T0178846

Lecture Notes in Artificial Intelligence   11489

Subseries of Lecture Notes in Computer Science

More information about this series at http://www.springer.com/series/1244

Marie-Jean Meurs · Frank Rudzicz (Eds.)

Advances in Artificial Intelligence

32nd Canadian Conference on Artificial Intelligence, Canadian AI 2019
Kingston, ON, Canada, May 28–31, 2019
Proceedings

 Springer

Editors
Marie-Jean Meurs 🆔
University of Quebec in Montreal
Montreal, QC, Canada

Frank Rudzicz 🆔
University of Toronto
Toronto, ON, Canada

ISSN 0302-9743 ISSN 1611-3349 (electronic)
Lecture Notes in Artificial Intelligence
ISBN 978-3-030-18304-2 ISBN 978-3-030-18305-9 (eBook)
https://doi.org/10.1007/978-3-030-18305-9

LNCS Sublibrary: SL7 – Artificial Intelligence

This Springer imprint is published by the registered company Springer Nature Switzerland AG
The registered company address is: Gewerbestrasse 11, 6330 Cham, Switzerland

Preface

We are pleased to assemble the papers presented at Canadian AI 2019: the 32nd Canadian Conference on Artificial Intelligence held May 28–31, 2019, at Queen's University in Kingston, Ontario. Canadian AI is an event organized by the Canadian Artificial Intelligence Association, and is collocated with the Canadian Graphics Interface and the Computer and Robot Vision conferences. These events (AI-GI-CRV 2019) bring together hundreds of leaders in research, industry, and government, as well as Canada's most accomplished students. This volume showcases Canada's ingenuity, innovation, and leadership in artificial intelligence research.

There were 132 submissions, representing an increase of 183% over 2018, undoubtedly reflecting the accelerating interest in artificial intelligence in Canada. We ran a double-blind review process, where each valid submission was reviewed by at least two Program Committee members. Based on the recommendations of the committee members, 27 submissions were accepted as long papers (20%) and 34 submissions as short papers (26%). In addition, the program included eight papers presented at the Graduate Student Symposium, chaired by Colin Bellinger and Elnaz Davoodi, which ran an independent review process, and four papers in the Industry Track, chaired by Eric Charton. We thank Colin, Elnaz, and Eric for organizing these sessions and for chairing the review processes.

We also thank all members of the Program Committee, and the external reviewers of the main conference, the Graduate Student Symposium, and the Industry Track reviewers, for their time and effort in providing valuable reviews in a timely manner. We thank all the authors for submitting their contributions and the authors of accepted papers for preparing the final version of their papers and presenting their work at the conference.

The conference was enriched by four keynote speakers, who are leaders in the field from academia and industry: Sven Dickinson (University of Toronto; Samsung AI Research), Chad Gaffield (University of Ottawa), Nancy Ide (Vassar College), and Maite Taboada (Simon Fraser University). We are grateful to them for their time and participation in this event.

The Canadian Conference on Artificial Intelligence is sponsored by the Canadian Artificial Intelligence Association (CAIAC). We especially thank and acknowledge the work of the executive committee of CAIAC, Ziad Kobti, Leila Kosseim, Xin Wang, Marina Sokolova, and Cory Butz, whose guidance was indispensable. We also thank Fabrizio Gotti, who designed and maintained the conference website with the utmost efficiency. We are grateful to Michael Greenspan, the general chair of AI/GI/CRV 2019, for his help with the organization of the broader conference.

Finally, we thank our financial sponsors, including Springer, the Vector Institute, HumanIA, and Layer6.

March 2018   Marie-Jean Meurs
                                                                                          Frank Rudzicz

Organization

Program Committee

Esma Aimeur	University of Montreal, Canada
Xiangdong An	University of Tennessee at Martin, USA
Ebrahim Bagheri	Ryerson University, Canada
Eric Beaudry	Université du Québec à Montréal, Canada
Colin Bellinger	National Research Council of Canada, Canada
Virendra Bhavsar	University of New Brunswick, Canada
Narjes Boufaden	KeaText Inc., Canada
Nizar Bouguila	Concordia University, Canada
Scott Buffett	National Research Council Canada, Canada
Cory Butz	University of Regina, Canada
Laurence Capus	Laval University, Canada
Laurent Charlin	HEC Montréal, Canada
Eric Charton	National Bank of Canada, Canada
Jackie Chi Kit Cheung	McGill University, Canada
Paul Cook	University of New Brunswick, Canada
Elnaz Davoodi	Concordia University, Canada
M. Ali Akber Dewan	Athabasca University, Canada
Abdoulaye Baniré Diallo	Université du Québec à Montréal, Canada
Chris Drummond	NRC Institute for Information Technology, Canada
Audrey Durant	McGill University, Canada
Ahmed Esmin	Federal University of Lavras, Brazil
Ali Etemad	Queen's University, Canada
Emma Frejinger	University of Montreal, Canada
Michel Gagnon	Polytechnique Montreal, Canada
Sebastien Gambs	Université du Québec à Montréal, Canada
Alice Gao	University of Waterloo, Canada
Yong Gao	The University of British Columbia, Canada
Nizar Ghoula	National Bank of Canada, Canada
Michael Guerzhoy	Princeton University, USA
Howard Hamilton	University of Regina, Canada
Jesse Hoey	University of Waterloo, Canada
Diana Inkpen	University of Ottawa, Canada
Ilya Ioshikhes	University of Ottawa, Canada
Aminul Islam	University of Louisiana at Lafayette, USA
Dhanya Jothimani	Indian Institute of Technology Delhi, India
Igor Jurisica	University of Toronto, Canada
Vlado Keselj	Dalhousie University, Canada
Fazel Keshtkar	St John's University, Canada

Kamyar Khodamoradi	University of Alberta, Canada
Richard Khoury	Laval University, Canada
Ziad Kobti	University of Windsor, Canada
Grzegorz Kondrak	University of Alberta, Canada
Leila Kosseim	Concordia University, Canada
Dmitry Kravchenko	Ben-Gurion University of the Negev, Israel
Adam Krzyzak	Concordia University, Canada
Majid Laali	Amazon, Canada
Sébastien Lallé	University of British Columbia, Canada
Luc Lamontagne	Laval University, Canada
Philippe Langlais	University of Montreal, Canada
Yves Lespérance	York University, Canada
Fuhua Lin	Athabasca University, Canada
Andrea Lodi	École Polytechnique de Montréal, Canada
Rongxing Lu	University of New Brunswick, Canada
Simone Ludwig	North Dakota State University, USA
Ali Mahdavi-Amiri	Simon Fraser University, Canada
Brad Malin	Vanderbilt University, USA
Stan Matwin	Dalhousie University, Canada
Robert Mercer	University of Western Ontario, Canada
Malek Mouhoub	University of Regina, Canada
Jian-Yun Nie	University of Montreal, Canada
Roger Nkambou	Université du Québec à Montréal, Canada
Iosif Viorel Onut	IBM CAS and University of Ottawa, Canada
Jeff Orchard	University of Waterloo, Canada
Fred Popowich	Simon Fraser University, Canada
Guillaume Rabusseau	McGill University, Canada
Sheela Ramanna	University of Winnipeg, Canada
Samira Sadaoui	University of Regina, Canada
Fatiha Sadat	Université du Québec à Montréal, Canada
Eugene Santos	Dartmouth College, USA
Weiming Shen	National Research Council Canada, Canada
Daniel L. Silver	Acadia University, Canada
Marina Sokolova	University of Ottawa, Canada
Bruce Spencer	University of New Brunswick, Canada
Stan Szpakowicz	University of Ottawa, Canada
Alain Tapp	Université de Montréal, Canada
Thomas Tran	University of Ottawa, Canada
Petko Valtchev	Université du Québec à Montréal, Canada
Roger Villemaire	Université du Québec à Montréal, Canada
Chun Wang	Concordia University, Canada
Xin Wang	University of Calgary, Canada
Dan Wu	University of Windsor, Canada
Yang Xiang	University of Guelph, Canada
Jingtao Yao	University of Regina, Canada
Harry Zhang	University of New Brunswick, Canada

Xiaodan Zhu	National Research Council Canada, Canada
Sandra Zilles	University of Regina, Canada
Farhana Zulkernine	Queen's University, Canada

Contents

Graduate Student Symposium Papers

Industrial Track Papers

Regular Papers

Categorizing Emails Using Machine Learning with Textual Features

Haoran Zhang[1], Jagadish Rangrej[1], Saad Rais[1(✉)], Michael Hillmer[1], Frank Rudzicz[2,3,4], and Kamil Malikov[1]

[1] Information Management, Data, and Analytics Division, Ministry of Health and Long-Term Care, Toronto, Ontario, Canada
`hran.zhang@utoronto.ca,`
`{Jagadish.Rangrej,Saad.Rais,Michael.Hillmer,Kamil.Malikov}@ontario.ca`
[2] Department of Computer Science, University of Toronto, Toronto, Ontario, Canada
[3] International Centre for Surgical Safety, Li Ka Shing Knowledge Institute, St. Michael's Hospital, Toronto, Ontario, Canada
[4] Vector Institute for Artificial Intelligence, Toronto, Ontario, Canada

Abstract. We developed an application that automates the process of assigning emails received in a generic request inbox to one of fourteen predefined topic categories. To build this application, we compared the performance of several classifiers in predicting the topic category, using an email dataset extracted from this inbox, which consisted of 8,841 emails over three years. The algorithms ranged from linear classifiers operating on n-gram features to deep learning techniques such as CNNs and LSTMs. For our objective, we found that the best-performing model was a logistic regression classifier using n-grams with TF-IDF weights, achieving 90.9% accuracy. The traditional models performed better than the deep learning models for this dataset, likely in part due to the small dataset size, and also because this particular classification task may not require the ordered sequence representation of tokens that deep learning models provide. Eventually, a bagged voting model was selected which combines the predictive power of the top eight models, with accuracy of 92.7%, surpassing the performance of any of the individual models.

Keywords: Machine learning · Deep learning · Natural language processing · Text classification · Email categorization

1 Introduction

The acceleration of communication through email, with all its benefits, has given birth to a laborious task: managing an overwhelming volume of incoming emails. The Ontario Ministry of Health and Long-Term Care, as a payer, steward, and provisioner of healthcare services in Ontario, through its reliance on email communication, faces this very challenge. One Ministry account, Data Decision Management (DDM) Support, is intended for data-related requests, and for access to

© Springer Nature Switzerland AG 2019
M.-J. Meurs and F. Rudzicz (Eds.): Canadian AI 2019, LNAI 11489, pp. 3–15, 2019.
https://doi.org/10.1007/978-3-030-18305-9_1

various data-related tools that the Ministry produces. The users of this account are not limited to Ministry staff – various healthcare organizations, including facilities, Local Health Integration Networks, Long-Term Care homes, and other partners of the healthcare sector within Ontario rely on this account to ask questions, obtain information, and submit requests.

Each year, DDM-support receives over 3000 emails, and, due to the diverse types of emails, is managed by multiple staff. Specifically, one person is responsible for categorizing each incoming email into one of fourteen pre-declared topic categories. Other dedicated staff periodically visit the inbox to check if any email has been flagged with a category that they are responsible for responding to. This manual email management process consumes valuable staff time, and therefore there was a desire to automate the process of categorizing emails with minimal human intervention. This led to the development of a data-driven application that automatically and seamlessly categorizes emails in the email client.

From a modelling perspective, emails contain a treasure trove of information, including the title and body. Features can also be derived from the metadata within the email header, which contains information such as the importance flag, sent time, and sender and recipient email addresses [1]. These can be further parsed to give the name and organization of each of the sender and recipients. In this paper, only the textual features were used for modelling, though adding the metadata is certainly something that could be explored in the future.

With the increase in computer memory, processing power, and more efficient training algorithms [2], linear classifiers such as Naïve Bayes, logistic regression, and support vector machines became more popular [2]. These classifiers have been shown to out-perform the rule-based approaches in email classification [3]. But more recently, deep learning approaches to text classification have been shown to surpass the performance of standard machine learning classifiers in many cases [4–9]. The goal of this paper is to present the performance of these aforementioned methods and models in order to develop a viable email classification system for internal use.

Email classification differs from a standard text classification problem in a few ways, however. First, there is generally a shift in email content over time, due to the constant changing of priorities and development of different projects. Thus, to achieve high performance, the classifier must give greater consideration to the most recent training data. Secondly, unlike many text classification datasets, emails have two distinct textual features - the title and the body. As these two features serve different purposes, they may have a different distribution of content, and therefore could potentially be treated differently in modelling. Lastly, domain-specific email inboxes such as DDM-Support will contain highly specific subject-matter terminology such as organization acronyms or hospital names, which are generally not present in large text corpora such as common crawl. This could present some issues for the use of pre-trained word embeddings.

2   Data

The dataset consisted of 8,841 emails received from 5 January 2015 to 6 March 2018. Each of these emails were manually flagged by subject expert staff into one of thirteen subject categories. An additional "other" category contained emails for rare requests not belonging in one of the thirteen categories. To prepare the email data for modelling, the sent date, email subject, email body, and associated category flag were extracted from each email. The subject was parsed by a simple regular expression to remove variations of "Re:" and "Fw:". The body of each email was stored as one long string of text containing the email text, and the headers that separated the intermediate messages in the email chain. These headers contained metadata such as sent times and subjects of the intermediate emails, and were removed by parsing the email body using regular expressions. The email body and title were then tokenized separately and stopwords were removed. The textual features that were used for modelling consisted of the parsed subject title, and the amalgam of all email-bodies in the email chain, excluding the headers.

3 Methodology

Initially, the data was split into training and test datasets by time to ensure that the test data was made up of more recent emails, which, due to proximity in time, would better reflect production data. Specifically, emails received on or after 1 December 2017 were assigned to the test set, whereas the remaining 8214 emails received before 1 December 2017 were used for training and validation. Similarly, time-dependent splitting of training-data by date was adopted during cross-validation, as discussed in Sect. 3.2. The distributions of categories in the

Table 1. Category distributions of the training and test sets

Group Name	Training set		Test set	
	N	%	N	%
OCC	3695	45.0%	7	1.1%
Bed census summary	2341	28.5%	505	80.5%
HDB Portal	859	10.5%	52	8.3%
OHRS	370	4.5%	24	3.8%
Data submission compliance reports	227	2.8%	1	0.2%
Other	145	1.8%	4	0.6%
SRI	129	1.6%	3	0.5%
HSFR	84	1.0%	21	3.3%
CDS-MH	91	1.1%	4	0.6%
IMSC	79	1.0%	7	1.1%
LTCH	86	1.0%	9	1.4%
Data quality	54	0.7%	9	1.4%
QIP	30	0.4%	1	0.2%
HIT	24	0.3%	0	0.0%
Total	8214		627	

training and test sets are shown in Table 1. The test set is composed of 7.1% of
the entire dataset. Table 1 reveals that the distributions of the email categories
in the two datasets are different, confirming our observation that there was a
shift in the distribution of the target variable, and therefore a shift in the content
of the emails, over time.

Python 3.6 was used for all modelling and data processing. The NLTK library
[10] was used for natural language processing tasks. Logistic regression was used
from the scikit-learn library [11]. The XGBoost [12] library was also used for classification modelling. Keras with Tensorflow was used for deep learning models.

3.1 Sample Weighting

The training data was weighted such that more recent data was given greater
weight [13] to help account for evolving email content. Specifically, each record
in the training set was weighted proportionally to the number of days the email
was received prior to 1 December 2017. The following function was used:

$$w(d_i) = \frac{1 - w_0}{f(D_s, D_e)} f(D_s, d_i) + w_0 \tag{1}$$

where w is a number between 0 and 1 corresponding to the weight, w_0 is a
number between 0 and 1 corresponding to the weight of the earliest date, d_i is
the date that an email was received, f is a function that returns the number of
calendar days between two dates, D_s is the start date (5 January 2015), and D_e
is the end date of the training set plus one day (1 December 2017).

3.2 Modified K-Folds

Since our particular dataset experienced a significant change in email content
over time, the performance of the model on old emails is much less relevant
than model performance on newer emails. Thus, a modified k-fold splitting algorithm was designed to ensure that the validation set resembled the test set (and
therefore will reflect the production data) as closely as possible, as shown in
Fig. 1. First, the training data was split into two equal sets by date, *old* and
new. The new half was split randomly into k folds, stratifying by the category to
ensure that each of the 14 categories were represented proportionally in each fold.

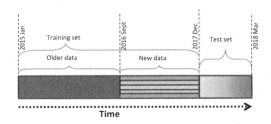

Fig. 1. Schematic for the modified K-fold splitting

The validation set was represented by one of the k-pieces, while the remaining pieces along with the *old*-Data formed the training set. Note that the *old* dataset was never used for validation – only training. For example, given $k = 5$, the algorithm produces five (training, validation) folds of size 90%, 10% respectively.

3.3 Models

Rule-Based. The rule-based approach searches for the pre-defined regular expressions in the email subject and body, and if one is found, assigns the email to the category that is associated to that regular expression. A set of 45 regular expression rules were created and mapped to the thirteen categories using subject matter knowledge, separately for the titles and bodies. Emails not matching any of the rules were put into the "Other" category. The motivation for using the rule-based method was to establish a baseline measure of performance that can be implemented in most email management programs.

Frequency-Based. On the training set, the n_b and n_t globally most frequent tokens were extracted from the body and the title respectively. Frequency counts of the selected words were computed for each category, converted into probabilities, and standardized for words in the category. Thus, 14 probability distributions were constructed from the body, and 14 were constructed from the title. To predict the category on new data, the algorithm processes each token and each category, and it acquires the probability for that token using the probabilities derived from the training data for each category. Summing the probabilities of each token in an email yields the probability that the email appears in each category. This procedure is done separately for both title tokens and email tokens. After two vectors of length 14 are computed for each email, a weighted average is computed across categories:

$$p = \alpha \cdot p_{body} + (1 - \alpha) \cdot p_{title} \tag{2}$$

Empirically, we found that $n_b = 2000, n_t = 80, \alpha = 0.65$ was an optimal combination for accuracy for our data. The category that corresponds to the highest probability is the one selected.

N-Grams. To derive the n-gram features, in addition to tokenization and stopword removal, any token that includes a numerical digit was also removed. We found that removing tokens with digits greatly increased the generalizability of the final model, as the classifier could potentially learn non-generalizable features such as numeric years or dates otherwise. From the training set, the 1400 and 3200 most frequent tokens were taken respectively from the title and the body. Similarly, bigrams were constructed, and the 800 and 1600 most frequent bigrams were taken from the title and the body respectively. In our n-grams approach, one-vs-all logistic regression (LR) and XGBoost [12] were used to predict email category. With both classifiers, a grid search was performed over the hyper-parameters using the modified k-folds ($k = 5, 3$ for LR, XGB respectively) method described in Sect. 3.2, optimizing on top-1 accuracy. A final classifier was then constructed with optimal hyper-parameters trained on the entire cross-validation set.

TF-IDF. TF-IDF [14] weighting was applied to the count matrix. Logistic regression and XGBoost were used to build a predictive model on this TF-IDF transformed matrix.

Noun-Phrase Extraction. The noun-phrase extraction method is an attempt to systematically engineer customized features that are more likely to correlate with the category, similar to what is described in [15]. After prepending the title tokens to the body tokens, part-of-speech tagging was applied, and the sentences were parsed into a tree structure. Nouns within noun phrases were extracted from the tree, and the collection of all noun phrases within the training set formed the feature space for the model. The values of the features corresponded to the depth of the phrase within the tree. A value of zero was assigned to a noun if it did not appear in the text. These features were then used to train a logistic regression model.

Character-Level CNN. A character-level convolutional neural network was used to predict email category. The architecture of the network was inspired by [5], except that a much smaller network was used due to the smaller training set. For the deep learning models, only the primary main body of the email was used (i.e. further bodies in the email chain were not used) in order to reduce noise and keep the text lengths to a reasonable level. The total size of the character set was 87, which consists of: all lower and upper case letters, standard ASCII punctuation, the space, new line, and tab characters. Digits were excluded. Unlike [5], who found that distinguishing between case of letters decreased model performance, we assumed that case is important in our dataset because of the prevalence of acronyms, which may correspond to unrelated words when uncased.

All of the characters in each email were one-hot encoded. Any character not included in the character list was assigned a vector of zeros. Each email is then passed through four sets of 1D convolution and max pooling layers, followed by global max and mean pooling, a dense layer with 128 units, and a 14 unit softmax output. A validation set was created for early stopping and hyperparameter tuning. K-fold cross-validation was not used for deep learning models due to resource limitations. Keeping with the idea of matching the validation set with the test set, the validation set was chosen to be the 646 emails between 1 September 2017 and 30 November 2017, and the training set was chosen to be the 7568 emails before 1 September 2017.

CNN with GloVe. The approach to word embedding that we use is GloVe [16], specifically the 840B tokens 300D cased vectors, trained on Common Crawl. The main advantage of using word vectors is that they have the ability to generalize across a large vocabulary from its pre-trained corpus [17], which allows the final model to take words which it has not seen during training into account. The architecture for this network was inspired by [18]. After each email is encoded as a sequence of word vectors, it is passed through four sets of 1D convolution and max pooling layers, followed by global max and mean pooling, a dense layer with 128 units, and a 14 unit softmax output.

LSTM with GloVe. LSTM networks have also been shown to give good performance on text classification tasks [19, 20]. For our LSTM model, the same text pre-processing procedure as the word level CNN model was used. Each encoded email was sent to a bidirectional LSTM with 32 units, followed by global max and mean pooling, a dense layer with 128 units, and a 14 unit softmax output.

Bagging. Bagging or voting techniques are used to combine the predictive power of multiple models in an attempt to surpass the performance of the best individual model [21]. One thing about our problem that differs from a standard multiclass classification problem is the presence of the "Other" category. We decided that emails which the bagged models are uncertain about will be assigned to this category, and will have to be manually reviewed by a human. Thus, there is a trade-off between potentially labelling an email incorrectly, versus assigning it to "Other". A bagging algorithm was used that allows emails where models disagree to be put into the "Other" category. Specifically, a certain threshold of models (c) have to agree on a certain category in order for it to be assigned there; otherwise, it will be assigned "Other". In addition, another parameter, n, the number of models to bag, can also be varied, where the models selected are the top n models by test set accuracy. This bagging algorithm was used with different combinations of n and c values.

Table 2. Metrics examined and their descriptions

Metric name	Description
Accuracy	Standard top-1 accuracy
Macro precision	Unweighted mean of precision scores for each label
Micro precision (without other)	Calculating precision globally by counting the total number of TP, over total of FP and TP. "Other" category is excluded; it would be equal to accuracy
Macro sensitivity	Unweighted mean of sensitivity scores for each label
Micro sensitivity (without other)	Calculating sensitivity globally by counting the total number of TP, over total of FN and TP. "Other" category is excluded
Micro distance to (1, 1)	Euclidean distance from the (micro sensitivity, micro precision) coordinate pair to (1, 1). i.e. $\sqrt{(1 - precision)^2 + (1 - sensitivity)^2}$. Considers the optimal point of trade-off between FN and FP

3.4   Metrics

A list of the metrics examined along with their descriptions can be found in
Table 2. Note that some of the metrics reflect the trade-off between labelling an
email incorrectly versus assigning it to "Other".

4   Results

Sample Weighting. A weighted n-gram logistic regression was ran for different
values of w_0, and the best cross-validation accuracy obtained from a grid search
of hyperparameters with the modified k-folds algorithm with $k = 5$ was recorded.
It was found that there is not a significant difference between varying values of
w_0, and the best performance occurs at $w_0 = 1$, corresponding to the unweighted
case. Thus, in all further modelling, the training samples were weighted equally.

Performance of Models. Table 3 summarizes the performance of the individual models on the test set, using the aforementioned metrics. The logistic regression TFIDF model has both the highest accuracy and the smallest distance to
$(1, 1)$.

Bagging. The individual models were ordered in descending order of accuracy
on the test set. The top n models by accuracy were selected, and the bagging
algorithm was used with a certain value of c. n from 3 to 10 were tried, in
combination with c from 2 to n. Table 4 shows the performance of these bagged
models on the test set, for the 10 combinations with the best accuracy. It can
be seen that the model with the best accuracy occurs at $n = 8, c = 4$, which
corresponds to using all models except the rule-based and the character-level
CNN. This model also has the lowest micro distance to $(1, 1)$.

Table 3. Performance of individual models on the test set, with the best-performing
model for each metric bolded

Method	Accuracy	Macro precision	Micro precision	Macro sensitivity	Micro sensitivity	Micro dist to $(1, 1)$
Rules	0.818	0.627	0.899	0.647	0.830	0.197
Freq	0.855	0.531	0.859	0.566	0.865	0.195
LR NGrams	0.896	0.630	0.905	0.633	0.909	0.132
XGB NGrams	0.898	0.676	0.901	**0.728**	0.912	0.133
LR TFIDF	**0.909**	**0.728**	**0.917**	0.608	**0.921**	**0.114**
XGB TFIDF	0.898	0.660	0.906	0.712	0.912	0.129
NLP	0.907	0.667	0.910	0.598	0.918	0.122
CharCNN	0.818	0.304	0.818	0.273	0.832	0.248
GloVeCNN	0.844	0.567	0.874	0.440	0.858	0.190
GloVeLSTM	0.889	0.618	0.895	0.677	0.903	0.143

Table 4. Performance of bagged models on the test set for the top ten models by accuracy

n	c	Accuracy	Macro precision	Micro precision	Macro sensitivity	Micro sensitivity	Micro dist to$(1, 1)$
8	4	0.927	0.761	0.939	0.716	0.940	0.086
8	3	0.927	0.730	0.932	0.720	0.942	0.090
8	2	0.927	0.730	0.930	0.720	0.942	0.091
10	3	0.926	0.728	0.929	0.708	0.940	0.093
10	2	0.926	0.728	0.927	0.708	0.940	0.094
9	4	0.926	0.758	0.936	0.712	0.939	0.089
9	3	0.926	0.732	0.930	0.716	0.940	0.092
9	2	0.926	0.731	0.929	0.716	0.940	0.093
7	2	0.924	0.705	0.927	0.741	0.939	0.095
10	4	0.924	0.741	0.933	0.703	0.939	0.091

Performance Decay. Fig. 2 shows the cumulative average of accuracy on the best bagged model ($n = 8, c = 4$) as a function of the email sent time. A general decrease in the performance of the model over time can be observed.

5   Discussion

Traditional Models. The traditional methods for feature extraction (Rules, Freq, N-Grams, TFIDF) are sufficient for simple algorithms like logistic regression, since categorizing an email to a particular category can be very well established based on the keywords alone. More nuanced NLP problems such as sentiment analysis, on the other hand, involve an understanding of the semantic similarity of tokens [22], and deriving the meaning of semantically different constructs of the sentence is necessary. Our noun phrase extraction method is a compromise which makes use of the grammatical structure but ultimately pulls out only syntactically sensible words relevant to the topic of interest with the feature that would correlate the words, i.e. depth of leaf at which word appears. When we ran stemming and lemmatization on the nouns that were identified in the noun-phrase based approach, our performance did not improve, demonstrating that the unification of lexically/semantically different tokens is not important to identify the category of the email. This observation concurs with the sentiment analysis performed on Amazon reviews data in [5]. With simple features such as n-grams and TFIDF, complex classifiers like XGB do not necessarily perform better than simpler classifiers such as logistic regression, as these features are word(s) that correlate with at most one neighbouring word.

Deep Learning Models. Overall, as seen in Table 3, all three of the deep learning models perform worse than the simpler n-grams or TF-IDF classifier based models. There are many possible reasons for this. First, the architecture and hyper-parameters of the neural network might not be well-optimized. Secondly, the size of the training set was small, thus it is possible that having only several thousand training samples is not sufficient to train a generalizable deep neural network model with hundreds of thousands of parameters. It was reported that the character-level CNNs only perform better than traditional methodologies (n-grams, TFIDF) when the size of the dataset is in the millions [5].

Thirdly, one reason why some deep learning methods might not perform well is that the pre-trained GloVe embeddings might not capture many of the knowledge-area specific words, such as acronyms, project names, or location names. Even for domain-specific terms that do appear in the word vectors, the meaning of a term in the training corpus might be different than its meaning in the context of healthcare, and so the particular word vector would not match the term in the email. This is one advantage of the character-level CNN model – it does not have to worry about out-of-vocabulary terms. However, there have been many strategies developed to deal with out-of-vocabulary words in word embeddings. One possible solution is to use an open vocabulary hybrid word-character model [23], such as fastText [7], which falls back to combining pre-learnt vectors for character-level n-grams in case of out-of-vocabulary words. Another strategy would be to generate word embeddings for out-of-vocabulary words on the fly while training [24].

Another possible reason for why the deep learning models performed no better than the other models is that this particular task is relatively simple. Often times, the category of the email was stated verbatim in the title or the body, so it could be easily identified by an n-grams model. The complex sequence representation offered by deep learning might not be necessary for this particular problem.

Performance of Bagged Models. Comparing Tables 3 and 4, the best bagged model had an accuracy of almost 2% more than the best individual model. Figure 3 shows the accuracy of each bagged model plotted against its c/n ratio. The best-performing model occurs at $c/n = 0.5$, and there is a rapid decrease in accuracy when c/n exceeds 0.5. For larger c/n, more models would have to agree on the category for it to be flagged as such, and it would otherwise be flagged as "Other", which would correspond to an incorrect classification in most cases. Similarly, as seen in Fig. 4, the Micro Distance to $(1, 1)$ also becomes worse for high c/n.

One trend that can be observed in the bagging approach is the trade-off between precision and sensitivity for increasing c, as shown in Fig. 5. As c decreases, precision always decreases and sensitivity always increases. This trade-off is well documented for binary classification [25]. In our case, however, the cause is a bit more subtle. For higher values of c, the bagged model is more likely to classify emails into the "Other" category, and will only categorize emails if the model surpasses a high threshold of confidence in that categorization. In calculating micro precision, the predicted "Other" category was excluded, resulting

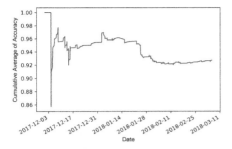

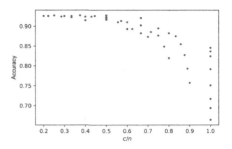

Fig. 2. Cumulative average of accuracy of best bagged model ($n = 8$, $c = 4$) on the test set as a function of sent time

Fig. 3. Accuracy of the bagged models as a function of c/n

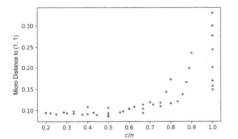

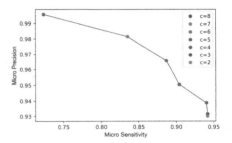

Fig. 4. Micro distance to $(1, 1)$ of the bagged models as a function of c/n

Fig. 5. Micro precision vs. micro sensitivity for bagging the top 8 models ($n = 8$) at different values of c

in a higher proportion of true positives, and therefore a higher micro precision. Similarly, for micro sensitivity, the actual "Other" category was removed, leaving very few emails, and thus all of the emails misclassified as "Other" counted towards false negatives, leading to a lower value in this metric. The proposed distance measure to $(1, 1)$ is a composite measure that takes into account the trade-off between micro-precision and micro-sensitivity at different values of c, and thus was used as one of the final measure in evaluating performance.

Performance Decay. Emails are time-sensitive, and the relevance of topics in emails can change over time. There were also a number of new terms introduced over time that were domain/topic specific. This change affected the model performance in two ways: first, by decreasing the relevance of the features that we developed in predicting the future emails, and second, by introducing new information that was not captured in past emails. Currently, this phenomenon of performance decay is being addressed through a batch offline retraining of the model with updated emails at regular intervals. However, a more robust approach would involve constant online knowledge acquisition and refinement, i.e.

generating new features that are weighted more heavily as new records of emails are received, and dropping or lowering the significance of out-dated features [26].

6 Conclusion

While our final model demonstrated a 92.7% accuracy, further improvements could be made to enhance the model's performance. For example, additional features from the email metadata could be used. Also, we used the NLTK library for NLP tasks, but the spaCy library has demonstrated high processing speed and accuracy relative to its competitors [27]. Our deep learning models could also be improved upon by using fastText instead of GLoVe, as fastText is better suited to industry-specific words that do not appear in its training data [28].

Notwithstanding some areas for improvement, our email categorization has had a positive impact on the Ministry's business operations: it has allowed two full time employees to re-distribute their workload to other more value generating tasks, and has significantly improved email response time. This success has attracted interest in expanding the application's use to other inboxes, and has even inspired new ideas to automate decision making processes that involve text data.

References

1. Yang, J., Park, S.-Y.: Email categorization using fast machine learning algorithms. In: Lange, S., Satoh, K., Smith, C.H. (eds.) DS 2002. LNCS, vol. 2534, pp. 316–323. Springer, Heidelberg (2002). https://doi.org/10.1007/3-540-36182-0_31
2. Sebastiani, F.: Machine learning in automated text categorization. ACM Comput. Surv. **34**(1), 1–47 (2001)
3. Provost, J.: Naïve-Bayes vs. rule-learning in classification of email. University of Texas at Austin, Artificial Intelligence Lab, CiteSeer (Ingebrigsten), pp. 1–4 (1999)
4. Zhou, C., Sun, C., Liu, Z., Lau, F.C.M.: A C-LSTM Neural Network for Text Classification. ArXiv e-prints, November 2015
5. Zhang, X., Zhao, J., LeCun, Y.: Character-level Convolutional Networks for Text Classification, pp. 1–9 (2015)
6. Lai, S., Xu, L., Liu, K., Zhao, J.: Recurrent convolutional neural networks for text classification. In: AAAI-29, pp. 2267–2273 (2015)
7. Joulin, A., Grave, E., Bojanowski, P., Mikolov, T.: Bag of Tricks for Efficient Text Classification (2016)
8. Johnson, R., Zhang, T.: Effective Use of Word Order for Text Categorization with Convolutional Neural Networks (2011, 2014)
9. Kim, T., Yang, J.: Abstractive Text Classification Using Sequence-to-convolution Neural Networks (2018)
10. Bird, S., Klein, E., Loper, E.: Natural Language Processing with Python (2009)
11. Pedregosa, F., Varoquaux, G., Gramfort, A.: Scikit-learn: machine learning in python. J. Mach. Learn. Res. **12**, 2825–2830 (2012)
12. Chen, T., Guestrin, C.: XGBoost: A Scalable Tree Boosting System (2016)
13. He, H., Garcia, E.A.: Learning from imbalanced data. IEEE Trans. Knowl. Data Eng. **21**(9), 1263–1284 (2009)

14. Manning, C.D., Raghavan, P., Schütze, H.: Introduction to Information Retrieval Introduction, vol. 35 (2008)
15. Lewis, D.D.: Feature selection and feature extraction for text categorization. Speech and natural language. In: Proceedings of a Workshop Held at Harriman, New York, 23–26 February 1992 (1992)
16. Pennington, J., Socher, R., Manning, C.: Glove: global vectors for word representation. In: Proceedings of the 2014 Conference on EMNLP, pp. 1532–1543 (2014)
17. Lecun, Y., Bengio, Y., Hinton, G.: Deep learning. Nature **521**(7553), 436–444 (2015)
18. Conneau, A., Schwenk, H., Le Cun, Y., Barrault, L.: Very deep convolutional networks for text classification. In: Proceedings of the 15th Conference of the EACL, vol. 1, pp. 1107–1116 (2017)
19. Jurafsky, D., Martin, J.: Speech & Language Processing, 2 edn., London (2014)
20. Ramos, J.: Using TF-IDF to determine word relevance in document queries. In: Proceedings of the First Instructional Conference on Machine Learning, pp. 1–4 (2003)
21. Breiman, L.: Bagging predictors. Mach. Learn. **24**(2), 123–140 (1996)
22. Maas, A.L., Daly, R.E., Pham, P.T., Huang, D., Ng, A.Y., Potts, C.: Learning word vectors for sentiment analysis. In: Proceedings of the 49th Annual Meeting of the ACL, pp. 142–150 (2011)
23. Luong, M.T., Manning, C.D.: Achieving Open Vocabulary Neural Machine Translation with Hybrid Word-Character Models (2016)
24. Bahdanau, D., Bosc, T.: Learning to Compute Word Embeddings on the Fly (2018)
25. Gordan, M., Kochen, M.: Recall-precision trade-off : a derivation. J. Am. Soc. Inf. Sci. **40** 145 (1989, 1998)
26. Fisher, D.: Knowledge acquisition via incremental clustering. Mach. Learn. **2**(1980), 139–182 (1987)
27. Choi, J.D., Tetreault, J., Stent, A.: It Depends: Dependency Parser Comparison Using a Web-based Evaluation Tool, pp. 387–396 (2015)
28. Bojanowski, P., Grave, E., Joulin, A., Mikolov, T.: Enriching word vectors with subword information. CoRR abs/1607.04606 (2016)

Weakly Supervised, Data-Driven Acquisition of Rules for Open Information Extraction

Fabrizio Gotti$^{(\boxtimes)}$ ⓘ and Philippe Langlais$^{(\boxtimes)}$

RALI, Université de Montréal, CP 6128 Succ. Centre-Ville,
Montreal H3C 3J7, Canada
{gottif,langlais}@iro.umontreal.ca

Abstract. We propose a way to acquire rules for Open Information Extraction, based on lemma sequence patterns (including potential typographical symbols) linking two named entities in a sentence. Rule acquisition is data-driven and requires little supervision. Given an arbitrary relation, we identify, in a large corpus, pairs of entities that are linked by the relation and then gather, score and rank other phrases that link the same entity pairs. We experimented with 81 relations and acquired 20 extraction rules for each by mining ClueWeb12. We devised a semi-automatic evaluation protocol to measure recall and precision and found them to be at most 79.9% and 62.4% respectively. Verbal patterns are of better quality than non-verbal ones, although the latter achieve a maximum recall of 76.5%. The strategy proposed does not necessitate expensive resources or time-consuming handcrafted resources, but does require a large amount of text.

Keywords: Open Information Extraction · Weak supervision · Natural Language Processing

1 Introduction

Open Information Extraction (OIE) is a research area in Natural Language Processing that seeks to acquire shallow semantic representation elements from unstructured texts [6]. Typically, this extracted knowledge takes the form of relational tuples like (*Einstein, was born in, Germany*), composed of arguments flanking a central relation. Contrarily to plain Information Extraction, and importantly, OIE is not bound by a closed list of pre-defined relations or patterns guiding extraction. On the contrary, an ideal OIE system can handle previously unseen ways of expressing information and cast it in tuples, whatever the domain and without supervision.

OIE has been used recently for question-answering [8], and for building domain-targeted knowledge bases [13], among various applications.

Most current OIE systems rely on sets of rules or patterns to perform extraction. These rules can be manually crafted and/or acquired automatically.

© Springer Nature Switzerland AG 2019
M.-J. Meurs and F. Rudzicz (Eds.): Canadian AI 2019, LNAI 11489, pp. 16–28, 2019.
https://doi.org/10.1007/978-3-030-18305-9_2

For the most part, they home in on verbs in the original text to produce an extraction. These heuristics are usually limited to producing extractions built from the tokens originally present in the source text. They rarely "conjure up" new tokens to produce an extraction, instead clinging to the tokens in the source text. Some systems overcome these limitations successfully for a limited number of cases (e.g. appositions, possessives), but overlook many others. Yet a recent handcrafted benchmark in OIE [11] shows that 39% of its triples' relations come from implicit semantics in the text, and that 54% of these triples contain tokens absent from the source text.

For instance, take the excerpt *Barcelona's surrender at the hands of the Nationalist forces of General Francisco Franco.* We understand at the very least that *(Barcelona, fall to, Franco)*, even though there are no verbs in the original text and the words *fall to* never occur within it. The hint that allows us humans to do this mental extraction is the presence of the token sequence starting with *'s surrender at the hands of* surrounded by two easily recognizable named entities. A machine could yet learn the association between that token sequence and the relation at hand, by mining large quantities of text for pairs of entities for which it *knows* that the relation *fall to* holds. The intervening tokens between these two named entities could be paraphrases of that relation.

In this paper, we propose a way of harnessing these lexical hints, in order to generate rules whose purpose is to extract and rephrase elements of meaning "buried" a little more deeply. We suggest a means of extending the expressivity of current extraction heuristics, without much supervision, while at the same time remaining as generic as possible.

Section 2 presents related work, focusing on current rule systems. In Sect. 3, we detail our acquisition methodology. We put it to the test in Sect. 4 and evaluate it in Sect. 5, before discussing the results in Sect. 6.

2   Related Work

As we mentioned in the introduction, extraction rules lie at the very heart of OIE systems. Extractors, starting with the seminal TextRunner [1], have intuitively focused on verbs to identify the relation at hand (but not always exclusively).

TextRunner first identifies interesting noun phrases and then uses a classifier to label the intervening words as part of the relation or not. ReVerb [7] seeks to sanitize the approach by introducing manually crafted regular expressions over the verb-based relational phrase and its arguments. Ollie [12] learns lexicalized and unlexicalized extraction rules using a large number of highly scored ReVerb extractions as seeds. It uses a dependency parser to delve deeper into the syntactic structure of sentences and find long-range dependencies between relations and arguments. Ollie can handle appositions and a few other non-verb-mediated relations, but up to a point. Its rule framework is nonetheless very powerful.

Other systems make use of rules applied to other text elements. ClausIE's hand-crafted rules [6] are applied to a sentence's parse tree to identify meaningful sentence clauses and to cast them into appropriate extraction tuples. Its successor MinIE [9] further processes ClausIE's output with manual rules to produce minimized, semantically annotated OIE tuples. Its authors use various types of heuristics, including rewrite rules for relations, non-verb-mediated extractors adapted from FINET [5], and word lists indicating a fact's polarity (positive or negative) and certainty. In a recent evaluation [11], MinIE was found to perform best among 7 extractors. Other simplifying extractors include the more recent Graphene [4], which implements a handcrafted simplification and decomposition of sentences in order to yield core tuples as well as accompanying contexts that are semantically linked by rhetorical relations (e.g. temporal).

Non-verb mediated OIE has been the focus of fewer systems. While Ollie or MinIE do handle some cases, more recent systems like RelNoun 1.1 and 2.2 [15] attempt to increase recall by nominal OIE, relying on a set of POS and NP-chunk patterns, as well as additional resources built specifically to handle capitalized relational nouns (e.g. *Paris Mayor Chirac*), demonyms (e.g. *Canadian*) and compound relational nouns (e.g. *health minister*).

Our approach bears some resemblance to PATTY [14], a large resource of textual patterns denoting relations between entities. The patterns are acquired on large corpora and include semantic types for their arguments. PATTY uses seeds from knowledge bases and links their named entities to tokens found in sentence parses in order to infer patterns. These patterns are generalized using sequence of words, part-of-speech tags, wildcards, and ontological types. In our case, we wanted to investigate whether we could do away with parsers, tackle larger corpora, and capture surface patterns including typographical marks like punctuation. Moreover, in this study, we do not rely on an external knowledge base: the process is entirely data-driven.

In a similar vein, the Coupled Pattern Learner (CPL) algorithm [3] of the "never-ending language learner" (NELL) [2] is a sophisticated algorithm designed to learn to extract both relation instances and argument categories (e.g. Movie, Athlete) from unstructured text. While related, it is quite different from our (more lightweight) approach, as it iteratively learns extraction patterns in a semi-supervised way, bootstrapped by a manually crafted input ontology and related seed instances and patterns.

We are also indebted to previous studies using external sources of structured knowledge to label a corpus in order to bootstrap a learning algorithm. See for instance the 2011 system MULTIR [10], which uses a probabilistic approach to handle overlapping relations. The authors match Freebase facts in a corpora from the *New York Times* to achieve weak supervision.

3   Semi-supervised Acquisition of Extraction Rules

3.1   Extraction Rules

In this work, an *extraction rule* is a lexical pattern between two named entities (NEs) that should trigger the creation of an OIE triple. Such a rule consists of two elements. The first one is a *pattern*, a sequence of consecutive lemmas flanked on both sides by NEs. The second one is a corresponding *template*, a shell of a triple with the relation already specified, and placeholders for each named entity matched by the pattern.

For instance, the extraction rule NE_1 's defeat by NE_2 → (NE_1, *fall to*, NE_2) will match against the excerpt *Constantinople's defeat by the Turks in 1453.* and trigger the production of the triple (*Constantinople, fall to, the Turks*). It is worth noting that the relation *fall to* expressed in the triple is not actually present in the original sentence, i.e. its tokens are absent, "conjured up" during the extraction process. Also worth mentioning is the presence of the non-word 's in the pattern. A pattern could in theory contain only typographical marks, e.g. an opening parenthesis.

3.2   Acquisition Process of Rules for an Arbitrary Relation

Principle. The method we propose to create such extraction rules rely on mining large corpora to find lexical hints that a given relation holds between two named entities. This acquisition proceeds with relatively little supervision. The method's steps are detailed below. Figure 1 illustrates the process for the relation *fall to*.

1. Select an arbitrary relation r for which extraction rules are to be acquired.
2. In a large corpus C, find pairs of named entities (NE_1, NE_2) such that the lemma sequence NE_1 r NE_2 is found verbatim within a sentence.
3. For each pair of named entities (NE_1, NE_2), find sentences in C where NE_1 and NE_2 are present, regardless of their respective positions in the sentences. Collect at most k different sentences matching these criteria, with k being a hyperparameter of the process. We use $k = 200,000$ in this study.
4. For each sentence found in the previous step, gather candidate patterns, i.e. the sequence of tokens positioned between NE_1 and NE_2. Collect and count the candidate patterns.
5. Filter and sort the candidate pattern list (see below for details). Pick the best patterns and associate them with the extraction template (NE_1, r, NE_2).

Filtering and Ranking Extraction Patterns. The noisy process leading up to Step 5 above produces a long list of candidate extraction patterns, many of which are irrelevant. Filtering and ranking these candidates is crucial.

Relying on mere frequency to rank these is fruitless, since the most frequent patterns are also the least specific, occurring for any given relation. For instance, for the relation *fall to*, the most frequent patterns are fall to (which is correct, but obviously tautological), to, and, by, and the comma.

We experimented with 3 algorithms in order to rank candidate patterns in decreasing order of specificity. We tested tf–idf, relative frequency and a chi-squared test. The latter was the most effective. This test allows us to compare the statistical distribution of candidate extraction patterns for a given relation r with their distribution in a generic corpus. In our case the latter is formed by the patterns for relations different than r. The contingency table for the chi-squared test is shown in Table 1, for a given relation r and a candidate pattern p for r. We use Yates's correction for continuity for counts < 5. We refer the reader to a statistics handbook for the complete formula for the chi-squared test.

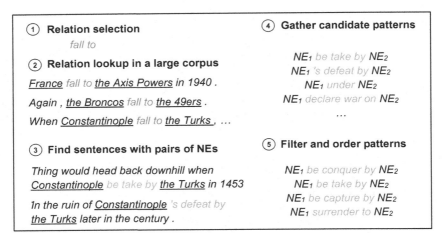

Fig. 1. Schema of the rule acquisition process for the relation *fall to*. Named entities are underlined in the figure. The best rule produced by the example shown would be NE_1 be conquer by $NE_2 \to (NE_1,$ *fall to*, $NE_2)$. All tokens are lemmatized.

We can then sort the candidate patterns in reverse order of their chi-squared score. To further sanitize the resulting list, we use simple additional filters:

- Remove all patterns with more than 7 tokens, e.g. , `r-texas , reveal his pick for a secret service nickname tuesday night on` for the relation *appear on*. In this case, the pattern is correct, but overly specific to appear in an extraction rule.
- Remove all patterns that appear in less than 2% of all relations studied. This indicates a pattern suspiciously specific to a relation, usually the result of a systematic extraction error for that relation.
- Remove the pattern that is identical to the relation at hand, like the pattern `fall to` for the relation *fall to*. Its presence in the candidate list could be viewed as auspicious nonetheless.
- Remove patterns that are stop words or contain a proper noun.

Table 1. Contingency table used for a chi-squared test assessing the specificity of a candidate pattern p for a relation r. Variables a, b, c, and d are frequencies (counts).

	In list of patterns for relation r	In list of patterns for all relations except r
Pattern p	a	b
Patterns different than p	c	d

4   Results for Rule Acquisition

4.1   Selection of Relations

Our acquisition method does not make any assumption about the kinds of relations we want extraction rules for. One starts by specifying them arbitrarily, which is not a form of supervision. In this work, the selection of these relations is dictated by the evaluation methodology we propose later on, in Sect. 5. In other words, we chose relations for which an evaluation is eventually possible. This means finding relations belonging to very high-quality OIE triples, whose veracity is indubitable. We will then be able to compute a recall measure over those triples, using the newly acquired extraction rules.

For the moment, suffice it to say that, to find these triples, we turned to a large collection of 15 million triples[1] extracted by the OIE system ReVerb [7] over the ClueWeb09 dataset. The latter contains 500M web pages collected in 2009. The triple collection was sanitized by its authors, using a threshold on a confidence score produced by ReVerb for each triple, and by additional reasonable heuristics. We then sorted these triples in reverse order of frequency, reasoning that frequent triples would prove more trustworthy.

Finding reliable triples in this list proved unexpectedly difficult. For instance, the most frequent triple is the odd (*Princeton, looks for, Lucy*) with a frequency of 3587. Other frequent examples include (*Red found in, strawberries*) and (*A, means to, an end*). This apparent cacophony has multiple explanations. First and foremost, ClueWeb09 is a relatively representative snapshot of the Internet circa 2009, including the inevitable repetitious and malicious content of defaced websites. Second, the extraction process is inherently noisy, even with a tool like ReVerb. Third, even a triple correctly extracted loses its meaning without its context, e.g. (*reprint, must include, byline*). Our search is further complicated by the fact that we seek triples with named entities for both arguments.

For these reasons, we had to sift through the top 2000 triples to find 128 triples, accounting for 43 different relations. Because 32 of these relations were all of the type *be a X* or *be the X* (e.g. *be a province of*), we explored a further 2000 triples to end up with a more varied final list of 179 triples, accounting for 81 different relations. Examples of such triples are (*ftp, stand for, file transfer protocol*), (*toronto, be the largest city in, canada*), and (*einstein, be born in, ulm*).

4.2   Finding Candidate Patterns in ClueWeb12

Once these 81 relations are selected, we can acquire associated extraction patterns. This entails the lookup of the relations in a large corpus, as previously explained in Sect. 3.2, starting with Step 2. Here, we decided to use ClueWeb12 to find the extraction patterns.[2] The dataset consists of 733M English web pages, collected between February 2012 and May 2012. By using this corpus, we wanted

---

[1] http://reverb.cs.washington.edu/.
[2] https://lemurproject.org/clueweb12/index.php.

to tackle unstructured text as is commonly found in the wild, and not only in curated texts like Wikipedia. This could allow more variety in the extraction rules ultimately yielded. Moreover, the sheer volume of ClueWeb12 increases our chances of finding relevant information in human-produced English documents.

We started by lemmatizing and indexing all 1.6×10^{10} sentences of the dataset, using Apache Lucene. We also performed part-of-speech tagging on all sentences to detect named entities. We distributed the task on 12 computers.

During **Step 2**, we looked up sentences with named entities flanking each relation, like so: $NE_1 \ r \ NE_2$. On average, we gathered 3400 pairs of named entities per relation (min = 61, max = 5000). The average frequency of any given NE pair is 1.1 (min = 1, max = 412 for the pair *Jesus* and *God* in the relation *be the son of*). Most pairs are found only once in co-occurrence with the relation, which is surprising given the large volume of text searched. A cursory examination indicates the pairs to be valid. This step is by far the longest and takes a day.

Step 3 consists in finding sentences in ClueWeb that contain the entity pairs selected during Step 2. We limited the total number of sentences for a given relation at 200k sentences, for performance reasons. We also quickly realized that we could not choose the most frequent entity pairs found in the previous step and find as many sentences as possible that contain them: A variety of entity pairs must be sampled for the overall rule acquisition pipeline to work correctly. Intuitively, this is necessary to gather as many different contexts as possible for these entity pairs, and to avoid oversampling entity pairs which may contain an extraction error. Therefore, we set a limit of 1000 total number of sentences sampled per entity pair. Per relation, we gathered 47k sentences on average (min = 6900 for *be the first wife of*, max = 200k for *be locate in*).

Step 4 sees the first candidate extraction patterns emerge. We obtain an average of 9794 extraction patterns, with a lot of variation ($\sigma = 10,820$). Here, we only consider patterns with a frequency of 5 or more, which gives us 409 extraction patterns on average ($\sigma = 369$).

Finally, **Step 5** filters and ranks extraction patterns. We only consider the top 20 patterns in this study, for a total of $81 \times 20 = 1620$ rules[3]. We show the top 10 patterns for 3 relations in Table 2. For instance, the top extraction rule for *die in* can be read off Table 2 as NE_1 `'s death in` $NE_2 \rightarrow (NE_1, die \ in, NE_2)$. We observe that these patterns are generally relevant and specific, but not always, e.g. the pattern `the great in` for the relation *die in*. Out of 1620 rules, we identified 809 rules (50%) whose pattern is non-verbal, like the previous example. We found 531 rules (32.7%) containing typographical marks, e.g. `- owner of`.

5   Evaluation

Evaluation is a delicate topic in OIE, as there are no gold standards or metrics that are agreed upon in the community. Some researchers will annotate the

---

[3] Download them here: http://rali.iro.umontreal.ca/rali/oie-pararules.

Table 2. Top 10 extraction patterns for 3 relations, in decreasing order of relevance. The top extraction rule for the relation *die in* can be read off column 2 as NE_1 's death in $NE_2 \rightarrow (NE_1,\ die\ in,\ NE_2)$. Non-verbal patterns are underlined.

fall to	*die in*	*be the owner of*
be conquer by	's death in	, owner of
be destroy by	the great in	, the owner of
be capture by	die on	own
at the hand of	society of	, who own
be take by	be assassinate in	(owner of
sink	's grave in	, which own
fall into the hand of	died in	- owner of
be occupy by	die last	be the president of
be defeat by	pass away in	a very
surrender to	be bury in	, who also own

output of their system for correctness, e.g. [12] or [16], which yields a precision measure without being able to offer a sense of the recall. Others have created benchmarks automatically, like the increasingly cited paper of Stanovsky and Dagan [17]. However such benchmarks are usually created with verb phrases as the focus of tuple extraction. They have other problems, described in [11].

In our case, we deemed it crucial to assess both precision and recall, while at the same time steer clear of any methodology that would only look at verbs in the source sentences. Indeed, the acquisition process we propose here can easily err on the side of recall at the expense of precision. If, for instance, we were to use a rule such as NE_1 at $NE_2 \rightarrow (NE_1,\ locate\ in,\ NE_2)$, then all occurrences of *at* would erroneously trigger the rule. The recall would be high for this relation, while at the same time extremely wanting in precision.

Moreover, verb-centric benchmarks like the one described in [17] would not be sensitive enough to measure the impact of extractions triggered by non-verbal clues, like the aforementioned NE_1 's death in $NE_2 \rightarrow (NE_1,\ die\ in,\ NE_2)$.

We propose an evaluation protocol taking these issues into account. Importantly, we did not create a full-fledged OIE system just for the purpose of evaluation. Rather, we used reasonable proxies to measure the performance of such a system were it to be programmed on the basis of the extraction rules found in the previous section.

5.1   Recall

Recall must be measured over a set of triples that are assuredly true and that each uses one of the 81 relations described earlier. We need only turn to Sect. 4.1 to find this test set: the 179 triples described are hand-validated and evidently use the relations at hand. Measuring recall then consists in matching a corpus's sentences against the extraction rules acquired and measuring what proportion of the 179 triples are found.

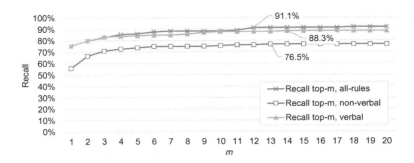

Fig. 2. Recall measure over 179 reference triples, for different values of m, the number of top rules retained for each relation. Three rule set are evaluated. **all-rules**: all the extraction rules acquired (1620 rules), **non-verbal** rules only (809 rules) and **verbal** rules only (811 rules). Each curve is labeled with the maximum recall value reached.

We applied our rules on ClueWeb12 and inspected the triples for recall over the 179 gold triples. Since we have 20 rules per relation, we also computed recall by considering only the top m rules per relation, yielding 20 additional recall metrics, Recall top-m, where $1 \leq m \leq 20$. We can further refine our metrics by considering only the subset of rules that are either verbal or non-verbal, as identified in Sect. 4.2. The complete results are shown in Fig. 2.

Unsurprisingly, the complete rule set (all-rules) achieves the highest recall, with 91.1% of all triples, a mark hit for $m \geq 12$. Not far behind is the rule set limited to verbal patterns, at 88.3% (starting at $m = 14$). The recall for the non-verbal rule set culminates at 76.5% (starting at $m = 13$). The best recall is quite high, which is encouraging. Even when only the top rule is kept for each relation ($m = 1$), the best recall is 75.4%. Because the recall starts at a relatively high value, further extraction rules have a limited effect, but they do help. Out of 179 reference triples, 16 triples cannot be found with our best system. Manual examination reveals that there are two causes for this.

The first difficulty is with triples whose relation yields relatively few candidate patterns (Step 4 in Fig. 1). For instance, the reference triple (*Aristotle, return to, Athens*) uses a relation that only generates 130 extraction patterns after filtering (the average is 409 patterns). The top patterns for this relation are dubious: `grip`, `may be from`, `profess his love for`, etc.

The second problem is actually an artefact of our evaluation procedure. For instance, we cannot find the reference triple (*britt gillette, be author of, the dvd report*) using our extraction rules because the only co-occurrence of *britt gillette* and *the dvd report* in ClueWeb sentences is in the token sequence *britt gillette be author of the dvd report*. Since we exclude the relation itself (*be author of* here) from the list of extraction rules evaluated, we miss this triple. If we compensate for this, the aforementioned 91.1% recall is bumped up to 92.7%.

The verbal patterns outperform the non-verbal ones by a significant margin of 11.8% absolute, even if their respective counts are almost the same.

This confirms the intuition that verb phrases are stronger hints of the presence of a relation, while at the same time shows that non-verbal patterns are fertile on their own, recalling at most 76.5% of triples.

5.2   Precision

Assessing precision in OIE is tricky, for the reasons put forward at the beginning of this section. We could not rely on an automatic metric like we did for recall and instead had to manually assess the quality of the output. Since our extraction rules are not yet part of a full-fledged OIE system, we had to simulate the processing performed by such a system manually.

For each relation r, we enumerated its 20 extraction patterns p_i and randomly selected from ClueWeb12 up to 10 sentences that matched the pattern NE_1 p_i NE_2. We then labeled the corresponding extractions (NE_1, r, NE_2) as either correct or incorrect. A correct extraction is one whose meaning is clearly stated in the original sentence. The task was very time-consuming, so we resorted to a random sample of 20 out of 81 relations to carry out the assessment.

For the 20 relations evaluated (and their associated 400 extraction rules), we labeled 3559 triples, an average of 8.9 triples per rule. Like we did for recall, we also computed precision by considering the top m rules per relation, yielding a precision measure for each m, where $1 \leq m \leq 20$. Figure 3 shows the results.

Precision decreases from 62.4% for $m = 2$ down to 36.4% for $m = 20$. This is expected, because the lower an extraction rule is in the list associated with a relation, the lower its quality will be. An F-measure cannot be computed here, since precision and recall are obtained from heterogeneous processes. Ultimately, for $m = 2$ (top 2 rules only), we have a recall of 79.9% and a precision of 62.4%. Non-verbal patterns' precision is disappointingly low.

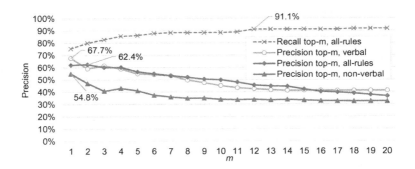

Fig. 3. Precision assessment over 3559 triples, for different values of m, the number of top rules retained for each relation. The recall measure is copied from Fig. 2 to provide a more complete picture of the evaluation.

Two types of error decrease precision. Firstly, erroneously acquired extraction patterns corrupt extraction. A rule like NE_1 `society of` $NE_2 \rightarrow (NE_1, \textit{die in}, NE_2)$ will always fail. This is the case for 50.7% of rules, a high figure, but it is crucial to point out that erroneous rules are typically ranked much lower, with a rank of 11.0 on average. Indeed, 97.8% of rules at rank 1 are valid. Secondly, some rules are correctly triggered in certain contexts, but not in others. For instance, the rule NE_1 `then part of` $NE_2 \rightarrow (NE_1, \textit{be ruled by}, NE_2)$ should be triggered in *Sicily, then part of the Roman Empire* but not for *Haifa, then part of British Palestine*. A manual examination of the non-verbal patterns show that they tend to be shorter on average (2.7 words vs. 3.0 words for verbal patterns) and tend to employ a more generic vocabulary (e.g. patterns like `mayor`, `within`), which may explain why they are triggered unnecessarily.

6   Discussion

In this work, we set out to extend the heuristic possibilities currently used in OIE by exploring the potential of arbitrary lemma patterns to trigger cogent extractions. We further wanted to acquire these additional rules without too much supervision, and in a generic framework consistent with the goals of OIE.

Our experiments show, at the very least, that it is possible to acquire these rules using data mining over a large corpus. Our manual evaluation shows that for 81 relations the top rule is sensible in 97.8% of the cases, comprising interesting non-verbal paraphrases like *Smith's death in Mali* $\rightarrow$ (*Smith, die in, Mali*). Our strategy can extract overlapping facts, a common occurrence according to [10].

While evaluation remains difficult in OIE, we managed to devise a proxy to a full-fledged extractor and apply it to measure recall and precision. The most precise configuration retains the top 2 rules for each relation, for a precision of 62.4% and a recall of 79.9%. The top precision and recall across all configurations is respectively 67.7% and 91.1%. Verbal patterns tend to outperform non-verbal ones in both respect, which is somewhat disappointing, but not unexpected.

While the results leave room for improvement, especially regarding precision, one must also consider how little supervision went into the acquisition process. By merely amplifying the signal given by a seed consisting of entity pairs linked by a relation, we can gather additional extraction rules by mining data.

We could nonetheless ameliorate our work by adding entity types (e.g. person, place) to our extraction rules, and by relying less on named entities, in order to extract desirable triples like (*meteors, occur in, the mesosphere*). It also remains to be seen how the strategy we propose can be extended to a very large number of relations. Data-mining approaches are sensitive to the amount of text at their disposal to accomplish their task. Here, we sifted through high-frequency ReVerb triples in order to find enough acceptable triples and their associated relations. It could prove more difficult to acquire rules in the long tail of rarer triples. This is important, because OIE strives to be generic.

References

1. Banko, M., Cafarella, M.J., Soderland, S., Broadhead, M., Etzioni, O.: Open information extraction from the web. In: Proceedings of the 20th International Joint Conference on Artificial Intelligence, IJCAI 2007, India, pp. 2670–2676 (2007)
2. Carlson, A., Betteridge, J., Kisiel, B., Settles, B., Hruschka Jr., E.R., Mitchell, T.M.: Toward an architecture for never-ending language learning. In: Proceedings of the Twenty-Fourth Conference on Artificial Intelligence (AAAI 2010), p. 3, July 2010
3. Carlson, A., Betteridge, J., Wang, R.C., Hruschka, Jr., E.R., Mitchell, T.M.: Coupled semi-supervised learning for information extraction. In: Proceedings of the Third ACM International Conference on Web Search and Data Mining, WSDM 2010, pp. 101–110. ACM, New York (2010)
4. Cetto, M., Niklaus, C., Freitas, A., Handschuh, S.: Graphene: semantically-linked propositions in open information extraction. In: Proceedings of the 27th International Conference on Computational Linguistics, pp. 2300–2311. ACL (2018)
5. Del Corro, L., Abujabal, A., Gemulla, R., Weikum, G.: FINET: context-aware fine-grained named entity typing. In: Proceedings of the 2015 Conference on Empirical Methods in Natural Language Processing, pp. 868–878. ACL (2015)
6. Del Corro, L., Gemulla, R.: ClausIE: clause-based open information extraction. In: Proceedings of the 22nd International Conference on World Wide Web, WWW 2013, pp. 355–366. ACM, New York (2013)
7. Fader, A., Soderland, S., Etzioni, O.: Identifying relations for open information extraction. In: Empirical Methods in Natural Language Processing, EMNLP 2011, pp. 1535–1545 (2011)
8. Fader, A., Zettlemoyer, L., Etzioni, O.: Open question answering over curated and extracted knowledge bases. In: Proceedings of the 20th ACM SIGKDD International Conference on Knowledge Discovery and Data Mining, KDD 2014, pp. 1156–1165. ACM, New York (2014)
9. Gashteovski, K., Gemulla, R., Del Corro, L.: MinIE: minimizing facts in open information extraction. In: Proceedings of the 2017 Conference on Empirical Methods in Natural Language Processing, pp. 2630–2640. ACL (2017)
10. Hoffmann, R., Zhang, C., Ling, X., Zettlemoyer, L., Weld, D.S.: Knowledge-based weak supervision for information extraction of overlapping relations. In: Proceedings of the 49th Annual Meeting of the ACL: Human Language Technologies - Volume 1, HLT 2011, pp. 541–550. ACL (2011)
11. Léchelle, W., Gotti, F., Langlais, P.: WiRe57 : A Fine-Grained Benchmark for Open Information Extraction. arXiv:1809.08962 [cs], September 2018
12. Mausam Schmitz, M., Bart, R., Soderland, S., Etzioni, O.: Open language learning for information extraction. In: Joint Conference on Empirical Methods in NLP and Computational Natural Language Learning, pp. 523–534 (2012)
13. Mishra, B.D., Tandon, N., Clark, P.: Domain-targeted, high precision knowledge extraction. Trans. ACL 5, 233–246 (2017)
14. Nakashole, N., Weikum, G., Suchanek, F.: PATTY: a taxonomy of relational patterns with semantic types. In: Joint Conference on Empirical Methods in NLP and Computational Natural Language Learning, pp. 1135–1145 (2012)
15. Pal, H., Mausam: demonyms and compound relational nouns in nominal open IE. In: AKBC@NAACL-HLT (2016)

16. Saha, S., Pal, H., Mausam: bootstrapping for numerical open IE. In: Proceedings of the 55th Annual Meeting of the Association for Computational Linguistics (Volume 2: Short Papers). pp. 317–323. ACL, Vancouver (2017)
17. Stanovsky, G., Dagan, I.: Creating a large benchmark for open information extraction. In: Proceedings of the 2016 Conference on Empirical Methods in Natural Language Processing, pp. 2300–2305. ACL, Austin (2016)

Measuring Human Emotion in Short Documents to Improve Social Robot and Agent Interactions

David Skillicorn[1(✉)], Nasser Alsadhan[1], Richard Billingsley[2], and Mary-Anne Williams[2]

[1] School of Computing, Queen's University, Kingston, Canada
`skill@cs.queensu.ca`
[2] Magic Lab, University of Technology Sydney, Ultimo, Australia

Abstract. Social robots and agents can interact with people better if they can infer their affective state (emotions). While they cannot yet recognise affective state from tone and body language, they can use the fragments of speech that they (over)hear. We show that emotions – as conventionally framed – are difficult to detect. We suggest, from empirical results, that this is because emotions are the wrong granularity; and that emotions contain submotions that are much more clearly separated from one another, and so are both easier to detect and to exploit.

1 Introduction

We consider the problem of detecting, from fragments of speech or text, those aspects of human affective state that change over short timeframes. (We consider only text, a minor limitation given the quality of speech-to-text provided by social-agent interfaces such as Siri, Alexa, Cortana and so on.) Our target application is social robots and social agents, where the ability of a robot/agent to detect the emotions of people with whom it is interacting provides significant added value.

For example, detecting a person's affective state enables a robot/agent to infer whether or not its actions are meeting the that person's expectations, and so improve the experience it provides. People are strong, but unconscious, mimics and so a social robot/agent will seem more natural if it reflects, in some way, the affective state of people around it.

More problematically, the robot/agent could act to change that person's emotional state. This latter case is almost unexplored – should a social robot/agent simply *reflect* the person's emotional state, or try to *improve* it by acting, say, more upbeat if it detects that the person is sad?

People recognise the affective state in others using channels such as appearance, movement patterns (body language), voice tonality, and aspects of their language such as choice of words, and this process forms a crucial part of social interactions. Social robots/agents are much more limited (at least for now) and so we consider only a limited channel (speech as text).

© Springer Nature Switzerland AG 2019
M.-J. Meurs and F. Rudzicz (Eds.): Canadian AI 2019, LNAI 11489, pp. 29–41, 2019.
https://doi.org/10.1007/978-3-030-18305-9_3

The contribution of this paper is to show that emotions, as they are conventionally framed, are not at the right granularity for effective detection. Emotions such as anger and disgust overlap substantially, based on the word usage that signals them. Thus a model built from lexical markers labelled by the emotions they convey is often ambiguous, and therefore provides a weak basis for response and action. We then show that conventional emotions contain subemotions that are much better separated from one another. Detecting the subemotion being expressed by a short utterance therefore provides a robust foundation for response.

2   Components of Affective State

Human affective state is a complex construct. It is not difficult to find disagreements within the psychology and linguistics community about particular issues, but there is widespread agreement about the main structures [2]. Human affective state can be considered as being made up of the moods and emotions. Emotions are a channel that parallels cognition, but focuses on the short-term significance associated with external situations. Studies of people with brain injuries that reduce emotional intensity indicate that such injuries cause difficulty making decisions, suggesting that a critical part of the function of emotions is to rank the instantaneous importance associated with potential goals and tasks. Emotions are primarily a reaction to external circumstances, and they are typically of short duration (minutes).

There are a number of categorization of basic emotions, but we will use anger, disgust, fear, and sadness (which are considered negatively associated emotions); and anticipation, and joy (which are considered positively associated emotions). Surprise reflects a reaction to change and so can be both positively or negatively associated.

3   Modelling Strategy

Emotions are usually regarded as unipolar from a psychological perspective (although in common usage, many emotions have named opposites, e.g. angry-calm). For example, the emotion of joy is unipolar: one can be joyful, but the opposite is a neutral state of not being joyful, rather than being sad. In this case, a count of words from a lexicon of joy-related words estimates how joyful the speaker or writer of an utterance is.

Our basic approach is to use lexicons associated with each emotion: the seven lexicons for anger, disgust, fear, sadness, anticipation, joy, and surprise from the NRC emotion lexicon [5]. Plutchik [6] listed these as primary emotions[1].

---

[1] Trust is included in many lists of basic emotions, including the NRC lexicon. However, this does not seem natural: 'I feel trust' seems to be more of an attitudinal statement than an affective one, especially as it must have an object. Compare this to 'I feel angry'.

Classification of emotions has been either physiological (for example, Ekman [3]) or semantic, based on the association between names of emotions. Many overlapping sets of 'basic' emotions have been defined e.g. Russell's circumplex model [7] itself derived from Wundt's work in the 19th Century. The substructure of emotions has also been considered, using clustering via MDS [4, Chapter 6].

The NRC lexicon is a binary matrix with 14,182 rows, each corresponding to a word, and (for us) 9 columns corresponding to positive and negative mood, and the seven emotions (anger, disgust, fear, sadness, anticipation, joy, and surprise). The entries were generated by crowd sourcing, using Amazon's Mechanical Turk. The entries are binary – either there is an association between a particular word and a particular emotion or not, but there is no estimate of the intensity of the association. Of the 14,182 words, only 3766 have an association with any of our 7 emotions. Of these 1898 are associated with 2 or more, 936 are associated with 3 or more, 345 with 4 or more, 80 with 5 or more, 7 with 6 or more, and 2 with all 7 emotions ('feeling' and 'treat'). It is surprising that so many words express multiple emotions (we shall see why), including words that straddle the boundary between positive and negative association. The words expressing multiple emotions also tend to be the frequent ones.

4   Scoring Documents Using the Lexicons

We use, as data, the first 100,000 tweets in the Microsoft Research Conversation Corpus (MRCC), a set of 4.46 million tweets from about 1.3 million conversations. Tweets are a reasonable surrogate for the kinds of short, informal utterances that a social robot/agent might (over)hear. The average length of a tweet is 9 words. These data are unlabelled with emotions.

We first construct a document-word matrix for the tweets. Variations in document length are an issue for all corpora, because of their effect on document-document similarity. Tweets had an imposed upper bound of 140 characters, and so have much more consistent lengths. However, they present an unusual problem for length normalization: repeated words are used as a mechanism for emphasis. A bag-of-words model of a tweet such as "I'm happy, happy, happy" contains more joy signals than "I'm happy, I'm ecstatic" based on words counts, but arguably the second required more cognitive processing to generate and so is more indicative of the emotion of joy. Normalizing to remove the effects of repetition is important.

Readers are cautioned that example tweets are presented intact, and some contain potentially offensive language.

4.1   Emotions

To measure emotion, the document-word matrix is weighted by each emotion column of the NRC lexicon matrix in turn (i.e. only word that signal that emotion are selected).

The simplest way to compute an emotion score is simply to sum each row of these matrices. However, this does not work well because of word repetitions. Replacing word counts by word presence or absence (mapping the document-word matrix to binary) improves the results a little. However, in a lexical model, the impact of each word as a signal of the desired property is not exactly the same – some words are stronger signals than others – and a simple summation cannot capture this. We turn therefore to a more sophisticated approach to scoring documents: compute the weighted MRCC document-word matrix (W) as before, but then apply a singular value decomposition to it. For example, if W is the anger-weighted matrix, with 100,000 rows and 2312 columns, the U matrix truncated at $k = 3$ is a 100,000 by 3 matrix and we compute the norm of each row and use this as its score for ranking. Tweets that are similar to many other tweets tend to be plotted close to the origin (i.e. their norms are small) in the truncated SVD, and so they are given low scores. Conversely, tweets that are unusual tend to be plotted far from the origin and so are given high scores.

The justification for this comes from the symmetry of the SVD. Words that play the same role as most other words tend to be placed close to the origin in the truncated SVD – such words are weak and/or redundant signals of the property of interest. In contrast, words placed far from the origin tend to be strong signals – so larger distances are associated with more significant words. (Words that lie in the same *direction* from the origin tend to be associated with one another, which we will also exploit.) Points corresponding to both words and tweets can be plotted in the same space, with unusual tweets aligning with unusual words. Hence, our high-scoring tweets will be those which contain high-scoring words, and these are words that vary strongly in usage across the corpus.

Table 1 show the top scoring tweets for each emotion, using presence/absence and distance from the origin in a truncated SVD embedding.

These tweets plausibly represent expression of the associated emotions, but there is still some ambiguity. For example, the same tweets are ranked first and second for both anger and disgust. Emotions are not well-separated based on the representation of the NRC lexicon. However, we suspect that this is not an artifact of this particular lexicon which, after all, was generated by humans. Rather, the conventional labelling of emotions captures differences that are not large to begin with, and are hard to separate based on small fragments of text. We suspect that this is also true for people – we are not able to estimate an emotion in another person from only a small piece of textual communication but rely on other, more sophisticated signals.

There are some other issues that arise from the way in which the lexicon was constructed: workers were asked to provide emotion labels on words without context. Unsurprisingly, a word like 'hope' is considered to be a signal for anticipation, but its actual force in a real context is usually quite weak.

4.2 Subemotions

Once each weighted document-word matrix has been projected into 3 dimensions[2] by an SVD, we can also examine the structure induced on the words (the V matrix). For example, there are 1247 words that signal anger, so the truncated V matrix is 1247 by 3; and plot of the rows of V shows the variation across this set of anger-signalling words. These plots reveal a strong and surprising structure.

Each emotion contains within it a set of orthogonally varying subemotions, each with an associated set of words that act as markers for the subemotion. For almost all of these subemotions, there is a single word that is a strong marker for it, and then a small set of other words that are also associated. The vast majority of words act as much weaker signals.

Figure 1 shows subemotion structure for the emotion of anticipation, as an example. The variation overall is small, and considerable scaling was required to make the structures visible.

Fig. 1. Mutual variation among words that are associated with the emotion of anticipation.

Table 2 shows representative words from each of the orthogonal clusters in the SVD of the document-word matrix for each emotion. These words were determined by hand from the relevant plots. The most common structure is a cross shape in two dimensions, producing four arms and so four clusters; for some emotions, five arms were present, roughly a cross with another arm into the plane of the cross. Thus each emotion contains 4 or 5 well-differentiated subclusters.

---

[2] Each document-word matrix is different, but, for anger, the amount of variation captured by truncating at $k = 3$ is 55%.

Table 1. Top ranked tweets for each emotion (presence and SVD)

Score	Tweet
Anticipation	
0.039	its been fun peoples but its time to catch that flight tomorrow...sis i hope u finding that flight
0.038	HAPPY BDAY UNNIE! mianhae for tday. but hope you guys will try to haave fun w/out me, i know it'll be hard. teeext
0.037	hey alex i hope you had a great time on your Birthday,Happy Birthday my friend,hope it been a good un for ya :-)
Joy	
0.040	Wishing you much Love, and a very Happy Birthday Chris. Hope you hav a good one *hugs*
0.039	aww paula you look soo pretty as always!! love ya! hope youre having fun!! :)
0.039	Hey Paula hope you had a cool fun and crazy (dont forget safe) 4th of July!! Love you!!
Surprise	
0.051	starting my day at work. I guess I will cheer for LA tonight. hope my son forgets that he forgot my b-day. classic
0.051	user4251 smh eeewww i guess lettuce smells bad & i hope wen u say camel's ass u mean cigarette butts. lol
0.051	I was 'round when the Chinaman pissed on the Dude's rug. Pleased to meet you, hope you guess my name. hoohoo
Anger	
0.051	Damn! Jay-Z jus went hard as hell! Kilt that shit!
0.050	OMG, I am so jealous of you. I want that damn brush so bad
0.050	RT user7381: DontYouHate when girls keep talking bout they Damn give it a rest ..I'm a bad nigga U don't see me bragging... Word we bad
Disgust	
0.051	Damn! Jay-Z jus went hard as hell! Kilt that shit!
0.050	OMG, I am so jealous of you. I want that damn brush so bad
0.050	dude. Bad week. Just wasn't possible last night. A damn shame. I trust it was ridiculous?
Fear	
0.053	i hate when you say shit to see my reaction... im so beating your tall ass down now.. watch
0.053	ahaha espn sucks no offense. i hate watching tv actually. i only can stand to watch certain things
0.053	Mornin tweeps! Hate when there's nothin to watch on tv. Gna have to run a Friends marathon I think.. :)
Sadness	
0.060	Oh hell i hate that too
0.060	35 min on ergometer and sweating like hell.. hate this moisty weather lately (still 70% humidity)
0.056	ill be fine soon as i figure out the hell a dm is i prolly kno 2 thats wuz bad lmao

Table 2. Words associated with the subcomponents of each emotion, tweets

Emotion	Word lists
anticipation	1. time —— top respect cash love precious analyst lover
	2. good —— celestial delightful score shining worship
	3. luck —— chocolate happiness goodness public sunny gift
	4. hope fun happy tomorrow wait long watch birthday ready
joy	5. good —— celestial worship save outstanding shining passionate gain saint miracle
	6. luck — holiday cheer sweetheart delicious celebrating freedom treat excitement
	7. hope fun happy birthday glad feeling
	8. love —— hilarious peace inspiration weight respect
surprise	9. good —— hilarious teach shout score sunny unique
	10. hope —— luck finally lovely exciting blast pray amazingly
	11. guess — money break feeling birthday catch leave chance surprise lucky
anger	12. bad —— lawyer confused frustrated smack unfair rage painful tease despise horrid hanging illegal
	13. shit —— rob soldier treat shoot lying tree punch demonic force
	14. damn hate hell hit bitch bout fight kick
	15. hot crazy mad jealous feeling ill hurting
disgust	16. hate damn hell finally hood fat
	17. shit —— rob treat lemon ugly tree lying demonic liar interested
	18. bad —— swine lawyer unfair hanging failure boil painful pathetic despise wasted dislike
	19. mad feeling jealous shame sick lie ill bloody
fear	20. hate —— god hell feeling ill fight bitch crazy
	21. bad —— flu worse lines confusion lawyer highest failure horrid fever
	22. watch —— cobra warning prowl giant vampire horror
	23. mad pain killing afraid lose kill problem
	24. slaughter unemployed fearful jail divorce venom monster escape
sadness	25. bad —— horrid impossible unfair failure hanging painful
	26. slaughter unemployed suicide mistake anthrax obnoxious sympathize rabid cutting
	27. hate — rape beating antisocial mad disappointed complain disaster nasty terrible pain
	28. hell —— music late feeling die bitch crazy bloody
	29. worse illegal hurt doubt debt falling

Table 3. Substructures from positively associated emotions

Subemotion name	cluster in emotion
time-related anticipation	1
religious-good	2, 5
luck	3, 6
fun/hope	4, 7, 10
love	8
conventional-good	9
possibility	11

Table 4. Substructures from negatively associated emotions

Subemotion name	Cluster in emotion
bad-state	12, 18, 21, 25
shit	13, 17
mad	15, 19, 23
(artificial) horror	22
hate	14, 16, 20, 27, 28
bad-events	24, 26
worsening	29

Almost always the head word is far more significant in the cluster than any of the other words (which we show by the length of the dash between it and the following words).

Although 29 subclusters of emotions appear in Table 2, it is clear that there is considerable overlap between the emotions to which these subclusters belong. At least as expressed in short texts such as tweets, the conventional separation into emotions does not properly match the variation observed in the use of words that are supposed to carry these emotions.

There is room for argument, but we suggest that these subclusters can be refactored into (sub)emotions, based on the observed empirical variation, as shown in Tables 3 and 4.

Thus seven emotions with a total of 29 subclusters, have been reduced to 14 subemotions. In other words, our empirically justified claim is that conventional emotions are the wrong granularity for detecting affective state from short documents – the signals overlap too much. Detecting subemotions instead is both more accurate, and more practical.

4.3   Analysis of Subemotions

Rather than using a lexicon of 14,182 words, we can determine the presence of subemotions using a lexicon of about 170 words. For example, the following rules are strong in the positively associated domain:

- 'time' → time-related-anticipation
- 'good' + 'celestial' → religious-good
- 'good' + 'hilarious' → conventional-good
- 'happiness' → luck
- 'hope' → fun/hope
- 'happy' → fun/hope

and these are strong in the negatively associated domain:

- 'lawyer' → bad-state
- 'rob' → shit
- 'hot' → mad
- 'crazy' → mad
- 'warning' → horror
- 'damn' → hate
- 'hell' → hate
- 'bitch' → hate
- 'worse' → worsening

In this framework, a social robot/agent need only listen for a small number of words, and use these to infer the current subemotion being displayed by the person.

Clearly some of the subemotions discovered here depend on the corpus from which they are inferred. For example, the association between 'watch' and horror presumably does not generalize to all other corpora. However, we have repeated the analysis here for the Facebook corpus collected by Stillwell and Kosinski [1]. It consists of 9918 posts with an average length of 10.9 words, very slightly longer than the Twitter corpus. The results are the same – strong overlapping subemotions are found within each emotion, and the structures are essentially the same as the ones described here, with small differences in vocabulary.

We now repeat the scoring procedure, using the words listed in Table 2, and using them to predict the presence of subemotions in tweets. The highest ranking tweets for each of the subemotions are shown in Tables 5 and 6. Since the subemotions are more or less orthogonal, there is no need to apply an SVD to the weighted document-word matrices but, as before, word counts are changed to presence or absence to reduce the effect of repeated words.

Although these subemotions are arguably more independent of one another than the standard set of emotions (because the words that signal them vary independently, shown by the orthogonality of the embeddings of each word group), there is no reason why an individual should not exhibit more than one of them at a time. Indeed, common co-occurrences of subemotions may be part of the explanation for our perceptions of the standard emotions.

Table 5. Top ranked tweets for each positive valence subemotion (presence)

Score	Tweet
Time-related anticipation	
2	@user138 give away a 3GS next time, people will LOVE you
2	Off to the love of my lifes house tonight for movie and sleep time
Religious good	
3	But is banning the schools a good idea? Just cut off the public money, & Let them worship the Celestial Teapot if they wisha
2	Watchin that spike lee movie miracle on st.anna....its good so far
Luck	
2	It's a chocolate cupcake that has zucchini bits in it! It's absolutely delicious
2	Good luck with your story. I know you'll treat the family with respect. You're not one of "those" reporters
Fun/hope	
4	Hiya Eddie! Glad to see you back on twitter Cant wait to see you tomorrow. Hope you like Lancaster and the sun's out for you
4	For toms birthday I got him a day out somewhere...going tomorrow. Yipeee!! I can't wait! Hope he enjoys it!! Xxxxx
Love	
1	Mad excited! Can't wait to go to the Tiffany and Co boutique. @user319 When????? reward for after our weight loss!
1	Mike rowe was on sesame street, hilarious!
Conventional good	
1	LOL Well when the house is built..I'll give ya a shout! XO
1	Ok someone told me who put it out Shout out to "Memory Man" so go and download it till the full course comes....on SEPT 8TH YALL
Possibility	
2	Prison Break - what an anti-climax sad ending - how could the hero dies at the end? A show not worth watching i guess
2	*Sighs* I Guess I'm Not Ready To Talk Then..... The Reason I'm This Upset Cause I Have In My Mind and I Need a Break

For example, a tweet such as 'Wishing you much Love, and a very Happy Birthday Chris. Hope you hav a good one *hugs*', which was ranked highly for joy contains aspects of the subemotions conventional-good ('good'), hope/fun ('hope', 'happy', 'birthday'), and love. Clearly these often go together.

Figure 2a shows how the positive valence subemotions fit together into the conventional positive valence emotions: anticipation and joy with four subemotions, and surprise with three. The only difference between anticipation and joy are that anticipation contains a subemotion with a time element, and joy

Table 6. Top ranked tweets for each negative valence subemotion (presence)

Score	Tweet
Bad state	
3	they're bloody impossible to peel if they're 2fresh. I boil mine @ a week old. Make egg salad, nobody can tell they look bad lol
3	i never sleep too.. its shit. yeah i been ill since wednesday night, been well bad, i been semi diagnosed with swine flu
Shit	
3	stop lying and get bacc to your good posture and shit dwn and shoot up (inside joke) ... sit down buddy.. stop lying
2	why you gotta be tryin to shoot me and shit
Mad	
2	work with me, i make coffees, ill throw hot milk at your pretty face. ooo yeah i know you want it now
2	Getting mad sick of iPhone, its boring after awhile
Horror	
2	Giant flying cockroach just flew over my head and disappeared into my desk. This has become a horror movie
1	Am i the only guy on earth that found Megan Fox's boobs really distracting whilst trying to watch giant robots fight!!!
Hate	
3	Wha the fuck is up with the carrots guys, i really hate them give me HAM god damn it
3	God I hate Grease, bloody music
Bad events	
2	@user13214 but how can someone hurt themselves? Besides suicide and cutting
1	i thought thwey all got shot in the great escape apart from steve hep cat mcqueen on his motor bike...oh and some people rowing
Worse	
1	has hurt his foot and is a bit worried about running with the bulls on Friday. Walking with the bulls doesn't have the same ring to it!
1	I never meant to hurt you. But i guess i did when i said those three words

contains a subemotion associated with love. Fun/hope is the only subemotion that is part of three conventional emotions. Rules for inferring emotions from subemotions are complex: love with any of fun/hope, luck, or religious-good signals joy, but any combination of fun/hope, luck and religious-good doesn't distinguish anticipation from joy. This, of course, explains why conventional explanations of emotions do not separate them cleanly either in theory or practice – as constructs, they are inherently overlapping.

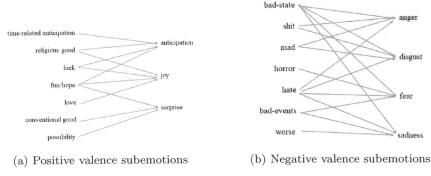

(a) Positive valence subemotions   (b) Negative valence subemotions

Fig. 2. Mapping subemotions to emotions

Figure 2b shows how the negative valence subemotions fit together into the conventional negative valence emotions. Note the greater density of edges – negative emotions overlap much more than their positive counterparts; and using subemotions is correspondingly more clarifying.

4.4   Benefits of Subemotions

A social robot/agent can use its awareness of the emotional state of a person in its vicinity to alter the way in which it responds and interacts with that person.

Suppose that a social robot/agent, working at the conventional level of emotions, detects that a person is angry. First, as we have seen, it is quite difficult for the social robot/agent to distinguish this from the other negative valence emotions. Is this really anger, or is it disgust, or something else? Second, what is the appropriate response? It may be difficult to tell whether the person is angry because of some internal dialogue, some action taken by the person that did not go as planned, or something in the external environment, including perhaps some action by the social robot/agent.

Contrast this with a social robot/agent that can detect that the person is in one of the subemotional states. Bad-state suggests that the source of the emotion is internal; bad-events suggests that it is external; shit again suggests an external focus, but on quite aggressive actions (probably not those of the social robot/agent); hate suggests an internal focus driven by values and attitudes; and worsening is a signal that whatever has been happening is not solving the problem the person faces. The ability to make these distinctions enables the social robot/agent to make better judgements about the domain of plausible actions it could take to address the current situation.

The same kind of analysis can be done with the positive valence emotions, although it is probably less critical that a social robot/agent get responses right in the direction.

The state of the art for social robots has not yet reached the level where emotional state detection is required. This is not true for social agents, however.

Technologies such as Siri, Alexa, and Cortana could already benefit from reading human (sub)emotions and tailoring their responses appropriately.

5 Conclusions

We have shown, empirically, that the relationship between a word and the emotion(s) it purports to signal, from the perspective of lexical approaches like that assumed by the NRC dataset, is more complex than it appears. Any given word tends to be an ambiguous signal of a particular emotion but this seems to be mostly because the conventional definitions of emotions are at the wrong granularity. Rather it seems as if there are subemotions, with less intuitive connotations. Conventional emotions are not mixtures of these subemotions; rather the subemotions appear as orthogonal components of conventional emotions (although multiple subemotions could be present at the same time).

Of course, the response problem has been made more difficult. Rather than having to answer the question "the person is *angry*, how should I respond?" the social robot/agent must answer questions such as "the person has made a fun-hope statement, how should I respond". This second class of questions seem harder to design from, partly because we are used to the intuition of emotions, and we have a large number of human-based or cultural rules about appropriate ways to respond to emotions in others. However, our analysis suggests that our view of emotions in others is perhaps rather fragile – as people we are probably not good at inferring emotions from short encounters with others, and we perhaps miss subtleties by immediately assuming our emotion constructs. Certainly the empirical investigation of short human communications does not support conventional emotions as the best representation of what people are trying to convey, consciously or unconsciously.

References

1. Costa Jr., P.T., McCrae, R.R.: Domains and facets: hierarchical personality assessment using the revised neo personality inventory. J. Pers. Assess. **64**(1), 21–50 (1995)
2. Davidson, R., Sherer, K., Goldsmith, H. (eds.): Handbook of Affective Sciences. Oxford University Press, Oxford (2009)
3. Ekman, P.: Basic emotions. In: Dalgleish, T., Power, M. (eds.) Handbook of Cognition and Emotion, pp. 45–60. John Wiley and Sons (1999)
4. Fillenbaum, S., Rapoport, A.: Structures in the Subjective Lexicon. Academic Press, New York (1971)
5. Mohammad, S.M., Turney, P.D.: Crowdsourcing a word-emotion association lexicon. Comput. Intell. **29**(3), 436–465 (2013)
6. Plutchik, R.: The nature of emotions. Am. Sci. **89**(4), 344–350 (2001)
7. Russell, J.: A circumplex model of affect. J. Pers. Soc. Psychol. **39**(6), 1161–1178 (1980)

Exploiting Symmetry of Independence
in d-Separation

Cory J. Butz[1]($\boxtimes$), André E. dos Santos[1], Jhonatan S. Oliveira[1],
and Anders L. Madsen[2,3]

[1] University of Regina, Regina, Canada
{butz,dossantos,oliveira}@cs.uregina.ca
[2] HUGIN EXPERT A/S, Aalborg, Denmark
anders@hugin.com
[3] Aalborg University, Aalborg, Denmark

Abstract. In this paper, we exploit the symmetry of independence in
the implementation of d-separation. We show that it can matter whether
the search is conducted from start to goal or vice versa. Analysis reveals
it is preferable to approach observed *v-structure* nodes from the bottom. Hence, a measure, called *depth*, is suggested to decide whether the
search should run from start to goal or from goal to start. One salient
feature is that depth can be computed during a pruning optimization
step widely implemented. An empirical comparison is conducted against
a clever implementation of d-separation. The experimental results are
promising in two aspects. The effectiveness of our method increases with
network size, as well as with the amount of observed evidence, culminating with an average time savings of 9% in the 9 largest BNs used in our
experiments.

Keywords: d-separation · Bayesian networks ·
Symmetry inference axiom · Conditional independence

1 Introduction

d-Separation [13] is perhaps the greatest contribution made in the founding of
Bayesian networks (BNs) [14]. d-Separation is a graphical test of those probabilistic conditional independencies encoded in the *directed acyclic graph* (DAG)
of a BN. One striking feature is that the independencies read from the DAG
are guaranteed to hold in the joint probability distribution defined by the BN.
Another salient characteristic is that d-separation can be implemented with linear complexity in the number of edges in the DAG. Consequently, d-separation
continues to be useful in a wide range of areas, including causal inference in
statistics [15], cause and correlation in biology [18], extrapolation across populations [16], cognition [11], handling missing data [9], bioinformatics [10], and deep
learning [4]. Many variations have been studied for d-separation [1,3,6,7,12,17].
Given disjoint pairwise sets X, Y, and Z of variables (nodes), the conditional

© Springer Nature Switzerland AG 2019
M.-J. Meurs and F. Rudzicz (Eds.): Canadian AI 2019, LNAI 11489, pp. 42–54, 2019.
https://doi.org/10.1007/978-3-030-18305-9_4

independence of X and Z given Y, denoted $I(X, Y, Z)$, holds in a DAG, if there does not exist an *active* path from X to Z with respect to Y [13]. It is known that if $I(X, Y, Z)$ holds by d-separation in the DAG, then $I(X, Y, Z)$ holds in joint probability distribution defined by the BN.

Lauritzen et al. [7] have established that said active paths can only involve nodes in XYZ and their ancestors. Hence, one fundamental optimization step is to restrict attention to the sub-DAG defined on these variables. However, all the above implementations find R_X, the reachable nodes from X along active paths with respect to Y, and then test whether

$$R_X \cap Z = \emptyset. \tag{1}$$

Thereby, all previous implementations ignore the fact that probabilistic conditional independence is *symmetric* [2].

In this paper, we propose SYMMETRIC D-SEPARATION as the first d-separation implementation that exploits the symmetry of probabilistic conditional independence. Since $I(X, Y, Z)$ is equivalent to $I(Z, Y, X)$ by symmetry, the test can be answered by computing R_Z, the nodes reachable from Z along active paths with respect to Y, and then testing whether

$$R_Z \cap X = \emptyset. \tag{2}$$

It is important to realize that R_X and R_Z can contain different nodes and may have distinct cardinalities. Our analysis of reachability highlights that *v-structure* [6] nodes play an important role in this regard. More specifically, it may be better to approach observed v-structure nodes from the bottom. Approaching an observed v-structure node from a child blocks an active path, whereas an active path can continue when approaching an observed v-structure node from a parent. Hence, a measure, called *depth*, is suggested to decide whether the search should run from start to goal or from goal to start. One salient feature is that depth can be computed during the pruning optimization step of finding the ancestor set of XYZ. An empirical comparison is conducted against a clever implementation of d-separation suggested by [1], which we refer to as DARWICHE. The experimental results are promising in two aspects. The effectiveness increases with network size, as well as with the amount of observed evidence, culminating with an average time savings of 9% in the 9 largest BNs used in our experiments.

The remainder of this paper is organized as follows. Section 2 contains background knowledge. d-Separation exploiting symmetry of independence is proposed in Sect. 3. Section 4 reports an empirical comparison of SYMMETRIC D-SEPARATION against DARWICHE. Conclusions are drawn in Sect. 5.

2   Background

A *Bayesian network* (BN) [13] is a *directed acyclic graph* (DAG) $\mathcal{D}$ on U together with *conditional probability tables* (CPTs) $P(v_1|Pa(v_1))$, $P(v_2|Pa(v_2))$, $\ldots, P(v_n|Pa(v_n))$. For example, Fig. 1i shows a BN, where CPTs $P(A)$, $P(B)$, $\ldots, P(J|G)$ are not provided. We call $\mathcal{D}$ a BN, if no confusion arises. The product of the CPTs for $\mathcal{D}$ on U is a *joint probability distribution* $P(U)$ [13]. The *conditional independence* [13] of X and Z given Y holding in $P(U)$ is denoted $I_P(X,Y,Z)$, where X, Y, and Z are pairwise disjoint subsets of U. It is known that if $I(X,Y,Z)$ holds in $\mathcal{D}$, then $I_P(X,Y,Z)$ holds in $P(U)$. We denote the descendants and ancestors of variable v by $De(v)$ and $An(v)$, respectively. For a set X of variables, we define $De(X)$ and $An(X)$ in the obvious way.

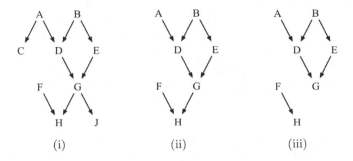

Fig. 1. When testing $I(D,G,H)$ in (i), active paths between D and H with respect to G can only involve variables in the sub-DAG in (ii) defined over $DGH \cup An(DGH)$. Darwiche's sub-DAG in (iii).

Definition 1. *[13] A variable v_k is called a v-structure [6] in a DAG $\mathcal{D}$, if $\mathcal{D}$ contains directed edges (v_i, v_k) and (v_j, v_k), but not a directed edge between variables v_i and v_j.*

For example, D, G, and H are the 3 v-structure nodes in Fig. 1i.

d-*Separation* [13] tests independencies in BNs and can be presented as follows [1]. Let X, Y, and Z be pairwise disjoint sets of variables in a BN $\mathcal{D}$. We say X and Z are d-*separated* by Y, denoted $I(X,Y,Z)$, if at least one variable on every undirected path from (any variable in) X to (any variable in) Z is closed. On a path, there are three kinds of variable v: (i) a *sequential* variable means v is a parent of one of its neighbours and a child of the other; (ii) a *divergent* variable is when v is a parent of both neighbours; and (iii) a *convergent* variable is when v is a child of both neighbours. A variable v is either open or closed. A sequential or divergent variable is *closed*, if $v \in Y$. A convergent variable is *closed*, if $(v \cup De(v)) \cap Y = \emptyset$. A path with a closed variable is *blocked*; otherwise, it is *active*.

A fundamental optimization step stems from Lemma 1.

Lemma 1. *[7] When testing $I(X, Y, Z)$ in a BN $\mathcal{D}$, an active path from X to Z with respect to Y can only involve nodes in $XYZ \cup An(XYZ)$.*

For example, consider testing $I(D, G, H)$ in the DAG shown in Fig. 1i. As

$$\{D, G, H\} \cup An(D, G, H) = \{A, B, D, E, F, G, H\}, \tag{3}$$

the independence $I(D, G, H)$ can be safely tested in the sub-DAG shown in Fig. 1ii.

Given an independence statement $I(X, Y, Z)$ to be tested in a BN $\mathcal{D}$. By R_X, we denote the set of nodes reachable from X along active paths with respect to Y. For instance, when testing $I(D, G, H)$ in the pruned DAG in Fig. 1ii, the reachable nodes from D are $R_D = \{A, B, E\}$.

2.1 Darwiche's Implementation of d-Separation

Here, we review a clever implementation of d-separation as suggested by [1], which we will henceforth refer to as DARWICHE.

Theorem 1. *[1] Testing whether X and Z are d-separated by Y in DAG $\mathcal{D}$ is equivalent to testing whether X and Z are disconnected in a new DAG $\mathcal{D}'$, which is obtained by pruning $\mathcal{D}$ as follows:*

- *delete any leaf node W from $\mathcal{D}$ as long as W does not belong to $X \cup Y \cup Z$. This process is repeated until no more nodes can be deleted.*
- *delete all edges outgoing from nodes in Y.*

This implementation is clever in that its connectivity test ignores edge directions, yet is equivalent to testing active paths [1].

Example 1. Consider testing $I(D, G, H)$ in the DAG $\mathcal{D}$ in Fig. 1i using DARWICHE. Leaf nodes C and J are removed in the first pruning step of Theorem 1. Directed edge (G, H) is deleted in the second pruning step, yielding the sub-DAG in Fig. 1iii. The nodes reachable from node D are

$$R_D = \{A, B, D, E\}.$$

Therefore, $I(D, G, H)$ holds, since node H is not reachable from D.

3 d-Separation Exploiting Symmetry

In this section, we present a novel implementation of d-separation that exploits the symmetry of probabilistic conditional independence.

Symmetry is known property of probabilistic conditional independence [2] and is included as one inference axiom in the semi-graphoid axioms [13]. Thus, $I(X, Y, Z)$ holds in $P(U)$ if and only if $I(Z, Y, X)$ holds in $P(U)$.

Given $I(X, Y, Z)$, the nodes R_X reachable from X will often be different from the nodes R_Z reachable from Z, since X and Z are disjoint sets of nodes.

Example 2. Recall testing $I(D, G, H)$ in Fig. 1i, where the pruning optimization step yields the sub-DAG on $\{D, G, H\} \cup An(DGH)$ in Fig. 1ii. Notice that $R_D = \{A, B, D, E\}$, while $R_H = \{F, H\}$.

We now analyze factors that affect reachability.

The set of reachable nodes can depend on the direction a node is approached from and on the kind of node. Consider a v-structure node v_i with parent set P_i and such that $v_i \in Y$ in $I(X, Y, Z)$. Then reaching any node in P_i makes all other non-evidence nodes in P_i reachable, since v_i is an open, convergent variable. On the contrary, P_i is not reachable from the children of v_i, since v_i is a closed, sequential variable.

Example 3. Consider the v-structure C in the DAG $\mathcal{D}$ in Fig. 2ii, where $C \in Y$ of $I(X, Y, Z)$. C does not block the path from A to B in Fig. 2i, since C is an open, convergent variable. On the contrary, C blocks the path from D to B (and A) in Fig. 2ii, since C is a closed, sequential variable.

(i)   (ii)

Fig. 2. Consider v-structure node $C \in Y$ in $I(X, Y, Z)$. (i) Approaching C from the top, say from A, renders B reachable. (ii) Approaching C from bottom, makes B unreachable.

Consequently, it may be better to compute reachability in DAGs from bottom-to-top rather than top-to-bottom. Given $I(X, Y, Z)$ to be tested in a DAG, the universal approach to implementing d-separation is to test whether

$$R_X \cap Z = \emptyset. \tag{4}$$

However, by symmetry of independence, one may equivalently test whether

$$R_Z \cap X = \emptyset. \tag{5}$$

The idea is to compute the smaller of R_X and R_Z. It is conjectured that the set with fewer nodes between R_X and R_Z corresponds to the set X or Z appearing "lower" in the DAG. In other words, if X is lower than Z in the DAG, then R_X likely has fewer elements than R_Z.

We now propose a measure of depth for a given set of nodes in a DAG.

Definition 2. *The* depth *of a set W of nodes in a DAG, denoted $depth(W)$, is defined as $|An(W)|$.*

Example 4. In Fig. 1ii, $depth(D) = 2$, since $An(D) = \{A, B\}$.

Given $I(X, Y, Z)$ to be tested in a DAG, if $depth(X) > depth(Z)$, we test

$$R_X \cap Z = \emptyset;$$

otherwise, test

$$R_Z \cap X = \emptyset.$$

Our main contribution, called SYMMETRIC D-SEPARATION, is formally given in Algorithm 1. For computing reachable nodes in lines 9 and 13, we use Algorithm 2.

Algorithm 1. Symmetric d-separation tests whether a given independence $I(X, Y, Z)$ holds in a given BN $\mathcal{D}$.

1: **procedure** SYMMETRIC D-SEPARATION(X,Y,Z,$\mathcal{D}$)
2:     ▷ Initialization
3:     $X^{up} \leftarrow X \cup An(X)$
4:     $Y^{up} \leftarrow Y \cup An(Y)$
5:     $Z^{up} \leftarrow Z \cup An(Z)$
6:     $XYZ^{up} \leftarrow X^{up} \cup Y^{up} \cup Z^{up}$ ▷ The relevant part of $\mathcal{D}$
7:     ▷ Exploit symmetry
8:     **if** $|An(X)| > |An(Z)|$ **then**   ▷ Check depth
9:         $R_X \leftarrow$ SYMMETRIC-REACHABLE($X, Y, Z, \mathcal{D}, XYZ^{up}$)
10:        **if** $R_X \cap Z = \emptyset$ **then**
11:            **return** true
12:    **else**   ▷ Test $I(X, Y, Z)$ as $I(Z, Y, X)$
13:        $R_Z \leftarrow$ SYMMETRIC-REACHABLE($Z, Y, X, \mathcal{D}, XYZ^{up}$)
14:        **if** $R_Z \cap X = \emptyset$ **then**
15:            **return** true
16:    **return** false

Example 5. Let us test $I(D, G, H)$ in Fig. 1i using Algorithm 1. In particular, line 3 computes

$$D^{up} = \{A, B, D\},$$

while line 5 determines

$$H^{up} = \{A, B, D, E, F, G, H\}.$$

Algorithm 2. Given an independence $I(X,Y,Z)$, SYMMETRIC-REACHABLE traverses all active paths from X within the relevant part of a BN $\mathcal{D}$.

```
1: procedure SYMMETRIC-REACHABLE(X,Y,Z,D,XYZ^up)
2:     V ← ∅                          ▷ (Variable,direction) marked as visited
3:     R ← ∅                          ▷ Variables reachable via an active path
4:     L ← ∅                          ▷ (Variable,direction) to be visited
5:     for v ∈ X do
6:         L ← L ∪ {(↑,v)}
7:     ▷ Starting from X traverse relevant paths that are active
8:     while L ≠ ∅ do                 ▷ While variables to be checked
9:         Select (d,v) in L
10:        L ← L − {(d,v)}
11:        if (d,v) ∉ V then          ▷ If v has not been visited from direction d
12:            if v ∉ Y then
13:                R ← R ∪ {v}        ▷ v is reachable
14:            V ← V ∪ {(d,v)}        ▷ Mark v as visited from direction d
15:            if d =↑ and v ∉ Y then
16:                for v_i ∈ Pa(v) do              ▷ v is open sequential
17:                    if v_i ∈ XYZ^up then        ▷ Only explore relevant paths
18:                        L ← L ∪ {(↑,v_i)}
19:                for v_i ∈ Ch(v) do              ▷ v is open divergent
20:                    if v_i ∈ XYZ^up then        ▷ Only explore relevant paths
21:                        L ← L ∪ {(↓,v_i)}
22:            else if d =↓ then
23:                if v ∉ Y then
24:                    for v_i ∈ Ch(v) do          ▷ v is open sequential
25:                        if v_i ∈ XYZ^up then    ▷ Only explore relevant paths
26:                            L ← L ∪ {(↓,v_i)}
27:                if v ∈ A then
28:                    for v_i ∈ Pa(v) do          ▷ v is open convergent
29:                        if v_i ∈ XYZ^up then    ▷ Only explore relevant paths
30:                            L ← L ∪ {(↑,v_i)}
31:    return R        ▷ All variables reachable from X via active paths within the
        relevant part of D
```

The computation of D^{up} and H^{up} is used in the key pruning step in line 6. Line 8 checks depth. Since $|An(D)| < |An(H)|$, we exploit symmetry to instead test $I(H,G,D)$ in line 12. In line 13,

$$R_H = \{F,H\}.$$

In line 14,

$$R_H \cap \{D\} = \emptyset.$$

Hence, $I(H,G,D)$ holds.

The key point in Example 5 is that it is beneficial to take advantage of symmetry in order to test $I(D, G, H)$ as $I(H, G, D)$. Recall how DARWICHE tests $I(D, G, H)$ in Example 1. First, DARWICHE prunes the DAG. Next, DARWICHE computes the nodes reachable from D,

$$R_D = \{A, B, D, E\}, \tag{6}$$

and then tests whether

$$R_D \cap \{H\} = \emptyset. \tag{7}$$

In contrast, consider how SYMMETRIC D-SEPARATION tested $I(D, G, H)$ in Example 5. By computing $An(D)$ and $An(H)$ during the pruning optimization step, it estimated that H is lower than D in the DAG of Fig. 1ii. Since it may be preferable to traverse bottom-up, symmetry is exploited to equivalently test $I(D, G, H)$ as $I(H, G, D)$. Hence, the nodes reachable from H are computed,

$$R_H = \{F, H\}, \tag{8}$$

and then it is checked whether

$$R_H \cap \{D\} = \emptyset. \tag{9}$$

Thereby, DARWICHE computed $R_D = \{A, B, D, E\}$ in (6), whereas SYMMETRIC D-SEPARATION computed $R_H = \{F, H\}$ in (8).

4   Experimental Results

We report on an empirical comparison of DARWICHE and SYMMETRIC-REACHABLE. Both methods were implemented in Python using the *NetworkX* library (see networkx.github.io). The experiments were conducted on a 2.9 GHz Intel Core i7 with 8 GB RAM. The evaluation was carried out on 18 real-world or benchmark BNs listed in the first columns of Table 1. The second column reports the number of variables in each BN. For each BN, 1000 independencies $I(X, Y, Z)$ were randomly generated and tested by DARWICHE and SYMMETRIC-REACHABLE. Following REACHABLE in [6], in our experiments X and Z are kept to being singleton sets. The average time in seconds required by DARWICHE and SYMMETRIC-REACHABLE are reported in the fourth and sixth columns, respectively. For completeness, the percentage of time taken by DARWICHE to prune is listed in the third column, while the percentage of time taken to exploit symmetry in SYMMETRIC-REACHABLE is given in the fifth column. In the last column, we show the time savings of SYMMETRIC-REACHABLE over DARWICHE.

Table 1. With 1% of evidence, comparison of DARWICHE and SYMMETRIC D-SEPARATION with 1000 randomly generated independencies in each BN.

BN	Size	Time pruning darwiche	Time (s) darwiche	Time swap	Time (s) sym sep	Time savings
child	20	4.77%	2.74E−02	1.21%	2.76E−02	−0.58%
insurance	27	4.77%	2.74E−02	1.15%	2.82E−02	−2.96%
water	32	5.17%	2.96E−02	1.16%	3.04E−02	−2.68%
mildew	35	5.04%	2.85E−02	1.08%	3.03E−02	−6.14%
alarm	37	4.87%	2.77E−02	1.25%	2.80E−02	−1.04%
barley	48	4.89%	3.00E−02	1.01%	3.16E−02	−5.48%
hailfinder	56	4.81%	2.91E−02	1.03%	3.05E−02	−4.92%
hepar2	70	4.64%	2.95E−02	1.22%	3.16E−02	−7.21%
win95pts	76	5.36%	3.12E−02	1.00%	3.33E−02	−6.57%
pathfinder	109	5.13%	3.90E−02	0.95%	4.21E−02	−7.91%
munin1	186	5.15%	3.38E−02	0.77%	3.95E−02	−16.70%
andes	223	3.97%	6.64E−02	0.49%	7.58E−02	−14.25%
pigs	441	5.01%	1.20E−01	0.37%	1.28E−01	−6.50%
link	724	4.26%	2.41E−01	0.20%	2.57E−01	−6.70%
munin2	1003	4.96%	4.44E−01	0.22%	4.76E−01	−7.26%
munin4	1038	4.55%	3.45E−01	0.19%	3.65E−01	−5.80%
munin	1041	3.97%	2.91E−01	0.15%	3.05E−01	−4.59%
munin3	1041	4.55%	3.76E−01	0.21%	4.10E−01	−8.96%
Average		4.77%	1.23E−01	0.76%	1.32E−01	−6.46%

Experiments are broken down into separate cases based on the percentage of evidence variables. Table 1 reports the case where 1% of the BN variables are randomly instantiated as evidence $Y \in I(X, Y, Z)$. Similarly, Tables 2, 3, and 4 show cases 5%, 10%, and 25%, respectively.

Table 1 shows that DARWICHE is faster than SYMMETRIC D-SEPARATION by an average of 6% when 1% evidence is considered. In Table 2, for 5% evidence, SYMMETRIC D-SEPARATION is faster in the 10 largest BNs and slower in the 8 smallest BNs. SYMMETRIC D-SEPARATION wins in 15 out of 18 BNs when 10% evidence is considered in Table 3. Finally, in Table 4, SYMMETRIC D-SEPARATION is faster than DARWICHE in all BNs by an average of 6%, where 25% evidence is examined.

Table 2. With 5% of evidence, comparison of DARWICHE and SYMMETRIC D-SEPARATION with 1000 randomly generated independencies in each BN.

BN	Size	Time pruning darwiche	Time (s) darwiche	Time swap	Time (s) sym sep	Time savings
child	20	5.24%	3.33E−02	1.15%	3.39E−02	−1.67%
insurance	27	5.11%	3.00E−02	1.09%	3.19E−02	−6.29%
water	32	5.12%	3.25E−02	1.19%	3.34E−02	−2.72%
mildew	35	5.04%	3.11E−02	1.03%	3.33E−02	−7.17%
alarm	37	5.32%	2.94E−02	1.18%	3.00E−02	−1.98%
barley	48	5.00%	6.45E−02	0.70%	6.83E−02	−5.97%
hailfinder	56	4.88%	5.82E−02	0.63%	5.98E−02	−2.73%
hepar2	70	4.57%	8.35E−02	0.47%	8.43E−02	−0.89%
win95pts	76	4.52%	8.31E−02	0.48%	8.30E−02	0.12%
pathfinder	109	3.90%	1.27E−01	0.32%	1.26E−01	0.87%
munin1	186	3.59%	2.26E−01	0.18%	2.25E−01	0.37%
andes	223	4.38%	3.62E−01	0.21%	3.59E−01	0.87%
pigs	441	3.56%	5.90E−01	0.09%	5.86E−01	0.56%
link	724	3.93%	1.30E+00	0.10%	1.29E+00	0.76%
munin2	1003	3.44%	1.52E+00	0.05%	1.48E+00	2.58%
munin4	1038	3.43%	1.55E+00	0.05%	1.50E+00	2.94%
munin	1041	3.41%	1.57E+00	0.05%	1.53E+00	2.91%
munin3	1041	3.34%	1.53E+00	0.05%	1.49E+00	2.31%
Average		4.32%	5.12E−01	0.50%	5.03E−01	−0.84%

Deeper analysis of experimental results reveals two trends. First note that the time savings offered by SYMMETRIC D-SEPARATION is proportional to the percentage of observed variables. The more evidence that is observed, the greater the time savings of SYMMETRIC D-SEPARATION over DARWICHE. DARWICHE is faster than SYMMETRIC D-SEPARATION by 6% when 1% evidence is considered, and by 1% when 5% evidence is examined. In contrast, SYMMETRIC D-SEPARATION, is faster than DARWICHE by 2% when 10% evidence is considered, and by 6% when 25% evidence is tested.

The second trend to note is that SYMMETRIC D-SEPARATION is especially effective on larger BNs. Consider those 9 BNs in Table 1 with more than 100 variables. It can be verified that DARWICHE is faster by 9% when 1% evidence is considered. SYMMETRIC D-SEPARATION, on the other hand, is faster in 5% by 2%, in 10% by 3%, and in 25% by 9%, respectively.

The important point in this section is that the experimental results show the usefulness of SYMMETRIC D-SEPARATION when implementing d-separation.

Table 3. With 10% of evidence, comparison of DARWICHE and SYMMETRIC D-SEPARATION with 1000 randomly generated independencies in each BN.

BN	Size	Time pruning darwiche	Time (s) darwiche	Time swap	Time (s) sym sep	Time savings
child	20	4.33%	5.20E−02	0.64%	5.27E−02	−1.52%
insurance	27	4.21%	5.21E−02	0.62%	5.21E−02	0.10%
water	32	4.45%	9.13E−02	0.50%	9.18E−02	−0.45%
mildew	35	4.10%	7.52E−02	0.45%	7.48E−02	0.60%
alarm	37	4.11%	7.45E−02	0.44%	7.35E−02	1.36%
barley	48	4.11%	1.06E−01	0.37%	1.05E−01	1.15%
hailfinder	56	3.83%	1.29E−01	0.34%	1.27E−01	2.27%
hepar2	70	3.66%	1.81E−01	0.24%	1.76E−01	2.93%
win95pts	76	4.30%	2.24E−01	0.28%	2.18E−01	2.66%
pathfinder	109	4.36%	3.00E−01	0.23%	2.95E−01	1.69%
munin1	186	3.98%	5.38E−01	0.14%	5.27E−01	2.14%
andes	223	3.41%	5.91E−01	0.09%	5.86E−01	0.90%
pigs	441	3.82%	1.46E+00	0.09%	1.48E+00	−1.88%
link	724	3.14%	2.22E+00	0.03%	2.21E+00	0.51%
munin2	1003	3.28%	3.94E+00	0.05%	3.73E+00	5.29%
munin4	1038	3.29%	3.90E+00	0.04%	3.65E+00	6.28%
munin	1041	3.08%	3.91E+00	0.03%	3.59E+00	8.11%
munin3	1041	3.20%	4.13E+00	0.04%	3.84E+00	6.87%
Average		3.81%	1.22E+00	0.26%	1.16E+00	2.17%

Table 4. With 25% of evidence, comparison of DARWICHE and SYMMETRIC D-SEPARATION with 1000 randomly generated independencies in each BN.

BN	Size	Time pruning darwiche	Time (s) darwiche	Time swap	Time (s) rp_sep	Time savings
child	20	4.09%	1.65E−01	0.35%	1.62E−01	1.74%
insurance	27	4.02%	1.69E−01	0.31%	1.62E−01	3.98%
water	32	4.14%	2.57E−01	0.29%	2.52E−01	1.85%
mildew	35	4.27%	2.41E−01	0.27%	2.37E−01	1.75%
alarm	37	4.01%	2.54E−01	0.24%	2.45E−01	3.62%
barley	48	3.85%	3.47E−01	0.20%	3.32E−01	4.15%
hailfinder	56	3.91%	4.19E−01	0.18%	4.05E−01	3.34%
hepar2	70	3.90%	4.97E−01	0.16%	4.82E−01	3.04%
win95pts	76	3.90%	5.66E−01	0.15%	5.62E−01	0.84%

(*continued*)

Table 4. (*continued*)

BN	Size	Time pruning darwiche	Time (s) darwiche	Time swap	Time (s) rp_sep	Time savings
pathfinder	109	3.78%	8.20E−01	0.12%	7.99E−01	2.57%
munin1	186	3.64%	1.49E+00	0.08%	1.44E+00	3.21%
andes	223	3.59%	1.80E+00	0.07%	1.74E+00	3.61%
pigs	441	3.09%	4.33E+00	0.05%	4.22E+00	2.53%
link	724	2.64%	8.76E+00	0.02%	8.38E+00	4.32%
munin2	1003	2.24%	1.36E+01	0.02%	1.12E+01	18.08%
munin4	1038	2.25%	1.41E+01	0.03%	1.20E+01	14.88%
munin	1041	2.26%	1.41E+01	0.02%	1.20E+01	15.04%
munin3	1041	2.26%	1.45E+01	0.02%	1.23E+01	15.02%
Average		3.44%	4.25E+00	0.14%	3.71E+00	5.76%

5 Conclusion

Although *d-separation* [13] has been extensively studied, all previous implementations, including [1,3,6,12,17], and [7], test independence $I(X,Y,Z)$ by computing R_X, the nodes reachable from X along active paths with respect to Y, and then test whether $R_X \cap Z = \emptyset$. Thereby, no previous study has examined exploiting the *symmetry* [2] of probabilistic conditional independence when implementing d-separation.

In this paper, we consider exploiting symmetry to answer $I(X,Y,Z)$ by testing $I(Z,X,Y)$. This involves computing R_Z, the nodes reachable from Z along active paths with respect to Y, and then checking whether $R_Z \cap X = \emptyset$. It is important to realize that R_X and R_Z can contain different variables and have distinct cardinalities. Our analysis in Sect. 3 reveals that v-structure nodes play a critical role in this respect. As shown in Example 3, it is better to approach an observed v-structure node from the bottom (from a child) rather than from the top (from a parent). When approached from the bottom, the parents are not reachable, since the v-structure node is a closed, sequential valve. In contrast, when approached from the top, all other non-evidence parents are reachable, since the v-structure node is an open, convergent valve. Then, more generally, it may be better to find reachable nodes from whichever of X and Z is "lower" in the DAG. We suggest a measure of depth in Definition 2. We incorporate symmetry into our implementation of d-separation, given in Algorithm 2, and called SYMMETRIC D-SEPARATION. We empirically compare our approach with the implementation of d-separation suggested by [1] with favourable results. The experimental results are promising in two aspects. The effectiveness increases with network size, as well as with the amount of observed evidence, culminating with an average time savings of 9% in the 9 largest networks.

This work was motivated in part by recent advances in graph search [5]. Classical search algorithms such as A^* are unidirectional and have been extended to search from goal to start and to search from the start and goal simultaneously, meeting somewhere in the middle. Future work can investigate a meet in the middle variation of d-separation. Other applications of our work may include *influence diagram* [6] and inference algorithms that test independencies, such as *Lazy Propagation* [8].

References

1. Darwiche, A.: Modeling and Reasoning with Bayesian Networks. Cambridge University Press, New York (2009)
2. Dawid, A.P.: Conditional independence in statistical theory. J. R. Stat. Soc. Ser. B (Methodol.) **41**, 1–31 (1979)
3. Geiger, D., Verma, T.S., Pearl, J.: d-separation: from theorems to algorithms. In: Proceedings of the Fifth Conference on Uncertainty in Artificial Intelligence, pp. 139–148 (1989)
4. Goodfellow, I., Bengio, Y., Courville, A.: Deep Learning. MIT Press, Cambridge (2016)
5. Holte, R.C., Felner, A., Sharon, G., Sturtevant, N.R., Chen, J.: MM: a bidirectional search algorithm that is guaranteed to meet in the middle. Artif. Intell. **252**, 232–266 (2017)
6. Koller, D., Friedman, N.: Probabilistic Graphical Models: Principles and Techniques. MIT Press, Cambridge (2009)
7. Lauritzen, S.L., Dawid, A.P., Larsen, B.N., Leimer, H.G.: Independence properties of directed Markov fields. Networks **20**, 491–505 (1990)
8. Madsen, A.L., Jensen, F.V.: Lazy propagation: a junction tree inference algorithm based on lazy evaluation. Artif. Intell. **113**(1–2), 203–245 (1999)
9. Mohan, K., Pearl, J.: On the testability of models with missing data. In: Proceedings of the Seventeenth International Conference on Artificial Intelligence and Statistics, vol. 33 (2014)
10. Neapolitan, R.E.: Probabilistic Methods for Bioinformatics: With an Introduction to Bayesian Networks. Morgan Kaufmann, San Francisco (2009)
11. Nobandegani, A.S., Psaromiligkos, I.N.: A rational distributed process-level account of independence judgment. arXiv preprint arXiv:1801.10186 (2018)
12. Pearl, J.: Fusion, propagation and structuring in belief networks. Artif. Intell. **29**, 241–288 (1986)
13. Pearl, J.: Probabilistic Reasoning in Intelligent Systems: Networks of Plausible Inference. Morgan Kaufmann, San Francisco (1988)
14. Pearl, J.: Belief networks revisited. Artif. Intell. **59**, 49–56 (1993)
15. Pearl, J.: Causal Inference in Statistics: A Primer. Wiley, Hoboken (2016)
16. Pearl, J., Bareinboim, E.: External validity: from do-calculus to transportability across populations. Stat. Sci. **29**(4), 579–595 (2014)
17. Shachter, R.D.: Bayes-ball: the rational pastime (for determining irrelevance and requisite information in belief networks and influence diagrams). In: Proceedings of the Fourteenth Conference on Uncertainty in Artificial Intelligence, pp. 480–487. Morgan Kaufmann Publishers Inc. (1998)
18. Shipley, B.: Cause and Correlation in Biology. Cambridge University Press, Cambridge (2016)

Personality Extraction Through LinkedIn

Frédéric Piedboeuf[1(✉)], Philippe Langlais[1], and Ludovic Bourg[2]

[1] RALI, Université de Montréal, Montreal, Canada
`frederic.piedboeuf@umontreal.ca, felipe@iro.umontreal.ca`
[2] LittleBIGJob, Montreal, QC H3B 4W5, Canada
`ludovic.bourg@lorenzandhamilton.com`
`http://rali.iro.umontreal.ca/rali/`

Abstract. LinkedIn is a professional social network used by many recruiters as a way to look for potential employees and communicate with them. In order to facilitate communication, it is possible to use personality models to gain a better understanding of what drives the person of interest. This paper first looks at the possibility of collecting a corpus on LinkedIn labelled with a personality model, which has never been done before, then looks at the possibility of extracting two different personalities from the user. We show that we can achieve results going from 73.7% to 80.5% of precision on the DiSC personality model and from 80.7% to 86.2% of precision on the MBTI personality model. These results are similar to what has been found on other social networks such as Facebook or Twitter, which is surprising given the more professional nature of LinkedIn. Finally, an analysis of the significance of the results and of the possible sources of errors is presented.

1 Introduction

With the advent of social networks and their more recent popularization through mainstream culture, the internet has now become a considerable source of information on people habits, opinions, and personalities. Research focusing on using this information has taken a great importance in the field of natural language processing, and more recently, personality extraction has started to grab the attention of researchers.

Personality extraction is the task of assessing the personality of a user through his social network interactions. The personality of an individual is defined as the set of responses to external stimuli [7]. It is usually described according to one of several models of personality that are used in psychology, such as the MBTI or the DiSC, both explained in Sect. 2. Personality extraction has many uses, from personalizing the user experience to using the personality type as input to other machine learning problems, e.g., by using it to better extract influential communities on Twitter [6].

The most popular social networks for such tasks are probably Twitter and Facebook, because they provide easy platforms for collecting the corpus and allow easy testing of the personality with the help of external web applications

© Springer Nature Switzerland AG 2019
M.-J. Meurs and F. Rudzicz (Eds.): Canadian AI 2019, LNAI 11489, pp. 55–67, 2019.
https://doi.org/10.1007/978-3-030-18305-9_5

that administer the tests. However, in the professional world, Facebook and Twitter are rarely consulted and most often, the professional social network LinkedIn will be favoured for evaluating a potential candidate when hiring [2]. To our knowledge, the only other study that has looked at personality assessment on LinkedIn is the one of [20], but the evaluation of the personality was made by humans and not by computers, even though the latter have been shown to achieve a better precision score when classifying personality based on digital footprints [21].

This paper looks at the possibility of automatically extracting the personality of a LinkedIn user and shows, somewhat surprisingly given the peculiar nature of the texts on LinkedIn, that our results are comparable to those reported on other types of texts. The automatic extraction of personality on LinkedIn could become an important tool for many companies, as it could diminish the cost of employee selection processes. It could also help improve communications with potential employees by understanding what drives them and what kind of communication they better respond to.

2   Personality Models

The personality model that has probably been the most studied in psychology is the Big-5 [4]. It characterizes a person with the help of 5 traits, usually on a scale of 1 to 5: Openness to experience, Conscientiousness, Extroversion, Agreeableness, and Neuroticism. Although this model has been used extensively in the scientific community, the continuous nature of the test results makes it difficult for the non-scientific community to discuss their results and so other, categorical models, such as the Myer-Briggs, have become more popular.

The Myer-Briggs model, also called MBTI, evaluates the personality along 4 different axes, representing the way one processes the surrounding information [11]. The four axes are:

- Introversion (I) vs. Extroversion (E), or does one focus on internal stimuli or external ones.
- Sensing (S) vs. Intuition (N), or does one process the events as facts happening or by trying to find patterns and meaning.
- Thinking (T) vs. Feeling (F), or does one make decisions by thinking over the consequences or by following their instincts.
- Judging (J) vs. Perceiving (P), or does one prefer the outside world to be structured or to be flexible and spontaneous.

The combination of these four characteristics form 16 distinct types of personality that are identified with a code such as "ENFP" or "ISFJ".

In our case, we were interested in DiSC, which is another categorical test that was developed as a way to describe how people interact with each other [15]. The model gives a score to 4 different traits: Dominance, Influence, Steadiness, and Conscientiousness, which indicate what drives the person in their professional environment. Dominant people are driven by results, Influent people by relationships with others and influencing others, Steady people by cooperation with other people, and Conscientious people by the quality and accuracy of the work done [17].

Since the DiSC personality model has not been as widely studied as some other personality models, our study also looks at the MBTI personality model, which is used extensively by companies and has also been considerably studied by researchers [8,9,12]. The goal of this research is to:

1. Assess whether a LinkedIn profile has enough information to successfully extract the personality of its author;
2. Compare the results of DiSC extraction with the results of MBTI extraction in order to determine how significant the DiSC extraction results are.

3   Related Work

Given the nature of personality models, most researchers break down the problem in as many traits of personality as the model has. For example, it is common practice to break the Big-5 model into 5 different classifiers, one for each of the traits, and the final result is then the prediction of all 5 classifiers.

In [14], the authors were among the firsts to enter the field by predicting the personality of Twitter users based on their number of following, followers, the number of times they had been listed, and the number of Facebook social contacts. They achieved a Normalized Root Square Error (NRSE) of 0.2 on average over the different traits of the Big-5 model, showing that we can successfully extract personality based on very little information - if that information is pertinent.

Also using the Big-5 model, the authors of [10] trained classifiers on a Facebook dataset of 250 profiles containing both texts from the users as well as some other information, such as the number of groups the person belongs to or the number of likes. The corpus was provided in the context of a shared task, and therefore 7 other teams also had access to the dataset [3]. They extracted a total of 725 features and performed feature selection with a Support Vector Machine (SVM) using the top 5 to 16 features, achieving better results for each of the traits than the other participating teams.

In [13], a way to obtain a corpus automatically by selecting users that voluntarily put their personality test results on social networks was proposed. The study was carried out using the MBTI model, which has the double advantage of being popular with the general population and easy to lookup using regular expressions, since the code associated with the different personality models can be searched for directly. Overall, they harvested 1.2 million tweets belonging to 1500 different users and used logistic regression with feature selection to predict each trait of the MBTI model, but achieved poor results, with a precision below the majority baseline for the Intuition-Sensing and Judging-Perception traits.

While these studies focused on the MBTI and Big-5 personality models, the two most popular ones, it is important to also study other personality models that are commonly used to avoid restricting the playing field. The authors of [1] chose Twitter for a data-mining study with the DiSC personality model. Using a list of keywords associated with each of the 4 dimensions of the model, they downloaded tweets containing these keywords. They then performed data analysis in order to extend the vocabulary related to each of the traits by finding the most common words used by each of the types, excluding stop words. Even if they did not perform classification on the data, the study still showed a correlation between the vocabulary used and the personality type, and laid the foundation for other personality studies using the DiSC personality model.

In [12] the starting dataset was a Dutch corpus of essays written by 145 students. Each essay was also labelled with the MBTI personality of the student. They used a centroid-based model to classify each essay according to each of the personality traits, achieving precision scores ranging between 76% and 85%.

The authors of [5] obtained a corpus labelled with the MBTI personality test results from the internet forum personalitycafe.com, a forum based on personality types. Working only with the text, they used a recurrent neural network with pre-trained word embeddings to predict the personality of the users. They broke down the MBTI model into its 4 dimensions and obtained precision figures ranging from 62% to 78%, which is lower than other studies on the MBTI model, suggesting that the use of neural networks on personality extraction has not been developed enough yet to outperform classic algorithms.

4   Corpus Preparation

Since there is no other study on automatic personality extraction on LinkedIn, no corpus was available. In order to obtain one, we followed the method of [13], who devised a way to extract social network profiles of users who had completed a personality test and posted the results on their social networks.

The GoogleSearch Library for Python allowed us to access approximately 81 million public LinkedIn profiles. We initially looked for all profiles containing the words "DiSC" and "personality", which gave us a total of 19,000 profiles. However, an examination of 100 randomly selected profiles found 4 of them having an actual DiSC personality and the rest of them being noise. The noise was introduced mostly by team-building coaches and musicians and although the words "DiSC" and "personality" were present on their LinkedIn page, there was no mention of the user personality test results.

Finding LinkedIn profiles with a DiSC personality by directly researching the personality types was not doable due to the fact that the DiSC personality model uses a very common vocabulary subset (e.g., the Dominant personality can be expressed as dominant, driver, achiever, inspirational, etc.) and because there are many ways for a user to express the same personality, as shown in Fig. 1.

In order to capture LinkedIn profiles labelled with the DiSC personality test results, we surmised that people who would write their MBTI personality would be more likely to also write their DiSC personality in their profiles. A corpus following that assumption would have both personality expressed in their profiles,

Disc Test: D55% I8% S10% C26%	Disc Assessment: (5|1|3|7)	Disc Profile: [Natural D Adaptive I]	My DiSC assessment reveals I am a persuader
High DI and above the line C on DISC	Disc: "SDIC"	Disc: (D) Driver followed by (C) Conscientious	DiSC: Result-Oriented Pattern

Fig. 1. Different ways to express the Dominant trait of the DiSC personality model.

and we would then be able to split them into 2 different corpora, one labelled with the MBTI personality model and the other labelled with the DiSC personality model. A search of the public LinkedIn profiles containing the words "DiSC", "personality", and one of the 16 MBTI codes yielded a total of 1253 profiles, that we then saved in JSON format. To eliminate the noise and keep only the profiles containing a DiSC personality test result, we eliminated all profiles containing the tokens "HR", "certified", "jockey", "administering", "workshop", "golf", "spine", and "coaching" close to the token "DiSC".

This produced a total of 841 profiles for the DiSC corpus, which is the set of profiles that we could label with one of the DiSC personalities, and we kept the original 1253 profiles for the MBTI corpus. The small size of the corpora can be explained by two factors: First, the use of Linkedin, which is a more professional social network than Twitter or Facebook, may discourage people from sharing their personality results on their profile and second, the use of DiSC, a less popular model, yields fewer results than if we had done corpus extraction solely based on MBTI.

An example of a profile, slightly modified for anonymity, is shown in Listing 1.2. We can see that the data will still need a lot of preprocessing before being usable by a classifier.

With the DiSC personality model, each user is expected to have one main trait and one or two supporting traits. However, on LinkedIn, users often express their personality test results as "I am high D and I", or "I am equal part I, S and C" and so it is hard to discern between the main trait and the secondary traits. We decided that we wanted a system that would return the personality test results the users say they have (e.g. DI for "I am high D and I"), and so our labeling did not discern between the main and secondary traits.

Initial attempts to capture the DiSC personality with regular expressions generated errors due to other types of personality tests that had a similar vocabulary to DiSC and to the fact that DiSC uses very common words to describe the different personalities. For example, "Achiever" and "Developer" are both in the vocabularies of the StrengthFinder personality test[1] and the DiSC personality model, which also includes common words such as "Result", "Creative",

---

[1] https://www.gallupstrengthscenter.com/home/en-us/cliftonstrengths-themes-domains.

and "Agent". The list of regular expressions for the Steadiness trait is reported in Listing 1.1 and shows the complexity of the task of capturing the personality type automatically. Although these were ultimately not used for the labeling process, they were used to remove the words that would give away the personality before classifying.

Listing 1.1. Regular expression for the Steadiness trait

```
S=["High[- ]S", "\"S\"", "[sS]econdary S", "[Ss]tead[a-zA-Z]*",
   " S ", "[/-] S ", " S[CIiD][CIiD ]", " [CIiD]S", " [CDIi][
   CIiD]S ", "S[CIiSD]?/", " S,","\([CIiD]?S\)", "\(S[CIiD]?\)",
   "\[S\]", "[Aa]chiever", "[Aa]gent", "[Cc]ounselor", "[Ii]
   nvestigator", "[pP]erfectionist", "[pP]ractitioner", "[sS]
   pecialist", "[sS]-oriente", "S[0-9]", "&S", "S&"]
```

Since we wanted to be sure to have a corpus that was correctly labelled to avoid any mistakes before starting the classification task, we looked at the $-150/+150$ window of characters around the word "DiSC" and labelled the profiles manually. This also gave us the opportunity to check that no noise was in the corpus. The labeling of the MBTI corpus was straightforward, since the MBTI codes could be looked up with simple regular expressions.

Listing 1.2. Linkedin profile slightly modified for anonymity, after being saved in JSON.

```
{
  "Summary": "I am recognized by my peers as a tech-savvy
      intuitive problem-solver. I thrive in environments where
      rapid change and the need to constantly adapt and learn new
      information are considered the norm. My DiSC personality
      is dominant",
  "Personal Branding Claim": "Problem Solver",
  "Connections": "500+",
  "Followers": "0",
  "Skills": ["Leadership 110 endorsements 99+ who are skilled 2
      colleagues", "Strategy 30 endorsements"],
  "Recommendation": "Received (12) Given (3)",
  "Voluntary experiences": {
    "Guest Speaker": {
      "To_date": "Feb 2005",
      "From_date": "Feb 2005"
      "Description": "Guest Speaking at East High School"}},
  "Educations": {
    "Northeastern College of Professional Studies":{
      "Field of study": "Organizational Leadership",
      "From_date": "1987",
      "To_date": "1993",
      "Description": "4 best invested years of my life"}},
  "Experiences": {"Teacher":{
    "To_date": "Present",
    "From_date": "2015",
    "Employer": "Neverends Education",
    "Description": "Teaching software to kids"}},
```

```
"Interests": ["Profyle Tracker 349 followers", "Leoprino Foods
    18802 followers"],
"Achievements": {
  "Course": {
    "Name": "Team management",
    "Description": "null"},
  "Test": {
    "Name": "MBTI",
    "Result": "ENFP"}}}
```

The distribution of the data is represented in Table 1. The classes are imbalanced, but at least for the MBTI results, the distribution is similar to the one found in [13] and to the statistics given by the Myers-Briggs Foundation[2]. Both are also shown in Table 1.

According to the DiSC Profile website[3], the main traits are expected to be roughly equally distributed, with 25% of users in each personality type. However, we found a radically different distribution. This could be explained by two factors. Firstly, the fact that we did not discern between main and supporting traits changes the statistics. For instance, it might be possible that the Dominant trait appears more often as a supporting trait than the Steadiness trait or the Conscientious one. Secondly, it is possible that Dominant and Influent users are more likely to put their personality test results on LinkedIn than Steady and Conscientious users.

Table 1. Percentage of the different traits for the MBTI and DiSC personality

MBTI	Linkedin	Plank 2015	MBTI foundation	DiSC	Linkedin	Disc Profile website (https://www.discprofile.com/)
I/E	35.7	36.0	50.7	D	60.5	24.8
N/S	72.0	73.0	73.3	I	57.2	25.1
T/F	55.6	58.0	59.8	S	27.3	25.7
P/J	34.0	41.0	45.9	C	36.9	24.4

5   Feature Engineering

For both corpora, we extracted textual and non textual features. Forty non textual features were hand-crafted and then extracted from the LinkedIn profiles.

---

[2] Found at https://www.myersbriggs.org/my-mbti-personality-type/my-mbti-results/how-frequent-is-my-type.htm?bhcp=1.
[3] Found at https://www.discprofile.com/what-is-disc/faq/.

5.1   Non-Textual Features

The non-textual features can be broken down in 7 categories.

1. **General information:** Information that concerns the user in general, such as the number of connections, whether the user has something written in the description, etc.
2. **Work:** Information about the work experience, such as the number of jobs, the average duration of those jobs, etc.
3. **Volunteer experience:** These features are related to the volunteer experiences of the user, which contains the number of volunteer experiences, the average duration of those, etc.
4. **Education:** Features related to number of diplomas, average length, etc. Very similar to the work and volunteer experience categories with the added feature of the highest level of education.
5. **Skills:** Features related to the skills section, such as the number of skills, average number of endorsements, number of endorsements from colleagues, etc.
6. **Accomplishments:** The total number of accomplishments, as well as the number of accomplishments broken down in categories (titles, languages, courses, etc.)
7. **Interests:** Features related to the user's interests, such as the number of interests, the number of followers of these interests, etc.

5.2   Textual Features

We also extracted general text from the LinkedIn profiles of each user, which was taken from 1- the introduction written by the user, and 2- the description of schools, jobs, and volunteering experience when these were available. The profiles have an average of 830.9 words, and each word is of an average length of 6.4 characters. We derived 415 textual features from this text. These features included 370 features from the General Inquirer[4] as well as the number of different POS (part of speech) tags used in the text. The General Inquirer is a Java program that takes a text as input and returns both percentages and total counts of words pertaining to each category included in its dictionary. Categories include, but are not limited to: Anxiety, Family, Health, Sadness, etc. [16]. We also used Term Frequency Inverse Document Frequency (Tf-Idf) in our experiments, using the 5000 most frequent words. For each classifier, we tried a combination of Tf-Idf and of the 415 features, with and without stopwords. The best results for each is reported in Sect. 6.

6   Experiments

Before classifying, we were careful to remove any words giving away directly the personality of either the DiSC or the MBTI models by using sets of regular

---

[4] http://www.wjh.harvard.edu/~inquirer/.

expressions, as shown in Listing 1.1. For the MBTI model, we simply looked for and removed the 16 personality codes.

The same procedure was used for the DiSC corpus and the MBTI corpus, and gave similar results. We trained a classifier for each trait, using 20-fold cross validation. We tested an SVM classifier with a feature ranking algorithm and optimization, a Random Forest with AdaBoost, and a simple Naives Bayes. Several architectures of neural networks were also tried, but due to the small size of the corpora, couldn't learn to generalize efficiently. The baseline chosen was a majority class rule, which seemed appropriate, since the classes are often imbalanced.

The feature ranking algorithm for the SVM started with all features and tried them k at a time, until there were only the k best features, discarding the worst c features at each iteration. Two different measures of efficiency for the features were tried: the difference between the precision with and without the feature, as well as simply the precision with the feature. The latter measure yielded better results and was therefore used here, k and c being meta-parameters of the algorithm. We set them at 30 and 15, respectively, which seemed to give the best results in a timely fashion. As with the final algorithms, we also used a 20-fold cross validation on each loop of the algorithm.

Without the feature ranking algorithm, the SVM performed at the same level or at best a few percents higher than the baseline. With the feature ranking algorithm, the SVM performed on average 11% better than the baseline, and with the fine tuning of the meta parameters through a grid search, the SVM slightly improved in precision. Adaboost wasn't used with the SVM, since it showed no improvements but instead a decrease in performance.

Naive Bayes (NB) and Random Forest (RF) were also tried. Random Forest worked best with a mix of Tf-Idf features and hand-crafted features, presented in Sect. 5. The mix contained only the n most frequent words, and the m hand crafted features used were those having the highest correlation with the trait being predicted. Different numbers of n and m were tried, but at the end, made little difference to the performance for the Random Forest due to its capacity to pick the most useful features for classification. For the Naive Bayes, we used a grid search to find the optimal number of n and m for each of the traits. Since it is a linear classifier, the features were ranked with the help of a Pearson correlation coefficient, and those that were the most correlated were introduced to the algorithm first. AdaBoost was then used to boost performance, allowing to gain on average 1% on precision for both NB and RF. Stop-words removal and lemmatization were also tried, but only hurt the performance and therefore were not kept.

The complete results can be seen in Table 2. An interesting thing to note here is that the SVM and the Random Forest capture some very distinct phenomenons. While the SVM finds itself, after the feature ranking algorithm, using mostly the non-textual features, as well as some of the POS tags, the RF mostly uses the textual features from the Tf-Idf. Despite several studies finding the features extracted by the General Inquirer very useful for classification [10,16], our algorithm learned to perform mostly without their help, maybe due to the fact that LinkedIn profiles are often written in a way which seems professional and not emotional.

Table 2. Precision (%) over the different traits for the MBTI (top) and the DiSC (bottom) personality models. NT stands for Non-Textual, T for Textual, NB for Naive Bayes, and RF for Random Forest.

Traits	Baseline	NB	RF NT	RF T	RF	SVM NT	SVM T	SVM
IE	64.3	72.5	75.2	82.6	**84.4**	77.5	71.8	77.9
NS	72.0	78.0	79.1	81.5	**86.2**	82.0	77.1	83.7
TF	55.6	65.5	68.9	78.3	**83.5**	69.4	68.3	71.7
JP	66.0	67.1	72.4	76.5	**80.7**	77.4	73.6	78.4
D	60.5	66.1	70.9	72.7	72.5	72.3	66.5	**73.7**
I	57.2	63.1	71.8	74.5	**75.1**	73.2	66.7	73.8
S	72.7	72.5	78.2	79.3	79.3	80.5	77.7	**80.5**
C	63.1	66.2	72.1	74.1	74.1	75.5	74.1	**76.7**

7   Analysis

Although there is no other study on personality classification on LinkedIn, we can still compare our results to other studies that have used the MBTI model for classification. Since the data is imbalanced, we also report the improvement over the majority baseline when available, which helps giving a better idea of the performance of the algorithm. In Table 3, we show the results of our MBTI classifier, of the study of [12] (Study 1) and the study of [5] (Study 2). In parentheses is the improvement over the majority baseline, which was available for Study 1, but not for Study 2.

Table 3. Comparison between our results and results of other studies for the MBTI personality model

Traits	Our results	Study 1	Study 2
IE	84.4 (+18.1)	76.4 (+21.2)	67.0
NS	86.2 (+14.2)	76.5 (+22.7)	62.0
TF	83.5 (+27.9)	80.5 (+8.1)	77.8
JP	80.7 (+14.7)	84.5 (+3.8)	63.7

We can see that, overall, we get rather good results over all traits. Although we cannot surpass Study 2 for the Judging-Perceiving trait, we get the best score overall for all other three traits. However, since the classes are very imbalanced, it is hard to say how much this really means.

The errors can be explained by several factors. Firstly, the method used, which takes profiles of people who have already passed the personality test, does not allow us to check how accurate the tests are. Personality tests often range from being given by trained psychologists to tests having been created on the internet by someone having no qualifications to do so. It is then very possible that some of these tests are not entirely accurate and introduce a bias in our data.

The choice of LinkedIn also makes the task more difficult. Since LinkedIn is a social network meant to display the users' professional qualities, not only will people tend to be more self conscious about what they publish [19], but they may also try to guess what type of personality an employer would want and "bend" their results towards the ideal personality. Even if the user did not actively change the results that they obtained on the personality test, the simple mental image of what a good employee is, and the desire to be a good employee at the moment of the test, may bias the outcome.

The results obtained here are nevertheless promising. The two goals of this study were to find out if LinkedIn had enough information to correctly classify personality, as well as if DiSC could be considered a valid personality model compared to those that have been more studied. Our initial assumption was that it would be harder to classify personalities on LinkedIn due to the peculiar nature of the social network, requiring that the user be more reserved in what is written since the user expects to be read by eventual employers.

The results show, interestingly, that it is possible to classify LinkedIn profiles to the same extent as profiles from other social networks. However, it should be mentioned that the LinkedIn profiles we obtained are profiles that had mostly been filled out thoroughly, and so it should be interesting to see how well this generalizes to profiles that lack information.

This study also gives some validity to the DiSC personality model. We collected a fairly small corpus and so it could be argued that the correlation found between profiles and personality traits are due to pure luck, but it would be highly improbable.

8   Conclusion and Future Work

LinkedIn is a professional social network used by many employers as a means to screen potential employees and communicate with them. In order to facilitate contacts between employers and employees, it would be easier to know the personality of the person contacted. However, it is very difficult, for a human, to detect the personality of someone on LinkedIn. This is due to the fact that descriptions on LinkedIn are often written in a way that the employees think the employers want, and so is not in accordance to their natural writing style [19].

Our study shows that a computer is able to extract the personality from a LinkedIn profile with a reliable precision. This was done using two personality models that are often used in professional settings as a way to better understand employees or coworkers: the MBTI personality model and the DiSC personality model.

Although studies have been done on correlating job success to personality traits [18], only neuroticism has been found to be negatively correlated with job performance, and so the application of personality extraction in the trimming of candidates must be done carefully and employers must work to avoid putting their personal biases in the selection process.

This study opens the door, however, to many interesting applications, such as facilitating conversations between employers and employees, helping recruiters find people that would fit the job better, or helping LinkedIn users to themselves find the next career step that would be best suited for their talents and personality.

In the future, studies on personality extraction through LinkedIn could focus on either getting a larger dataset so that better machine learning algorithms may be applied, finding a way to transpose with accuracy the linguistic manual features to other languages, or working on the anonymization of a LinkedIn dataset so that researchers may compare their results on a common benchmark.

References

1. Ahmad, N., Siddique, J.: Personality assessment using Twitter tweets. Proc. Comput. Sci. **112**, 1964–1973 (2017)
2. Caers, R., Castelyns, V.: Linkedin and Facebook in Belgium: the influences and biases of social network sites in recruitment and selection procedures. Soc. Sci. Comput. Rev. **29**(4), 437–448 (2011)
3. Celli, F., Pianesi, F., Stillwell, D., Kosinski, M.: Workshop on computational personality recognition (shared task). In: Proceedings of the Workshop on Computational Personality Recognition (2013)
4. Goldberg, L.R.: The structure of phenotypic personality traits. Am. Psychol. **48**(1), 26 (1993)
5. Hernandez, R.K., Scott, I.: Predicting Myers-Briggs type indicator with text (2017)
6. Kafeza, E., Kanavos, A., Makris, C., Vikatos, P.: T-PICE: Twitter personality based influential communities extraction system. In: 2014 IEEE International Congress on Big Data (BigData Congress), pp. 212–219. IEEE (2014)
7. Kaushal, V., Patwardhan, M.: Emerging trends in personality identification using online social networks' a literature survey. ACM Trans. Knowl. Discov. Data (TKDD) **12**(2), 15 (2018)
8. Lima, A.C.E., de Castro, L.N.: Predicting temperament from Twitter data. In: 2016 5th IIAI International Congress on Advanced Applied Informatics (IIAI-AAI), pp. 599–604. IEEE (2016)
9. Ma, A., Liu, G.: Neural networks in predicting Myers Brigg personality type from writing style (2017)
10. Markovikj, D., Gievska, S., Kosinski, M., Stillwell, D.: Mining Facebook data for predictive personality modeling. In: Proceedings of the 7th International AAAI Conference on Weblogs and Social Media (ICWSM 2013), Boston, MA, USA, pp. 23–26 (2013)

11. Myers, I.B., McCaulley, M.H., Quenk, N.L., Hammer, A.L.: MBTI Manual: A Guide to the Development and Use of the Myers-Briggs Type Indicator, vol. 3. Consulting Psychologists Press, Palo Alto (1998)

12. Noecker Jr., J., Ryan, M., Juola, P.: Psychological profiling through textual analysis. Literary Linguistic Comput. **28**(3), 382–387 (2013)

13. Plank, B., Hovy, D.: Personality traits on Twitter—or—how to get 1,500 personality tests in a week. In: Proceedings of the 6th Workshop on Computational Approaches to Subjectivity, Sentiment and Social Media Analysis, pp. 92–98 (2015)

14. Quercia, D., Kosinski, M., Stillwell, D., Crowcroft, J.: Our Twitter profiles, our selves: predicting personality with Twitter. In: 2011 IEEE Third International Conference on Privacy, Security, Risk and Trust (PASSAT) and 2011 IEEE Third International Conference on Social Computing (SocialCom), pp. 180–185. IEEE (2011)

15. Reynierse, J.H., Ackerman, D., Fink, A.A., Harker, J.B.: The effects of personality and management role on perceived values in business settings. Int. J. Value Based Manage. **13**(1), 1–13 (2000)

16. Stone, P.J., Bales, R.F., Namenwirth, J.Z., Ogilvie, D.M.: The general inquirer: a computer system for content analysis and retrieval based on the sentence as a unit of information. Behav. Sci. **7**(4), 484–498 (1962)

17. Sugerman, J.: Using the disc® model to improve communication effectiveness. Ind. Commercial Training **41**(3), 151–154 (2009)

18. Tett, R.P., Jackson, D.N., Rothstein, M.: Personality measures as predictors of job performance: a meta-analytic review. Pers. Psychol. **44**(4), 703–742 (1991)

19. Van Dijck, J.: 'You have one identity': performing the self on Facebook and Linkedin. Media Cult. Soc. **35**(2), 199–215 (2013)

20. van de Ven, N., Bogaert, A., Serlie, A., Brandt, M.J., Denissen, J.J.: Personality perception based on Linkedin profiles. J. Managerial Psychol. **32**(6), 418–429 (2017)

21. Youyou, W., Kosinski, M., Stillwell, D.: Computer-based personality judgments are more accurate than those made by humans. Proc. Nat. Acad. Sci. **112**(4), 1036–1040 (2015)

Solving Influence Diagrams with Simple Propagation

Anders L. Madsen[1,2(✉)], Cory J. Butz[3], Jhonatan Oliveira[3],
and André E. dos Santos[3]

[1] HUGIN EXPERT A/S, Aalborg, Denmark
anders@hugin.com
[2] Department of Computer Science, Aalborg University, Aalborg, Denmark
[3] Department of Computer Science, University of Regina, Regina, Canada

Abstract. Recently, Simple Propagation was introduced as an algorithm for belief update in Bayesian networks using message passing
in a junction tree. The algorithm differs from other message passing algorithms such as Lazy Propagation in the message construction process. The message construction process in Simple Propagation
identifies relevant potentials and variables to eliminate using the *one-in,
one-out*-principle. This paper introduces Simple Propagation as a solution algorithm for influence diagrams with discrete variables. The *one-in,
one-out*-principle is not directly applicable to influence diagrams. Hence,
the principle is extended to cope with decision variables, utility functions,
and precedence constraints to solve influence diagrams. Simple Propagation is demonstrated on an extensive example and a number of useful
and interesting properties of the algorithm are described.

Keywords: Influence diagrams · Simple propagation ·
Discrete variables

1 Introduction

An influence diagram [1] is a natural representation of a decision problem where
a single decision maker has to make a sequence of decisions under uncertainty. An
influence diagram is essentially a Bayesian network [2,3] augmented with facilities (decision variables, utility functions and precedence constraints) to support
decision making under uncertainty. In essence, the solution to a decision problem
represented as an influence diagram consists of an optimal strategy specifying
which decision option the decision maker should take at each decision depending
on the information known to her at that point in time.

Various extensions of traditional influence diagrams have been introduced.
These include limited-memory influence diagrams [4], unconstrained influence
diagrams [5], and influence diagrams with mixed variables [6]. The focus of this
paper is on traditional influence diagrams with discrete variables. The solution of
an influence diagram is extended to consist of the optimal strategy, the expected

© Springer Nature Switzerland AG 2019
M.-J. Meurs and F. Rudzicz (Eds.): Canadian AI 2019, LNAI 11489, pp. 68–79, 2019.
https://doi.org/10.1007/978-3-030-18305-9_6

utility of adhering to the optimal strategy and the probability of future decisions. The probability of future decisions under a (optimal) strategy can be determined by encoding the decision policy for each decision as a conditional probability distribution (CPD) [7]. We consider this part of the process of solving an influence diagram.

Some of the popular algorithms for solving the traditional influence diagram include [8–13]. One of the first methods for solving influence diagrams was based on Cooper's technique [14] of transforming the influence diagram into a Bayesian network and using a Bayesian network inference algorithm to solve the influence diagram. Other methods include message passing in a tree and approximation algorithms based on sampling. A recent review of methods for solving influence diagrams can be found here [15] and a survey of probabilistic decision graphs can be found in this paper [16]. Recent work on solving influence diagrams include finding bounds for influence diagrams using join graph decomposition [17] and an improved method for solving hybrid influence diagrams [18].

Simple Propagation [19] is a new algorithm for belief update in Bayesian networks. It proceeds by message passing in a junction tree representation of the Bayesian network taking advantage of a decomposition of clique and separator potentials. In Simple Propagation, message construction is based on a *one-in, one-out*-principle meaning that a potential relevant for a message has at least one non-evidence variable in the separator and at least one non-evidence variable not in the separator. The advantage of Simple Propagation is its simplicity compared to, for instance, Lazy Propagation [20]. This paper extends Simple Propagation to the solution of traditional influence diagrams, where the solution consists of an optimal strategy, the expected utility of adhering to this strategy and the probability of future decisions under the strategy.

Simple Propagation is extended to cover solutions of traditional influence diagrams by identifying an optimal strategy computing the expected utility, which involves computing the probability of future decisions under the optimal strategy. To achieve this, the *one-in, one-out*-principle is extended such that it considers both probability and utility potentials. Our approach supports exploitation of probabilistic barren variables [9] and the decomposition of utility potentials reducing the number of calculation operations.

2   Background

A traditional influence diagram over discrete variables is a triple $N = (G, \Phi, \Psi)$ where $G = (V, E)$ is a DAG (directed, acyclic graph) over vertices V and edges $E \subseteq V \times V$. Each vertex $v \in V$ represents a random (or chance) variable, a decision variable, or a utility function. Let $\mathcal{D} = \{D_1, \ldots, D_n\}$ be the set of decision variables and let $\mathcal{X}$ be the set of random variables and $\mathcal{U}$ be the set of utility nodes. Each random variable $X \in \mathcal{X}$ has a CPD $P(X \mid \text{pa}(X))$, where $\text{pa}(X)$ are the parent vertices of X in G, and each utility node $u \in \mathcal{U}$ represents a utility function $u(\text{pa}(u))$ (we use u to denote both the utility node and the utility function). Finally, we denote $\Phi = \{P(X \mid \text{pa}(X)) \in \mathcal{X}\}$, $\Psi = \{u(\text{pa}(u)) \mid u \in \mathcal{U}\}$, and $\text{dom}(\phi)$ (or $\text{dom}(\psi)$) the domain of ϕ (ψ).

The influence diagram $N = (G, \Phi, \Psi)$ is an efficient representation of a joint expected utility function:

$$EU(\mathcal{X}, \mathcal{D}) = \prod_{X \in \mathcal{X}} P(X \,|\, \text{pa}(X)) \times \sum_{u \in \mathcal{U}} u(\text{pa}(u)). \tag{1}$$

The regularity constraints of the traditional influence diagram state that there must be a total order on the decisions and that the decision maker has perfect recall. Assuming decisions are ordered according to index, the random variables can be partitioned into information sets using a partial precedence ordering $\prec$: $\mathcal{I}_0 \prec D_1 \prec \mathcal{I}_1 \prec \cdots \prec \mathcal{I}_{n-1} \prec D_n \prec \mathcal{I}_n$. The set $\mathcal{I}_i$ is the set of random variables observed (by the decision maker) after making decision D_i and before decision $\mathcal{I}_{i+1}$, i.e., $\mathcal{I}_0$ is the random variables initially observed and $\mathcal{I}_n$ is the set of random variables never observed or observed after the last decision.

In a Bayesian network, a variable X is a barren node when X is never observed, $P(X)$ is of no interest, and the same holds true for each of X's descendants (if any) [9]. A variable X in an influence diagram is a probabilistic barren node when X is a barren node if only the set of probability potentials are considered [9,21].

2.1 Strategies, Decision Policies and Future Decisions

A decision policy $\delta_{D_i}(\text{rel}(D_i))$ is a mapping from $\text{rel}(D_i)$ to D_i where $\text{rel}(D_i) \subseteq \bigcup_{j<i} \mathcal{I}_j \cup \{D_j\}$ is defined below. A strategy is a collection of decision policies $\Delta = \{\delta_D \,|\, D \in \mathcal{D}\}$, one for each decision. A decision policy $\delta_D(\text{rel}(D))$ is encoded as a CPD $P(D \,|\, \text{rel}(D))$ as follows:

$$P_\Delta(D = d \,|\, \text{rel}(D) = z) = \begin{cases} 1 & \text{if } \delta_D(\text{rel}(D) = z) = d \\ 0 & \text{otherwise.} \end{cases} \tag{2}$$

The strategy Δ induces a joint probability distribution:

$$P_\Delta(\mathcal{X} \cup \mathcal{D}) = \prod_{X \in \mathcal{X}} P(X \,|\, \text{pa}(X)) \times \prod_{D \in \mathcal{D}} P_\Delta(D \,|\, \text{rel}(D)). \tag{3}$$

This factorization can be used to compute the probability of future decisions [7]. Notice that the probability of future decisions can be computed under any strategy D.

The expected utility $EU(\Delta)$ of a strategy Δ is the expectation of the total utility $U(\bigcup_{u \in \mathcal{U}} \text{pa}(u)) = \sum_{u \in \mathcal{U}} u(\text{pa}(u))$ under the probability distribution $P_\Delta(\mathcal{X} \cup \mathcal{D})$, $\sum_{Y \in \mathcal{X} \cup \mathcal{D}} U(\bigcup_{u \in \mathcal{U}} \text{pa}(u)) \times P_\Delta(\mathcal{X} \cup \mathcal{D})$ [22]. A strategy that maximizes the expected utility is an optimal strategy, denoted $\hat{\Delta}$, and has the property $EU(\hat{\Delta}) \geq EU(\Delta)$, for all Δ.

An optimal strategy $\hat{\Delta}$ and its expected utility of $EU(\hat{\Delta})$ can be computed as:

$$EU(\hat{\Delta}) = \sum_{\mathcal{I}_0} \max_{D_1} \sum_{\mathcal{I}_1} \cdots \sum_{\mathcal{I}_{n-1}} \max_{D_n} \sum_{\mathcal{I}_n} EU(\mathcal{X}, \mathcal{D}). \tag{4}$$

Under the regularity constraints, the decision policy δ_{D_i} for decision variable D_i can, in principle, be a mapping from all past observations and decisions $\bigcup_{j<i} \mathcal{I}_j \cup \{D_j\}$ onto D_i making the policy exponentially large in the number of observations. Luckily, this is often not the case though. That is, not every past observation or decision variable is *requisite* for a decision, i.e., not all past observations may impact the choice of the decision maker at a decision D. A variable $Y \in \mathrm{pa}(D) \subseteq \mathcal{X} \cup \mathcal{D}$, $D \in \Delta$ is non-requisite, if $(\mathcal{U} \cap \mathrm{de}(D)) \perp \{Y\} \mid (\mathrm{fa}(D) \setminus \{Y\})$ [4,23,24]. This condition states that Y is non-requisite, if it is separated from the utility nodes that are descendents of D given the family of D except Y, i.e., the value of Y does not affect the optimal choice at D. If a variable is not non-requisite, then it is requisite and belongs to the relevant past [7], denoted $\mathrm{rel}(D)$.

2.2   Strong Junction Tree Construction

Simple propagation is performed in a strong junction tree representation. The strong junction tree ensures that variables are eliminated in the reverse order of $\prec$ during evaluation. In addition, it serves as a caching structure through the message passing process.

The construction of a strong junction tree representation $T = (\mathcal{C}, \mathcal{S})$ with cliques $\mathcal{C}$ and separators $\mathcal{S}$ of influence diagram $N = (G, \Phi, \Psi)$ proceeds in four steps [12]:

Minimalization where information arcs from non-requisite parents of decision nodes are removed.

Moralization where each pair of parents X_i and X_j with a common child (representing either a random variable or a utility function) are connected by an undirected edge, all edges are made undirected and utility nodes are removed to produce G^M.

Strong Triangulation where G^M is triangulated to produce G^T using an elimination order σ such that $\sigma(\mathcal{I}_{i+1}) < \sigma(D_{i+1}) < \sigma(\mathcal{I}_i)$, for all $i \in [0, n-1]$.

Strong Junction Tree Structure where the cliques $\mathcal{C}$ are connected by separators $\mathcal{S}$ producing a tree $T = (\mathcal{C}, \mathcal{S})$ with strong root $R \in \mathcal{C}$.

Initialization of Junction Tree is the process of assigning $\phi \in \Phi$ and $\psi \in \Psi$ to cliques such that $\mathrm{dom}(\phi) \subseteq C$ and $\mathrm{dom}(\psi) \subseteq C$.

Notice that in the last step, the initial clique potential $\pi_A = (\Phi_A, \Psi_A)$ consists of the set of probability distributions $\Phi_A = \{P_1, \ldots, P_l\}$ and the set of utility functions $\Psi_A = \{u_1, \ldots, u_m\}$ assigned to A. Simple Propagation proceeds by message passing between neighboring cliques of T. The message from clique B to clique A is denoted $\pi_{B \to A} = (\Phi_{B \to A}, \Psi_{B \to A})$. The combination of a clique potential $\pi_A = \{\Phi_A, \Psi_A\}$ and the message $\pi_{B \to A} = (\Phi_{B \to A}, \Psi_{B \to A})$ is denoted $\otimes$ and simply amounts to set union, i.e., $\pi_A \otimes \pi_{B \to A} = (\Phi_A \cup \Phi_{B \to A}, \Psi_A \cup \Psi_{B \to A})$.

If $C \in \mathcal{C}$, then $\mathcal{X}(C)$ denotes the variables of C and $\mathrm{adj}(C)$ denotes the cliques adjacent to C. Let $A \in \mathrm{adj}(B)$ with A closer to R, then $\mathrm{pa}(B) = A$ denotes the parent clique and $S = A \cap B$ is denoted the parent separator.

3   Simple Propagation

Simple Propagation was introduced by [19] as a message passing algorithm for belief update in Bayesian networks. In this section, Simple Propagation is extended to cover solutions of traditional influence diagrams by identifying an optimal strategy $\hat{\Delta}$, computing $EU(\hat{\Delta})$, and computing $P_{\hat{\Delta}}(D)$, for each $D \in \mathcal{D}$ under $\hat{\Delta}$.

When constructing the message $\pi_{A \to B}$ from clique A to clique B over separator S, Simple Propagation uses the *one-in, one-out*-principle to order the computations. This means that Simple Propagation is not driven by identifying an elimination order. Instead the elimination order is induced by the order in which potentials satisfying the *one-in, one-out*-principle are processed. The next potential to consider is selected randomly among the potentials satisfying the principle. In the case of influence diagrams, the induced elimination order must satisfy the partial order $\prec$. This is achieved by applying an extended *one-in, one-out*-principle relative to the decision variables in clique A, not in clique B, in reverse order of $\prec$ when computing a message up the junction tree.

Let A and B be adjacent cliques with $B = \mathrm{pa}(A)$ and $\pi_A \otimes_{C \in \mathrm{adj}(A) \setminus \{B\}} \pi_{C \to A} = (\Phi, \Psi)$. The *one-in, one-out*-principle is extended such that it considers both probability and utility potentials, i.e., $\Phi \cup \Psi$, when selecting a potential. When constructing the message $\pi_{A \to B}$, the principle is applied relative to each decision variable $D \in A \setminus B$ in reverse order of $\prec$. For decision variable $D_i \in A \setminus B$ any potential ω with $D_i \in \mathrm{dom}(\omega)$ such that ω has a $Y \in \mathrm{dom}(\omega) \cap \mathcal{I}_i$ is selected. If, for decision variable $D_i \in \mathcal{D}$, there is no potential ω such that $D_i \in \mathrm{dom}(\omega)$, then D has no descendants and any policy can be assigned to D. Each time a variable must be eliminated, Variable Elimination is applied as the marginalization operation (see Algorithm 3).

Let $\mathcal{D}(A) = \{D_{A_1}, \dots, D_{A_m}\} \subseteq \mathcal{D}$ be the set of decision variables in clique A. The variables $\mathcal{X}(A)$ of clique A can be partitioned according to the partial precedence ordering $\prec$ and must be processed in reverse order to produce correct results. However, when sending a message from clique A to B, we only have to consider the order of decisions in $A \setminus B$ that must be eliminated when constructing the message $\pi_{A \to B}$. Once the decisions $A \setminus B$ have been eliminated, the *one-in, one-out*-principle is applied on the result relative to parent separator $S = A \cap B$ between A and B.

Algorithm 1 shows the general Simple Propagation algorithm for influence diagrams. It solves the influence diagram to identify $\hat{\Delta}$ and computes $EU(\hat{\Delta})$ as well as constructs a Bayesian network N^* to compute $\{P_{\hat{\Delta}}(D) \mid D \in \mathcal{D}\}$. The Bayesian network N^* is constructed from N by changing decision variables to random variables, removing utility functions, and keeping the structure of G induced by the random and decision variables. The algorithm performs a collect operation where messages are passed from the leaf cliques towards R. Next, a Bayesian network is constructed where each decision policy $\delta_D(\mathrm{rel}(D))$ is encoded as a CPD $P_{\hat{\Delta}}(D \mid \mathrm{rel}(D))$. This is followed by a distribute operation where messages are passed from R towards the leaf cliques. This operation only

---

Procedure *SP-ID(N = (G, P, U), T)*

1   Perform a collect operation on the strong root R of T to identify $\hat{\Delta}$ using
     Algorithm 2

2   Compute $EU(\hat{\Delta})$ at R

3   Create BBN N^* from N
     for $D \in \mathcal{D}$ *represented in* N^* **do**
       | Encode decision policy $\delta_D(\mathrm{rel}(D))$ as $P(D \,|\, \mathrm{rel}(D))$
     end

4   Perform a distribute operation on R in N^*

5   **for** $D \in \mathcal{D}$ *represented in* N **do**
       | Compute $P_{\hat{\Delta}}(D)$
     end

6   **return** $\hat{\Delta}$ and $\{P_{\hat{\Delta}}(D) \,|\, D \in \mathcal{D}\}$

Algorithm 1. The Simple Propagation algorithm for Influence Diagrams.

involves elimination of random variables and proceeds as in Simple Propagation for Bayesian networks [19].

All probability and utility potentials satisfying the *one-in, one-out*-principle have variables to be eliminated. In addition, any potential ϕ (or ψ) with $\mathrm{dom}(\phi) \subseteq S$ (or $\mathrm{dom}(\psi) \subseteq S$) must be included in the message.

Algorithm 2 is the Simple Message Computation algorithm applied during the collect operation (SMC-CE). It applies the extended *one-in, one-out*-principle when processing $\mathcal{D}(A)$ in reverse order of $\prec$.

To complete the collect operation, the potential $\pi_R \otimes \bigotimes_{A \in \mathrm{adj}(R)} \pi_{A \to R}$ of the strong root R of T is marginalized to $\emptyset$ to determine $EU(\hat{\Delta})$. After the last decision variable to consider in the solution process, i.e., D_1, is eliminated, the remaining potentials are processed to compute $EU(\hat{\Delta})$.

Algorithm 2 applies Algorithm 3 to eliminate a variable (decision or chance) in the message construction process. It is essentially Variable Elimination.

In Algorithm 3, $\Phi^{\downarrow - Y}$ denotes marginalization of Y over the set Φ using $\sum$ for random variables, i.e., $Y \in \mathcal{X}$ and max for decision variables, i.e., $Y \in \mathcal{D}$. In the process of performing a max-marginalization of $D \in \mathcal{D}$, the maximizing alternatives for each parent configuration are recorded as the optimal decision rule for D.

Theorem 1. *Simple Propagation in an Influence Diagram N computes:*

1. an optimal strategy $\hat{\Delta}$ for the decision problem represented as N,
2. the expected utility $EU(\hat{\Delta})$ of $\hat{\Delta}$, and
3. the probability of future decisions $\{P_{\hat{\Delta}}(D) \,|\, D \in \mathcal{D}\}$ under $\hat{\Delta}$.

Proof. Simple Propagation processes the variables in reverse order of $\prec$. Each variable (random or decision) is eliminated using Algorithm 3, which is equivalent to Variable Elimination. Message passing proceeds as in Lazy Propagation. Hence, the calculations performed during the collect operation correspond to solving the influence diagram with Lazy Propagation using the induced elimination

---

Procedure $SMC\text{-}CE(A,\ B,\ (\Phi,\Psi) = \pi_A \otimes \bigotimes_{C \neq B} \pi_{C \to A})$

1 **for** $D_i \in \mathcal{D}(A) \setminus B$ *in reverse $\prec$-order* **do**
2   **while** $\exists \omega \in \Phi \cup \Psi$ *satisfying the extended one-in, one-out-principle relative to D_i* **do**
3   Select variable $Y \in \mathrm{dom}(\omega)$ such that $Y \in \mathcal{I}_i$
4      $\Phi^*, \Psi^* = \mathrm{MARGINALIZATION}(Y, \Phi, \Psi)$
5   Set $\Phi = \Phi^*$ and $\Psi = \Psi^*$
  end
6    $\Phi^*, \Psi^* = \mathrm{MARGINALIZATION}(D_i, \Phi, \Psi)$
7   Set $\Phi = \Phi^*$ and $\Psi = \Psi^*$
end
8 **while** $\exists \omega \in \Phi \cup \Psi$ *satisfying the extended one-in, one-out-principle relative to parent separator S* **do**
9   Select variable $Y \in \mathrm{dom}(\omega)$ such that $Y \in A \notin B$
10   $\Phi^*, \Psi^* = \mathrm{MARGINALIZATION}(Y, \Phi, \Psi)$
11   Set $\Phi = \Phi^*$ and $\Psi = \Psi^*$
end
12 Set $\Phi = \Phi^*$ and $\Psi = \Psi^*$
13 **return** $\pi_{A \to B} = (\Phi, \Psi)$

Algorithm 2. The Simple Message Computation algorithm under collect.

Procedure $\mathrm{MARGINALIZATION}(Y,\ \Phi,\ \Psi)$

1 Set $\Phi_Y = \{\phi \,|\, Y \in \mathrm{dom}(\phi)\}$
2 Set $\Psi_Y = \{\psi \,|\, Y \in \mathrm{dom}(\psi)\}$
3 Compute $\phi_Y = (\prod_{\phi \in \Phi_Y} \phi)^{\downarrow - Y}$
4 Compute $\psi_Y = (\prod_{\phi \in \Phi_Y} \phi \times \sum_{\psi \in \Psi_Y} \psi)^{\downarrow - Y}$
5 Set $\Phi^* = \{\phi_Y\} \cup \Phi \setminus \Phi_Y$
6 Set $\Psi^* = \{\frac{\psi_Y}{\phi_Y}\} \cup \Psi \setminus \Psi_Y$
7 **return** Φ^*, Ψ^*

Algorithm 3. The $\mathrm{MARGINALIZATION}$ algorithm.

---

ordering [20]. The Bayesian network for computing the probability of future decisions is constructed as described by [7]. The distribute operation proceeds as Simple Propagation in a Bayesian network [19].

4 Analysis

This section considers solving two different influence diagrams with Simple Propagation illustrating the process and the properties of the algorithm.

4.1 Influence Diagram with Four Decisions

Figure 1 shows a commonly used example of an influence diagram N_{JJD} with four decisions introduced by [12]. The figure does not include non-forgetting information links in order to reduce the clutter to a minimum.

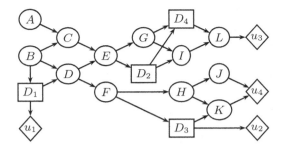

Fig. 1. An influence diagram N_{JJD} with four decisions [12].

Figure 2 shows a strong junction tree representation T_{JJD} of N_{JJD} with strong root $R = BD_1ACDEF$. This is not an optimal junction tree in terms of, for instance, total clique weight, but it serves to illustrate important properties of Simple Propagation for solving influence diagrams. For simplicity, we refer to the two leaf cliques as A and B as indicated in the subscripts of the clique potentials π_A and π_B in the figure.

$$\pi_R = (\{P(A), P(B), P(C\,|\,A, B), P(D\,|\,B, D_1), P(E\,|\,C, D), P(F\,|\,D)\}, \{u_1(D_1)\})$$

$$\pi_A = (\{P(G|E), P(I\,|\,G, D_2), P(L\,|\,I, D_4)\}, \{u_3(L)\})$$

$$\pi_B = (\{P(H\,|\,F), P(J\,|\,H), P(K\,|\,H, D_3)\}, \{u_2(D_3), u_4(J, K)\})$$

Fig. 2. A junction tree representation T_{JJD} of N_{JJD}.

Collect Operation. Simple Propagation solves N_{JJD} by first performing a collect operation on R passing messages $\pi_{A \to R}$ and $\pi_{B \to R}$ from the leaf cliques A and B, respectively. The message $\pi_{A \to R}$ is computed from $\pi_A = (\{P(G\,|\,E), P(I\,|\,G, D_2), P(L\,|\,I, D_4)\}, \{u_3(L)\})$, while the message is computed from $\pi_B = (\{P(H\,|\,F), P(J\,|\,H), P(K\,|\,H, D_3)\}, \{u_2(D_3), u_4(J, K)\})$.

Consider the construction of the message $\pi_{A \to R}$ passed from A to R during the collect operation. As $D_2, D_4 \in \mathcal{D}(A) \setminus R$ and $D_2 \prec D_4$, the process of constructing $\pi_{A \to R}$ starts with D_4. In this case, the potential including D_4 satisfying the *one-in, one-out*-principle is $P(L\,|\,D_4, I)$, where $I, L \in \mathcal{I}_4$ must be eliminated. The vanilla version of Simple Propagation selects randomly between $I, L \in \mathcal{I}_4$.

Assume L is selected first. The elimination of L (Algorithm 3) produces $\phi(\cdot \mid D_4, I) = \sum_L P(L \mid D_4, I) = 1_{D_4, I}$ and $\psi_X(D_4, I) = \sum_L P(L \mid D_4, I) u_3(L)$, where $1_{D_4, I}$ is the unity potential indexed by D_4, I, i.e., L is probabilistic barren and we do not need to compute this potential nor the division in Algorithm 3. Hence, $\pi_A^{\downarrow -L} = (\{P(G \mid E), P(I \mid G, D_2)\}, \{\psi(D_4, I)\})$. The next set of potentials satisfying the *one-in, one-out*-principle is $\{P(I \mid G, D_2), \psi(D_4, I)\}$. Here, I is the only *out*-variable and must be eliminated producing $\pi_A^{\downarrow -\{L, I\}} = (\{P(G \mid E)\}, \{\psi(D_4, G, D_2)\})$ as I is probabilistic barren. Now, the variables $\mathcal{I}_4$ have been eliminated. Decision variables D_4 is eliminated from $\psi(D_4, G, D_2)$ and the decision policy $\delta_{D_4}(G, D_2)$ is identified producing $(\{P(G \mid E)\}, \{\psi(G, D_2)\})$. Finally, variables G and D_2 are eliminated to produce the message $\pi_{A \to R} = \pi_A^{\downarrow E} = (\{\}, \{\psi(E)\})$. The induced elimination order is $\sigma = (L, I, D_4, G, D_2)$.

Now assume I is selected first. The elimination of I (Algorithm 3) produces $\phi(L \mid D_2, G, D_4) = \sum_I P(I \mid G, D_2) P(L \mid I, D_4)$. That is, $\pi_A^{\downarrow -I} = (\{P(G \mid E), P(L \mid D_2, G, D_4)\}, \{\psi(L)\})$. The elimination of I is followed by the elimination of L as $P(L \mid D_2, G, D_4)$ is the only potential satisfying the *one-in, one-out*-principle and $L \in \mathcal{I}_4$. The process continues as above producing the message $\pi_{A \to R} = \pi_A^{\downarrow E} = (\{\}, \{\psi(E)\})$ with the induced elimination order $\sigma = (I, L, D_4, G, D_2)$. For the message $\pi_{A \to R}$ the only random element is the selection of the variable to eliminate from $P(L \mid D_4, I)$.

Next, the construction of the message $\pi_{B \to R}$ passed from B to R during the collect operation is constructed. The message is $\pi_{B \to R} = (\{\}, \{\psi(F)\})$. Following the collect operation, the root R is processed to identify $\delta_{D_1}(B)$ and compute $EU(\Delta)$ for $\Delta = \{\delta_{D_1}(B), \delta_{D_2}(E), \delta_{D_3}(F), \delta_{D_4}(D_2, G)\}$).

The root node is processed similarly.

Distribute Operation. The purpose of the distribute operation is to compute the probability of future decisions under the optimal strategy $\hat{\Delta}$, i.e., $P_{\hat{\Delta}}(D)$, for each $D \in \mathcal{D}$. Following the collect operation, the Bayesian network N_{JJD}^* is constructed over $\mathcal{X} \cup \mathcal{D}$, where each decision policy $\delta_D(\mathrm{rel}(D))$ of the optimal strategy $\hat{\Delta}$ is encoded as a CPD $P(D \mid \mathrm{rel}(D))$ and D is turned into a random variable. Subsequently, the distribute operation is performed to compute $P(D)$ for each $D \in \mathcal{D}$. This process proceeds as the distribute operation in a Bayesian network using Simple Propagation [19].

4.2   Distributive Law

Figure 3(a) shows the structure of a simple influence diagram N with one decision $\mathcal{D} = \{D\}$, three random variables $\mathcal{X} = \{C_1, C_2, C_3\}$ and two utility functions $\mathcal{Y} = \{u_1(C_1, C_3), u_2(D, C_2)\}$ and Fig. 3(b) shows a strong junction tree T with root $R = C_1 D C_3$.

$$\pi_{C_1 DC_3} = (\{P(C_1 \,|\, C_3), P(C_3)\}, \{u_1(C_1, C_3)\})$$

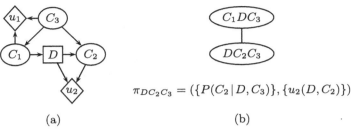

$$\pi_{DC_2 C_3} = (\{P(C_2 \,|\, D, C_3)\}, \{u_2(D, C_2)\})$$

(a)   (b)

Fig. 3. An influence diagram and its junction tree representation.

Simple Propagation solves N by first performing a collect operation on R passing the message from clique $DC_2 C_3$ to clique $C_1 DC_3$. The collect message $\pi_{DC_2 C_3 \to C_1 DC_3}$ from clique $DC_2 C_3$ to clique $C_1 DC_3$ is computed as:

$$\pi_{DC_2 C_3 \to C_1 DC_3} = (\emptyset, \{\sum_{C_2} P(C_2 \,|\, D, C_3) u_2(D, C_2)\})$$
$$= (\emptyset, \{\psi(D, C_3)\}),$$

where C_2 is probabilistic barren making $\Phi_{DC_2 C_3 \to C_1 DC_3} = \emptyset$. At the root, we need to determine the policy δ_D of D and process all potentials to compute $EU(\hat{\Delta})$. The identification of δ_D starts from $\psi(D, C_3)$ as this is the only probability or utility potential associated with R and incoming messages that contains D, i.e., $\pi_R \otimes \pi_{DC_2 C_3 \to C_1 DC_3}$.

The decomposition of clique probability and utility potentials gives a number of advantages. One advantage is the option to exploit the distributive law of algebra when eliminating variables [20, 21]. The distributive law can be exploited when eliminating C_3:

$$P(C_1) \times (\psi(C_1) + \psi(C_1, D))$$
$$= \sum_{C_3} P(C_3) P(C_1 \,|\, C_3)(u_1(C_1, C_3) + \psi(D, C_3))$$
$$= \sum_{C_3} (P(C_3) P(C_1 \,|\, C_3) u_1(C_1, C_3)) + \sum_{C_3} (P(C_3) P(C_1 \,|\, C_3) \psi(D, C_3)).$$

This approach supports exploitation of probabilistic barren variables and the decomposition of utility potentials reducing the number of calculation operations.

5   Conclusion

This paper has introduced Simple Propagation as a method for solving influence diagrams with discrete variables exactly and with optimality. The main advantage of Simple Propagation over other methods for finding an optimal strategy and computing the probability of future decisions under an optimal strategy is its simplicity through the absence of the need for graph theoretical considerations to identify the potentials relevant for a message.

The paper has extended the *one-in, one-out*-principle to cope with decision variables and utility functions to solve influence diagrams. It has also illustrated and described properties of solving influence diagrams with Simple Propagation.

Future work includes considering other algorithms than VE as the variable elimination algorithm as well as solving limited-memory and mixed influence diagrams using Simple Propagation. It also includes considered other computational structures than the strong junction tree as well as any-time approximation.

References

1. Howard, R.A., Matheson, J.E.: Influence diagrams. In: Readings in Decision Analysis, pp. 763–771. Strategic Decisions Group, Menlo Park (1981)
2. Kjærulff, U.B., Madsen, A.L.: Bayesian Networks and Influence Diagrams: A Guide to Construction and Analysis, 2nd edn. Springer, New York (2013). https://doi.org/10.1007/978-0-387-74101-7
3. Koller, D., Friedman, N.: Probabilistic Graphical Models — Principles and Techniques. MIT Press, Cambridge (2009)
4. Lauritzen, S.L., Nilsson, D.: Representing and solving decision problems with limited information. Manage. Sci. **47**, 1238–1251 (2001)
5. Jensen, F.V., Vomlelova, M.: Unconstrained influence diagrams. In: Proceedings of Uncertainty in Artificial Intelligence Conference, pp. 234–241 (2002)
6. Madsen, A.L., Jensen, F.: Mixed influence diagrams. In: Nielsen, T.D., Zhang, N.L. (eds.) ECSQARU 2003. LNCS (LNAI), vol. 2711, pp. 208–219. Springer, Heidelberg (2003). https://doi.org/10.1007/978-3-540-45062-7_17
7. Nilsson, D., Jensen, F.V.: Probabilities of future decisions. In: Bouchon-Meunier, B., Yager, R.R., Zadeh, L.A. (eds.) Information, Uncertainty and Fusion. The Springer International Series in Engineering and Computer Science, vol. 516, pp. 161–171. Springer, Boston (2000). https://doi.org/10.1007/978-1-4615-5209-3_12
8. Olmsted, S.M.: On representing and solving decision problems. Ph.D. thesis, Department of Engineering-Economic Systems, Stanford University, Stanford, CA (1983)
9. Shachter, R.D.: Evaluating influence diagrams. Oper. Res. **34**(6), 871–882 (1986)
10. Shachter, R.D., Peot, M.A.: Decision making using probabilistic inference methods. In: Proceedings of the Eighth Conference on Uncertainty in Artificial Intelligence, pp. 276–283. Morgan Kaufmann Publishers, San Mateo (1992)
11. Shenoy, P.P.: Valuation-based systems for Bayesian decision analysis. Oper. Res. **40**(3), 463–484 (1992)
12. Jensen, F., Jensen, F.V., Dittmer, S.: From influence diagrams to junction trees. In: Proceedings of Uncertainty in Artificial Intelligence Conference, pp. 367–373. Morgan Kaufmann Publishers, San Francisco (1994)
13. Madsen, A.L., Jensen, F.V.: Lazy evaluation of symmetric Bayesian decision problems. In: Proceedings of Uncertainty in Artificial Intelligence Conference, pp. 382–390 (1999)
14. Cooper, G.F.: A method for using belief networks as influence diagrams. In: Proceedings of Uncertainty in Artificial Intelligence Conference, pp. 55–63 (1988)

15. Yang, M., Zhou, L., Ruan, H.: The methods for solving influence diagrams: a review. In: International Conference on Information Technology and Applications, pp. 427–431 (2013)
16. Jensen, F.V., Nielsen, T.D.: Probabilistic decision graphs for optimization under uncertainty. 4OR-Q J. Oper. Res. **9**(1), 1–28 (2011)
17. Lee, J., Ihler, A., Dechter, R.: Join graph decomposition bounds for influence diagrams. In: Proceedings of Uncertainty in Artificial Intelligence Conference, pp. 1053–1062 (2018)
18. Yet, B., Neil, M., Fenton, N., Constantinou, A., Demetiev, E.: An improved method for solving hybrid influence diagrams. IJAR **95**, 93–112 (2018)
19. Butz, C.J., de Oliveira, J.S., dos Santos, A.E., Madsen, A.L.: Bayesian network inference with simple propagation. In: Proceedings of Florida Artificial Intelligence Research Society Conference, pp. 650–655 (2016)
20. Madsen, A.L., Jensen, F.V.: Lazy propagation: a junction tree inference algorithm based on lazy evaluation. Artif. Intell. **113**(1–2), 203–245 (1999)
21. Cabañas, R., Cano, A., Gómez-Olmedo, M., Madsen, A.L.: On SPI-lazy evaluation of influence diagrams. In: van der Gaag, L.C., Feelders, A.J. (eds.) PGM 2014. LNCS (LNAI), vol. 8754, pp. 97–112. Springer, Cham (2014). https://doi.org/10.1007/978-3-319-11433-0_7
22. Madsen, A.L., Nilsson, D.: Solving influence diagrams using HUGIN, Shafer-Shenoy and lazy propagation. In: Proceedings of Uncertainty in Artificial Intelligence Conference, pp. 337–345 (2001)
23. Nielsen, T.D., Jensen, F.V.: Welldefined decision scenarios. In: Proceedings of Uncertainty in Artificial Intelligence Conference, pp. 502–511 (1999)
24. Nielsen, T.D.: Decomposition of influence diagrams. In: Benferhat, S., Besnard, P. (eds.) ECSQARU 2001. LNCS (LNAI), vol. 2143, pp. 144–155. Springer, Heidelberg (2001). https://doi.org/10.1007/3-540-44652-4_14

Uncertain Evidence for Probabilistic Relational Models

Marcel Gehrke[(✉)], Tanya Braun, and Ralf Möller

Institute of Information Systems, University of Lübeck, Lübeck, Germany
{gehrke,braun,moeller}@ifis.uni-luebeck.de

Abstract. Standard approaches for inference in probabilistic relational models include lifted variable elimination (LVE) for single queries. To efficiently handle multiple queries, the lifted junction tree algorithm (LJT) uses a first-order cluster representation of a model, employing LVE as a subroutine in its steps. LVE and LJT can only handle certain evidence. However, most events are not certain. The purpose of this paper is twofold, (i) to adapt LVE, presenting LVE^{evi}, to handle uncertain evidence and (ii) to incorporate uncertain evidence for multiple queries in LJT, presenting LJT^{evi}. With LVE^{evi} and LJT^{evi}, we can handle uncertain evidence for probabilistic relational models, while benefiting from the lifting idea. Further, we show that uncertain evidence does not have a detrimental effect on completeness results and leads to similar runtimes as certain evidence.

1 Introduction

Areas such as health care or logistics involve probabilistic data with relational aspects where many objects are in relation to each other with uncertainties about object existence, attribute value assignments, or relations between objects. E.g., health care systems involve electronic health records (EHRs) (the relational part) for many patients (the objects) and uncertainties [22] due to, e.g., missing information. Additionally, evidence or events are not always certain due to different sensors or where the tests are performed. Automatically analysing EHRs can improve the care of patients and save time for medical professionals to spend on other important tasks.

Answering queries in probabilistic, relational environments, like predicting outcomes of treatments, needs efficient exact inference algorithms, which is particularly true for health care as approximations might not be good enough [24]. Probabilistic databases (PDBs) can answer queries for relational models with uncertainties [8,19]. However, each query can contain redundant information, resulting in huge queries. In contrast to PDBs, we build more expressive and compact models (offline) enabling efficient answering of more compact queries (online). Currently, these compact models cannot handle uncertain evidence.

This research originated from the Big Data project being part of Joint Lab 1, funded by Cisco Systems Germany, at the centre COPICOH, University of Lübeck.

© Springer Nature Switzerland AG 2019
M.-J. Meurs and F. Rudzicz (Eds.): Canadian AI 2019, LNAI 11489, pp. 80–93, 2019.
https://doi.org/10.1007/978-3-030-18305-9_7

Therefore, in this paper, we study the problem of exact inference in relational temporal probabilistic models with uncertain evidence.

Research in the field of lifted inference has lead to efficient algorithms for relational models. lifted variable elimination (LVE), first introduced in [16] and expanded in [12,17,21], saves computations by reusing intermediate results for isomorphic subproblems when answering a query. The lifted junction tree algorithm (LJT) sets up a first-order junction tree (FO jtree) to handle multiple queries efficiently [2] using LVE as a subroutine. Van den Broeck et al. apply lifting to weighted model counting and knowledge compilation [6]. To scale lifting, Das et al. use graph databases storing compiled models to count faster [9]. Lifted belief propagation (BP) provides approximate solutions to queries, often using lifted representations, e.g. [1]. But, to the best of our knowledge, none of the approaches handle uncertain evidence.

We focus on *exact* inference for multiple queries and present an efficient algorithm for based on LJT, called LJT^{evi}, handling uncertain evidence. For Bayesian networks, work on uncertain evidence exists, sometimes called soft evidence in contrast to certain, i.e., hard evidence [7,10,13–15]. In this paper, we interpret uncertain evidence in the sense of a priori distributions, which is closely related to Pearl's method of virtual evidence [13]. This paper includes two main contributions, (i) an algorithm, LVE^{evi}, handling uncertain evidence for probabilistic relational models and (ii) LJT^{evi}, handling uncertain evidence for multiple queries. Additionally, we show soundness and completeness results for LVE^{evi} and LJT^{evi} and a brief empirical case study.

The remainder of this paper is structured as follows: First, we introduce basic notations and recap LVE and LJT. Then, we show how to handle uncertain evidence and present LJT^{evi}, followed by a discussion. We conclude with upcoming work.

2   Preliminaries

This section specifies notations and recaps LJT. Based on [17], a running example models the interplay of natural or man-made disasters, an epidemic, and people being sick, travelling, and being treated. Parameters represent disasters, people, and treatments.

2.1   Parameterised Probabilistic Models

Parameterised models compactly represent models using logical variables (logvars) to parameterise random variables (randvars), so called parameterised randvars (PRVs).

Definition 1. *Let* **L**, *Φ, and* **R** *be sets of logvar, factor, and randvar names respectively. A PRV* $R(L_1, \ldots, L_n)$, $n \geq 0$, *is a syntactical construct with* $R \in$ **R** *and* $L_1, \ldots, L_n \in$ **L** *to represent a set of randvars. For PRV A, the term* $range(A)$ *denotes possible values. A logvar L has a domain* $\mathcal{D}(L)$. *A constraint*

$(\mathbf{X}, C_{\mathbf{X}})$ *is a tuple with a sequence of logvars* $\mathbf{X} = (X_1, \dots, X_n)$ *and a set* $C_{\mathbf{X}} \subseteq \times_{i=1}^{n} \mathcal{D}(X_i)$ *restricting logvars to values. The symbol* $\top$ *marks that no restrictions apply and may be omitted. For some construct* P, *the term* $lv(P)$ *refers to its logvars, the term* $rv(P)$ *to its PRVs with constraints, and the term* $gr(P)$ *to all instances of* P, *i.e.* P *grounded w.r.t. constraints.*

For the example, we build the boolean PRVs *Epid*, *Sick*(X), and *Travel*(X) from $\mathbf{R} = \{Epid, Sick, Travel\}$ and $\mathbf{L} = \{X\}$, $\mathcal{D}(X) = \{alice, eve, bob\}$. *Epid* holds if an epidemic occurs. *Sick*(X) holds if a person X is sick, *Travel*(X) holds if X travels. With a constraint $C = (X, \{eve, bob\})$, $gr(Sick(X)_{|C}) = \{Sick(eve), Sick(bob)\}$. $gr(Sick(X)_{|\top})$ also contains *Sick*($alice$). Parametric factors (parfactors) link PRVs. A parfactor describes a function, identical for all argument groundings, mapping argument values to real values (potentials), of which at least one is non-zero.

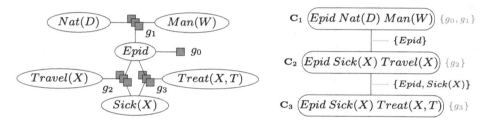

Fig. 1. Parfactor graph for G_{ex} **Fig. 2.** FO jtree for G_{ex}

Definition 2. *Let* $\mathbf{X}$ *be a set of logvars,* $\mathcal{A} = (A_1, \dots, A_n)$ *a sequence of PRVs with* $lv(\mathcal{A}) \subseteq \mathbf{X}$, C *a constraint on* $\mathbf{X}$, *and* $\phi : \times_{i=1}^{n} range(A_i) \mapsto \mathbb{R}^+$ *a function with name* $\phi \in \Phi$, *identical for* $gr(\mathcal{A}_{|C})$. *We denote a* parfactor g *by* $\forall \mathbf{X} : \phi(\mathcal{A})_{|C}$. *We omit* $(\forall \mathbf{X} :)$ *if* $\mathbf{X} = lv(\mathcal{A})$ *and* $|C$ *if* $C = \top$. *A set of parfactors forms a* model $G := \{g_i\}_{i=1}^{n}$.

We define a model G_{ex} as our running example. Let $\mathbf{L} = \{D, W, M, X\}$, $\Phi = \{\phi_0, \phi_1, \phi_2, \phi_3\}$, and $\mathbf{R} = \{Epid, Nat, Man, Sick, Travel, Treat\}$. We build three more boolean PRVs. $Nat(D)$ holds if a natural disaster D occurs, $Man(W)$ if a man-made disaster W occurs. $Treat(X, T)$ holds if a person X is treated with treatment T. The other domains are $\mathcal{D}(D) = \{earthquake, flood\}$, $\mathcal{D}(W) = \{virus, war\}$, and $\mathcal{D}(T) = \{vaccine, tablet\}$. The model reads $G_{ex} = \{g_i\}_{i=0}^{3}$, $g_0 = \phi_0(Epid)$, $g_1 = \phi_1(Epid, Nat(D), Man(W))_{|\top}$, $g_2 = \phi_2(Epid, Sick(X), Travel(X))_{|\top}$, and $g_3 = \phi_3(Epid, Sick(X), Treat(X, T))_{|\top}$. Parfactors g_1 to g_3 have eight input-output pairs, g_0 has two (omitted here). Figure 1 depicts G_{ex} as a graph with six variable nodes for the PRVs and four factor nodes for the parfactors with edges to arguments.

Evidence displays symmetries if observing the same value for n instances of a PRV [21]. In a parfactor $g_E = \phi_E(R(\mathbf{X}))_{|C_E}$, a potential function ϕ_E and constraint C_E encode the observed values and instances for PRV $R(\mathbf{X})$. Assume we observe the value *true* for ten randvars of the PRV $Sick(X)$. The corresponding parfactor is $\phi_E(Sick(X))_{|C_E}$. C_E represents the domain of X restricted to the 10 instances and $\phi_E(true) = 1$ and $\phi_E(false) = 0$. A technical remark: To *absorb* evidence, we split all parfactors g_i that cover $R(X)$, called shattering [17], restricting C_i to those tuples that contain $gr(R(X)_{|C_E})$ and a duplicate of g_i to the rest. g_i absorbs g_E (cf. [21]).

The *semantics* of a model G is given by grounding and building a full joint distribution P_G. With Z as the normalisation constant, G represents $P_G = \frac{1}{Z} \prod_{f \in gr(G)} f$. The query answering (QA) problem asks for a marginal distribution of a set of randvars or a conditional distribution given events, which boils down to computing marginals w.r.t. a model's joint distribution, eliminating non-query terms. Formally, $P(\mathbf{Q}|\mathbf{E})$ denotes a query with $\mathbf{Q}$ a set of grounded PRVs and $\mathbf{E} = \{E_i = e_i\}_{i=1}^n$ a set of events. An example query for G_{ex} is $P(Epid|Sick(eve) = true)$. Next, we look at LJT, a lifted QA algorithm, which seeks to avoid grounding and building a full joint distribution.

2.2   Query Answering Algorithms

LVE and LJT answer queries for probability distributions. LJT uses an FO jtree with LVE as a subroutine. We briefly recap LVE and LJT.

Lifted Variable Elimination: In essence, LVE computes variable elimination for one case and exponentiates the result for isomorphic instances (lifted summing out), avoiding duplicate calculations. Taghipour et al. implement LVE through an operator suite (cf. [21] for details). Its main operator *sum-out* realises lifted summing out. An operator *absorb* handles evidence in a lifted way. The remaining operators (*count-convert*, *split*, *expand*, *count-normalise*, *multiply*, *ground-logvar*) aim at enabling a lifted summing out, transforming part of a model. All operators have pre- and post-conditions to ensure computing a result equivalent to one computed on $gr(G)$. Algorithm 1 shows an outline. To answer a query, LVE eliminates all non-query randvars from the model.

Algorithm 1. Outline of LVE	**Algorithm 2.** Outline of LJT
1: LVE(Model G, Query $\mathbf{Q}$, Evidence $\mathbf{E}$)	1: LJT(M. G, Queries $\{\mathbf{Q}_i\}_{i=1}^m$, Evidence $\mathbf{E}$)
2: Absorb $\mathbf{E}$ in G	2: Construct FO jtree J
3: **while** G has non-query PRVs **do**	3: Enter $\mathbf{E}$ into J
4: **if** PRV A fulfils *sum-out* prec. **then**	4: Pass messages on J
5: Eliminate A using *sum-out*	5: **for** each query $\mathbf{Q}_i$ **do**
6: **else**	6: Find subtree J_i for $\mathbf{Q}_i$
7: Apply transformator	7: Extract submodel G_i from J_i
8: Multiply and normalise G	8: LVE(G_i, $\mathbf{Q}_i$, $\emptyset$)

Lifted Junction Tree Algorithm: Algorithm 2 outlines LJT for a set of queries $\{\mathbf{Q}_i\}_{i=1}^m$ given a model G and evidence $\mathbf{E}$. First, LJT builds an FO jtree, which clusters a model into submodels that LJT uses to answer queries after preprocessing. We define a minimal FO jtree with parameterised clusters (parclusters) as nodes, which are sets of PRVs connected by parfactors, as follows.

Definition 3. *Let $\mathbf{X}$ be a set of logvars, $\mathbf{A}$ a set of PRVs with $lv(\mathbf{A}) \subseteq \mathbf{X}$, and C a constraint on $\mathbf{X}$. Then, $\forall \mathbf{X}{:}\mathbf{A}_{|C}$ denotes a parcluster. We omit $(\forall \mathbf{X}{:})$ if $\mathbf{X} = lv(\mathbf{A})$ and $|\top$. An FO jtree for a model G is a cycle-free graph $J = (V, E)$, where V is the set of nodes, i.e., parclusters, and E the set of edges. J must satisfy three properties: (i) $\forall \mathbf{C}_i \in V$: $\mathbf{C}_i \subseteq rv(G)$. (ii) $\forall g \in G$: $\exists \mathbf{C}_i \in V$ s.t. $rv(g) \subseteq \mathbf{C}_i$. (iii) If $\exists A \in rv(G)$ s.t. $A \in \mathbf{C}_i \wedge A \in \mathbf{C}_j$, then $\forall \mathbf{C}_k$ on the path between $\mathbf{C}_i$ and $\mathbf{C}_j$: $A \in \mathbf{C}_k$ (running intersection property). J is minimal if by removing a PRV from any parcluster, J ceases to be an FO jtree, i.e., no longer fulfills at least one of the three properties. The parameterised set $\mathbf{S}_{ij}$, called separator of edge $\{i, j\} \in E$, is given by $\mathbf{C}_i \cap \mathbf{C}_j$. Each $\mathbf{C}_i \in V$ has a local model G_i and $\forall g \in G_i$: $rv(g) \subseteq \mathbf{C}_i$. The G_i's partition G.*

In a minimal FO jtree, no parcluster is a subset of another parcluster. Figure 2 shows a minimal FO jtree for G_{ex} with parclusters $\mathbf{C}_1 = \{Epid, Nat(D), Man(W)\}$, $\mathbf{C}_2 = \{Epid, Sick(X), Travel(X)\}$, and $\mathbf{C}_3 = \{Epid, Sick(X), Treat(X, T)\}$. $\mathbf{S}_{12} = \{Epid\}$ and $\mathbf{S}_{23} = \{Epid, Sick(X)\}$ are the separators. Parfactor g_0 appears at $\mathbf{C}_1$ but could be in any local model as $rv(g_0) = \{Epid\} \subset \mathbf{C}_i \; \forall \, i \in \{1, 2, 3\}$.

During construction, LJT assigns the parfactors in G to local models (cf. [2]). LJT enters $\mathbf{E}$ into each parcluster $\mathbf{C}_i$ where $rv(\mathbf{E}) \subseteq \mathbf{C}_i$. Local model G_i at $\mathbf{C}_i$ absorbs $\mathbf{E}$ as described above. Message passing distributes local information within the FO jtree. Two passes from the periphery to the center and back suffice [11]. If a node has received messages from all neighbours but one, it sends a message to the remaining neighbour (*inward* pass). In the *outward* pass, messages flow in the opposite direction. Formally, a *message m_{ij}* from node i to node j is a set of parfactors, with arguments from $\mathbf{S}_{ij}$. LJT computes m_{ij} by eliminating $\mathbf{C}_i \setminus \mathbf{S}_{ij}$ from G_i and the messages of all other neighbours with LVE. A minimal FO jtree enhances the efficiency of message passing. Otherwise, messages unnecessarily copy information between parclusters. To answer a query $\mathbf{Q}_i$, LJT finds a subtree J' covering $\mathbf{Q}_i$, compiles a submodel G' of local models in J' and messages from outside J', and sums out all non-query terms in G' using LVE.

3 LVE for Uncertain Evidence

Currently, evidence in LVE and therefore LJT is always certain. However, often sensors are not completely reliable or some test may be more precisely performed in a hospital compared to a test in a general practice [18]. Before we incorporate uncertain evidence in LVE, we take a closer look at how LVE handles certain evidence.

3.1   Evidence in LVE

In Algorithm 1, handling evidence appears as "absorb evidence". Thus, let us now have a look at lifted absorption, which is outlined in Algorithm 3, without including counting PRVs for ease of explanation. The operator uses a count function defined as follows.

Definition 4. *Given a constraint* $C = (\mathbf{X}, C_{\mathbf{X}})$*, for any* $\mathbf{Y} \subseteq \mathbf{X}$ *and* $\mathbf{Z} \subseteq \mathbf{X} \backslash \mathbf{Y}$*, the function* $\mathrm{COUNT}_{\mathbf{Y}|\mathbf{Z}} : C_{\mathbf{X}} \to \mathbb{N}$ *is defined by*

$$\mathrm{COUNT}_{\mathbf{Y}|\mathbf{Z}}(t) = |\pi_{\mathbf{Y}}(C_{\mathbf{X}} \bowtie_{\mathbf{Z}} \pi_{\mathbf{Z}}(t))|.$$

i.e., for a tuple $t \in C_{\mathbf{X}}$*, it outputs how many constants for* $\mathbf{Y}$ *co-occur with the value of* $\mathbf{Z}$ *in* t*. We define* $\mathrm{COUNT}_{\mathbf{Y}|\mathbf{Z}}(t) = 1$ *for* $\mathbf{Y} = \emptyset$*.* $\mathbf{Y}$ *is count-normalised w.r.t.* $\mathbf{Z}$ *in* C *iff*

$$\exists n \in \mathbb{N} : \forall t \in C_{\mathbf{X}} : \mathrm{COUNT}_{\mathbf{Y}|\mathbf{Z}}(t) = n.$$

If n *exists, we call it the conditional count of* $\mathbf{Y}$ *given* $\mathbf{Z}$ *in* C*, denoted by* $\mathrm{COUNT}_{\mathbf{Y}|\mathbf{Z}}(C)$*.*

Before we take a closer look at the operator, we illustrate the count function. Consider the constraint $C = ((X, T), \{(eve, tablet), (alice, vaccine), (alice, tablet), (bob, vaccine), (bob, tablet)\})$. With $\mathbf{X} = \{X, T\}$, $\mathbf{Y} = \{T\}$, and $\mathbf{Z} = \{X\}$, the count function calculates the following for tuple $(eve, tablet)$: First, it projects $(eve, tablet)$ onto $\{X\}$, which leaves (eve). Then, it joins eve with the tuples from C, i.e., $(eve, tablet)$, and projects the tuples onto $\{T\}$, which results in a set with one element, $(tablet)$. Last, it outputs the cardinality of the set, here 1. For $(alice, vaccine)$, the first projection yields $(alice)$, with the join resulting in $(alice, vaccine)$ and $(alice, tablet)$ and the second projection resulting in $(vaccine)$ and $(tablet)$, yielding a cardinality of 2. Thus, there does not exist a unique n for all tuples in C, that is, M is not count-normalised. Now, assume the constraint $C' = ((X, T), \{(alice, vaccine), (alice, tablet), (bob, vaccine), (bob, tablet)\})$. Here, each tuple leads to a count of 2 given $\mathbf{X} = \{X, T\}$, $\mathbf{Y} = \{T\}$, and $\mathbf{Z} = \{X\}$ and thus, M is count-normalised w.r.t. X in C'. The conditional count of T given X in C' is 2. In case, $alice$ and bob would additionally receive another treatment, the count would be 3. The count in this case is important as absorbing evidence eliminates as many instances as the count function yields, and thus, LVE needs to exponentiate the result with the count.

ABSORB has as inputs an evidence parfactor g_E with evidence for a PRV A_i and a parfactor g, which contains A_i. As a precondition, A_i covers at most the randvars of g_E in g. Thus, LVE often performs a shattering before absorption to split parfactors into parts with evidence and without evidence. The other precondition relates to logvars being eliminated during absorption. For the output parfactor, the operator deletes A_i from g, reducing the dimensions in g. The operator also projects the constraint C of g onto the remaining logvars. Lastly, it collects all potentials that agree with the evidence, i.e., where $A_i = o$, and

Algorithm 3. Lifted Absorption [21].

Operator ABSORB
Inputs:
 (1) $g = \phi(\mathcal{A})_{|C}$: a parfactor in G
 (2) $A_i \in \mathcal{A}$ with $A_i = R(\mathbf{X})$
 (3) $g_E = \phi_E(R(\mathbf{X}))_{|C_E}$: an evidence parfactor
 Let $\mathbf{X}^{excl} = \mathbf{X} \setminus lv(\mathcal{A} \setminus A_i)$;
 $L' = lv(\mathcal{A}) \setminus \mathbf{X}^{excl}$;
 o = the observed value for $R(\mathbf{X})$ in g_E
Preconditions:
 (1) $gr(A_{i|C_i}) \subseteq gr(A_{i|C_E})$
 (2) $\mathbf{X}^{excl}$ is count-normalised w.r.t. L' in C.
Output: $\phi'(\mathcal{A}')_{|C'}$, with
 (1) $\mathcal{A}' = \mathcal{A} \setminus A_i$
 (2) $C' = \pi_{lv(C) \setminus \mathbf{X}^{excl}}(C)$
 (3) $\phi'(\ldots, a_{i-1}, a_{i+1}, \ldots) = \phi(\ldots, a_{i-1}, o, a_{i+1}, \ldots)^r$ with $r = \text{COUNT}_{\mathbf{X}^{excl}|L'}(C)$
Postcondition: $G \cup \{g_E\} \equiv G \setminus \{g\} \cup \{g_E, \text{ABSORB}(g, A_i, g_E)\}$

---

exponentiates them accordingly. As the operator performs a dimension reduction by deleting A_i from the argument sequence, rather than keeping the argument and setting all potentials where $A_i \neq o$ to 0, LVE has to apply the absorption operator to each parfactor that contains A_i.

To illustrate the ABSORB operator, assume that eve is sick, i.e. $Sick(eve) = true$. LVE builds an evidence parfactor $g_E = \phi_E(Sick(X))_{|C_E}$, with $C_E = (X, \{eve\})$. As g_2 and g_3 also contain $Sick(X)$, both need to absorb g_E. To absorb g_E in g_2, LVE first splits g_2 into g_2' for eve and g_2'' for all other instances, i.e., $alice$ and bob. With g_E and g_2' as inputs, the first precondition holds as both g_E and g_2' have X restricted to eve. Since $\mathbf{X}^{excl} = X \setminus lv(Epid, Travel(X)) = \emptyset$, i.e., no logvars are eliminated, $\mathbf{X}^{excl}$ is count-normalised and $r = 1$. Hence, the operator can proceed. It removes $Sick(X)$ from g_2'. The constraint remains unchanged. Lastly, all potentials that agree with $Sick(eve) = true$ remain and are exponentiated to the power of 1. Similarly g_E gets absorbed in g_3. One could also perform lifted absorption by multiplying g_E into g, which leads to potentials of 0 whenever $A_i \neq o$. Afterwards, one could drop the mappings with potentials of 0 and then eliminate A_i from the argument sequence as after dropping the mappings, $A_i = o$ in all remaining mappings, holding no further information. However, absorption as in Algorithm 3 only works for certain evidence.

3.2   Uncertain Evidence in LVE^{evi}

The main differences to certain evidence and its handling are in specifying evidence, constructing evidence parfactors, and handling evidence parfactors within LVE. Currently, one event has a potential of 1, while all others have a potential of 0 in evidence. With uncertain evidence, we need to be able to specify potentials different from 0 for possible events of a PRV. However, evidence should not incur a scaling factor. Therefore, individual events of a PRV A have assigned a potential p with $p \in [0,1]$ and the potentials of all possible events of A add up to 1. We allow for two options to specify potentials for events. The first option is to specify the potential for each possible event of a PRV A_i with the sum of the potentials being 1. LVE^{evi} then constructs an evidence parfactor $g_E = \phi_E(A_i)_{|C_E}$ accordingly. The second option is to specify a subset of the events with the sum of the potentials s being at most 1. LVE^{evi} constructs an evidence parfactor $g_E = \phi_E(A_i)_{|C_E}$, distributing the residual potential $1 - s$ on the remaining range values in a max-entropy style [23]. Constructing evidence parfactors in such a way ensures that all range values have a potential and that the potentials add up to 1.

Assume the potential of eve being sick is 0.9. We may specify the evidence using a complete distribution, $Sick(eve) = ((true, 0.9), (false, 0.1))$. The other option is to only specify $Sick(eve) = (true, 0.9)$, a subset of the distribution. Then, LVE^{evi} would distribute the remaining 0.1 max-entropy alike on the remaining range values, while constructing the evidence parfactor. With $Sick(X)$ being boolean, there is only one other range value, namely $false$, which would be assigned a potential of 0.1. In case of another range value, e.g., $immune$, then both would be assigned a potential of 0.05. Assigning a distribution to evidence still allows for specifying certain evidence. Given $Sick(eve) = ((true, 1))$, all other range values would be assigned the potential 0, which is identical to the evidence so far in LVE.

We now present LVE^{evi} to handle uncertain evidence while answering a query. The workflow of LVE^{evi} is identical to LVE as given in Algorithm 1 with line 2 changing. Instead of absorbing all evidence $\mathbf{E}$ in affected parfactors, a case distinction occurs, which is specified in Algorithm 4, for each evidence parfactor g_E constructed for $\mathbf{E}$. If g_E contains certain evidence, g_E is absorbed in G as before. If g_E is uncertain evidence, g_E is added to G. During query answering, the uncertain evidence is then properly accounted for since g_E is multiplied into the model at one point and therefore, influences a queried distribution accordingly. Next, we consider soundness and completeness of LVE^{evi}.

Algorithm 4. Evidence Handling in LVE^{evi}

1: **procedure** ADDEVIDENCE(G, g_E)
2:    **if** g_E is uncertain **then**
3:        Add g_E to G
4:    **else**
5:        Absorb g_E in G

Theorem 1. *LVE^{evi} is sound, i.e., computes a correct result for a query* $\mathbf{Q}$ *given an input model G and evidence* $\mathbf{E}$.

Proof of Sketch. Since both LVE^{evi} and LVE handle certain evidence in the same way and LVE is correct [21], LVE^{evi} is correct w.r.t. certain evidence. We interpret uncertain evidence as an a priori distribution for events. LVE^{evi} simply adds evidence parfactors of uncertain evidence to a model. During query answering, LVE^{evi} then handles these parfactors as part of the model, multiplying evidence parfactors into other parfactors accordingly, thus, accounting for evidence as a form of a priori distribution.   □

Theorem 2. *LVE^{evi} is complete for unary evidence, i.e., the time complexity is polynomial in the domain sizes of the model logvars.*

Proof of Sketch. Given certain, unary evidence, i.e., evidence which can be represented in a parfactor with an evidence PRV using one-logvar, LVE is complete [5,20]. Replacing certain, unary evidence with uncertain, unary evidence with a given distribution leads to the same number of splits during shattering and the number of splits is linear per evidence PRV and model parfactor [21]. Thus, LVE^{evi} still has a time complexity polynomial in the domain sizes of the model logvars given uncertain, unary evidence and the completeness results for unary evidence from LVE also hold for LVE^{evi}.   □

4   LJT for Uncertain Evidence

LVE^{evi} handles uncertain evidence efficiently for single queries. To handle multiple queries efficiently, we incorporate uncertain evidence into LJT based on the same principles that have guided the adaptation of LVE to handle uncertain evidence. Before we present LJT^{evi}, we first take a closer look at how LJT handles evidence.

4.1   Evidence in LJT

Evidence handling in LJT generally works by performing the following steps: (i) Construct evidence parfactors. (ii) Enter evidence parfactors into FO jtree. (iii) Shatter local models on entered evidence parfactors. (iv) Absorb evidence parfactor in local models. Basically, LJT handles evidence in each local model as LVE does in its input model. In each parcluster that covers an evidence PRV, LJT tests each parfactor for evidence absorption. If a parfactor in a local model covers the evidence PRV, LJT shatters the parfactor on the evidence and lets the affected parfactor absorb the evidence parfactor. Whenever a separator no longer covers an evidence PRV, LJT can omit checking the subtree behind the neighbour associated with the separator based on the running intersection property.

Again, assume that *eve* is sick as certain evidence with an evidence parfactor $g_E = \phi_E(Sick(X))_{|C_E}$, with $C_E = (X, \{eve\})$. Then, LJT enters g_E in J_{ex}, which is shown in Fig. 1. The PRV $Sick(X)$ occurs in $\mathbf{C}_2$ and $\mathbf{C}_3$. LJT shatters the local models G_2 and G_3, i.e., g_2 and g_3. LJT splits g_2 into g_2' for *eve* and g_2'' for all other instances, in this case, *alice* and *bob*. Analogously, LJT splits g_3 into g_3' and g_3''. Finally, LJT absorbs g_E in g_2' and g_3'. After the absorption, all local models encode information about certain evidence and the overall model.

Algorithm 5. Evidence Handling in LJT^{evi}

1: **procedure** ENTEREVIDENCE(J, g_E)
2:    **if** g_E is uncertain **then**
3:        Add g_E to the local model of *one* parcluster, which contains the PRV of g_E
4:        Shatter local model (optional)
5:        Multiply g_E into local model (optional)
6:    **else**
7:        Enter g_E in *all* parclusters, which contain the PRV of g_E
8:        Shatter local models
9:        Absorb g_E

4.2 Uncertain Evidence in LJT

LJT^{evi} is based on LJT and is able to handle uncertain evidence as well. Evidence may be specified in the same manner as for LVE^{evi}, which allows for certain evidence as well as uncertain evidence, partially or fully specified with distributions, whose potentials add up to 1. LJT^{evi} has the same workflow as LJT, outlined in Algorithm 2. Only line 3 now references steps that incorporate uncertain evidence. Algorithm 5 describes the steps to enter an evidence parfactor g_E in an FO jtree J. Again, a case distinction occurs. If g_E encodes certain evidence, LJT^{evi} works as LJT, absorbing g_E in the affected parfactors of all parclusters that cover the evidence PRV. If g_E encodes uncertain evidence, LJT^{evi} adds g_E to one local model of a parcluster that covers the evidence PRV. During message passing, the information about the evidence is distributed to all other parclusters, which makes it apparent, why uncertain evidence should only be added to one local model. In case LJT^{evi} would add the uncertain evidence parfactor to all parclusters containing the evidence PRV, then the evidence would be distributed during message passing and accounted for multiple times. One could directly shatter a local model of the chosen parcluster and multiply g_E into it. But, the operations are optional: LJT^{evi} uses LVE for its calculations, which is able to handle g_E accordingly and multiply g_E into other parfactors when necessary, resulting in more efficient multiplications.

Let us consider uncertain evidence and LJT^{evi}. Assume that *eve* is sick with a potential of 0.9. So, LJT^{evi} builds an evidence parfactor $g_E = \phi_E(Sick(X))_{|C_E} = ((true, 0.9), (false, 0.1))$, with $C_E = (X, \{eve\})$, as would LVE^{evi}. Now, LJT^{evi} only needs to find one parcluster containing $Sick(X)$, instead of all parclusters containing $Sick(X)$. Both parclusters $\mathbf{C}_2$ and $\mathbf{C}_3$ contain $Sick(X)$. LJT^{evi} randomly chooses to add g_E to $\mathbf{C}_3$. As the remaining part is optional, we choose against it for efficiency reasons. Evidence entering now is complete.

During message passing, LJT^{evi} sends a message m_{32} from $\mathbf{C}_3$ to $\mathbf{C}_2$. To calculate m_{32}, LJT^{evi} splits g_3 into g'_3 for *eve* and g''_3 for all other instances. Then, LJT^{evi} eliminates $Treat(X, T)$ from g'_3 and g''_3. Afterwards, LJT^{evi} sends m_{32}, which contains g_E, g'_3, and g''_3, to $\mathbf{C}_2$. In m_{32}, we can easily see that LJT^{evi} propagates evidence to all parclusters containing the PRV of the evidence parfactor as it is an explicit part of the message. Next, we consider soundness and completeness of LJT^{evi}.

Theorem 3. *LJT^{evi} is sound, i.e., computes a correct result for a query **Q** given an input model G and evidence **E**.*

Proof of Sketch. For certain evidence, LJT^{evi} computes the same result as LJT since they perform the same steps. Given that LJT is correct [3], LJT^{evi} is correct. For uncertain evidence, LJT^{evi} adds evidence parfactors once to a local model of one parcluster. During message passing and query answering, LJT^{evi} then properly accounts for the evidence as an a priori distribution for the given events.   □

Theorem 4. *LJT^{evi} is complete for unary evidence, i.e., the time complexity is polynomial in the domain sizes of the model logvars.*

Proof of Sketch. The completeness results for unary evidence and LVE [5,20] extend also to LJT. Following the same argument as in the proof sketch of completeness for LVE^{evi}, the change from certain to uncertain evidence over one distribution does not lead to groundings, which means the runtime complexity is still polynomial in the domain sizes of the model logvars and the completeness results extend to LJT^{evi}.   □

5   Empirical Case Study

We have implemented a prototype version of LJT^{evi} and adapted an LVE implementation by Taghipour (https://dtai.cs.kuleuven.be/software/lve) for uncertain evidence. Given the changes from certain to uncertain events in LVE and LJT and their effects on completeness, we expect implementations of the algorithms to accomplish similar runtimes for certain and uncertain evidence given certain evidence does not cancel out a majority of the model. If certain evidence exists for a majority of the PRVs in a model, the dimension reduction during absorption leaves a very small model, enabling fast query answering. Thus, we use the running example with a domain size of 1000 and add certain evidence $Sick(X) = ((true, 1))$ as well as uncertain evidence $Sick(X) = ((true, 0.8), (false, 0.2))$, covering 0% to 100% of $gr(Sick(X))$ in 10% steps. The query randvar is $Travel(x_{1000})$. We look at two aspects, (i) runtimes for answering a single query with LVE^{evi} and LJT^{evi} and (ii) runtimes of the LJT^{evi} steps.

Figure 3 shows runtimes in milliseconds [ms] for answering a single query with LVE^{evi} (triangles) and LJT^{evi} (circles) with evidence coverage ranging from 0% to 100% on the x-axis. The filled symbols show runtimes for certain evidence. The hollow symbols show runtimes for uncertain evidence. As expected, LJT runtimes are shorter than LVE runtimes since LJT is able to use a smaller submodel compared to the original input model. For LJT^{evi}, certain evidence leads to shorter runtimes than uncertain evidence due to the dimension reduction as well as its preprocessing. Evidence is already handled when LJT^{evi} starts answering the query. And as the submodel for query answering is rather small, the dimension reduction has a comparatively large impact. For LVE, certain

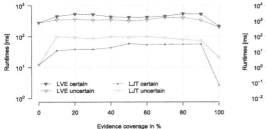 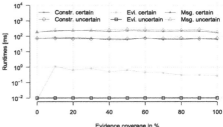

Fig. 3. Runtimes for query answering   **Fig. 4.** Runtimes for LJT steps

evidence leads to larger runtimes as the overall impact of the dimension reduction is not as large and absorption in itself is a rather expensive operation, even though is leads to faster runtimes afterwards. The increase in runtimes from 0% to 10% evidence as well as the decrease in runtimes from 90% to 100%, which occurs for both certain and uncertain evidence, comes from the shattering on the evidence. With 0% and 100% evidence, no splits are necessary, which means smaller models in terms of the number of parfactors to handle.

Figure 4 shows runtimes in milliseconds [ms] of the steps construction (diamond), evidence entering (squares), and message passing (triangles) of LVE^{evi} with evidence coverage ranging from 0% to 100% on the x-axis (filled = certain, hollow = uncertain). Evidence has no influence on construction. Therefore, runtimes are nearly the same for certain and uncertain evidence. Certain evidence leads to larger runtimes as LJT^{evi} absorbs the evidence during this step. Uncertain evidence is simply added to a local model and thus, entering uncertain evidence does not depend on evidence coverage. Message passing with uncertain evidence takes slightly longer than with certain evidence as the dimension reduction also helps during message calculation.

Overall, the case study shows that uncertain evidence leads to similar runtimes for LVE^{evi} and LJT^{evi} compared to certain evidence with a limited scope. Comparing runtimes for domain sizes of 10 to domain sizes of 1000 shows that even though the domain sizes rise by a factor of 100, runtimes only rise by a factor of 2.7 to 8.6 for uncertain evidence and LJT^{evi}. As uncertain evidence basically leads to an additional parfactor and a limited number of splits, we expect further empirical results from [3,4] to also hold for LVE^{evi} and LJT^{evi}, with both algorithms outperforming the ground case.

6   Conclusion

We present LVE^{evi} and LJT^{evi}, versions of LVE and LJT, which incorporate uncertain evidence and allow for similar runtimes as before. We specify how to construct and handle uncertain evidence. LVE^{evi} and LJT^{evi} close the gap to PDBs to also allow for uncertain evidence in probabilistic relational models.

We currently work on learning lifted models. Other interesting algorithm extensions include parallelisation, construction using hypergraph partitioning, and different message passing strategies. Additionally, we look into areas of application to see its performance on real-life scenarios.

References

1. Ahmadi, B., Kersting, K., Mladenov, M., Natarajan, S.: Exploiting symmetries for scaling loopy belief propagation and relational training. Mach. Learn. **92**(1), 91–132 (2013)
2. Braun, T., Möller, R.: Lifted junction tree algorithm. In: Friedrich, G., Helmert, M., Wotawa, F. (eds.) KI 2016. LNCS (LNAI), vol. 9904, pp. 30–42. Springer, Cham (2016). https://doi.org/10.1007/978-3-319-46073-4_3
3. Braun, T., Möller, R.: Counting and conjunctive queries in the lifted junction tree algorithm - extended version. In: Croitoru, M., Marquis, P., Rudolph, S., Stapleton, G. (eds.) GKR 2017. LNCS (LNAI), vol. 10775, pp. 54–72. Springer, Cham (2018). https://doi.org/10.1007/978-3-319-78102-0_3
4. Braun, T., Möller, R.: Parameterised queries and lifted query answering. In: Proceedings of IJCAI 2018, pp. 4980–4986 (2018)
5. Van den Broeck, G., Davis, J.: Conditioning in first-order knowledge compilation and lifted probabilistic inference. In: Proceedings of the Twenty-Sixth AAAI Conference on Artificial Intelligence, pp. 1–7. AAAI Press (2012)
6. van den Broeck, G., Taghipour, N., Meert, W., Davis, J., Raedt, L.D.: Lifted probabilistic inference by first-order knowledge compilation. In: IJCAI-11 Proceedings of the 22nd International Joint Conference on AI (2011)
7. Chan, H., Darwiche, A.: On the revision of probabilistic beliefs using uncertain evidence. Artif. Intell. **163**(1), 67–90 (2005)
8. Dalvi, N., Suciu, D.: Efficient query evaluation on probabilistic databases. VLDB J. Int. J. Very Large Data Bases **16**(4), 523–544 (2007)
9. Das, M., Wu, Y., Khot, T., Kersting, K., Natarajan, S.: Scaling lifted probabilistic inference and learning via graph databases. In: Proceedings of the SIAM International Conference on Data Mining, pp. 738–746 (2016)
10. Jeffrey, R.C.: The Logic of Decision. University of Chicago Press, Chicago (1990)
11. Lauritzen, S.L., Spiegelhalter, D.J.: Local computations with probabilities on graphical structures and their application to expert systems. J. Royal Stat. Soc. Ser. B Methodol. **50**, 157–224 (1988)
12. Milch, B., Zettlemoyer, L.S., Kersting, K., Haimes, M., Kaelbling, L.P.: Lifted probabilistic inference with counting formulas. In: AAAI-08 Proceedings of the 23rd Conference on AI, pp. 1062–1068 (2008)
13. Pearl, J.: Probabilistic Reasoning in Intelligent Systems: Networks of Plausible Reasoning. Morgan Kaufmann, San Mateo (1988)
14. Pearl, J.: On two pseudo-paradoxes in Bayesian analysis. Ann. Math. Artif. Intell. **32**(1–4), 171–177 (2001)
15. Peng, Y., Zhang, S., Pan, R.: Bayesian network reasoning with uncertain evidences. Int. J. Uncertainty Fuzziness Knowl. Based Syst. **18**(05), 539–564 (2010)
16. Poole, D.: First-order probabilistic inference. In: IJCAI-03 Proceedings of the 18th International Joint Conference on AI (2003)
17. de Salvo Braz, R., Amir, E., Roth, D.: Lifted first-order probabilistic inference. In: IJCAI-05 Proceedings of the 19th International Joint Conference on AI (2005)

18. Steinhäuser, J., Kühlein, T.: Role of the general practitioner. In: Patient Blood Management, pp. 61–65. Thieme, Stuttgart (2015)
19. Suciu, D., Olteanu, D., Ré, C., Koch, C.: Probabilistic databases. Synth. Lect. Data Manage. **3**(2), 1–180 (2011)
20. Taghipour, N., Fierens, D., van den Broeck, G., Davis, J., Blockeel, H.: Completeness results for lifted variable elimination. In: Proceedings of the 16th International Conference on AI and Statistics, pp. 572–580 (2013)
21. Taghipour, N., Fierens, D., Davis, J., Blockeel, H.: Lifted variable elimination: decoupling the operators from the constraint language. J. AI Res. **47**(1), 393–439 (2013)
22. Theodorsson, E.: Uncertainty in measurement and total error: tools for coping with diagnostic uncertainty. Clin. Lab. Med. **37**(1), 15–34 (2017)
23. Thimm, M., Kern-Isberner, G.: On probabilistic inference in relational conditional logics. Logic J. IGPL **20**(5), 872–908 (2012)
24. Wemmenhove, B., Mooij, J.M., Wiegerinck, W., Leisink, M., Kappen, H.J., Neijt, J.P.: Inference in the promedas medical expert system. In: Bellazzi, R., Abu-Hanna, A., Hunter, J. (eds.) AIME 2007. LNCS (LNAI), vol. 4594, pp. 456–460. Springer, Heidelberg (2007). https://doi.org/10.1007/978-3-540-73599-1_61

Enhanced Collaborative Filtering Through User-Item Subgroups, Particle Swarm Optimization and Fuzzy C-Means

Ayangleima Laishram$^{(\boxtimes)}$ and Vineet Padmanabhan

School of Computer and Information Sciences,
University of Hyderabad, Hyderabad, India
ayang.laishram@gmail.com, vcpnair73@gmail.com

Abstract. Recommender systems are information filtering systems that assist users to retrieve relevant information from massive amounts of data. Collaborative filtering (CF) is the most widely used technique in recommender systems for predicting the interests of a user on particular items. In traditional CF preferences of all items from many users are collected in the prediction process and this may include items that are irrelevant to the *active user* (the user for whom the prediction is for). Recently, subgroup based methods have emerged which take into account correlation of users and a set of items to rule out consideration of superfluous items. In this paper our objective is to explore CF that considers only user-item subgroups which consist of only similar subset of users based on a subset of items. We propose a novel hybrid framework based on Particle Swarm Optimization and Fuzzy C-Means clustering that optimizes the searching behaviour of user-item subgroups in CF. The proposed algorithm is experimented and compared with several state-of-the-art algorithms using benchmark datasets. Accuracy metrics such as precision, recall and mean average precision is used to find the top N recommended items.

Keywords: Recommender system · Collaborative filtering ·
Particle Swarm Optimization · Fuzzy C-Means · Co-clustering

1 Introduction

It is quite evident these days that the world wide web has occupied an important place in our day to day life and it can be said without a doubt that the most evolving applications in the web are enabled with the recommender system (RS) technology. RS technology is widely applied by several enterprises that deal with huge amount of information such as Twitter, Netflix, Google, Amazon etc. The main objective behind a recommender system is to suggest unknown and suitable information to the users based on his/her preferences by getting rid of irrelevant information. This saves time for the user who otherwise has to search through a

© Springer Nature Switzerland AG 2019
M.-J. Meurs and F. Rudzicz (Eds.): Canadian AI 2019, LNAI 11489, pp. 94–106, 2019.
https://doi.org/10.1007/978-3-030-18305-9_8

huge amount of information which eventually leads to the problem called *infor-mation overloading* wherein a user is provided with abundant information but he/she fails in taking an appropriate decision. Two main techniques that are used by the recommender system technology is that of *Collaborative Filtering (CF)* and *Content Based filtering (CB)*. The basic difference between the two approaches is that, CB solely depends on the features/contents of the items that are liked in the past for making a recommendation whereas CF relies on past behaviour of the users and involves preferences of other users as well in order to make a recommendation. In CF strategy, the idea is to learn hidden features of the user's past activities so as to build an accurate model for recommendation and therefore CF is more accurate and widely used.

In traditional CF algorithms, a group of similar users are found if a match can be ascertained among the preferences the users have given for all the items they have consumed. Based on this matching of preferences unknown items that are most liked by the group can be recommended. However, this framework is not always tenable because the users having strong correlation on a subset of items may have completely different preferences on another subset of items. Moreover, it is natural for a user to have a concentrative yet diverse set of preferences rather than diverse preferences over all the items available. So, in traditional CF, one might miss the chance to capture the strongly correlated group of users based on a subset of items while searching for a group of similar users over All the items. This kind of mismatch of preferences arising in a group while taking into consideration the preferences given to all items could degrade the similarity degree and can further degrade the quality of recommendation. Thus, we focus on filtering irrelevant information by discovering only correlated user-item subgroups. In this paper we refer to a group of similar users based on a set of items as a *user-item* subgroup. We discover highly correlated user-item subgroups by using a hybrid framework that consists of a population-based search algorithm like *Particle Swarm Optimization (PSO)*, and a co-clustering technique to compute the prediction score matrix of user and item interaction. The co-clustering technique [1] is an extension of the traditional clustering technique which performs clustering of the data points simultaneously on both the dimensions of a matrix to capture the local similarity structure instead of grouping the data-points separately in each dimension. The co-clustering technique generates a subgroup of the rows and columns of the matrix which exhibits similar behaviour. We employ *Fuzzy C-Means (FCM)* clustering to perform the co-clustering and this in turn discovers highly correlated subgroups. Our objective is to exploit the globalised behaviour of PSO in discovering optimal centroids to be initialised as the initial centroids for the FCM. Therefore FCM discovers meaningful user-item subgroups which in a combined manner computes a unified prediction score matrix for collaborative filtering to make recommendation. The proposed framework refrains Fuzzy C-Means clustering from premature convergence that happens due to the sensitivity of randomly initialized centroid of the clusters.

The paper is organized in the following order. Section 2 describes about the background information. Section 3 presents proposed method. The experimental settings are given in Sect. 4 followed by the results and discussion in Sect. 5. We conclude the paper with Sect. 6.

2 Background

Co-clustering technique, also known as bi-clustering was first introduced by Hartigan [1], and is usually termed as direct clustering. Co-clustering technique is well studied in microarray data analysis and text mining for handling sparsity and dimensionality reduction problem. Several researchers in the area of collaborative filtering also explored co-clustering for solving different aspects of recommender systems. In Symeonidis et al. [2] it was shown how to find the partial matching information that existed among the users by discovering bi-clusters from the user-item rating matrix. Since the algorithm was limited to find overlapping between the biclusters, it was not possible to capture different preferences of the users. Pablo [3] applied immune inspired algorithm to discover the bi-clusters, and also allowed overlapping of the biclusters to capture more similar local structures of the matrix. Alqadah et al. [4] obtained personalised bi-clusters on demand by mapping a user to a bi-cluster, and find the nearest bi-cluster based on the relationship between the target user and other similar users. Kant and Mahara [5] applied biclustering algorithm to find subgroups which are later assigned to user/item based on the similarity degree. [5] fused the item-based and user-based CF to get predictions. Honda [6] reported the application of co-cluster structure analysis by using fuzzy c-means clustering. Fuzzy c-means clustering has also been widely used in collaborative filtering since the last decade. It has been applied in user-based [7] and item-based [8] CF to improve the quality of recommendations. Yang and Zhang [9] optimized fuzzy c-means clustering for traditional CF by hybridising with hidden semantic algorithm. Chen and Ludwig [10] find an optimal number of clusters for fuzzy c means clustering by using PSO. However, to the best of our knowledge other than the work of Xu et al. [11,12], no one has exploited the soft clustering benefit of fuzzy c-means in co-clustering technique with the aim to find correlated user-item subgroups. Xu et al. [11] formulated multi-class co-clustering problem to improve traditional CF models. Later Bu et al. [12] extended the multi-class co-clustering framework by analysing the user-item interactions from three different angles such as user-user, user-item and item-item interactions.

In Assad Abhas et al. [13] several researches that focus on addressing the problems of CF by computational intelligence techniques have been discussed. Similarly, the role of nature-inspired algorithms and its application to mitigate the challenges in CF has been discussed at length in [14–19]. Wasid and Kant [20] employed PSO to learn the preference weights of the users for various item features and applied fuzzy set to build user profiles. Recently, a user-item subgroup based CF model through genetic algorithm was presented by [21] wherein promising results were achieved.

2.1   Co-Clustering

The objective of Co-Clustering in CF is to simultaneously partition both the users and items into c co-clusters/subgroups from $Y \in \mathbb{R}^{n \times m}$ user-item rating matrix in which there are n users and m items. Each subgroup represents the similarity of a set of users' preferences based on a subset of items. The Multiclass Co-Clustering (MCoC) technique [11] is adopted as base co-clustering technique in this paper. A partition matrix, $P \in [0,1]^{(n+m) \times c}$, is used to represent membership degree of the users in the respective subgroups. The partition matrix P can be represented as given by (1):

$$P = \begin{bmatrix} Q \\ R \end{bmatrix} \tag{1}$$

where $Q \in [0,1]^{n \times c}$ and $R \in [0,1]^{m \times c}$ represent the partition matrices of users and items respectively. The partition matrix P represents solution of the base co-clustering technique. The property of the partition matrix is explained by (2).

$$P_{i,j} = \begin{cases} P_{i,j} > 0, & \text{if the } i^{th} \text{ entry belongs to the } j^{th} \text{ subgroup} \\ P_{i,j} = 0, & \text{Otherwise} \end{cases} \tag{2}$$

Each element $P_{i,j}$ denotes membership degree of an entry i of the partition matrix P to the jth subgroup. The membership degree of an entry i of the matrix P for each subgroup sums to one. Each entry belong to a fix number of subgroups such that $1 \leq k \leq c$ wherein that each entry is allowed to belong to k subgroups. Note that when $k = 1$, the technique works as the traditional clustering algorithm. The aim of the base co-clustering is to map all the users and items to low dimensional space while preserving the original information, and search correlated user-item subgroups to discover relevant items for recommendation. The base co-clustering technique consists of two main stages which are explained below.

Step 1: *Application of Fuzzy C-Means clustering.*

Step 1.1: *Map all the users and items to low dimenensional space.*
      The optimal z-dimensional solution X that preserves preference information can be obtained from the optimization function [11] which is given by (3):

$$\min_X \quad Tr(X^T N X) \tag{3}$$

such that

$$X \in \mathbb{R}^{(n+m) \times z}, X^T X = I, \qquad N = \begin{bmatrix} I_n & -S \\ -S^T & I_m \end{bmatrix}, \qquad S = (G^{row})^{-1/2} Y (G^{col})^{-1/2}, \tag{4}$$

where I_n and I_m correspond to the identity matrices of size $n \times n$ and $m \times m$ whereas $G^{row} \in \mathbb{R}^{n \times n}$ and $G^{col} \in \mathbb{R}^{m \times m}$ correspond to diagonal degree matrices of users and items with $G_{i,i}^{row} = \sum_{j=1}^{m} Y_{i,j}$ and $G_{j,j}^{col} = \sum_{i=1}^{n} Y_{i,j}$. The optimization function given by (3) is minimized by the optimal solution X which is the solution of eigenvalue problem $NX = \lambda X$ where $X = [x_1, x_2, \cdots, x_a]$ such that $x_1, \cdots, x_a$ are the least eigenvectors of matrix N for the respective sorted eigenvalues.

Step 1.2: *Find correlated user-item subgroups.*

The optimal partition matrix P can be obtained by finding correlated subgroups from the unified user-item data X. The user-item subgroups can be discovered in two ways: single class/multi class based on the factor of overlapping among the subgroups. The k-means clustering is used to discover the subgroups from the data X in singleclass co-clustering ($SCoC$). Each entry of P in $SCoC$ belongs to only one subgroup. Whereas fuzzy c-means (FCM) clustering is used to find subgroups for multiclass co-clustering. FCM works by assigning fuzzy membership value to each data object corresponding to each cluster and updates the membership value until the algorithm converges to an optimal solution. The optimization problem of FCM is a minimization function [11] which is given by (5).

$$J_m(P, V) = \sum_{i=1}^{m+n} \sum_{j=1}^{c} P_{ij}^l d(x_i, v_j)^2 \tag{5}$$

where P_{ij} represents membership value of the entry x_i to the jth cluster, v_j represents cluster centroid while V represents matrix of the cluster centroids. l represents the weighting exponent that controls the fuzziness of the partition and d is distance function wherein in this paper we use the Euclidean distance. The partition matrix and cluster centroids are updated in every iteration by making use of the equation as given by (6).

$$P_{ij} = \frac{d(x_i, v_j)^{\frac{2}{1-l}}}{\sum_{k=1}^{c}(d(x_i, v_k))^{\frac{2}{1-l}}}, \quad v_j = \frac{\sum_{i=1}^{m+n} P_{ij}^l x_i}{\sum_{i=1}^{m+n} P_{ij}^l}, \quad \forall i = 1, \cdots, (n+m); \forall j = 1, \cdots, c. \tag{6}$$

The algorithm is terminated when an improvement in the value of the optimization function at two consecutive iterations is smaller than the threshold ϵ.

Step 2: *Recommendation.*

Once correlated subgroups are discovered, any existing CF algorithm can be applied in each subgroup. The results from each subgroup are merged to get a unified prediction score for recommendation. The unified framework [11] copes with all the cases including the overlapping of the subgroups as given in (7). Let $Pred(u_i, y_j, k)$ represent prediction score obtained by a user i on item j in a subgroup k and Z_{ij} represent the unified prediction score of user i on item j.

$$Z_{i,j} = \begin{cases} \sum_k Pred(u_i, y_j, k).\delta_{ik} & \text{if } u_i \text{ and } y_j \text{ belong to one or more subgroups,} \\ 0 & \text{otherwise} \end{cases} \tag{7}$$

where δ_{ik} is an indicator value of user i that corresponds to the most interesting subgroup k among the overlapped subgroups with item j. Now, δ_{ik} can be described as in (8).

$$\delta_{ik} = \begin{cases} 1 & \text{if } P_{ik} \text{ is } \max(Q_{u_i} \cap R_{y_j}), \\ 0 & \text{otherwise} \end{cases} \tag{8}$$

3   Proposed Method

The base co-clustering technique extracts correlated user-item subgroups by employing Fuzzy C-Means (FCM) clustering. The FCM algorithm randomly initializes initial centroid of the clusters and subsequently learns the centroids in further iterations through clustering. However, the performance of FCM algorithm has high impact on the initial position of the centroids being selected and may get trapped in local optima if it was not chosen right. Thus, we employ a population based search optimization algorithm, namely Particle Swarm Optimization (PSO), to discover the nearest optimal centroid for each cluster. PSO algorithm inherits intelligence exhibited by the collective behaviours of the particles when they interact with local environment and thereby leads to convergence of globally optimal solution. Thus, PSO algorithm discovers the nearest optimal centroids of the clusters by exploring high dimensional search space with a set of possible solutions which eventually converges to global optima.

In this section, we propose a novel hybrid framework of Particle Swarm Optimization and Co-Clustering, named as *PSO-CoC*. The goal of the proposed algorithm is to find highly correlated user-item subgroups from the whole user-item rating matrix by initializing the clusters' centroids of FCM to the nearest optimal centroids in collaborative filtering. The block diagram of the proposed framework is shown in Fig. 1. As shown in Fig. 1, PSO module is fed with known user-item rating matrix, Y, and number of clusters, c, as inputs. The PSO module discovers nearest optimal centroids of the clusters effectively by utilizing collective intelligence exhibited by the multiple particles and feeds the nearest optimal centroids as initialized inputs to the co-clustering module. The FCM technique of the co-clustering module, in turn, starts from the position of nearest optimal centroids discovered by the PSO module to co-cluster the similar data and further optimizes the position of centroids by using the minimization function defined in (5) in order to find highly correlated subgroups. Finally after the convergence of FCM, a traditional CF algorithm can be executed in each subgroup to predict the unknown ratings of the items in the subgroup by utilizing the rating information of the highly correlated users in the subgroup. A unified prediction score of the rating matrix is obtained by integrating results from different subgroups according to the method given by (7).

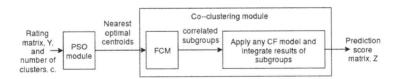

Fig. 1. Block diagram of the proposed hybrid method, PSO-CoC

PSO module: In our proposed algorithm, each particle is represented by a matrix $\{v_1, v_2, \cdots, v_c\}$, where c is the number of clusters and v_i denotes centroid vector of ith cluster wherein a is the dimension of the optimal solution X. The representation of a particle is given below.

$$particle = \begin{pmatrix} v_1 \\ v_2 \\ \cdots \\ v_c \end{pmatrix} = \begin{pmatrix} v_{11} & v_{12} & \cdots \cdots & v_{1a} \\ v_{21} & v_{22} & \cdots \cdots & v_{2a} \\ \cdots & \cdots & \cdots \cdots & \cdots \\ v_{c1} & v_{c2} & \cdots \cdots & v_{ca} \end{pmatrix}$$

The fitness degree of each particle is evaluated by finding the average distance between a cluster and a user, which is given by (9). Thus, smaller value of the fitness function signifies stronger correlation of the cluster [22].

$$f = \frac{\sum_{i=1}^{c} \frac{\sum_{j=1}^{q_i} d(v_i, u_{ij})}{q_i}}{c} \tag{9}$$

where q_i denotes the number of users which belongs to cluster i; u_{ij} represents the jth user vector that belongs to cluster i; v_i is the centroid vector of ith cluster; $d(v_i, u_{ij})$ represents distance between centroid vector of the cluster i and user vector j that belongs to the cluster i.

Let the position of a particle be $particle_i(t)$ where i denotes the particle i at time step t. The position of the particle can be updated by adding a velocity vector, $vel_i(t)$, in which the velocity vector drives the particle towards the optimal solution. The position update rule and velocity vector is given by (10) and (11) respectively.

$$particle_i(t+1) = particle_i(t) + vel_i(t+1) \tag{10}$$

$$vel_{ij}(t+1) = w.vel_{ij}(t) + c_1 rand1_{1j}(t)(y_{ij}(t) - particle_{ij}(t)) + c_2 rand2_{2j}(t)(\hat{y}_j(t) - particle_{ij}(t)) \tag{11}$$

where c_1 and c_2 are acceleration constants that control degree of participation of cognitive and social components in the velocity update, the constants $rand1_{1j}(t)$ and $rand2_{2j}(t)$ are uniformly distributed random values between $[0, 1]$ and w denotes the inertia. The best position visited by each particle i in the swarm so far is recorded as personal best position which is represented by y_i. The termination condition of the PSO module is user-defined number of iterations.

4   Experimental Settings

We experimented our proposed algorithm on a benchmark dataset, *Movielens*, in which there are 100,000 ratings (100k) given by 945 users on 1682 items. The dataset is a sparse matrix by having 93.71% sparsity, thus missing ratings are filled with zero whereas the existing ratings range from 1 to 5 such that 1 being the lowest and 5 being the highest. The project, Grouplens[1], at University of Minnesota collected the dataset. The best model is achieved by fine tuning the

---

[1] http://grouplens.org/datasets/movielens/.

Algorithm 1. PSO-CoC

Input: User-item rating matrix Y, and number of clusters c.
Output: Prediction score matrix Z
//Start PSO Module//
Initialize swarm with randomly selected k user vectors from matrix Y as cluster centroids;
while *(max-iteration is not met)* **do**
    for *each particle* **do**
        Assign each user vector to the nearest cluster centroid;
        Evaluate fitness function by using (9);
        Update position and velocity of the particle based on (10, 11) to generate more optimal solution;

//Start Co-clustering Module//
Initialize initial centroids of FCM with the nearest optimal centroids found by PSO module;
while *(termination condition is not satisfied)* **do**
    Compute objective function, J, which is given by (5);
    Update partition matrix, P, and centroid of the clusters, V, given by (6).

Apply any CF algorithm to each subgroup, and predict missing values in each subgroup;
Apply the unified framework given by (7) to compute the prediction score matrix, Z ;
Find top N item recommendations and compute accuracy of the algorithm;

---

parameters of the problem. A parameter controls the effectiveness of an algorithm by tuning towards the optimal values. The swarm size is set to 50. The parameters values of $rand1$ and $rand2$ are set to 0.1270 and 0.0975 respectively. The acceleration coefficients c_1, c_2 and inertia weight w are set to 1.42, 1.42 and 0.72 respectively. The user-defined termination condition is set to 50. The weighting exponent of FCM, l is set to 2, while the threshold ε to 0.00001. The user-defined termination condition of PSO and Co-clustering modules are set to 50. The number of subgroups, k, that a user is allowed to overlap is $k = log_2(c)$ where c is the total number of subgroups [11]. The user-defined termination condition is set to 50. The popular accuracy metrics of top N recommendation such as precision, recall, F1-score and Mean Average Precision (MAP) are utilized to measure our proposed algorithm.

To measure the effectiveness of our proposed algorithm, we selected popular user-item subgroup based algorithms called Multiclass Co-Clustering ($MCoC$), Single-class Co-Clustering algorithms ($SCoC$) [11] and an extended version of the Multiclass Co-Clustering ($MCoC$) algorithm which gives more insight about the user-item relationship [12]. The extended algorithm is named as $MCoC$-UU-II-UI. The mentioned algorithms aims at utilising user-item subgroups in the prediction of score matrix for top N recommendation. Moreover, the $MCoC$ algorithms allows the users to fall into more than one subgroups whereas overlapping of the subgroups is not allowed in SCoC. We employ a few state-of-the-art collaborative filtering algorithms such as Maximum Margin Matrix Factorization ($MMMF$) [23], Probabilistic Matrix Factorization (PMF) [24] and Item-Based (IB) [25] algorithms as base CF models in our proposed algorithm.

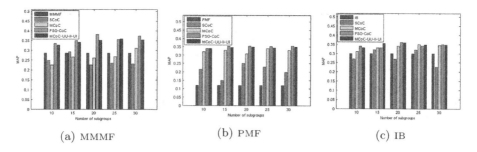

(a) MMMF   (b) PMF   (c) IB

Fig. 2. Performance of proposed algorithm (PSO-CoC) with three well known CF models across different number of subgroups.

5   Experimental Results and Discussions

The comparisons of proposed algorithm PSO-CoC with different recommendation algorithms such as $MMMF$, PMF and IB as base CF models and application of different base CF models with popular subgroup based models such as $MCoC$, $SCoC$, $MCoC$-UU-II-UI [11,12] for top N recommendations are shown in Tables 1, 2, 3 and Fig. 2. In Fig. 2, the comparisons are made in terms of *Mean Average Precision* (MAP) across different *number of subgroups* for all the algorithms. The performance of the base CF models are represented by dark blue bar in the figure. We observe from the figure that single class co-clustering model performs better when the number of subgroups are lesser and the performance level falls with an increase in the number of subgroups. As the number of subgroups increases, the number of items contained in a subgroup decreases. This results in overspecification of the features in the subgroups. Thus, accuracy drops in the case of single class model with higher number of subgroups due to the restriction of overlapping among the subgroups that consequently lead to inability of discovering meaningful subgroups. On the otherhand, the multi-class subgroup based models such as our proposed algorithm (PSO-CoC), $MCoC$ and $MCoC$-UU-II-UI perform better with higher number of subgroups by addressing the preference overspecification problem. It is also the case that, the incorporation of flexibility of preferences to match among multiple subgroups helps the models to find accurate preferences of the users. The advantage of overlapping among the subgroups is to boost the possibility of capturing user's preferences by removing irrelevant information. The reason why multi class models do not perform well with lesser number of subgroups is because of the incapability of extraction of highly similar interests from the large number of choices available in the subgroup. From the Tables 1, 2, 3 and Fig. 2, we observe that the multi class models improve both the single class models and the base algorithms in all the cases. Our findings suggest that the correlated user-item subgroups influence the accuracy of the recommender system in prediction. The superiority of the proposed algorithm from the existing well known subgroup based CF model is the intialization of the centroids of fuzzy c-means clustering to nearest optimal solutions instead of initializing randomly on any point of the hyperdimensional search space. The initialization of centroids to optimal positions lead the proposed framework to better prediction.

Apart from *Mean Average Precision*, the proposed algorithm (PSO-CoC) is further measured by using popular metrics for top N recommendation accuracy such as *Precision, Recall* and *F1-score*. The proposed algorithm is compared for two values of k, the number of subgroups, and the results are shown in Tables 1, 2 and 3. It is evident from the results that the models with higher number of subgroups are able to capture more accurate fuzzy weights in clustering than the lesser number of subgroups, and the recommender system benefits from the participation of higher number of correlated subgroups in prediction. The Tables 1, 2 and 3 show that the subgroup based algorithms outperform the corresponding base CF algorithm, and also the multi-class algorithms outperforms the single class algorithms. Further, our proposed algorithm (PSO-CoC) improves the popular subgroup based algorithm $MCoC$ by directing FCM clustering from optimized initial centroids which are discovered by PSO. The proposed algorithm also outperforms the extended multiclass co-clustering model $MCoC$-UU-II-UI with the impact of proper initialization of the centroids for FCM clustering.

Table 1. Comparisons of proposed algorithm (PSO-CoC) with base$MMMF$ and execution of $MMMF$ with $MCoC$, $MCoC$-UU-II-UI and $SCoC$

Base MMMF	Precision		Recall		F1-score	
	0.158		0.0355		0.058	
Subgroup based	*10 subgroups*			*20 subgroups*		
CF models	Precision	Recall	F1-score	Precision	Recall	F1-score
SCoC	0.152	0.0198	0.035	0.155	0.0347	0.0564
MCoC	0.168	0.022	0.039	0.171	0.0399	0.0635
MCoC-UU-II-UI	**0.18**	0.0242	0.0426	0.192	0.0276	0.0481
PSO-CoC	0.178	**0.0244**	**0.043**	**0.21**	**0.0662**	**0.052**

Table 2. Comparisons of proposed algorithm (PSO-CoC) with base PMF and execution of PMF with $MCoC$, $MCoC$-UU-II-UI and $SCoC$

Base PMF	Precision		Recall		F1-score	
	0.05		0.02		0.020	
Subgroup based	*10 subgroups*			*20 subgroups*		
CF models	Precision	Recall	F1-score	Precision	Recall	F1-score
SCoC	0.104	0.02	0.03	0.15	0.021	0.032
MCoC	0.174	0.026	0.045	0.175	0.027	0.046
MCoC-UU-II-UI	0.18	**0.0262**	**0.0457**	0.198	0.0276	0.0482
PSO-CoC	**0.182**	0.024	0.043	**0.202**	**0.035**	**0.06**

Table 3. Comparisons of proposed algorithm (*PSO-CoC*) with base *IB* and execution of *IB* with *MCoC*, *MCoC-UU-II-UI* and *SCoC*

Base IB	Precision		Recall		F1-score	
	0.8		0.0355		0.058	
Subgroup based	*10 subgroups*			*20 subgroups*		
CF models	Precision	Recall	F1-score	Precision	Recall	F1-score
SCoC	0.106	0.018	0.0313	0.11	0.019	0.032
MCoC	0.178	0.024	0.041	0.186	0.029	0.042
MCoC-UU-II-UI	0.19	0.0241	0.0439	0.20	0.0315	0.0544
PSO-CoC	**0.19**	**0.025**	**0.044**	**0.204**	**0.037**	**0.06**

The results of the Tables 1, 2 and 3 suggest that the subgroups retrieved are highly correlated and hence influence the recommender system.

The Fig. 3a, b represent the fitness graphs of the proposed algorithm for different base CF models with different number of subgroups. The fitness funtion is a minimization function and thus the graphs have a decreasing slope across the subsequent iterations. We observe a common characteristic in both the graphs of the figures that there is a rapid drop of the graphs in the beginning of the iterations, then gradually converge to the optimal solution and become stable. This means that the fitness function which is an average distance between the centroid of the cluster and the user vector is able to get to the nearby subspace of the optimal centroids after a few iterations which subsequently lead to the nearest optimal centroids of the clusters. We observe different patterns of fitness graphs for different number of subgroups in the figures. The performance of *PMF* behaves similarly with Fig. 3a due to the similar characteristics of the matrix factorization. The graph with 10 number of subgroups achieved the lowest fitness value in the Fig. 3a suggesting that the proposed algorithm

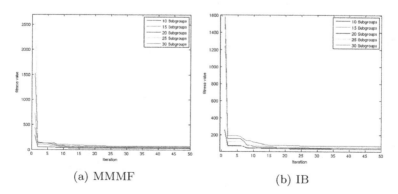

(a) MMMF   (b) IB

Fig. 3. Performance of proposed algorithm with different number of subgroups for base CF models.

converged immaturely with lower number of subgroups since the subgroups have large number of items which make it hard to capture only highly similar preferences. Figure 3b suggests that the proposed algorithm is not able to capture optimum fuzzy weights for larger number of subgroups in the beginning of iteration but subsequently converged to the optimal solution. Thus, trade-off between the number of subgroups and fitness value is a factor in achieving an accurate recommender system.

6 Conclusions

We presented a hybrid framework of particle swarm optimization and fuzzy c-means clustering to discover meaningful user-item subgroups for CF. The significance of utilization of user-item subgroups in CF for computing prediction score is demonstrated by comparing with base CF models such as maximum margin matrix factorization, probabilistic matrix factorization and item-based models. The particle swarm optimization technique is used to drive the searching direction of the centroids for fuzzy c-means to the optimal subspace by exploiting the globalized searching behaviour of PSO. Then, fuzzy c-means is used to find highly correlated user-item subgroups wherein any original collaborative filtering is adopted to find prediction score of each user in the subgroup. A final prediction score of each user evaluated from all the subgroups that the user belongs to is used to find top N recommendation items. In our proposed algorithm, the population based optimization algorithm acts as a boost to improve the fuzzy c-means clustering in discovering highly correlated user-item subgroups by initializing the initial centroid of the clusters to the nearest optimal solutions. Our experimental results have shown a promising way of user-item subgroups in helping to capture highly similar user preferences on a subset of items. As continuation of the research, we focus to explore alternate subgroup based models for the base CF algorithm that capture accurate partial preferences.

Acknowledgment. The first author would like to express gratitude to Council of Scientific and Industrial Research (CSIR) Government of India for the funding support in the form of a Senior Research Fellowship.

References

1. Hartigan, J.A.: Direct clustering of a data matrix. J. Am. Stat. Assoc. **67**(337), 123–129 (1972)
2. Symeonidis, P., et al.: Nearest-biclusters collaborative filtering based on constant and coherent values. Inf. Retrieval **11**(1), 51–75 (2008)
3. de Castro, P.A.D., et al.: Applying biclustering to perform CF. In: Seventh International Conference on Intelligent Systems Design and Applications, pp. 421–426 (2007)
4. Alqadah, F., et al.: Biclustering neighborhood-based collaborative filtering method for top-n recommender systems. Knowl. Inf. Syst. **44**(2), 475–491 (2015)

5. Kant, S., Mahara, T.: Nearest biclusters collaborative filtering framework with fusion. Int. J. Comput. Sci. **25**, 204–212 (2018)
6. Honda, K.: Fuzzy co-clustering and application to CF. In: Huynh, V.-N., Inuiguchi, M., Tran, D.H., Denoeux, T. (eds.) Integrated Uncertainty in Knowledge Modelling and Decision Making. LNCS, vol. 7027, pp. 16–23. Springer, Cham (2016). https://doi.org/10.1007/978-3-642-24918-1
7. Koohi, H., Kiani, K.: User based CF using FCM. Measurement **91**, 134–139 (2016)
8. Birtolo, C., et al.: Improving accuracy of recommendation system by means of item-based fuzzy clustering CF. In: 11th International Conference on Intelligent Systems Design and Applications, pp. 100–106 (2011)
9. Yang, Y., Zhang, Y.: CF recommendation model based on fuzzy clustering algorithm, May 2018
10. Chen, M., Ludwig, S.: Fuzzy clustering using automatic particle swarm optimization. In: IEEE International Conference on Fuzzy Systems, pp. 1545–1552, September 2014
11. Xu, B., et al.: An exploration of improving collaborative RS via user-item subgroups. In: Proceedings of the 21st International Conference on WWW, pp. 21–30. ACM (2012)
12. Bu, J., et al.: Improving collaborative recommendation via user-item subgroups. IEEE Trans. Knowl. Data Eng. **28**(9), 2363–2375 (2016)
13. Abbas, A., et al.: A survey on context-aware recommender systems based on computational intelligence techniques. Computing **97**(7), 667–690 (2015)
14. Devi, V.S., et al.: Collaborative filtering by PSO-based MMMF. In: IEEE International Conference on Systems, Man and Cybernetics (SMC), pp. 569–574. IEEE (2014)
15. Navgaran, D.Z., et al.: Evolutionary based matrix factorization method for CF systems. In: 21st Iranian Conference on Electrical Engineering (ICEE), pp. 1–5 (2013)
16. Laishram, A., et al.: CF, MF and population based search: the nexus unveiled. In: 23rd International Conference on Neural Information Processing, ICONIP, pp. 352–361 (2016)
17. Katarya, R., Verma, O.P.: Effectual recommendations using artificial algae algorithm and FCM. Swarm Evol. Comput. **36**, 52–61 (2017)
18. da Silva, E.Q., et al.: An evolutionary approach for combining results of RS techniques based on CF. Expert Syst. Appl. **53**, 204–218 (2016)
19. Bobadilla, J., et al.: Improving collaborative filtering recommender system results and performance using GA. Knowl. Based Syst. **24**(8), 1310–1316 (2011)
20. Wasid, M., Kant, V.: A particle swarm approach to collaborative filtering based RS through fuzzy features. Proc. Comput. Sci. **54**, 440–448 (2015)
21. Laishram, A., et al.: Analysis of similarity measures in user-item subgroup based CF via GA. Int. J. Inf. Technol. **10**(4), 523–527 (2018)
22. Cui, X., Potok, T.E.: Document clustering analysis based on hybrid PSO+K-means algorithm. Spec. Issue 27–33 (2005)
23. Rennie, J.D.M., Srebro, N.: Fast MMMF for collaborative prediction. In: Proceedings of the 22nd International Conference on ML, pp. 713–719. ACM (2005)
24. Salakhutdinov, R., Mnih, A.: Probabilistic MF. In: Proceedings of the 20th International Conference on Neural Information Processing Systems, pp. 1257–1264. NIPS, USA (2007)
25. Sarwar, B., et al.: Item-based CF recommendation algorithms. In: Proceedings of the 10th International Conference on WWW, pp. 285–295. ACM (2001)

TextKD-GAN: Text Generation Using Knowledge Distillation and Generative Adversarial Networks

Md. Akmal Haidar$^{(\boxtimes)}$ and Mehdi Rezagholizadeh$^{(\boxtimes)}$

Huawei Noah's Ark Lab, Montreal Research Centre, Montreal, Canada
{md.akmal.haidar,mehdi.rezagholizadeh}@huawei.com

Abstract. Text generation is of particular interest in many NLP applications such as machine translation, language modeling, and text summarization. Generative adversarial networks (GANs) achieved a remarkable success in high quality image generation in computer vision, and recently, GANs have gained lots of interest from the NLP community as well. However, achieving similar success in NLP would be more challenging due to the discrete nature of text. In this work, we introduce a method using knowledge distillation to effectively exploit GAN setup for text generation. We demonstrate how autoencoders (AEs) can be used for providing a continuous representation of sentences, which is a smooth representation that assign non-zero probabilities to more than one word. We distill this representation to train the generator to synthesize similar smooth representations. We perform a number of experiments to validate our idea using different datasets and show that our proposed approach yields better performance in terms of the BLEU score and Jensen-Shannon distance (JSD) measure compared to traditional GAN-based text generation approaches without pre-training.

Keywords: Text generation · Generative adversarial networks · Knowledge distillation

1 Introduction

Recurrent neural network (RNN) based techniques such as language models are the most popular approaches for text generation. These RNN-based text generators rely on maximum likelihood estimation (MLE) solutions such as teacher forcing [11] (i.e. the model is trained to predict the next item given all previous observations); however, it is well-known in the literature that MLE is a simplistic objective for this complex NLP task [15]. MLE-based methods suffer from exposure bias [20], which means that at training time the model is exposed to gold data only, but at test time it observes its own predictions.

However, GANs which are based on the adversarial loss function and have the generator and the discriminator networks suffers less from the mentioned problems. GANs could provide a better image generation framework comparing

© Springer Nature Switzerland AG 2019
M.-J. Meurs and F. Rudzicz (Eds.): Canadian AI 2019, LNAI 11489, pp. 107–118, 2019.
https://doi.org/10.1007/978-3-030-18305-9_9

to the traditional MLE-based methods and achieved substantial success in the field of computer vision for generating realistic and sharp images. This great success motivated researchers to apply its framework to NLP applications as well.

GANs have been exploited recently in various NLP applications such as machine translation [24,25], dialogue models [15], question answering [26], and natural language generation [7,13,19,20,28,29]. However, applying GAN in NLP is challenging due to the discrete nature of the text. Consequently, back-propagation would not be feasible for discrete outputs and it is not straightforward to pass the gradients through the discrete output words of the generator. The existing GAN-based solutions can be categorized according to the technique that they leveraged for handling the problem of the discrete nature of text: Reinforcement learning (RL) based methods, latent space based solutions, and approaches based on continuous approximation of discrete sampling. Several versions of the RL-based techniques have been introduced in the literature including Seq-GAN [27], MaskGAN [5], and LeakGAN [8]. However, they often need pre-training and are computationally more expensive compared to the methods of the other two categories. Latent space-based solutions derive a latent space representation of the text using an AE and attempt to learn data manifold of that space [13]. Another approach for generating text with GANs is to find a continuous approximation of the discrete sampling by using the Gumbel Softmax technique [14] or approximating the non-differentiable argmax operator [28] with a continuous function.

In this work, we introduce TextKD-GAN as a new solution for the main bottleneck of using GAN for text generation with knowledge distillation: a technique that transfer the knowledge of softened output of a teacher model to a student model [9]. Our solution is based on an AE (Teacher) to derive a smooth representation of the real text. This smooth representation is fed to the TextKD-GAN discriminator instead of the conventional one-hot representation. The generator (Student) tries to learn the manifold of the softened smooth representation of the AE. We show that TextKD-GAN outperforms the conventional GAN-based text generators that do not need pre-training. The remainder of the paper is organized as follows. In the next two sections, some preliminary background on generative adversarial networks and related work in the literature will be reviewed. The proposed method will be presented in Sect. 4. In Sect. 5, the experimental details will be discussed. Finally, Sect. 6 will conclude the paper.

2   Background

Generative adversarial networks include two separate deep networks: a generator and a discriminator. The generator takes in a random variable, z following a distribution $P_z(z)$ and attempt to map it to the data distribution $P_x(x)$. The output distribution of the generator is expected to converge to the data distribution during the training. On the other hand, the discriminator is expected to discern real samples from generated ones by outputting zeros and ones, respectively. During training, the generator and discriminator generate samples and classify them, respectively by adversarially affecting the performance of each other. In this regard, an adversarial loss function is employed for training [6]:

$$\min_{G} \max_{D} V(D, G) = E_{x \sim P_x(x)}[\log D(x)] + E_{z \sim P_z(z)}[\log(1 - D(G(z)))] \quad (1)$$

This is a two-player minimax game for which a Nash-equilibrium point should be derived. Finding the solution of this game is non-trivial and there has been a great extent of literature dedicated in this regard [22].

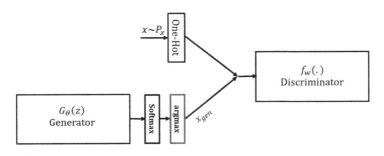

Fig. 1. Simplistic text generator with GAN

As stated, using GANs for text generation is challenging because of the discrete nature of text. To clarify the issue, Fig. 1 depicts a simplistic architecture for GAN-based text generation. The main bottleneck of the design is the *argmax* operator which is not differentiable and blocks the gradient flow from the discriminator to the generator.

$$\min_{G} E_{z \sim P_z(z)}[\log(1 - D(\text{argmax}(\text{softmax}(G(z)))))] \quad (2)$$

2.1 Knowledge Distillation

Knowledge distillation has been studied in model compression where knowledge of a large cumbersome model is transferred to a small model for easy deployment. Several studies have been studied on the knowledge transfer technique [9,21]. It starts by training a big teacher model (or ensemble model) and then train a small student model which tries to mimic the characteristics of the teacher model, such as hidden representations [21], it's output probabilities [9], or directly on the generated sentences by the teacher model in neural machine translation [12]. The first teacher-student framework for knowledge distillation was proposed in [9] by introducing the softened teacher's output. In this paper, we propose a GAN framework for text generation where the generator (Student) tries to mimic the reconstructed output representation of an auto-encoder (Teacher) instead of mapping to a conventional one-hot representations.

2.2   Improved WGAN

Generating text with pure GANs is inspired by improved Wasserstein GAN (IWGAN) work [7]. In IWGAN, a character level language model is developed based on adversarial training of a generator and a discriminator without using any extra element such as policy gradient reinforcement learning [23]. The generator produces a softmax vector over the entire vocabulary. The discriminator is responsible for distinguishing between the one-hot representations of the real text and the softmax vector of the generated text. The IWGAN method is described in Fig. 2. A disadvantage of this technique is that the discriminator is able to tell apart the one-hot input from the softmax input very easily. Hence, the generator will have a hard time fooling the discriminator and vanishing gradient problem is highly probable.

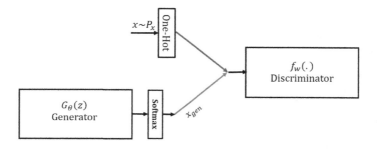

Fig. 2. Improved WGAN for text generation

3   Related Work

A new version of Wasserstein GAN for text generation using gradient penalty for discriminator was proposed in [7]. Their generator is a CNN network generating fixed-length texts. The discriminator is another CNN receiving 3D tensors as input sentences. It determines whether the tensor is coming from the generator or sampled from the real data. The real sentences and the generated ones are represented using one-hot and softmax representations, respectively.

A similar approach was proposed in [20] with an RNN-based generator. They used a curriculum learning strategy [2] to produce sequences of gradually increasing lengths as training progresses. In [19], RNN is trained to generate text with GAN using curriculum learning. The authors proposed a procedure called teacher helping, which helps the generator to produce long sequences by conditioning on shorter ground-truth sequences.

All these approaches use a discriminator to discriminate the generated softmax output from one-hot real data as in Fig. 2, which is a clear downside for them. The reason is the discriminator receives inputs of different representations: a one-hot vector for real data and a probabilistic vector output from the generator. It makes the discrimination rather trivial.

AEs have been exploited along with GANs in different architectures for computer vision application such as AAE [17], ALI [4], and HALI [1]. Similarly, AEs can be used with GANs for generating text. For instance, an adversarially regularized AE (ARAE) was proposed in [13]. The generator is trained in parallel to an AE to learn a continuous version of the code space produced by AE encoder. Then, a discriminator will be responsible for distinguishing between the encoded hidden code and the continuous code of the generator. Basically, in this approach, a continuous distribution is generated corresponding to an encoded code of text.

4   Methodology

AEs can be useful in denoising text and transferring it to a code space (encoding) and then reconstructing back to the original text from the code. AEs can be combined with GANs in order to improve the generated text. In this section, we introduce a technique using AEs to replace the conventional one-hot representation [7] with a continuous softmax representation of real data for discrimination.

4.1   Distilling Output Probabilities of AE to TextKD-GAN Generator

As stated, in conventional text-based discrimination approach [7], the real and generated input of the discriminator will have different types (one-hot and softmax) and it can simply tell them apart. One way to avoid this issue is to derive a continuous smooth representation of words rather than their one-hot and train the discriminator to differentiate between the continuous representations. In this work, we use a conventional AE (Teacher) to replace the one-hot representation with softmax reconstructed output, which is a smooth representation that yields smaller variance in gradients [9]. The proposed model is depicted in Fig. 3. As seen, instead of the one-hot representation of the real words, we feed the softened reconstructed output of the AE to the discriminator. This technique would makes the discrimination much harder for the discriminator. The GAN generator (Student) with softmax output tries to mimic the AE output distribution instead of conventional one-hot representations used in the literature.

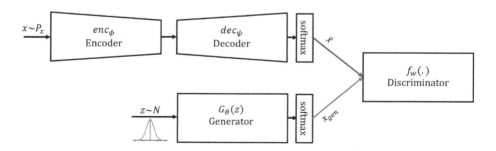

Fig. 3. TextKD-GAN model for text generation

4.2 Why TextKD-GAN Should Work Better Than IWGAN

Suppose we apply IWGAN to a language vocabulary of size two: words x_1 and x_2. The one-hot representation of these two words (as two points in the Cartesian coordinates) and the span of the generated softmax outputs (as a line segment connecting them) is depicted in the left panel of Fig. 4. As evident graphically, the task of the discriminator is to discriminate the points from the line connecting them, which is a rather simple very easy task.

Now, let's consider the TextKD-GAN idea using the two-word language example. As depicted in Fig. 4 (Right panel), the output locus of the TextKD-GAN decoder would be two red line segments instead of two points (in the one-hot case). The two line segments lie on the output locus of the generator, which will make the generator more successful in fooling the discriminator.

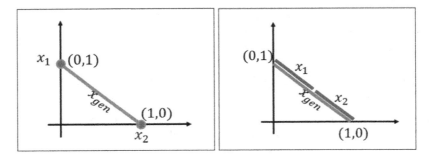

Fig. 4. Locus of the input vectors to the discriminator for a two-word language model; Left panel: IWGAN, Right panel: TextKD-GAN.

4.3 Model Training

We train the AE and TextKD-GAN simultaneously. In order to do so, we break down the objective function into three terms: (1) a reconstruction term for the AE, (2) a discriminator loss function with gradient penalty, (3) an adversarial cost for the generator. Mathematically,

(1) $\min\limits_{(\phi,\psi)} L_{AE}(\phi,\psi) = \min\limits_{(\phi,\psi)} ||x - \mathrm{softmax}(\mathrm{dec}_\psi(\mathrm{enc}_\phi(x)))||^2$

(2) $\min\limits_{w \in W} L_{\mathrm{discriminator}}(w) =$

$\min\limits_{w \in W} -E_{x \sim P_x}[f_w(\mathrm{dec}_\psi(\mathrm{enc}_\phi(X)))] + E_{z \sim P_z}[f_w(G(z))] + \lambda_2 E_{\hat{x} \sim P_{\hat{x}}}[(||\nabla_{\hat{x}} f_w(\hat{x})||_2 - 1)^2]$

(3) $\min\limits_{\theta} L_{\mathrm{Gen}}(\theta) = -\min\limits_{\theta} E_{z \sim P_z}[f_w(G(z))].$

$$(3)$$

These losses are trained alternately to optimize different parts of the model. We employ the gradient penalty approach of IWGAN [7] for training the discriminator. In the gradient penalty term, we need to calculate the gradient norm of random samples $\hat{x} \sim P_{\hat{x}}$. According to the proposal in [7], these random samples can be obtained by sampling uniformly along the line connecting pairs of generated and real data samples:

$$[\hat{x} \sim P_{\hat{x}}] \leftarrow \alpha [x \sim P_x] + (1 - \alpha) [x_{gen} \sim G(z)] \tag{4}$$

The complete training algorithm is described in 1.

Algorithm 1. TextKD-GAN for text generation.

Require: The Adam hyperparameters α, β_1, β_2, the batch size m. Initial AE parameters (encoder (ϕ_0), decoder ψ_0), discriminator parameters w_0 and initial generator parameters θ_0

1: **for** number of training iterations **do**
      AE Training:
2:    Sample $\{x^{(i)}\}_{i=1}^{m} \sim P_x$ and compute code-vectors $c^i = enc_\phi(x^i)$
3:    and reconstructed text $\{\tilde{x}^i\}_{i=1}^{m}$.
4:    Backpropagate reconstruction loss $L_{AE}(\phi, \psi)$.
5:    Update with $(\phi, \psi) \leftarrow Adam(L_{AE}(\phi, \psi), \alpha, \beta_1, \beta_2)$.
      Train the discriminator:
6:    **for** k times **do:**
7:        Sample $\{x^{(i)}\}_{i=1}^{m} \sim P_x$ and Sample $\{z^{(i)}\}_{i=1}^{m} \sim N(0, I)$.
8:        Compute generated text $\{x_{gen}^{(i)}\}_{i=1}^{m} \sim G(z)$
9:        Backpropagate discriminator loss $L_{discriminator}(w)$.
10:      Update with $w \leftarrow Adam(L_{discriminator}(w), \alpha, \beta_1, \beta_2)$.
      end for
      Train the generator:
11:    Sample $\{x^{(i)}\}_{i=1}^{m} \sim P_x$ and Sample $\{z^{(i)}\}_{i=1}^{m} \sim N(0, I)$.
12:    Compute generated text $\{x_{gen}^{(i)}\}_{i=1}^{m} \sim G(z)$
13:    Backpropagate generator loss $L_{Gen}(\theta)$.
14:    Update with $\theta \leftarrow Adam(L_{Gen}(\theta), \alpha, \beta_1, \beta_2)$.
    end for

---

5 Experiments

5.1 Dataset and Experimental Setup

We carried out our experiments on two different datasets: Google 1 billion benchmark language modeling data[1] and the Stanford Natural Language Inference (SNLI) corpus[2]. Our text generation is performed at character level with a sentence length of 32. For the Google dataset, we used the first 1 million sentences and extract the most frequent 100 characters to build our vocabulary. For the SNLI dataset, we used the entire preprocessed training data[3], which contains

---

[1] http://www.statmt.org/lm-benchmark/.
[2] https://nlp.stanford.edu/projects/snli/.
[3] https://github.com/aboev/arae-tf/tree/master/data_snli.

714667 sentences in total and the built vocabulary has 86 characters. We train the AE using one layer with 512 LSTM cells [10] for both the encoder and the decoder. We train the autoencoder using Adam optimizer with learning rate 0.001, $\beta_1 = 0.9$, and $\beta_2 = 0.9$. For decoding, the output from the previous time step is used as the input to the next time step. The hidden code c is also used as an additional input at each time step of decoding. The greedy search approach is applied to get the best output [13]. We keep the same CNN-based generator and discriminator with residual blocks as in [7]. The discriminator is trained for 5 times for 1 GAN generator iteration. We train the generator and the discriminator using Adam optimizer with learning rate 0.0001, $\beta_1 = 0.5$, and $\beta_2 = 0.9$.

We use the *BLEU-N* score to evaluate our techniques. *BLEU-N* score is calculated according to the following equation [3, 16, 18]:

$$BLEU\text{-}N = BP \cdot \exp(\sum_{n=1}^{N} w_n \log(p_n)) \tag{5}$$

where p_n is the probability of n-gram and $w_n = \frac{1}{n}$. We calculate BLEU-n scores for n-grams without a brevity penalty [29]. We train all the models for 200000 iterations and the results with the best *BLEU-N* scores in the generated texts are reported. To calculate the *BLEU-N* scores, we generate ten batches of sentences as candidate texts, i.e. 640 sentences (32-character sentences) and use the entire test set as reference texts.

5.2   Experimental Results

The results of the experiments are depicted in Tables 1 and 2. As seen in these tables, the proposed TextKD-GAN approach yields significant improvements in terms of *BLEU-2*, *BLEU-3* and *BLEU-4* scores over the IWGAN [7], and the ARAE [13] approaches. Therefore, softened smooth output of the decoder can be more useful to learn better discriminator than the traditional one-hot representation. Moreover, we can see the lower *BLEU*-scores and less improvement for the Google dataset compared to the SNLI dataset. The reason might be the sentences in the Google dataset are more diverse and complicated. Finally, note that the text-based one-hot discrimination in IWGAN and our proposed method are better than the traditional code-based ARAE technique [13].

Table 1. Results of the *BLEU-N* scores using 1 million sentences from 1 billion Google dataset

Model	BLEU-2	BLEU-3	BLEU-4
IWGAN	0.50	0.27	0.11
ARAE	0.13	0.02	0.00
TextKD-GAN	**0.51**	**0.29**	**0.13**

Table 2. Results of the *BLEU-N* scores using SNLI dataset

Model	BLEU-2	BLEU-3	BLEU-4
IWGAN	0.57	0.44	0.30
ARAE	0.37	0.27	0.17
TextKD-GAN	**0.62**	**0.50**	**0.38**

Some examples of generated text from the SNLI experiment are listed in Table 3. As seen, the generated text by the proposed TextKD-GAN approach is more meaningful and contains more correct words compared to that of IWGAN [7].

Table 3. Example generated sentences with model trained using SNLI dataset

IWGAN	TextKD-GAN
The people are laying in angold	Two people are standing on the s
A man is walting on the beach	A woman is standing on a bench
A man is looking af tre walk aud	People have a ride with the comp
A man standing on the beach	A woman is sleeping at the brick
The man is standing is standing	Four people eating food
A man is looking af tre walk aud	The dog is in the main near the
The man is in a party	A black man is going to down the
Two members are walking in a hal	These people are looking at the
A boy is playing sitting	The people are running at some l

We also provide the training curves of Jensen-Shannon distances (JSD) between the n-grams of the generated sentences and that of the training (real) ones in Fig. 5. The distances are derived from SNLI experiments and calculated as in [7]. That is by calculating the log-probabilities of the n-grams of the generated and the real sentences. As depicted in the figure, the TextKD-GAN approach further minimizes the JSD compared to the literature methods [7,13]. In conclusion, our approach learns a more powerful discriminator, which in turn generates the data distribution close to the real data distribution.

5.3   Discussion

The results of our experiment shows the superiority of our TextKD-GAN method over other conventional GAN-based techniques. We compared our technique with those GAN-based generators which does not need pre-training. This explains why we have not included the RL-based techniques in the results. We showed the power of the continuous smooth representations over the well-known tricks to work around the discontinuity of text for GANs. Using AEs in TextKD-GAN adds another important dimension to our technique which is the latent space,

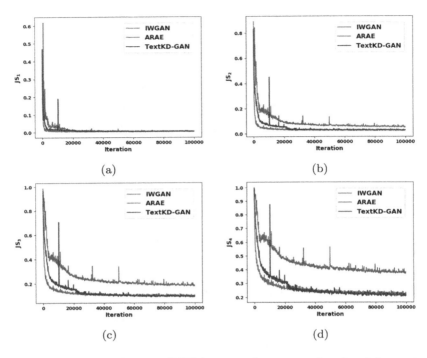

Fig. 5. Jensen-Shannon distance (JSD) between the generated and training sentences n-grams derived from SNLI experiments. (a) js_1, (b) js_2, (c) js_3, and (d) js_4 represent the JSD for 1, 2, 3, and 4-grams respectively

which can be modeled and exploited as a separate signal for discriminating the generated text from the real data. It is worth mentioning that our observations during the experiments show training text-based generators is much easier than training the code-based techniques such as ARAE. Moreover, we observed that the gradient penalty term plays a significant part in terms of reducing the mode-collapse from the generated text of GAN. Furthermore, in this work, we focused on character-based techniques; however, TextKD-GAN is applicable to the word-based settings as well. Bear in mind that pure GAN-based text generation techniques are still in a newborn stage and they are not very powerful in terms of learning semantics of complex datasets and large sentences. This might be because of lack of capacity of capturing the long-term information using CNN networks. To address this problem, RL can be employed to empower these pure GAN-based techniques such as TextKD-GAN as a next step.

6   Conclusion and Future Work

In this work, we introduced TextKD-GAN as a new solution using knowledge distillation for the main bottleneck of using GAN for generating text, which is the discontinuity of text. Our solution is based on an AE (Teacher) to derive a continuous smooth representation of the real text. This smooth representation is distilled to the GAN discriminator instead of the conventional one-hot representation. We demonstrated the rationale behind this approach, which is to make the discrimination task of the discriminator between the real and generated texts more difficult and consequently providing a richer signal to the generator. At the time of training, the TextKD-GAN generator (Student) would try to learn the manifold of the smooth representation, which can later on be mapped to the real data distribution by applying the argmax operator. We evaluated TextKD-GAN over two benchmark datasets using the *BLEU-N* scores, JSD measures, and quality of the output generated text. The results showed that the proposed TextKD-GAN approach outperforms the traditional GAN-based text generation methods which does not need pre-training such as IWGAN and ARAE. Finally, We summarize our plan for future work in the following:

1. We evaluated TextKD-GAN in a character-based level. However, the performance of our approach in word-based level needs to be investigated.
2. Current TextKD-GAN is implemented with a CNN-based generator. We might be able to improve TextKD-GAN by using RNN-based generators.
3. TextKD-GAN is a core technique for text generation and similar to other pure GAN-based techniques, it is not very powerful in generating long sentences. RL can be used as a tool to accommodate this weakness.

References

1. Belghazi, M.I., Rajeswar, S., Mastropietro, O., Rostamzadeh, N., Mitrovic, J., Courville, A.: Hierarchical adversarially learned inference. arXiv preprint arXiv:1802.01071 (2018)
2. Bengio, Y., Louradour, J., Collobert, R., Weston, J.: Curriculum learning. In: Proceedings of the 26th Annual International Conference on Machine Learning, pp. 41–48. ACM (2009)
3. Cer, D., Manning, C.D., Jurafsky, D.: The best lexical metric for phrase-based statistical MT system optimization. In: Human Language Technologies: The 2010 Annual Conference of the North American Chapter of the Association for Computational Linguistics, pp. 555–563. Association for Computational Linguistics (2010)
4. Dumoulin, V., et al.: Adversarially learned inference. arXiv preprint arXiv:1606.00704 (2016)
5. Fedus, W., Goodfellow, I., Dai, A.M.: Maskgan: Better text generation via filling in the _. arXiv preprint arXiv:1801.07736 (2018)
6. Goodfellow, I., et al.: Generative adversarial nets. In: Advances in Neural Information Processing Systems, pp. 2672–2680 (2014)
7. Gulrajani, I., Ahmed, F., Arjovsky, M., Dumoulin, V., Courville, A.: Improved training of wasserstein gans. arXiv preprint arXiv:1704.00028 (2017)
8. Guo, J., Lu, S., Cai, H., Zhang, W., Yu, Y., Wang, J.: Long text generation via adversarial training with leaked information. arXiv preprint arXiv:1709.08624 (2017)
9. Hinton, G., Vinyals, O., Dean, J.: Distilling the knowledge in a neural network. arXiv preprint arXiv:1503.02531 (2015)

10. Hochreiter, S., Schmidhuber, J.: Long short-term memory. Neural Comput. **9**(8), 1735–1780 (1997)
11. Williams, R.J., Zipser, D.: A learning algorithm for continually running fully recurrent neural networks. Neural Comput. **1**(2), 270–280 (1989)
12. Kim, Y., Rush, A.M.: Sequence-level knowledge distillation. In: EMNLP, pp. 1317–1327 (2016)
13. Kim, Y., Zhang, K., Rush, A.M., LeCun, Y., et al.: Adversarially regularized autoencoders for generating discrete structures. arXiv preprint arXiv:1706.04223 (2017)
14. Kusner, M.J., Hernández-Lobato, J.M.: Gans for sequences of discrete elements with the gumbel-softmax distribution. arXiv preprint arXiv:1611.04051 (2016)
15. Li, J., Monroe, W., Shi, T., Ritter, A., Jurafsky, D.: Adversarial learning for neural dialogue generation. arXiv preprint arXiv:1701.06547 (2017)
16. Liu, C.W., Lowe, R., Serban, I.V., Noseworthy, M., Charlin, L., Pineau, J.: How not to evaluate your dialogue system: an empirical study of unsupervised evaluation metrics for dialogue response generation. arXiv preprint arXiv:1603.08023 (2016)
17. Makhzani, A., Shlens, J., Jaitly, N., Goodfellow, I., Frey, B.: Adversarial autoencoders. arXiv preprint arXiv:1511.05644 (2015)
18. Papineni, K., Roukos, S., Ward, T., Zhu, W.: Bleu: a method for automatic evaluation of machine translation. In: ACL, pp. 311–318 (2002)
19. Press, O., Bar, A., Bogin, B., Berant, J., Wolf, L.: Language generation with recurrent generative adversarial networks without pre-training. arXiv preprint arXiv:1706.01399 (2017)
20. Rajeswar, S., Subramanian, S., Dutil, F., Pal, C., Courville, A.: Adversarial generation of natural language. arXiv preprint arXiv:1705.10929 (2017)
21. Romero, A., Ballas, N., Kahou, S.E., Chassang, A., Gatta, C., Bengio, Y.: Fitnets: Hints for thin deep nets. In: ICLR (2015)
22. Salimans, T., Goodfellow, I., Zaremba, W., Cheung, V., Radford, A., Chen, X.: Improved techniques for training gans. In: Advances in Neural Information Processing Systems, pp. 2234–2242 (2016)
23. Sutton, R.S., McAllester, D.A., Singh, S.P., Mansour, Y.: Policy gradient methods for reinforcement learning with function approximation. In: NIPS, pp. 1057–1063 (1999)
24. Wu, L., et al.: Adversarial neural machine translation. arXiv preprint arXiv:1704.06933 (2017)
25. Yang, Z., Chen, W., Wang, F., Xu, B.: Improving neural machine translation with conditional sequence generative adversarial nets. arXiv preprint arXiv:1703.04887 (2017)
26. Yang, Z., Hu, J., Salakhutdinov, R., Cohen, W.W.: Semi-supervised qa with generative domain-adaptive nets. arXiv preprint arXiv:1702.02206 (2017)
27. Yu, L., Zhang, W., Wang, J., Yu, Y.: Seqgan: Sequence generative adversarial nets with policy gradient. In: AAAI, pp. 2852–2858 (2017)
28. Zhang, Y., et al.: Adversarial feature matching for text generation. arXiv preprint arXiv:1706.03850 (2017)
29. Zhu, Y., et al.: Texygen: A benchmarking platform for text generation models. arXiv preprint arXiv:1802.01886 (2018)

SALSA-TEXT: Self Attentive Latent Space Based Adversarial Text Generation

Jules Gagnon-Marchand, Hamed Sadeghi, Md. Akmal Haidar,
and Mehdi Rezagholizadeh[✉]

Huawei Noah's Ark Lab, Montreal Research Centre, Montreal, Quebec, Canada
jgagnonmarchand@gmail.com, haamed.sadeghi@gmail.com,
{md.akmal.haidar,mehdi.rezagholizadeh}@huawei.com

Abstract. Inspired by the success of self attention mechanism and Transformer architecture in sequence transduction and image generation applications, we propose novel self attention-based architectures to improve the performance of adversarial latent code-based schemes in text generation. Adversarial latent code-based text generation has recently gained a lot of attention due to its promising results. In this paper, we take a step to fortify the architectures used in these setups, specifically AAE and ARAE. We benchmark two latent code-based methods (AAE and ARAE) designed based on adversarial setups. In our experiments, the Google sentence compression dataset is utilized to compare our method with these methods using various objective and subjective measures. The experiments demonstrate the proposed (self) attention-based models outperform the state-of-the-art in adversarial code-based text generation.

Keywords: Text generation · Self-attention ·
Generative adversarial network

1 Introduction

Text generation is of particular interest in many natural language processing (NLP) applications such as dialogue systems, machine translation, image captioning and text summarization. Recent deep learning-based approaches to this problem can be categorized into three classes: auto-regressive or maximum likelihood estimation (MLE)-based, generative adversarial network (GAN)-based and reinforcement learning (RL)-based approaches.

MLE-based methods (such as [26]) model the text (language) as an auto-regressive process, commonly using RNNs. RNNs compactly represent the samples history in the form of recurrent states. In these models, text is generated by predicting next token (character, word, etc) based on the previously generated ones [10].

One of the main challenges involved with auto-regressive methods is exposure bias [4]. This problem arises due to discrepancy between the training and generation phases. In fact, ground-truth samples from the past are used in training,

© Springer Nature Switzerland AG 2019
M.-J. Meurs and F. Rudzicz (Eds.): Canadian AI 2019, LNAI 11489, pp. 119–131, 2019.
https://doi.org/10.1007/978-3-030-18305-9_10

while past generated ones are used in generation. A number of solutions have been proposed to address this problem by modifying the training procedure including scheduled sampling [4], Gibbs sampling [25], and Professor forcing [17].

Over the past few years, researchers have extensively used GANs [9] as a powerful generative model for text [6,29], inspired by the great success in the field of image generation. GANs are believed to be capable of solving the exposure bias problem in text generation raised from using MLE. The reason is that they solved a similar issue of blurry image generation in MLE-based variational autoencoders (VAEs). It is belived that the discriminator is able to guide the text generator, through their training exchange, how to generate samples similar to real (training) data. However, there are other challenges involved in GAN-based text generation.

A few of these challenges in text generation are inherent to GANs themselves, such as mode collapse and training instability. The mode collapse problem happens when the adversarially trained generator does not produce diverse texts. These issues can be mitigated by using well-known techniques such as feature matching [31], and entropy regularization [24]. Another challenge is due to the discrete nature of text, which causes the generator sampling to be non-differentiable over the categorical distribution of the words.

In this paper, we take advantage of Transformer self-attention mechanism [27] and incorporate it in two state-of-the-art adversarial latent code-based schemes proposed for text generation. More specifically:

- We incorporate the Transformer structure in the design of encoder and decoder blocks of AAE [18] and ARAE [13] setups for text generation.
- Blocks closely inspired from the Transformer's encoder layers, incorporating self-attention and element-wise fully-connected layers in a residual configuration and with positional encodings, are used along with spectral normalization to propose a novel GAN (both generator and discriminator) structure for AAE and ARAE setups.
- The performance improvement obtained from the proposed architectures is demonstrated via objective and subjective measures used in extensive experiments.

2   Related Work

2.1   Spectral Normalization

Spectral normalization [19] is a weight normalization method proposed to stabilize the training of GANs. The authors show that the Lipshitz norm of a neural networks can be bounded by normalizing the spectral norm of layer weight matrices. As opposed to local regularizations used in WGAN-GP, etc., the network-wide spectral regularization stabilizes the GAN training, produces more diverse outputs and results in higher inception scores. We use spectral normalization in our adversarial setups for the same reasons.

2.2   Attention Models

In sequence modeling literature, attention was initially proposed by [3]. It recognizes the fixed-length latent representation of the input sequence as the main performance bottleneck in the seq-to-seq models and proposed using soft-attention in the decoder. Using attention, the decoder can also attend to any desired token in the input sequence besides consuming the compressed representation resulting at the end of encoding operation.

Self-attention was initially proposed for language inference in [21]. The authors named it as "intra-attention" and showed that their structure can be an effective alternative for LSTMs in the task of natural language inference [5], at the time achieving state of the art performance with much fewer parameters as well as requiring a training time an order of magnitude shorter. Self-attention structures have since been used to set the state of the art in a number of different tasks [1, 8, 22, 27, 28]. They drastically reduce the path length between any two sequence inputs, making the learning of long term dependencies much easier [27]. They are considerably easier to parallelize, reducing the number of operations that are required to be sequential.

Recently, [30] applied self attention along with spectral normalization to the task of image generation using GANs. It showed by visualization that using attention, the generator can attend to far neighborhoods of any shape rather than close-by fixed-shape ones at each level in a hierarchical generation. The authors claim that applying spectral normalization to generator as well as discriminator helps training dynamics (stability). Similarly, we also adopt self attention and spectral normalization in our architecture designs.

Transformer [27] extended the use of self attention mechanism and was proved to be the state-of-the-art in sequence transduction applications such as machine translation. It dispenses convolutional and recurrent layers and relies entirely on attention-only layers and element-wise feed forward layers.

2.3   Latent Space-Based Text Generation

One of the main challenges of the language generation task originates from the discrete nature of text. Similarly to generating other discrete tokens, the back propagation of error through argmax operator is not well-defined. To address this problem, various approaches have been proposed in the literature including continuous approximation of discrete sampling [11, 16], using policy gradient from reinforcement learning [12, 24], etc. One of the most successful solutions is based on autoencoders with continuous latent spaces (i.e. latent code-based methods). Various training setups have been proposed for training these autoencoders including adversarial [13] and variational [15] setups.

A recent paper [7] performs a thorough review of the state-of-the-art latent code-based text generation methods. It studies the performance of a number of code-based text generation schemes and uses a unified rigorous evaluation protocol to evaluate them. We got inspired by their evaluation protocol to demonstrate the strength of our self attention-based approach in the context. They use a broad set of measures to perform a comprehensive study. We adopt forward and reverse perplexity as well as BLEU from their objective measures and fluency from the subjective ones.

2.4   Adversarial Text Generation

In this section, we briefly explain two prominent baseline methods using adversarial latent code-based generation techniques and present the technical details in Sect. 3.

AAE. Adversarial autoencoder (AAE) [18] proposes an adversarial setup to train probabilistic autoencoders. It matches the aggregated posterior of the encoder output (latent codes) to an arbitrary distribution that can be easily sampled from. Although authors demonstrate the applications of their setup in semi-supervised learning, style and content disentanglement, etc, AAE decoder can be effectively used as a generative model, converting samples of the arbitrary distribution (noise) to real-like outputs. From application perspective, authors only evaluated AAE performance in vision-related applications. In this paper, we tailor AAE for text generation, following guidelines proposed by [7] and incorporate self attention and Transformer as novel parts in the model.

ARAE. The adversarially regularized autoencoder (ARAE) [14] learns an autoencoder with continuous contracted codes that highly correlate with discrete inputs. That is, similar inputs get encoded (mapped) to nearby continuous codes. ARAE aims at exploiting GAN's ability to force the generator to output continuous codes corresponding to the code space obtained from encoding the real text data. By matching the outputs of generator and encoder, ARAE provides an implicit latent code GAN that serves as a generative model for decoding text.

3   Self Attentive Latent Code-Based Models

In this section, we explain the details of our self attention-based models following the ARAE and AAE setups proposed in [7]. These setups have shown comparable to the state-of-the-art results in text generation. We select similar setups to provide fair comparisons and report the best techniques/parameters based on our experiments.

In our architectures, Transformer [27] is used in designing all autoencoders. In both encoder and decoder, we use three blocks of Transformer. 'Block' and 'layer' names are used, respectively, instead of 'layer' and 'sub-layer' in the original paper.

Layer normalization [2] is applied on every layer (multi-head attention, masked multi-head attention and feed forward layers) within each Transformer

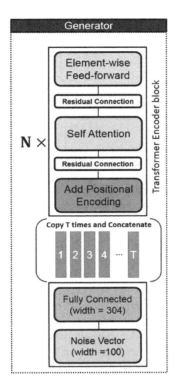
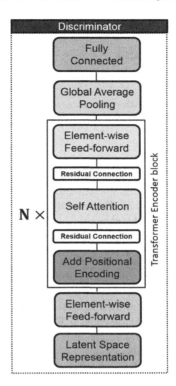

Fig. 1. SALSA-TEXT generator and discriminator architecture designed using Transformer encoder structure

block. Multi-head attentions have eight heads and embedding layers are of size 304 (a multiple of eight). Similarly to [27], positional encoding is used at the very first layer of the encoder and decoder. The dimensions and encoding place were found empirically for the best objective and subjective performance.

For GAN structures, i.e. the generator and discriminator architectures, we use modified Transformer encoder layers combined with spectral normalization, as depicted in Fig. 1 ($N = 3$). As in the regular transformer blocks, all connections are residual. Inspired by spectral normalization successes in the GAN-based image generation, especially proved in SAGAN [30], we apply it to the weights of the discriminator and the generator in our network. We did not find layer normalization (used in original Transformer) to be useful, when applied along with spectral normalization in the generator and discriminator architectures. Hence, only use spectral normalization in our GAN structures.

3.1   Adversarial Techniques

We use self attention-based structures in two well-known adversarial setups [13,18].

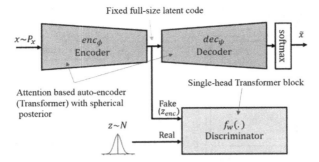

Fig. 2. SALSA-AAE for text generation. As explained in the figure, Transformer-based encoder, decoder and discriminator (with self attention) architectures are used. Decoder uses fixed-size full codes.

AAE. We use the AAE-SPH setup used in [7]. It is based on the original setup proposed in [18]. The discriminator forces encoder outputs to follow a uniform distribution on the unit sphere. Similarly to [18], a two-phase training is used, where there is regular alternation between minimizing reconstruction and adversarial (regularization) costs. The trade-off factor (λ) between reconstruction and adversarial costs is 20 (as in [7]). All over the encoder, decoder and discriminator, input and attention heads are dropped with a probability of 0.1. The general architecture and the proposed (self) attention-based changes are depicted in Fig. 2.

ARAE. We use the original setup from [13] with fixed-size full codes as inputs to the decoder. Inside the encoder and decoder, word and attention head dropout is performed with a probability of 0.1 and a maximum of 3-word shift is applied

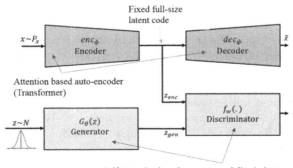

Fig. 3. SALSA-ARAE for text generation. Similarly to SALSA-AAE, all blocks are Transformer-based and decoder uses fixed-size full codes. Generator and discriminator comprise of self-attention layers.

to input words. The general architecture and the proposed (self) attention-based changes are depicted in Fig. 3.

4   Experiments

We study the performance of our self attentive (SALSA) architectures and compare it with that of the code-based setups studied in [7].

4.1   Experimental Setup

The performance of the models is evaluated in sentence generation (sampling), on the Google Sentence Compression (GSC) dataset[1] (as in [7]). Training on this dataset is very challenging as the sentences are relatively long (average of 24.8 words) and diverse in terms of content, grammar, etc. GSC comprises $200,000$ training and $9,995$ test sentences. For all the trained models, we use Google's SentencePiece[2] tokenizer using byte-pair encoding (BPE [23]) as in [7].

We filter the dataset to only include sentences with a maximum of 50 PBE tokens. This only lowers the average number of words per sentence and total number of sentences to 23.1 and 183739, respectively in the training set. The test dataset is also reduced to 9254 lines with an average of 22.7 words per sentence. Samples of generated sentences from all models is listed in Sect. 4.3.

The input noise to the generator is of size 100 (as in [7]). We upsample the noise to the embedding size of 304 by using a fully connected layer. The same upsampled noise is copied a number of times equal to the maximum number of steps in the sentence. We use $T = 50$ times in our experiments. The noise is then fed to the generator, where positional encodings are added to each step. The previously mentioned fully connected layer also serves to allow the model to learn to protect the information of the positional encodings from the noise. Positional encodings are also added at the start of each transformer encoder block. As we use fixed size sequences, the attention depth is always fixed (T bpe). Positional encodings are also added to the input of each transformer encoder block, inside of the generator.

4.2   Evaluation Metrics

We use various objective and subjective measures to evaluate the models. As objective measures, we use BLEU [20], Self-BLEU [32], forward and reverse perplexity.

BLEU [20] is a widely used metric to compute the similarity of a set of generated sentences with a reference dataset. The results are described in Table 2.

Self BLEU [32] (Table 3) is a measure of diversity for generated texts. In Self-BLEU, for each generated sentence, we compute the BLEU using the sentence as hypothesis and the rest of the generated sentences as the reference.

---

[1] https://github.com/google-research-datasets/sentence-compression.

[2] https://github.com/google/sentencepiece.

Table 1. Samples from models trained on GSC dataset

Model	Sample sentences
AAE	– on Thursday for an undisclosed amount of US Senate
	– .com is launching a new album on "The Idol's"
	– .
	– Tuesday for a US dollar in the US, according to the US Department
	– US US markets on Wednesday as US stocks rose to a US dollar, dealers said
	– The US will open its first-day visit to next year
SALSA-AAE	– The world's largest car maker, said it will buy back a new US$4 million fine for the first time in three years
	– London, May 6A 48-year-old man has been charged with raping a woman in the face of livelihood on Tuesday
	– In a bid to save money, the US government's most expensive land
	– The Governors of the city of Caterpillar, who was in talks with the world's most expensive officials
	– Israel has launched a new website that would help the Gaza Strip, and the first such incident in a deadly attack on the West Bank and the United States
	– Harrington has been found guilty of two counts of driving under the influence of a semi-final at a local court
ARAE	– A man has been arrested and charged with sexual assault on an alleged assault and assault for allegedly assaulting him to death and a child
	– A man is accused of stealing a child in her own home case of her husband
	– A man, who was accused of killing the two-year-old girl in connection with several of them who died from injuries last week
	– A man is facing charges of sexual assault for allegedly assaulting a woman and her wife
	– A man accused of killing the man in his home is being sentenced to life and is expected to be on the way
	– A man has been arrested in connection with her death and then sex with her husband, who was in critical condition and will be released on Monday
SALSA-ARAE	– Former PSV Ethanolve said he has "two and the title of his contract," the first-round pick of the season
	– The Queen will present a "Curprone of the Donegal Plan" in a new biopic of the forthcoming musical, and "Kalways," the school said in a statement
	– This week between the "Daily Show" flights, a year after the airing, a day before the launch of a summer vacation
	– However, the Myssenon, is planning to open its first wedding, with a move that will include a number of its customers, but they are not planning to sell their services
	– The Dallas Morning News reported that a Houston man is in "a new, a state that is leaving the Houston Rockets
	– The Dallas Morning News Corp. said it is to open a subsidiary of Houston, the largest newspaper and its staff, to be in the city

When averaged over all the references, it gives us a measure of how diverse the sentences are. Lower Self-BLEU scores are better, as high BLEU scores indicate great similarity.

In **Perplexity** evaluation (Table 4), the goal is to measure the individual quality of the sentences generated. We train an LSTM language model on the WMT News 2017 Dataset[3] filtered for lines of a maximum of 50 BPE tokens (a total of 200000 sentences). The perplexity of the language model is computed over 100000 generated sentences for each model.

Reverse perplexity evaluation (Table 4) aims to measure variety of the generated sentences. For each model, we train an LSTM-based language model based on 100000 generated sentences, and then evaluate the perplexity on the GSC test dataset, filtered to a maximum length of 50 BPE. Diverse generated sentences that cover dataset to a good extent would result in better (lower) reverse perplexity measures resulting from the trained LSTM network (language model).

For the **subjective** evaluation (Table 5), we use Amazon mechanical Turk[4] online platform. 18 sentences are sampled from each model, i.e. a total of 162 sentences. We assign 81 randomly selected sentences to 50 native English speakers (among Mechanical Turk Masters with hit approval ratings greater than 75%). The remaining 81 are assigned to another group of 50 people with the same qualifications. Each person was asked to evaluate the assigned 81 sentences in one and a half hours. In the evaluation, the 5-point Likart scale is used to measure grammaticality, semantic consistency and overall (Fluency). The overall reflects both grammar and semantic consistency in addition to other human-specific factors. Hence, it is a good representative of "Fluency" measure used in [7].

Objective and subjective evaluation of the studied models on the GSC dataset

Table 2. BLEU

	AAE	SALSA-AAE	ARAE	SALSA-ARAE
BLEU-1	0.671	**0.905**	0.924	0.865
BLEU-2	0.500	**0.638**	0.698	0.585
BLEU-3	0.279	**0.367**	0.402	0.294
BLEU-4	0.149	**0.192**	0.212	0.137
BLEU-5	0.086	**0.101**	0.116	0.071

---

[3] http://www.statmt.org/wmt17/.

[4] https://www.mturk.com/.

Table 3. Self-BLEU

	AAE	SALSA-AAE	ARAE	SALSA-ARAE
Self BLEU-1	0.950	**0.897**	0.973	**0.896**
Self BLEU-2	0.759	**0.738**	0.921	**0.731**
Self BLEU-3	0.604	**0.573**	0.843	**0.549**
Self BLEU-4	0.412	**0.410**	0.751	**0.367**
Self BLEU-5	0.270	**0.275**	0.653	**0.226**

Table 4. Perplexity

	AAE	SALSA-AAE	ARAE	SALSA-ARAE
Reverse perplexity	10309	**822**	8857	**1008**
Perplexity	88	**61**	37	106

Table 5. Human Evaluations

	AAE	SALSA-AAE	ARAE	SALSA-ARAE
Grammaticality	2.756	**3.09**	2.980	2.898
Semantic consistency	2.575	**2.597**	2.856	2.617
Fluency	2.604	**2.700**	2.851	2.652

4.3 Samples of Generated Sentences

In Table 1, we list six generated sentences for each model. As seen, AAE generates rather short sentences, while the corresponding SALSA version (SALSA-AAE) has alleviated the issue to a good extent. Finally, ARAE suffers from extreme mode collapse as opposed to its SALSA counterpart.

4.4 Results and Discussion

The results of objective and subjective evaluations are presented in Tables 2, 3, 4 and 5. As seen, the proposed self attention-based (SALSA) architectures consistently outperform the non-attention-based benchmarks in terms of diversity (measured by reverse perplexity). Moreover, they often show better performance in terms of output quality (measured by BLEU, self BLEU, preplexity and human evaluations) on the long and complicated sentences of the GSC dataset.

As seen in the generated samples (Table 1), human evaluation (Table 5) and objective metrics (Tables 2, 3 and 4), the original AAE and ARAE setups perform very poorly on GSC with long sentences. With reverse perplexities of over 8000 and high self-BLEU scores close to 0.9, they suffer from a high level of mode collapse (repeated sentences).

Human evaluations do not account for lack of diversity. The reason is humans are presented with a number of shuffled sentences and asked to evaluate them independently (without knowing which sentence coming from which model).

Hence, in our experiments for the original AAE and ARAE, a model can generate similar sentences (maybe due to mode collapse) and still receives high subjective scores.

It seems that, in our experiments, the original ARAE model suffers from mode collapse. We can see that it has slightly higher human evaluation scores, but extremely poor diversity metrics, i.e. very high reverse perplexity and self-BLEU scores. It can also be seen in the randomly selected generated sentences (Table 1), where all the sentences start with "A man" and invariably mention he is being arrested or accused of grievous crimes. This is likely because the sentences in the GSC dataset are long and that their structure is elaborate. SALSA-ARAE on the other hand reliably produces sentences of quality with great diversity.

SALSA-AAE has both considerably higher individual quality metrics than the original AAE and much better diversity metrics. It is the strongest pure adversarial text model. As seen in Table 5, SALSA-AAE provides the best grammaticality, semantic consistency and Fluency performance.

5   Conclusion and Future Work

In this paper, we introduced SALSA-TEXT, a Transformer-based architecture for adversarial code-based text generation. It incorporates self-attention mechanism by utilizing Transformer architecture in autoencoder and GAN setups. Our extensive experiments demonstrate the better performance of our models compared to the state-of-the-art in adversarial code-based text generation (without self-attention). The proposed architectures provide diverse, long and high quality output sentences as confirmed by objective metrics and human evaluations in extensive experiments.

As a future direction, it is beneficial to study the performance of self attention in other text generation methods including variational code-based and reinforcement learning-based approaches. Another interesting direction is to experiment with deeper Transformer-based autoencoders to better capture the underlying language model and perform unsupervised pre-training inspired by the success of [1,22].

References

1. Al-Rfou, R., et al.: Character-level language modeling with deeper self-attention. arXiv preprint arXiv:1808.04444 (2018)
2. Ba, J.L., Kiros, J.R., Hinton, G.E.: Layer normalization. arXiv preprint arXiv:1607.06450 (2016)
3. Bahdanau, D., Cho, K., Bengio, Y.: Neural machine translation by jointly learning to align and translate. arXiv preprint arXiv:1409.0473 (2014)
4. Bengio, S., Vinyals, O., Jaitly, N., Shazeer, N.: Scheduled sampling for sequence prediction with recurrent neural networks. In: Advances in Neural Information Processing Systems, pp. 1171–1179 (2015)

5. Bowman, S.R., Angeli, G., Potts, C., Manning, C.D.: A large annotated corpus for learning natural language inference. In: Proceedings of the 2015 Conference on Empirical Methods in Natural Language Processing (EMNLP). Association for Computational Linguistics (2015)
6. Che, T., et al.: Maximum-likelihood augmented discrete generative adversarial networks. arXiv preprint arXiv:1702.07983 (2017)
7. Cífka, O., Severyn, A., Alfonseca, E., Filippova, K.: Eval all, trust a few, do wrong to none: comparing sentence generation models. arXiv preprint arXiv:1804.07972 (2018)
8. Dehghani, M., Gouws, S., Vinyals, O., Uszkoreit, J., Kaiser, Ł.: Universal transformers. arXiv preprint arXiv:1807.03819 (2018)
9. Goodfellow, I., et al.: Generative adversarial nets. In: Advances in Neural Information Processing Systems, pp. 2672–2680 (2014)
10. Graves, A.: Generating sequences with recurrent neural networks. arXiv preprint arXiv:1308.0850 (2013)
11. Gulrajani, I., Ahmed, F., Arjovsky, M., Dumoulin, V., Courville, A.: Improved training of wasserstein gans. arXiv preprint arXiv:1704.00028 (2017)
12. Guo, H.: Generating text with deep reinforcement learning. CoRR abs/1510.09202 (2015). http://arxiv.org/abs/1510.09202
13. Haidar, Md.A., Rezagholizadeh, M., Omri, A.D., Rashid, A.: Latent code and text-based generative adversarial networks for soft-text generation. In: NAACL-HLT 2019 (2019)
14. Haidar, Md.A., Rezagholizadeh, M.: TextKD-GAN: text generation using knowledge distillation and generative adversarial networks. In: Canadian AI 2019 (2019)
15. Hu, Z., Yang, Z., Liang, X., Salakhutdinov, R., Xing, E.P.: Toward controlled generation of text. arXiv preprint arXiv:1703.00955 (2017)
16. Jang, E., Gu, S., Poole, B.: Categorical reparameterization with gumbel-softmax. arXiv preprint arXiv:1611.01144 (2016)
17. Lamb, A.M., Goyal, A.G.A.P., Zhang, Y., Zhang, S., Courville, A.C., Bengio, Y.: Professor forcing: a new algorithm for training recurrent networks. In: Advances In Neural Information Processing Systems, pp. 4601–4609 (2016)
18. Makhzani, A., Shlens, J., Jaitly, N., Goodfellow, I., Frey, B.: Adversarial autoencoders. arXiv preprint arXiv:1511.05644 (2015)
19. Miyato, T., Kataoka, T., Koyama, M., Yoshida, Y.: Spectral normalization for generative adversarial networks. arXiv preprint arXiv:1802.05957 (2018)
20. Papineni, K., Roukos, S., Ward, T., Zhu, W.J.: Bleu: a method for automatic evaluation of machine translation. In: Proceedings of the 40th Annual Meeting on Association for Computational Linguistics. ACL 2002, pp. 311–318. Association for Computational Linguistics, Stroudsburg (2002). https://doi.org/10.3115/1073083.1073135
21. Parikh, A.P., Täckström, O., Das, D., Uszkoreit, J.: A decomposable attention model for natural language inference. arXiv preprint arXiv:1606.01933 (2016)
22. Radford, A., Narasimhan, K., Salimans, T., Sutskever, I.: Improving language understanding by generative pre-training. https://s3-us-west-2.amazonaws.com/openai-assets/research-covers/language-unsupervised/language_understanding_paper.pdf
23. Sennrich, R., Haddow, B., Birch, A.: Neural machine translation of rare words with subword units. CoRR abs/1508.07909 (2015). http://arxiv.org/abs/1508.07909
24. Shi, Z., Chen, X., Qiu, X., Huang, X.: Towards diverse text generation with inverse reinforcement learning. arXiv preprint arXiv:1804.11258 (2018)

25. Su, J., Xu, J., Qiu, X., Huang, X.: Incorporating discriminator in sentence generation: a GIBBS sampling method. arXiv preprint arXiv:1802.08970 (2018)
26. Sutskever, I., Vinyals, O., Le, Q.V.: Sequence to sequence learning with neural networks. In: Advances in Neural Information Processing Systems, pp. 3104–3112 (2014)
27. Vaswani, A., et al.: Attention is all you need. In: Advances in Neural Information Processing Systems, pp. 5998–6008 (2017)
28. Yu, A.W., et al.: Qanet: combining local convolution with global self-attention for reading comprehension. arXiv preprint arXiv:1804.09541 (2018)
29. Yu, L., Zhang, W., Wang, J., Yu, Y.: Seqgan: sequence generative adversarial nets with policy gradient. In: AAAI, pp. 2852–2858 (2017)
30. Zhang, H., Goodfellow, I., Metaxas, D., Odena, A.: Self-attention generative adversarial networks. arXiv preprint arXiv:1805.08318 (2018)
31. Zhang, Y., et al.: Adversarial feature matching for text generation. arXiv preprint arXiv:1706.03850 (2017)
32. Zhu, Y., et al.: Texygen: a benchmarking platform for text generation models. In: The 41st International ACM SIGIR Conference on Research & Development in Information Retrieval, SIGIR 2018, Ann Arbor, MI, USA, 08–12 July 2018, pp. 1097–1100 (2018). https://doi.org/10.1145/3209978.3210080. http://doi.acm.org/10.1145/3209978.3210080

Automatically Learning
a Human-Resource Ontology
from Professional Social-Network Data

David Alfonso-Hermelo[1](✉), Philippe Langlais[1], and Ludovic Bourg[2]

[1] Université de Montréal, Montreal, Quebec H3C 3J7, Canada
david.alfonso.hermelo@umontreal.ca, felipe@iro.umontreal.ca
[2] LittleBIGJob, Montreal, Quebec H3B 4W5, Canada
ludovic.bourg@lorenzandhamilton.com
http://rali.iro.umontreal.ca/rali/

Abstract. In this work, we build an ontology (automatically learned) in the domain of Human Ressources by using a simple, efficient and undemanding procedure. Our principal challenge is to tackle the problem of automatically grouping human-provided job titles into a hierarchy and by similarity (as they are presented in human-made HR ontologies). We use the Louvain algorithm, a greedy optimization method that, given a sufficient amount of data, interconnects domain-specific jobs that have more skills in common than jobs from different domains. In our case, we used publicly available profiles from LinkedIn (written in English by users in France). An automatic evaluation was performed and shows that the resulting ontology is similar in size and structure to ESCO (one of the most complete human-made ontology for HR). The whole procedure allows recruitment professionals to easily generate and update this ontology with virtually no human intervention.

Keywords: Automatic Ontology Learning · E-recruitment ·
Occupations and skills ontology · Community detection ·
Relational model · Natural language processing · Taxonomy ·
Data mining · Artificial intelligence

1 Introduction

In computer science and according to the authors of [9], an ontology is a formal, explicit specification of a shared conceptualization that is characterized by high semantic expressiveness. It typically encompasses sectors of knowledge (more or less limited) in a fast computer-accessible way. With the development of AI technology mimicking human understanding of natural language, the number of technologies that can benefit from ontology-based artificial understanding has increased greatly. The problem is that, in addition to requiring multiple domain specialists, the elaboration of an ontology is labor and cost intensive.

© Springer Nature Switzerland AG 2019
M.-J. Meurs and F. Rudzicz (Eds.): Canadian AI 2019, LNAI 11489, pp. 132–145, 2019.
https://doi.org/10.1007/978-3-030-18305-9_11

In spite of these obstacles, multiple ontologies have been manually created in the domain of Human-Resources (HR), as, for instance, the ESCO ontology that we describe in Sect. 2.2. Those ontologies encode a hierarchy of categories and relations between occupations and skills. They serve as administrative tools in the recruitment domain to conduct operations such as grasping the skills needed for a given occupation, pinpointing the right jobs for a given set of skills, or guiding post assignation.

Still, even the most covering and detailed of these ontologies is incomplete and even if it is an important tool for HR workers and guidance counsellors, there are usually information gaps: brands are usually absent from skills (e.g., using *"text processor program"* instead of *"Microsoft Word"*), some occupation categories are constantly evolving, occupations are generally limited to one standardized denomination and so on. This was our first motivation to use a different strategy.

The fact that governments and institutions have gathered the necessary resources to produce human-made HR ontologies leads us to believe that there is great interest in having, updating and expanding this domain-specific resource.

In this paper, we explore the first version of a system designed to build an HR Ontology using the publicly available user data of a professional social network (LinkedIn in this case). This system is named HOLA for **HR O**ntology **L**earned **A**utomatically. In Sect. 2, we take a quick look at the most relevant related works. We present our system in Sect. 3. We describe the evaluation we conducted in Sect. 4 and conclude in Sect. 5.

2   Related Work

2.1   Automatic Ontology Learning

Every Automatic Ontology Learning (AOL) research approach we have encountered in the literature is divided in two main parts: a semantic triple[1] extractor from free text, and an ontology generator (e.g., the works of [13], or [5]). The appeal of unstructured data is evident: it is widely available, abundant and far richer than structured data.

2.2   HR Ontologies

As mentioned in Sect. 1, there have been some notable efforts to build human-made taxonomies and ontologies in the HR sector. We will mention two of the largest and publicly available ontologies: ROME and ESCO.

ROME is the Operational Directory of Trades and Jobs (Répertoire Opérationnel des Métiers et des Emplois), an HR ontology in French whose last version dates to 2009. It contains 532 job titles and 12,099 skills (divided in competences, activities and environments).

---

[1] Three entities that codify a statement in the form of subject-predicate-object expressions.

ESCO is the European Skills/Competences, qualifications and Occupations ontology project developed by the European Commission since 2010 and described in [17]. In its current state, ESCO stands as the most comprehensive free-of-charge ontology with regard to the number of languages (26), the number of job titles (2,942), the number of skills/competences and knowledges (13,485) and the number of job title variants (23,281 in English). ESCO is intended to serve as a multicultural and multilingual unifying resource between job seekers and job providers.

We now briefly present the structure of ESCO since our system design was mostly inspired by it. ESCO shows three different types of concepts: *occupation*, *knowledge* and *skill/competence*. All concepts have a preferred label, a list of alternative labels, a human-understandable description and are uniquely identified by a Uniform Ressource Identifier (URI). The occupation concepts are classified with a deep hierarchy based on the ISCO (International Standard Classification of Occupations) structure[2]. An excerpt of ESCO is provided in Table 1.

Table 1. Excerpt of ESCO entries.

Concept type	ISCO group	Preferred label	Alternative labels	Description
Occupation	2512	Software analyst	Software analysts, software requirements analyst, etc.	Software analysts elicit and prioritise user [...]
Skill		Load animals for transportation	Load animals safely, load animal, etc.	Load and unload animals safely into containers or [...]
Knowledge		Animal transport regulations	Animal welfare during transportation, etc.	The legal requirements relating to safe and [...]

2.3   Automatic HR Ontology Learning

The authors of [12] use automatic extraction methods on a large corpus of job offers to obtain job titles and their corresponding skills. After the extraction, the job titles are grouped into 27 clusters that represent the domain-specific categories. The final result is an ontology generated without human supervision, containing 440 job titles (128 in English, 312 in French) and 6,226 skills (4,059 in English, 2,167 in French). This system describes very interesting methods for HR AOL, but because of the unstructured nature of job offer texts, the number of reliable extracted data is relatively small and the resulting ontology is therefore incomplete.

---

[2] The hierarchy for the skill/competence and knowledge concepts is not yet available in full.

Google has also been developing an occupation and skill ontology for their recently released service Google Jobs [15]. According to the author of the blog, this ontology is a machine learning enrichment of the O*NET Standard Occupational Classification combined with a proprietary skill ontology containing 50,000 skills. The result is an HR ontology of 30 job categories, 1,100 job 'families' and 250,000 job titles; all connected to specific sets of skills from the skill ontology. Unfortunately, this ontology is not available to the public.

2.4 Ontology Evaluation

One of the problems in AOL is the evaluation of the resulting ontology. Building on the works of the authors of [4], we present the following grouping of the evaluation approaches usually used in AOL, for which there are four main strategies.

A *human inspection* is the most commonly used strategy. According to some authors (like [14] and [10]) three human annotators (not necessarily experts) can offer a good precision score if the sample is representative and if a protocol is uniformly followed.

A *comparison to a gold standard* measures the overlap between a candidate ontology and a reference one, at the risk of being penalized by an incomplete reference. Some researchers (see, for instance, [8] and [7]) go further and compare the *structural* similarity between ontologies. This measure depends on having an ontology and a gold standard whose structure is similar enough to be compared.

When an ontology is designed with a task in mind, a *benchmark assessment* measures its ability to perform this task (see, for instance, [18]).

It is also possible to use *quality scores* based on a collection of indicative proxies. The authors of [6] introduce a tool to detect inconsistencies, redundancies and incompleteness in taxonomies. In [1] the authors present a ranking system that uses 4 metrics (class match, density, semantic similarity, and betweenness) to compare multiple ontologies and rank them according to a quality score.

One such system, OntoQA [16] caught our attention. It is based on a structure-oriented score that uses 8 different metrics. We will analyze it in some details in Sect. 4.

3 System Design

The aim of HOLA is to automatically learn a domain-specific ontology for the HR sector. The human-made ontology already encompass the organizational knowledge HR specialists can bring forth. In those ontologies however, it is common for the more specific information to go unnoticed (i.e. the latest professional tools; the rare, new or specialized job titles; the diversification of skills; etc.).

Our claim is that professionals have a greater understanding of their own occupational fields and, by extension, of the kind of detailed knowledge our ideal ontology seeks to capture. One way to do this is to use professional social networks. Just like HR ontologies, these social networks are meant as a link between hiring entities and professionals. They use semi-structured data to allow an easy data query and they allow unlimited modifications and updates in order to be constantly up-to-date.

One such network is LinkedIn, for which we have collected the occupation and skill data from publicly available profiles made in France in 2016 and written in English. It should be fairly simple to continuously update this LinkedIn data collection and transform HOLA into a dynamic ontology. This is however left as future work.

3.1 Input Data, Term and Triple Extraction

As we discussed in Sect. 2, the state-of-the-art AOL projects we have encountered mine unstructured data. Because mining free text is a noisy process, we use the semi-structured data of LinkedIn user profiles.

Some HR ontologies make a distinction between different types of skills (i.e., skill, competence, qualification, knowledge, etc., see Table 1). To simplify the task for its users, LinkedIn allows listing skills in a separate section of their profile. LinkedIn does not expect its users to be HR specialists capable of differentiating the distinct types of skills, and neither do we. This is why, of all the available LinkedIn profile data, we only extract the *job title* and *skill* data for our use in HOLA.

The HOLA system depends on a fairly simple ontology design inspired by ESCO described in Subsect. 2.2. The ontology is encoded as a graph characterized as:

- **Node**: Fundamental unit of the graph. In HOLA, every concept, whether it is a job title, a skill or a community is represented as a node.
- **Community**: Cluster or group of nodes densely connected. Each node is assigned to the cluster with which it shares the most connections.
- **Edge**: Ordered pair of nodes. In HOLA, there are four different types of edges: job title - skill, community - job title, community - skill, community - community.
- **Edge weight**: Numerical value assigned to an edge. In HOLA, the edge weight is either null (for community - community, community - job title and community - skill edges) or it represents the skill frequency (for job title - skill edges).

At this point, our 2016 corpus has 5,110,768 profiles. Each profile contains an average of 1.1 job title and 2.1 skills. We can see an example of a LinkedIn profile's data in Table 2.

Of these LinkedIn profiles, more than half (55.9% of our corpus) have no job titles[3] and 18.8% have multiple job titles. LinkedIn profiles are presented with a Resumé structure that allows and encourages the chronological listing of each job title, instead of replacing it with the most current. The profiles with

---

[3] These profiles most probably correspond to LinkedIn users who have not yet any experience in the professional world or who have not updated their profile.

Table 2. Excerpt of a LinkedIn profile.

Country code	FR	Language code	en
Personal Branding Pitch	Prove organizational, analytical and process oriented skills and can demonstrate the ability to serve clients in a professional manner [...]		

Experience	– Experience 2
	  • **Function** : SAP SMB Delivery Manager
	  • **Start Date** : 2014-04
	  • **Missions** : SAP ByD and B1 Projects Delivery Improve Team [...]
	– Experience 1
	  • **Function** : SAP Technical Lead
	  • **Start Date** : 2010-07 • **End Date** : 2013-04
	  • **Missions** : Functionnal and Technical implementation of [...]		

Skills	SAP, ERP, Microsoft Office, Business Analysis, IT Strategy [...]		

multiple job titles, as the one in Table 2, pose a problem because of the difficulty to pinpoint what skills correspond to which job title. For this first version of HOLA, we chose to limit ourselves to the profiles containing exactly one job title and at least one skill. The implicit assumption is that the skills given in these profiles are pertinent to that single job title. By doing so, we are left with 1,293,082 profiles (25.3% of our initial corpus).

3.2 Filtering

Many of the job titles and skills appear several times in our profiles. After suppressing all duplicates, we are left with 57,655 job title nodes and 50,557 skill nodes, and 795,044 relationship edges linking them. Unfortunately, as it is often the case with real user data, we observe a rather high level of noise in both job title and skill nodes. A careful inspection led us to distinguish five types of noise:

- Lack of usefulness: the skill is not useful to the job title to which it is associated (e.g., skill *"powerpoint"* for job title *"Sushi Chef"*).
- Foreign words: a word or set of words written in a different language than the one specified in the metadata (e.g., 中国市场负责人 *Marketing Development"*[4]).
- Occupational over-specification: the job title is exclusively used inside a particular company and is not recognized by a wider group (e.g., *"sandwich artist at Subway Boulevard Voltaire"* instead of *"fast-food cook"*).
- Incoherent content: nonsensical or misplaced job titles or skills (e.g., *"XXxxxXX"*).
- Spelling error: (e.g., *"Research Ingenieer"*).

---

[4] Loosely translated as: "head of China market/person who is in charge of the Chinese market".

To filter out noisy job titles and skills, we use five heuristics[5] that are very different in nature.

- **Language detector**: By using Google's language detection library *langdetect*[6] we detect the language for each job title and skill and select those who correspond to our targeted language (English). This heuristic leaves out 56.8% of the job titles and 46.69% of the skills.
- **Gibberish detector**: We remove meaningless strings of characters (e.g., *XXXXXX*, *****). This heuristic filters out 0.08% of the job titles and 1.5% of the skills.
- **Token counter**: In human-made ontologies, job titles and skills rarely exceed 5 tokens (e.g., less than 1% of ESCO job titles have 6 or more tokens). We speculate that filtering out the nodes containing 6 or more tokens (not counting stop-words), will leave out most over-specific job titles (e.g., *"Internship as Assistant Product Manager and Business Developer (3D Printing Systems)"*, *"Consultant food and beverage and culinary for hotels and restaurants opening"*, *"Senior Administrative Assistant European Sales Management"*). This heuristic leaves out 5% of the job titles and 2.31% of the skills.
- **2 in 1 label detector**: As we explained in Subsect. 3.1, we select the profiles containing only one job title. Nevertheless, we sometimes observe that those still contain two (or more) job titles in the form of an enumeration (e.g., *"Owner/Partner & Managing Director"*, *"director / film researcher"*), which defeats our goal. If we detect this type of job title, we remove it and its connected skill nodes. This heuristic eliminates 8.27% of the job titles and 2.91% of the skills.
- **Isolated trees eliminator**: We want to avoid capturing job titles whose skills are exclusive to one job title and that are so specialized that they only appear once in the whole dataset. We thus filter out a) the trees isolated from the rest of the ontology forest, and b) the job titles and skills appearing in only one or two profiles. This heuristic leaves out 4.98% of the job titles and 35.78% of the skills.

By applying the filters, we obtain a data reduction of 75.16% of the job title, 89.26% of the skill nodes, and 80.60% of the edges, leaving our graph with 14,326 job title nodes, 5,430 skill nodes and 154,258 edges. Figure 1 shows in gray the concepts and relations that are removed by our filters and, in bold and red, the ones remaining.

Evaluating such a pipeline of filters is, of course, tricky. But since our goal was to eradicate not only aberrant noise but also spelling errors, gender versus singular-plural form variations and the like, we can measure the percentage of nodes in the filtered resource that have a label close to the label of another node. Intuitively, a good filter would significantly reduce the number of close labels. We used a simple Levenshtein edit-distance at the character level to detect,

---

[5] We did not include a spelling checker because the data contains terminology, acronyms, neologisms and variations that do not correspond to dictionary forms.

[6] http://code.google.com/p/language-detection/.

Fig. 1. Excerpt of job title and skill nodes in the graph before (in gray) and after applying the filters (in bold and red). (Color figure online)

for each node, neighbors within a distance between 1 and 3. Before filtering, 39% of the nodes have a neighbor. After filtering, this percentage falls to less than 33% (as Table 3 shows). This represents a small decrease after eliminating so many nodes during filtering, which goes against out initial intuition. After analyzing the captured neighbors, we observe that the greater part of what we managed to capture are node labels with distant meanings who happen to have similar spellings (e.g., "*heating*" and "*healing*", "*Geologist*" and "*Urologist*", "*Tutor*" and "*Tenor*") and a rare amount of the variations we aimed (and failed) to eradicate (e.g., "*Systems engineer*" and "*System Engineer*", "*Navy Officer*" and "*Naval Officer*"). We have no automatic way to determine which is which but since our heuristics focused on eliminating the noisy nodes and not the paronymic nodes, the decrease of neighbors from 39 to 33% suggests that there actually is a measurable reduction of noise.

3.3 Community Identification Process

At this point we have obtained a graph composed of skill and job title nodes connected between them. This graph still lacks the necessary structure to be a functional ontology. The nodes still need to be grouped according to their similarities and those groups must be placed in a hierarchy. This can be achieved with the Louvain algorithm, a community identification algorithm presented in [3]. It is a non-deterministic algorithm that is considered one of the fastest (with a complexity of $O(n \log n)$ according to the analysis presented in [11]).

Table 3. Number of nodes in HOLA at a given edit-distance of a neighbor after filtering.

Edit distance	Nb of nodes after filtering	Percentage of nodes after filtering	Example	
1	2365	12%	**Heating**	Healing
1-2	4208	21.3%	**Systems engineer**	System engineer
1-3	6505	32.9%	**Java engineer**	Naval engineer

In a nutshell, each node of the graph is assigned a different community label so there are as many communities as there are nodes. Then, for each node i we consider the neighbors j of i. We calculate the modularity gain of removing i from its community and placing it in the community of j. The node i is then placed in the community for which the modularity gain is maximum and positive. If no positive gain is possible, i stays in its original community.

This reduces the total amount community labels from one per node to one per community cluster. This process is applied repeatedly and sequentially for all nodes until there is no more modularity gain. After calculating the modularity gain for all the nodes, we iterate the whole process but, this time, reassigning a community label to each cluster of nodes instead of each node. For each iteration over i, we obtain a hierarchical ordering of the communities where the communities after the first iteration can be found inside the communities found after the second iteration and so forth.

After applying the Louvain algorithm to our graph, we obtain a 3-level classification for each node. At the first, less restrictive level we have 6 *parent* communities (level-1), these can be further divided into 8 *children* communities (level-2), which can be even further divided into 88 *grand-children* communities (level-3).

The community detection algorithm does not offer a perfect match to the usual domain segmentation and hierarchy. The main reason is that the current classification of occupations is the result of centuries of arbitrary categorization while the community detection is purely based on the shared adjacent nodes (the shared skills to determine job title categories, and vice versa). Nevertheless, if we observe the nodes composing each community and sub-community we can see in Table 4 that they show certain affinities.

Table 4. Excerpt of HOLA nodes and edges grouped in the 3rd-level community 6.6.17.

Job Titles	Yoga Teacher, Health Coach, complementary therapist, homeopath, Ayurveda Therapist, Mandataire Syner J Health, body psychotherapist, Osteopath
Skills	Physical therapy, smoking cessation, exercise physiology, energy healing, craniosacral therapy, wellness coaching, body massage, hatha yoga, holistic health
Edges	Ayurveda Therapist - wellness coaching, Health Coach - wellness coaching, Osteopath - craniosacral therapy, Ayurveda Therapist - holistic health

At this point, the outcome we obtain is a graph of 14,325 job title nodes, 5,431 skills, a 3-level node hierarchy and 154,259 job title-skill edges. Even after filtering, in pure numbers of nodes and edges, this is still more than ESCO or ROME (as we can see in Table 5).

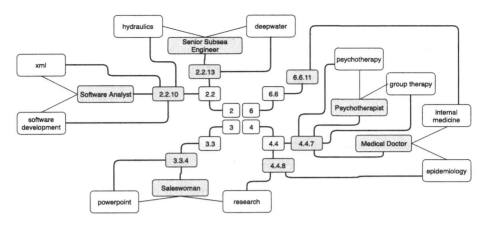

Fig. 2. Excerpt of the HOLA ontology.

In Fig. 2, we observe that the job title and skill nodes in our graph are connected to multiple community nodes, which are ordered in a hierarchical way. In this figure, the nodes in white represent skills, in blue are job titles, in red are level-3 communities, in orange are level-2 communities, in yellow are level-1 communities. We feel comfortable calling this graph an ontology[7].

4   Evaluation

As mentioned in Sect. 2.4 we opt for an evaluation based on quality scores and retain the OntoQA System described in [16] for comparing the ESCO ontology and the HOLA one. The OntoQA system is based on 8 metrics, but only 5 are actually useful in our setting (that is, for comparing ontologies):

- **Relationship Richness**: ratio of the number of non-inheritance relationships, divided by the total number of relationships defined in the schema. The closer this normalized score is to 1.0, the more diverse types of relations in the ontology. Compared to HOLA, ESCO uses three more non-inheritance relationships (*Alternative Label, Description, Skill Type*), which we do not implement yet. This is why we expect ESCO to surpass HOLA for this score. Yet, since we can successfully emulate many non-inheritance relationships we anticipate a high score approaching 1.0.
- **Class Richness**[8]: ratio of the number of classes in the ontology divided by the total number of classes defined in the ontology schema (non-empty classes/all classes). HR ontologies usually get the maximum score of 1.0 since

---

[7] The complete HOLA graph is available for a dynamic consultation at http://www-etud.iro.umontreal.ca/~alfonsda/project/holaOntology/index.html.

[8] In our case, the classes correspond to the automatically detected communities.

the HR classification evolves slowly and, therefore, it is rare to see a schema class that is not represented in the ontology[9].

- **Attribute Richness**: average number of attributes (nodes and edges) per class. The assumption behind this metric is that more attributes equals more knowledge conveyed. Since we have quite a high number of nodes and edges for an HR ontology, we expect to get a higher score than the human-made ontology.
- **Inheritance Richness**: average number of subclasses per class. A low Inheritance Richness score indicates a deep (or vertical) ontology that covers a specific domain in a detailed manner, while a high value denotes a shallow (or horizontal) ontology indicating a lower level of details. LinkedIn profiles tend to denominate specific job titles and skills but cover many professional sectors. We aim to get an ontology that would mirror these qualities and produce a corresponding middle ground score: not so high as to indicate a broad and general classification and not so low as to indicate a restrictive coverage.
- **Cohesion**: number of connected components. If the ontology is a connected graph, then the Cohesion value should be 1, if not, it should have a value of 2 or more. Since we specifically eliminated isolated data (as described in Subsect. 3.2) we expect to get a score of 1.

Evidently, there is still room for improvement (particularly for the Relationship Richness score) but, in general, we are quite pleased with the results obtained and reported in Table 5. These depict an ontology coinciding with the properties we aim to obtain: whose structure profile is (metrically) comparable to one of a human-made ontology, conveying more knowledge than its counterparts and indicating a compromise in inheritance coverage.

Table 5. OntoQA metric scores obtained for ESCO, ROME and HOLA.

Ontology name	Nodes	Edges	Relationship Richness	Class Richness	Attribute Richness	Inheritance Richness	Cohesion
ESCO	16,427	114,406	**0.929**	1.0	11,894	54	1
ROME	12,631	31,099	0.65	1.0	2,082	5	1
HOLA	**19,756**	**154,259**	0.909	1.0	**29,002**	16	1

Another point of comparison between the two ontologies is the number of job titles and skills they share. Actually, we were surprised to measure that only 286 job titles and 300 skills are shared among those ontologies, which represents only 3% of HOLA and 3.6% of ESCO nodes. We can see some examples in Table 6.

---

[9] Some ontologies from other domains evolve so quickly that conventional classes at the time of the schema conception might become obsolete, e.g., the smartphone sensor network ontologies analyzed in [2], like OntoSensor or CESN have a Class Richness score of 0.59 and 0.71 respectively.

While we anticipated a larger intersection, we feel this is a good sign. Our aim is not to replicate the ESCO ontology (or any other ontology) but to imitate its structure while using more dynamic data in nature[10].

Table 6. Excerpt of job titles and skills appearing in ESCO or HOLA. The nodes common to both ontologies appear in bold.

ESCO		HOLA	
Job Title	Skills	Job Title	Skills
webmaster	**css**, **php**, python (computer programming), pascal (computer programming), javascript, etc.	**webmaster**	html, xhtml, html5, **css**, css3, **php**, xml, python, mysql, flash, etc.
specialised doctor	general surgery, plastic surgery, radiology, **emergency medicine**, clinical biology, cardiology, etc.	medical doctor	medicine, surgery, emr, **emergency medicine**, clinical research, etc.

5 Conclusion and Future Work

We proposed a procedure to create an HR Ontology automatically, using a selection of the public semi-structured information found in LinkedIn. Even though there is still much to improve, this is a first step towards a system that resembles human-made HR ontologies but needs less time and effort. We report encouraging results both in terms of number of nodes, and level of structuring. While using user profiles to populate an ontology gives the hope to obtain a more up-to-date ontology for various activities in the recruitment domain, we showed that dealing with real user data is in itself a challenge that needs to be addressed.

There are several avenues of this work we plan to address in future research. First, we proposed filters in the HOLA pipeline that are not without limits: we do not detect all the noise in the data, and we do remove profiles that could contribute to populate the HOLA ontology. In particular, we discarded profiles with more than one professional experience, in order to associate skills to the mentioned job title. We should definitely leverage the data we discarded in this first version of our system. Second, we believe we could automatically identify descriptions and variants of job titles and skills, very similarly to what is manually done in the ESCO ontology. Third, the community detection algorithm is not naming a found community, a challenge we want to look at. Last, one motivation in conducting this work was to be able to update and increase the ontology with new job titles or skills emerging from professional social networks. This remains to be investigated.

---

[10] Even thought the enrichment of the ESCO ontology was not among this work's objectives, we do not discard this possible application of the HOLA procedure.

References

1. Alani, H., Brewster, C., Shadbolt, N.: Ranking ontologies with AKTiveRank. In: Cruz, I., et al. (eds.) ISWC 2006. LNCS, vol. 4273, pp. 1–15. Springer, Heidelberg (2006). https://doi.org/10.1007/11926078_1
2. Ali, S., Khusro, S., Ullah, I., Khan, A., Khan, I.: Smartontosensor: ontology for semantic interpretation of smartphone sensors data for context-aware applications. J. Sens. **2017** (2017)
3. Blondel, V.D., Guillaume, J.L., Lambiotte, R., Lefebvre, E.: Fast unfolding of communities in large networks. J. Stat. Mech: Theory Exp. **2008**(10), P10008 (2008)
4. Brank, J., Grobelnik, M., Mladenić, D.: A survey of ontology evaluation techniques. In: Slovenian KDD Conference (2005)
5. Cimiano, P., Völker, J.: Text2Onto. In: Montoyo, A., Muñoz, R., Métais, E. (eds.) NLDB 2005. LNCS, vol. 3513, pp. 227–238. Springer, Heidelberg (2005). https://doi.org/10.1007/11428817_21
6. Corcho, Ó., Gómez-Pérez, A., González-Cabero, R., Suárez-Figueroa, M.C.: ODEval: a tool for evaluating RDF(S), DAML+OIL, and OWL concept taxonomies. In: Bramer, M., Devedzic, V. (eds.) AIAI 2004. IIFIP, vol. 154, pp. 369–382. Springer, Boston, MA (2004). https://doi.org/10.1007/1-4020-8151-0_32
7. Dasgupta, S., Padia, A., Maheshwari, G., Trivedi, P., Lehmann, J.: Formal ontology learning from English is-a sentences. arXiv preprint arXiv:1802.03701 (2018)
8. Dellschaft, K., Staab, S.: On how to perform a gold standard based evaluation of ontology learning. In: Cruz, I., et al. (eds.) ISWC 2006. LNCS, vol. 4273, pp. 228–241. Springer, Heidelberg (2006). https://doi.org/10.1007/11926078_17
9. Feilmayr, C., Wöß, W.: An analysis of ontologies and their success factors for application to business. Data Knowl. Eng. **101**, 1–23 (2016)
10. Gupta, A., Piccinno, F., Kozhevnikov, M., Pasca, M., Pighin, D.: Revisiting taxonomy induction over wikipedia. In: Proceedings of COLING 2016, the 26th International Conference on Computational Linguistics: Technical Papers, Osaka, Japan, December 11–17 2016, pp. 2300–2309. EPFL-CONF-227401 (2016)
11. Kanawati, R.: Détection de communautés dans les grands graphes d'interactions (multiplexes): état de l'art. In: HAL archives ouvertes (2013)
12. Kessler, R., Lapalme, G.: Agohra: Génération d'une ontologie dans le domaine des ressources humaines. Traitement Automatique des Langues **58**(1), 39–62 (2017)
13. Mukherjee, S., Ajmera, J., Joshi, S.: Unsupervised approach for shallow domain ontology construction from corpus. In: Proceedings of the 23rd International Conference on World Wide Web. WWW 2014 Companion, pp. 349–350. ACM, New York (2014). https://doi.org/10.1145/2567948.2577350
14. Oliveira, H., Lima, R., Gomes, J., Ferreira, R., Freitas, F., Costa, E.: A confidence–weighted metric for unsupervised ontology population from web texts. In: Liddle, S.W., Schewe, K.-D., Tjoa, A.M., Zhou, X. (eds.) DEXA 2012. LNCS, vol. 7446, pp. 176–190. Springer, Heidelberg (2012). https://doi.org/10.1007/978-3-642-32600-4_14
15. Posse, C.: Cloud jobs API: machine learning goes to work on job search and discovery (2016). https://cloud.google.com/blog/big-data/2016/11/cloud-jobs-api-machine-learning-goes-to-work-on-job-search-and-discovery
16. Tartir, S., Arpinar, I.B., Sheth, A.P.: Ontological evaluation and validation. In: Poli, R., Healy, M., Kameas, A. (eds.) Theory and Applications of Ontology: Computer Applications, pp. 115–130. Springer, Dordrecht (2010). https://doi.org/10.1007/978-90-481-8847-5_5

17. le Vrang, M., Papantoniou, A., Pauwels, E., Fannes, P., Vandensteen, D., De Smedt, J.: ESCO: boosting job matching in europe with semantic interoperability. Computer **47**(10), 57–64 (2014)
18. Wandmacher, T., Ovchinnikova, E., Krumnack, U., Dittmann, H.: Extraction, evaluation and integration of lexical-semantic relations for the automated construction of a lexical ontology. In: Proceedings of the Third Australasian Workshop on Advances in Ontologies, vol. 85, pp. 61–69. Australian Computer Society, Inc. (2007)

Efficient Transformer-Based Sentence Encoding for Sentence Pair Modelling

Mahtab Ahmed[✉] and Robert E. Mercer[✉]

Department of Computer Science, The University of Western Ontario,
London, ON, Canada
mahme255@uwo.ca, mercer@csd.uwo.ca

Abstract. Modelling a pair of sentences is important for many NLP tasks such as textual entailment (TE), paraphrase identification (PI), semantic relatedness (SR) and question answer pairing (QAP). Most sentence pair modelling work has looked only at the local context to generate a distributed sentence representation without considering the mutual information found in the other sentence. The proposed attentive encoder uses the representation of one sentence generated by a multi-head transformer encoder to guide the focussing on the most semantically relevant words from the other sentence using multi-branch attention. Evaluating this novel sentence encoder on the TE, PI, SR and QAP tasks shows notable improvements over the standard Transformer encoder as well as other current state-of-the-art models.

Keywords: Transformer · Sentence encoder · Attention ·
Semantic similarity · Paraphrase identification ·
Question answer pairing

1 Introduction

Modelling the relationship of sentence pairs typically involves some comparison of the sentences: semantic relatedness [1], textual entailment between premise and hypothesis [2], true-false question-answer selection [3] and paraphrase identification [4]. Previous work tends to use the representation of each sentence separately overlooking the impact of combining the information from the two sentences. It is contrary to what humans do, we usually look at the keywords of both sentences when doing a comparison. For example, when comparing the sentence *"There is no biker jumping in the air"* with the sentence *"A lone biker is jumping in the air"* for the semantic relatedness task, attention should be given on the word *"no"* if we consider both sentences at the same time, whereas if we look at them independently, then the attention shifts to the most important portion of the sentences, *"biker is jumping in the air"*, which may not have any contribution to the semantics. This example shows the need for an architecture that builds a representation of one sentence by looking at both.

© Springer Nature Switzerland AG 2019
M.-J. Meurs and F. Rudzicz (Eds.): Canadian AI 2019, LNAI 11489, pp. 146–159, 2019.
https://doi.org/10.1007/978-3-030-18305-9_12

Recurrent neural networks (RNNs) [5] are the most widely used neural network model for sequence data. To overcome issues such as the "vanishing gradient problem", variants have been proposed: LSTM [6] and GRU [7]. They have been successful in a variety of applications such as machine translation, sequence labelling, speech recognition, and question answering. These models are difficult to parallelize because of a built-in state dependency and they are also inclined to overfit.

A popular LSTM framework is the sequence-to-sequence model using an encoder-decoder architecture where the encoder encodes the entire source sequence and the decoder generates the target sequence. For longer sequences, this model tends to generate asymmetrical sequences. To solve this, "Attention" was proposed, especially for the machine translation task: the decoder looks at each encoder word before generating a single output at each time step [8,9]. However this kind of Attention has a high computational overhead and it is affected by the state dependency problem. As a variant to this, Parikh et al. [10] proposed a decomposable attention mechanism which first decomposes the entire problem into sub-problems and then calculates attention via a soft alignment. Utilizing this, Vaswani et al. [11] proposed the "Transformer" model, which doesn't have any state dependencies, for the machine translation task. The main building blocks of their model are positional encoding [12], multi-head attention, and layer normalization [13].

In this paper, we propose an improved "Transformer Encoder" model to encode a pair of sentences for the sentence pair modelling task. To model a sentence pair, each sentence is viewed individually via multi-head attention. The semantic keywords in both sentences are looked at using a multi-branch cross attention which takes the representation of one sentence to put attention on the words of the other sentence. Evaluation done on the four aforementioned tasks shows notable improvements over the standard Transformer encoder on all of the tasks and best results on the textual inference and question-answer pair tasks when compared against all state-of-the-art encoder models including the best models especially designed for the specific tasks.

2   Related Work

Cer et al. [14] propose a sentence encoder model based on the encoder portion of Transformer [11] and perform an element-wise sum of the encoded representations at each word position to get a fixed length sentence representation. They evaluate their model mostly on sentence classification tasks and proved the effectiveness of transfer learning at the sentence level compared to the word level. Conneau et al. [15] also propose a universal sentence representation model based on LSTMs where it is first trained on the Stanford Natural Language Inference task and then uses transfer learning. Evaluation is on a range of tasks including QAP, PI, and SR. Zhou et al. [16] propose a sentence pair ranking model where they encode attention in the tree structure of the hypothesis based on the sequential representation of the premise. Lin et al. [17] propose a self-attentive sentence encoding model where they replace the pooling block with a self attention block on top of an LSTM encoder. Instead of extracting a vector representation, the authors use a matrix as the sentence representation where each row attends to different portions of the sentence. Zhao et al. [18] propose a self-adaptive hierarchical model which first extracts an intermediate representation of all possible phrases and finally takes the convex combination of them through gating. Yang et al. [19] use a hierarchical attention network for document classification where the first BiLSTM gets the word level attention and the second BiLSTM extracts the sentence level attention. Socher et al. [20] propose a recursive auto-encoder-based paraphrase identification model that first reconstructs each of the phrases from the tree representation of both sentences. Finally, this model extracts the sentence representation by doing a min-pooling operation over the transformed differences between the sentences. Yin et al. [3] propose an attention based convolutional sentence encoding model where they infer the attention matrix over a sentence pair through a simple matching function and then apply weighted pooling followed by another set of convolutions to get the final sentence representation. Mueller et al. [21] first encode the sentence pair to be compared using a siamese BiLSTM encoder and then use a simple Manhattan distance based matching function to infer the similarity score. Finally, they use an additional non-parametric regression block to further calibrate their model's prediction.

3   The Model

The standard pipeline architecture for sentence pair modelling starts with encoding both the hypothesis and premise sentences as vectors using a neural sentence encoder followed by a matching layer where the corresponding vector representations are compared with a similarity metric. Our sentence encoder architecture for modelling pairs of sentences is based on the "Transformer" [11] and the "Universal Sentence Encoder" [14]. Unlike LSTMs where the information at the current time step is dependent on the previous one, Transformer can encode all the words of a sentence using only linear tensor multiplications making it easily parallelizable. Figure 1 shows a sketch of our Transformer based sentence encoder which is built using stacked self-attention, point-wise fully connected layers, and branch attention for both of the encoders, respectively.

First our encoder inputs both the hypothesis and premise as sequences, $H = [h_1, h_2, \ldots, h_a]$ and $P = [p_1, p_2, \ldots, p_b]$ respectively where a denotes number of words in the hypothesis and b denotes number of words in the premise. We then initialize a pre-trained word embedding layer which transforms each of the words in those sequences into vectors and gives a new input representation in terms of tensors $H_{a \times d}$ and $P_{b \times d}$. However, these tensors only have bag of words information without any explicit knowledge of the word positions. In order to encode the positional information with the input, we initialize a positional encoding layer, where the positions are assigned by a unique 300 dimensional vector with each dimension being a sinusoid. Out of many possible initialization, we choose to initialize the odd dimensions by a sine function and the even dimensions by a cosine function as follows,

$$PE\,(pos, 2i) = \texttt{sin}(pos/10000^{(2i/d_{model})})$$
$$PE\,(pos, 2i + 1) = \texttt{cos}(pos/10000^{(2i/d_{model})}) \tag{1}$$

where pos denotes the position and i denotes the dimension. These wavelengths form a geometric progression from 2π to $10000 \cdot 2\pi$ having an interesting property that for any fixed offset k, $\texttt{sin}(k/z)$ and $\texttt{cos}(k/z)$ are constant and PE_{pos+k} can be represented by some linear function of PE_{pos}. Following this, we add a deep "Multi-Head Self Attention" layer which allows each position of our encoder to attend to all relative positions in the previous layer enabling it to easily learn long-range dependencies. In order to avoid repetition, we will explain this layer only for the hypothesis as for the premise it is exactly same. We start by initializing three learned parameters $\mathbf{W}_q$, $\mathbf{W}_k$, $\mathbf{W}_v \in \mathbb{R}^{n \times d \times l}$ where n denotes the number of heads and d and l represent the hidden dimensions having the constraint $d = n \times l$. Next, we copy the hypothesis tensor, H, n times creating a new tensor $\tilde{H}$ of size $n \times p \times d$ and multiply it with the three parameter tensors $\mathbf{W}_q$, $\mathbf{W}_k$, $\mathbf{W}_v$ independently as follows,

$$Q = \tilde{H}^T \mathbf{W}_q \quad K = \tilde{H}^T \mathbf{W}_k \quad V = \tilde{H}^T \mathbf{W}_v \tag{2}$$

This produces three new tensors called query (Q), key (K), and value (V) with each of them having dimension $n \times p \times l$. Next we apply batch-wise tensor multiplication on query and key, normalize it and pass the result through $\texttt{Softmax}$ which gives us the attention probabilities, $A_{n \times p \times p}$ as follows,

$$A = \texttt{Softmax}\left(\frac{QK^T}{\sqrt{d}}\right) \tag{3}$$

Following this, we multiply this attention A with value V which gives us the transformed hypothesis tensor $\hat{H}_{n \times p \times l}$ ($\hat{H} = AV$) where each of the n heads captures how much attention should be given on different phrases within the sentence and finally, we combine all the decisions by doing a concatenation over the cardinal dimension of the resulting tensor. Next we add a residual block followed by a layer normalization block where we do the normalization of inputs across the features. This permits each input to have a different normalization operation thereby allowing arbitrary mini-batch sizes to be used easily. This gives the final self attentive representation of the hypothesis. We follow the same procedures as above and derive the self attentive representation of the premise $\hat{P}_{q \times d}$. Instead of passing these representations to a matching layer we utilize the mutual information between the sentences through another layer of attention. Our idea is inspired by Rocktaschel et al. [22], Hermann et al. [23] and Zhou et al. [16]. Contrary to choosing "Multi-Head Attention", we choose "Multi-Branch Attention" as the composition function of this layer. According to Ahmed et al. [24], these branches mimic the same principles of multiple heads with an additional advantage of inferring a different representation at a time and the model automatically learns to combine these representations during training. Next, we explain this layer in terms of the hypothesis only in order to

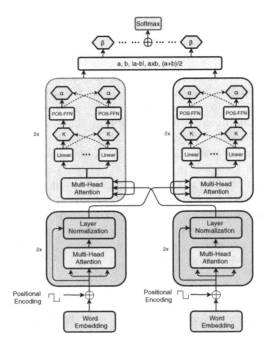

Fig. 1. Model architecture

avoid needless repetition. Initially, we utilize the same equations as above from Multi-Head attention with the exception that the query is calculated on the hypothesis whereas the key and value are calculated on the premise as follows,

$$\hat{Q} = \hat{H}^T \hat{\mathbf{W}}_q \quad \hat{K} = \hat{P}^T \hat{\mathbf{W}}_k \quad \hat{V} = \hat{P}^T \hat{\mathbf{W}}_v \tag{4}$$

Here $\hat{\mathbf{W}}_q$, $\hat{\mathbf{W}}_k$, $\hat{\mathbf{W}}_v$ are new model parameters defined especially for this layer. Next we apply Eq. 3 with these new $\hat{Q}$ and $\hat{K}$ to get a cross attention probability matrix $\hat{A}_{n \times p \times q}$ and following this, we extract an intermediate representation through $\hat{H} = \hat{A}\hat{V}$. We then initialize a learned parameter $\mathbf{W}_z \in \mathbb{R}^{n \times d \times d}$ and perform a batch-wise tensor multiplication with $\hat{H}$ to get a cross attentive hypothesis representation $\overline{H}_{n \times p \times d}$ as follows,

$$\overline{H} = \hat{H} \mathbf{W}_z \tag{5}$$

Here n depicts the branch id. Intuitively, this $\overline{H}$ contains n different transformed representations of its input where each of them resides in its individual branch. Following this, we scale each of these branch representations with a new learned parameter $\kappa \in \mathbb{R}^n$ and add a layer normalization block on top of it,

$$\overline{H}_i = \texttt{layerNorm}(\overline{H}_i \kappa_i) \tag{6}$$

However, we require $\sum \kappa_i = 1$.

Next we add a separate and identical position-wise feed forward module on each position. Vaswani et al. [11] use two convolutions with kernel size 1 and a RELU activation function. However, in this work we use two feed forward layers with feature expansion dimension of size 2048. Equation 7 summarizes these as follows,

$$FFN(x) = \texttt{max}(0, xW_1 + b_1)W_2 + b_2, \quad \text{where } x \in \{\overline{H}_1, \overline{H}_2 \dots, \overline{H}_n\} \qquad (7)$$

This gives a new set of transformed hypothesis tensors $\bar{\bar{H}}_1, \bar{\bar{H}}_2, \dots, \bar{\bar{H}}_n \in \mathbb{R}^{p \times d}$. Finally we perform another scaling operation on this set of tensors with a new learned parameter $\boldsymbol{\alpha} \in \mathbb{R}^n$ and then add all these resulting tensors to get the final attentive hypothesis representation as follows,

$$H^f = \sum_{i=1}^{n} \bar{\bar{H}}_i \boldsymbol{\alpha}_i \qquad (8)$$

Again, we require $\sum \alpha_i = 1$. We use the same procedures as above with slightly changed notation to get the final attentive premise representation P^f. Specifically, to achieve this, in Eq. 4, query is calculated on the premise whereas key and value are calculated on the hypothesis and following this all instances of H are replaced with P. Next, we add a matching layer where both of these representations are compared and a class decision is made according to the underlying task. In this layer, both the hypothesis (H^f) and premise (P^f) representations are passed through a position-wise feed forward block as depicted in Eq. 7. Next, these self- and cross-context-aware word representations are converted to a fixed length sentence encoding vector by computing the element-wise sum of the representations at each word position. A nonlinearity, $\texttt{tanh}$, is introduced in order to make this vector compatible with the underlying objective functions.

$$a = \texttt{tanh}(\texttt{elmentwiseSum}(H^f))$$
$$b = \texttt{tanh}(\texttt{elmentwiseSum}(P^f)) \qquad (9)$$

Next, we compute a range of features such as difference, element-wise product and average for the tuple $< a, b >$. We anticipate that such an operation could help to capture the inference between components in the tuples and catch induction relationships such as logical inconsistency. These features are then concatenated with the original representations, a and b, and we create a new tensor representation from this as follows,

$$F = \texttt{makeTensor}([a, b, |a - b|, a * b, \frac{(a + b)}{2}]) \qquad (10)$$

Next, we perform an element-wise multiplication of this feature tensor with a learned parameter vector $\boldsymbol{\beta} \in \mathbb{R}^n$ having the constraint $\sum \beta_i = 1$. The final inference relationship vector results from an element-wise weighted addition of each feature position is as follows,

$$\overline{F} = \sum_{i=1}^{4} F_i \boldsymbol{\beta}_i \qquad (11)$$

Next, an `MLP` maps this $\overline{F}$ to the required number of classes $Y_{1\times c} = \texttt{MLP}(\overline{F})$, where c represents the number of classes. Finally we use `Softmax` to turn this into class probabilities and our model predicts a corresponding label y^* as follows,

$$p(\mathbf{y}|\mathbf{x},\boldsymbol{\theta}) = \texttt{Softmax}(\texttt{MLP}(\overline{F}))$$
$$y^* = \arg\max_y p(\mathbf{y}|\mathbf{x},\boldsymbol{\theta}) \tag{12}$$

4   Experimental Setup

In this section, we detail the experimental setup used for evaluation. We first describe our training corpora as well as all of the benchmarks used in other standard sentence pair modelling studies. Following this, we explain the technical details of our proposed architectures along with their hyper-parameter settings.

First, we conduct our experiment on the Sentences Involving Compositional Knowledge (SICK) dataset [1]. The sentences are derived from video and image annotations. The first task is to classify a given sentence pair into three classes: Entailment, Neutral and Contradiction. The standard evaluation metric used for this inference task is accuracy. In addition to this, each of the SICK sentence pairs is also annotated with a relatedness label $\in [1,5]$ corresponding to the average relatedness as judged by different individuals. The standard evaluation metrics used for this relatedness measurement task are Pearson's r and mean squared error (MSE). We experiment with two different forms of this problem. While treating it as a regression problem, we use Manhattan distance similarity metric [21], and when converting it into a distribution mapping task, we measure the continuous distance between the predicted and ground truth distribution [16,25]. We use weighted cross entropy as the loss function for the inference task and for the relatedness task, we use MSE when it's a regression problem and KL divergence loss when it's a distribution mapping problem. The next task used for evaluation is paraphrase identification using the Microsoft Research Paraphrase Corpus (MSRP) [26]. Given a pair of sentences, the task is to identify whether or not they are paraphrases of each other. As it is a classification task, we use accuracy as the evaluation metric. Since the sentences in this dataset are mostly stock market related and contain many numbers, during preprocessing we replace all these numbers with a ⟨number⟩ tag. The final task on which we evaluate our model is the true-false question selection task from the AI2-8grade dataset [27]. Each data sample consists of a pair of sentences with one being the question and the other being the evidence formed by replacing the wh in the question by the answer. These evidence sentences in the sentence pairs are mainly collected from CK12 textbooks. As it is a binary classification task, we use accuracy as the evaluation metric. Detailed statistics for each of these datasets are summarized in Table 1.

Table 2 shows the hyper-parameter settings used during the experiments. We train our models on a GeForce GTX 1080Ti GPU with 'Adam', 'AdaDelta', 'AdaGrad' and 'SGD' optimizers. Results in the next section are reported using 'SGD' as it was giving comparatively better results. We use PyTorch 0.4.1 for implementing the models under the Linux environment.

Table 1. Dataset Description. Number of sentences in each split.

	SICK	MSRP	AI2-8grade
Train	4500	4076	12689
Valid	500	N/A	2483
Test	4927	1725	11359

Table 2. Hyper-parameters used for the experiments (in boldface) and the ranges that were searched during tuning.

Config	Value	Config	Value
Learning rate	**0.1**/0.05/0.001	Max Norm	5
Batch size	10/15/**25**	Learning rate decay	**0.99**
No. of layers	1/2/3	Dropout FC	**0.1**
Hidden dimension	**300**	No. of Head/Branches	8
Word embedding	Glove 300D	W_q, W_k, W_v dimension	**64**

5   Experimental Results and Analysis

In this section, we present the detailed results obtained with our Transformer encoder and compare with some of the top performing models for the four sentence pair modelling tasks. Additionally, we give a qualitative analysis by showing the predictions of our models on some random test samples for all of the tasks. Finally, we conclude this section by giving an insight about the performance of our model by analyzing the attention weights through heatmap visualization.

Table 3 displays our model's overall evaluation on the four corpora in terms of task specific evaluation metrics. The first group contains the results of our model in two different configurations, each one having two semi versions: **1.** Head attention for both self and cross, **2.** Head attention for self and Branch attention for cross. In configuration 1, we keep the self and cross attention blocks separate for the hypothesis and premise whereas in the siamese version, we have only one copy of self and cross attention block for both hypothesis and premise as we keep the parameters shared. In order to analyze the potency of our model, we implement some currently available top performing sentence encoders with our fixed hyper-parameter settings and report their results in the second group of this table. The last group contains the results of some preeminent models on each task as well as the ones that report their evaluation on one or more corpora. For the paraphrase identification task on the MSRP dataset, our model achieves the second best accuracy falling just short of RAE [20] which is specifically designed for this task. However, our model with head attention outperforms all the SOTA sentence encoders by quite a good margin of 1.94%. We also get better accuracy than all of the tree structured models from [18] which have an implicit advantage of having access to the sentence structure as parse trees. For the true-

Table 3. Performance comparison of our model on different tasks against some existing top performing models. We mark models that we implemented as †.

Model		MSRP Acc.	AI2-8grade Acc.	SICK-E Acc.	SICK-R r/MSE
Transformer Non Siamese	Head - Head	75.90	75.29	83.17	72.75/0.5129
	Head - Branch	75.80	74.78	82.63	73.04/0.5042
Transformer Siamese	Head - Head	76.40	**77.29**	84.80	72.78/0.5127
	Head - Branch	75.29	74.20	**85.22**	73.18/0.4969
InferSent [15] †		74.46	77.29	84.62	85.63/0.2732
LSTM [15] †		70.74	76.97	76.80	82.91/0.3244
BiLSTM Projection Layer [15] †		74.24	74.88	85.20	80.37/0.3667
BiGRU Last Encoder [15] †		70.46	74.76	81.47	83.17/0.3147
Inner Attention [17] †		69.74	74.77	72.01	78.63/0.3944
ConvNet Encoder [18] †		73.96	75.43	83.82	85.20/0.2806
Seq-LSTMs [16]		71.70	63.30	-	0.8528/0.2831
Seq-GRUs [16]		71.80	62.40	-	0.8595/0.2689
Tree LSTM [16]		73.50	69.10	-	0.8664/0.2610
Tree LSTM + Attn. [16]		75.80	72.50	-	**0.8730**/0.2426
Tree GRU [16]		73.96	70.60	-	0.8672/0.2573
Tree GRU + Attn. [16]		74.80	72.10	-	0.8701/0.2524
RAE [20]		**76.80**	-	-	-
Combine-skip + feats [28]		75.80	-	-	-
RNN [27]		-	36.1	-	-
CNN [27]		-	38.4	-	-
RNN-CNN [27]		-	37.6	-	-
Attn1511 [27]		-	35.8	-	-
Ubu.RNN [27]		-	44.1	-	-
Illinois-LH [29]		-	-	84.60	-
UNAL NLP [30]		-	-	83.10	-
SNLI-Transfer 3-class LSTM [2]		-	-	80.80	-
MaLSTM features + LSTM [21]		-	-	84.20	-
ECNU [31]		-	-	83.60	0.8414/-
Child sum Tree LSTM [25]		-	-	-	0.8676/0.2532
combine skip + COCO [28]		-	-	-	0.8655/0.2561
ConvNet [4]		-	-	-	0.8686/0.2606

false question-answer pairing task, our model with head attention achieves the best results. It clearly exceeds all of the tree structured models [18] by a very good margin of approximately 5–7% points, but it ties with InferSent [15] on accuracy: 77.29% However, it achieves better results on two other tasks. It is to be noted that the standard RNN and CNN based methods [27] perform very poorly on this dataset. One reason could be there is a lot of repetitiveness in the questions which makes the number of unique samples to be around 1/7th of the real data. For the SICK entailment task (SICK-E), the siamese version of our model with branch attention achieves the best result among all of the reported results as well as the ones that we re-implemented. Among the reported ones, Illinois-LH performs the best, but it employs many task specific features such as the count of words like 'no', and 'not' and also hypernyms. Among the models that we re-implemented, BiLSTM with a projection layer comes very close to our results but could not surpass. For SICK relatedness task (SICK-R), our model performs very badly compared to all others. To overcome this, we replace our matching layer with a Manhattan distance similarity function [21] but got almost the same results. One reason could be that the vectors produced by Transformer based models have to go through a linear summation followed by a normalization at the end which shrinks the norms of the vector drastically. This transformation makes it unsuitable for regression as well as the distribution mapping problem. However, for classification problems, another linear transformation with a `tanh` activation overcomes this.

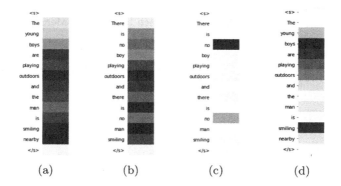

$$(a) \qquad (b) \qquad (c) \qquad (d)$$

Fig. 2. Visualization of attention weights. (a) $Q = H$, $K = V = H$ (b) $Q = P$, $K = V = P$ (c) $Q = H$, $K = V = P$ (d) $Q = P$, $K = V = H$ (**H** = Hypothesis and **P** = Premise)

Table 4 displays some example predictions produced by our Transformer encoder model. The SICK-E examples show that our model is very good at predicting the textual entailment class given a sentence pair. However on the same dataset but with a different task (SICK-R), we noticed some deviation between our model's predicted score and the actual score. We argue that this behavior is incongruous because the sentences in the SICK dataset are fairly simple and there are fewer named entities and uncommon words in the vocabulary.

Table 4. Example predictions from the test set. **GT**: ground truth, **Pr**: predicted.

Dataset	Sentence 1	Sentence 2	GT	Pr
SICK-E	A biker is riding away from a fence	A man is dancing on the road	N	N
	Two white dogs are quickly running together	Two white dogs are running together	E	E
	The man is going into the water	The person is going into the sea	E	N
SICK-R	A few men in a competition are running outside	A few men are running competitions outside	3.9	3.7
	A girl in white is dancing	A girl is wearing white clothes and is dancing	4.9	4.5
	A man is talking on the radio	A man is spitting	1.7	1.4
MSRP	I'm never going to forget this day	I am never going to forget this throughout my life	1	1
	You can reach George Hunter at (313) 222-2027 or ghunter@detnews.com	Reach her at (248) 647-7221 or email lberman@detnews.com	0	0
	Orange shares jumped as much as 15%	France Telecom shares dropped 3.6% while Orange surged 13%	0	1
AI2-8grade	Venus has a warmer average surface temperature than Earth?	Even though it is farther from the Sun, Venus is even much hotter than Mercury	1	1
		As a result, Venus is the hottest planet	1	1
		Venus's clouds are a lot less pleasant	1	0
		Clouds on Earth are made of water vapor	1	0

We observe that our model is able to grasp the word ordering deviation correctly as in the first example, but fails in cases where both sentences do not share the same words as happened in the second and third examples. The third group displays the example predictions on the MSRP dataset. We find that our model is able to predict the correct paraphrases in most of the cases even though the sentence pairs have different numbers but sometimes gives false positive predictions as happens in the third example where both sentences are talking about share prices. The AI2-8grade group shows that our model is very good at selecting false questions in almost every case and true questions in most of the cases but sometimes it makes false negative predictions if the evidence is quite different.

Figure 2 visualizes the amount of attention our model gives on different segments of the sentence pair when doing the comparison. As our model involves two self attentions and two cross attentions it outputs four attention vectors with different combinations of key, query, and value matrices. In the self attention case where the key, query, and value matrices are the same, our model puts more attention on keywords such as activities like 'playing' and 'smiling' and places like 'outdoors' and 'nearby'. However the effectiveness of our model is more discernible with the results of cross attention. When we have the hypothesis as query, the premise as key, and the relationship being Contradiction, our model gives all of its attention to the word 'no' from the premise. On the other hand, when we have the premise as query and the hypothesis as key, our model puts all the attention on "young boys are playing outdoors" and 'smiling' which contain all of the information from the hypothesis, showing that adding multi-branch cross attention makes our model think more like a human when comparing two sentences thus making it very effective for any sentence pair modelling task.

6 Conclusion

Being able to independently model the context aware representations of words in a sentence with correct orderings, Transformer-based encoders are perfect candidates for parallelization. We have introduced a way to incorporate cross attention in the Transformer-based sentence encoders capturing the shared mutual information between sentences much like humans do. The attention heatmap clearly shows where attention is put when comparing two sentences. We evaluate our proposed models on four sentence pair modelling tasks, achieving state-of-the-art performance on two of them. Our sentence encoders do not take any task specific features into account making it more generalized. A thorough comparison against the available SOTA sentence encoders shows that, for classification tasks, Transformer-based encoders are superior to the LSTM and CNN-based ones. Adding cross attention on top of self attention usually makes it more robust in identifying the underlying relationship between the sentence pairs.

References

1. Marelli, M., et al.: Semeval-2014 task 1: evaluation of compositional distributional semantic models on full sentences through semantic relatedness and textual entailment. In: Proceedings of the 8th International Workshop on Semantic Evaluation, pp. 1–8 (2014)
2. Bowman, S.R., Angeli, G., Potts, C., Manning, C.D.: A large annotated corpus for learning natural language inference. arXiv preprint arXiv:1508.05326 (2015)
3. Yin, W., et al.: ABCNN: attention-based convolutional neural network for modeling sentence pairs. Trans. Assoc. Comput. Linguist. **4**, 259–272 (2016)
4. He, H., Gimpel, K., Lin, J.: Multi-perspective sentence similarity modeling with convolutional neural networks. In: Proceedings of the 2015 Conference on Empirical Methods of Natural Language Processing, pp. 1576–1586 (2015)
5. Rumelhart, D.E., Hinton, G.E., Williams, R.J.: Learning representations by back-propagating errors. Nature **323**(6088), 533 (1986)

6. Hochreiter, S., Schmidhuber, J.: Long short-term memory. Neural Comput. **9**(8), 1735–1780 (1997)
7. Chung, J., et al.: Empirical evaluation of gated recurrent neural networks on sequence modeling. arXiv preprint arXiv:1412.3555 (2014)
8. Luong, M.T., Pham, H., Manning, C.D.: Effective approaches to attention-based neural machine translation. arXiv preprint arXiv:1508.04025 (2015)
9. Bahdanau, D., Cho, K., Bengio, Y.: Neural machine translation by jointly learning to align and translate. arXiv preprint arXiv:1409.0473 (2014)
10. Parikh, A.P., et al.: A decomposable attention model for natural language inference. arXiv preprint arXiv:1606.01933 (2016)
11. Vaswani, A., et al.: Attention is all you need. In: Advances in Neural Information Processing Systems, pp. 5998–6008 (2017)
12. Gehring, J., Auli, M., Grangier, D., Yarats, D., Dauphin, Y.N.: Convolutional sequence to sequence learning. arXiv preprint arXiv:1705.03122 (2017)
13. Ba, J.L., et al.: Layer normalization. arXiv preprint arXiv:1607.06450 (2016)
14. Cer, D., et al.: Universal sentence encoder. arXiv preprint arXiv:1803.11175 (2018)
15. Conneau, A., et al.: Supervised learning of universal sentence representations from natural language inference data. arXiv preprint arXiv:1705.02364 (2017)
16. Zhou, Y., Liu, C., Pan, Y.: Modelling sentence pairs with tree-structured attentive encoder. arXiv preprint arXiv:1610.02806 (2016)
17. Lin, Z., et al.: A structured self-attentive sentence embedding. arXiv preprint arXiv:1703.03130 (2017)
18. Zhao, H., Lu, Z., Poupart, P.: Self-adaptive hierarchical sentence model. In: IJCAI, pp. 4069–4076 (2015)
19. Yang, Z., et al.: Hierarchical attention networks for document classification. In: Proceeding of the 2016 Conference of NAACL: HLT, pp. 1480–1489 (2016)
20. Socher, R., et al.: Dynamic pooling and unfolding recursive autoencoders for paraphrase detection. In: Advances in Neural Inf. Processing Systems, pp. 801–809 (2011)
21. Mueller, J., Thyagarajan, A.: Siamese recurrent architectures for learning sentence similarity. In: AAAI, vol. 16, pp. 2786–2792 (2016)
22. Rocktäschel, T., et al.: Reasoning about entailment with neural attention. arXiv preprint arXiv:1509.06664 (2015)
23. Hermann, K.M., et al.: Teaching machines to read and comprehend. In: Advances in Neural Information Processing Systems, pp. 1693–1701 (2015)
24. Ahmed, K., Keskar, N.S., Socher, R.: Weighted transformer network for machine translation. arXiv preprint arXiv:1711.02132 (2017)
25. Tai, K.S., et al.: Improved semantic representations from tree-structured long short-term memory networks. arXiv preprint arXiv:1503.00075 (2015)
26. Dolan, B., Quirk, C., Brockett, C.: Unsupervised construction of large paraphrase corpora: exploiting massively parallel news sources. In: Proceedings of the 20th International Conference on Computational Linguistics, p. 350 (2004)
27. Baudis, P., Stanko, S., Sedivy, J.: Joint learning of sentence embeddings for relevance and entailment. arXiv preprint arXiv:1605.04655 (2016)
28. Kiros, R., et al.: Skip-thought vectors. In: Advances in Neural Information Processing Systems, pp. 3294–3302 (2015)
29. Lai, A., et al.: Illinois-LH: a denotational and distributional approach to semantics. In: Proceedings of the 8th International Workshop on Semantic Evaluation, pp. 329–334 (2014)

30. Jimenez, S., et al.: UNAL-NLP: combining soft cardinality features for semantic textual similarity, relatedness and entailment. In: Proceedings of the 8th International Workshop on Semantic Evaluation (SemEval 2014), pp. 732–742 (2014)
31. Zhao, J., et al.: ECNU: one stone two birds: ensemble of heterogenous measures for semantic relatedness and textual entailment. In: Proceedings of the 8th International Workshop on Semantic Evaluation (SemEval 2014), pp. 271–277 (2014)

Instance Ranking and Numerosity Reduction Using Matrix Decomposition and Subspace Learning

Benyamin Ghojogh[(✉)] [iD] and Mark Crowley[(✉)] [iD]

Machine Learning Laboratory, Department of Electrical and Computer Engineering,
University of Waterloo, Waterloo, ON, Canada
{bghojogh,mcrowley}@uwaterloo.ca

Abstract. One way to deal with the ever increasing amount of available data for processing is to rank data instances by usefulness and reduce the dataset size. In this work, we introduce a framework to achieve this using matrix decomposition and subspace learning. Our central contribution is a novel similarity measure for data instances that uses the basis obtained from matrix decomposition of the dataset. Using this similarity measure, we propose several related algorithms for ranking data instances and performing numerosity reduction. We then validate the effectiveness of these algorithms for data reduction on several datasets for classification, regression, and clustering tasks.

Keywords: Instance ranking · Numerosity reduction ·
Matrix decomposition · Subspace learning

1 Introduction

Data processing and pattern analysis can be challenging because of the excessively large mass of data [1]. The data can be reduced either for the goal of better data representation and discrimination of classes or for the sake of storage efficiency [2]. Two main approaches are used for data reduction, i.e., dimensionality reduction and numerosity reduction in which the number of *features* (or dimensions) and the number of *instances* (or samples, points, prototypes) [3,4] are reduced, respectively. This paper proposes several related algorithms for numerosity reduction using matrix decomposition and subspace learning.

Problem Definition: Assume there exists a set of n instances $\{\boldsymbol{x}_j\}_{j=1}^n$ in a d-dimensional Euclidean space, $\forall j \in \{1, \ldots, n\} : \boldsymbol{x}_j \in \mathbb{R}^d$. The instances together form a matrix $\boldsymbol{X} = [\boldsymbol{x}_1, \ldots, \boldsymbol{x}_n] \in \mathbb{R}^{d \times n}$. In supervised learning, we have $\{(\boldsymbol{x}_j, \boldsymbol{y}_j)\}_{j=1}^n$ meaning that every instance $\boldsymbol{x}_i$ has a corresponding label $\boldsymbol{y}_j \in \mathbb{R}^\ell$, where ℓ is the dimensionality of the label. We can then form the label matrix $\boldsymbol{Y} = [\boldsymbol{y}_1, \ldots, \boldsymbol{y}_n] \in \mathbb{R}^{\ell \times n}$. In classification, each instance belongs to one of $|\mathcal{C}|$ classes where $\mathcal{C}$ is the set including labels of classes. The cardinality of the

© Springer Nature Switzerland AG 2019
M.-J. Meurs and F. Rudzicz (Eds.): Canadian AI 2019, LNAI 11489, pp. 160–172, 2019.
https://doi.org/10.1007/978-3-030-18305-9_13

set of instances in class c is denoted by n_c. The goal of our work is to reduce the cardinality of a dataset from n to $m \in (0, n]$ by ranking the instances from the most to least important in terms of representation, discrimination, etc. In other words, we want to have $\hat{X} \in \mathbb{R}^{d \times m}$ and $\hat{Y} \in \mathbb{R}^{\ell \times m}$ (if supervised), where $\hat{X}$ and $\hat{Y}$ include the best m instances and their labels in terms of representation of data and/or discrimination of classes.

Related Work: Numerosity reduction is also known by other names such as *instance selection, instance ranking* and *prototype selection* [2,3,5]. In *Edited Nearest Neighbor (ENN)* [6], instances surrounded by a majority of neighbors from other classes are removed. *Decremental Reduction Optimization Procedure (DROP)* [7] removes instances one by one if the number of neighbor instances which are correctly classified improves after omitting the instance. Among DROP1 to DROP5 algorithms [7], DROP3 has the best accuracy-time trade-off [5]; however, its time complexity is not good. There are also some heuristic prototype selection methods such as *Random Mutation Hill Climbing (RMHC)* [8] which finds the best instances using mutations and testing the accuracy fitness. Heuristic methods usually take a noticeable amount of time to run. Some methods focus on the boundary and median points, such as *Stratified Ordered Selection (SOS)* [2] which concentrates on selecting boundary instances and recursively finds the median instances. *Shell Extraction (SE)* [9] defines a reduction sphere and removes the instances in it, resulting in a shell of boundary points. *Principal Sample Analysis (PSA)* [10] is another method which makes use of the scatter of data. Our proposed algorithms perform ranking of instances as well as data reduction. They are task agnostic, so they can be used for all unsupervised (clustering), classification, and regression tasks. Our algorithms are also very fast in run-time as the experiments demonstrate.

2   Background

In this section, we review some background on matrix decomposition and subspace learning, required in this paper.

2.1   Matrix Decomposition

Matrix decomposition factorizes the matrix $X \in \mathbb{R}^{d \times n}$ into multiplication of two matrices $X = UV^{\top}$ where the columns of $U \in \mathbb{R}^{d \times k}$ and $V \in \mathbb{R}^{n \times k}$ can be interpreted as bases and corresponding coefficients, respectively, and $k \in \mathbb{Z}_{+}$, usually $k := \min(d, n)$. There exist many different types of matrix decomposition such as *EigenValue Decomposition (EVD), Singular Value Decomposition (SVD), Nonnegative Matrix Factorization (NMF), PLU Decomposition, QR Decomposition, Cholesky Decomposition,* and *Dictionary Learning (DL)*. We will not cover EVD and Cholesky Decomposition as they decompose a square matrix and in our work $d = n$ does not necessarily hold. The rest of the above methods are used here for presenting our algorithm; however, note that our algorithm can be used with almost any other decomposition technique.

Singular Value Decomposition: SVD decomposes the matrix as $\boldsymbol{X} = \widetilde{\boldsymbol{U}} \boldsymbol{\Lambda} \widetilde{\boldsymbol{V}}^{\top}$ where columns of $\widetilde{\boldsymbol{U}} \in \mathbb{R}^{d \times k}$ are eigenvectors $(\widetilde{\boldsymbol{u}})$ of $\boldsymbol{X}^{\top} \boldsymbol{X}$, columns of $\widetilde{\boldsymbol{V}} \in \mathbb{R}^{n \times k}$ are eigenvectors $(\widetilde{\boldsymbol{v}})$ of $\boldsymbol{X} \boldsymbol{X}^{\top}$, $\boldsymbol{\Lambda} \in \mathbb{R}^{k \times k}$ is a diagonal matrix, containing singular values σ, with entries as square roots of eigenvalues of $\boldsymbol{X}^{\top} \boldsymbol{X}$, and $k = \min(d, n)$ [11]. Note that $\boldsymbol{X}^{\top} \boldsymbol{X}$ and $\boldsymbol{X} \boldsymbol{X}^{\top}$ are linear kernels over columns and rows of $\boldsymbol{X}$; therefore, $\widetilde{\boldsymbol{U}}$ and $\widetilde{\boldsymbol{V}}$ can be interpreted as measures of similarity between instances and features, respectively. One of the algorithms to compute $\widetilde{\boldsymbol{U}}$, $\boldsymbol{\Lambda}$, and $\widetilde{\boldsymbol{V}}$ in SVD is Jordan's algorithm [12], also called the power method. This algorithm extracts the leading singular vectors and values first. It continues iteratively by subtracting $\widetilde{\boldsymbol{u}} \sigma \widetilde{\boldsymbol{v}}^{\top}$ from $\boldsymbol{X}$ in every iteration.

Nonnegative Matrix Factorization: NMF targets decomposition of a matrix with nonnegative entries $\boldsymbol{X} \in \mathbb{R}_{\geq 0}^{d \times n}$ into $\boldsymbol{X} = \boldsymbol{U} \boldsymbol{V}^{\top}$ where $\boldsymbol{U} \in \mathbb{R}_{\geq 0}^{d \times k}$ and $\boldsymbol{V} \in \mathbb{R}_{\geq 0}^{n \times k}$ are also nonnegative [13]. This decomposition is beneficial especially if we want to have nonnegative basis and coefficients for better interpretation in image processing for example. Except in some exceptional methods [14], this decomposition is mostly treated as an optimization problem: $\min_{\boldsymbol{U}, \boldsymbol{V}} ||\boldsymbol{X} - \boldsymbol{U} \boldsymbol{V}^{\top}||_F^2$ subject to: $\boldsymbol{U} \succeq 0$, $\boldsymbol{V} \succeq 0$ which is NP-hard [15] and thus is usually solved heuristically.

PLU Decomposition: PLU decomposition is a method for solving linear systems of equations based on Gaussian elimination using elementary matrices [16]. It decomposes matrix $\boldsymbol{X} \in \mathbb{R}^{d \times n}$ as $\boldsymbol{X} = \boldsymbol{P} \widetilde{\boldsymbol{L}} \widetilde{\boldsymbol{U}}$ where $\boldsymbol{P} \in \mathbb{R}^{d \times d}$ is the permutation matrix. The matrices $\widetilde{\boldsymbol{L}} \in \mathbb{R}^{d \times k}$ and $\widetilde{\boldsymbol{U}} \in \mathbb{R}^{k \times n}$ are lower and upper triangular matrices, respectively.

QR Decomposition: The QR decomposition factorizes the matrix $\boldsymbol{X} \in \mathbb{R}^{d \times n}$ as $\boldsymbol{X} = \boldsymbol{Q} \boldsymbol{R}$ where $\boldsymbol{R} \in \mathbb{R}^{k \times n}$ is upper triangular and $\boldsymbol{Q} \in \mathbb{R}^{d \times k}$ is an orthogonal matrix whose columns, as basis vectors, span the same space as the columns of $\boldsymbol{X}$ [16]. The $k = \min(d, n)$ is named the reduced or economy size here. We refer the reader to [16,17] for more details.

Dictionary Learning: The DL tries to decompose matrix into $\mathbb{R}^{d \times n} \ni \boldsymbol{X} = \boldsymbol{D} \boldsymbol{R}$ where $\boldsymbol{D} \in \mathbb{R}^{d \times k}$ is the dictionary whose columns are basis vectors also called atoms, and $\mathbb{R}^{k \times n} \ni \boldsymbol{R} = [\boldsymbol{r}_1, \dots, \boldsymbol{r}_n]$ is the representation (components) [18]. Every column of $\boldsymbol{X}$ should be represented as a linear combination of a few atoms: $\min_{\boldsymbol{r}_j \in \mathbb{R}^k} ||\boldsymbol{x}_j - \boldsymbol{D} \boldsymbol{r}_j||_2^2 + \lambda ||\boldsymbol{r}_j||_1$, where λ is the penalty weight and the ℓ_1-norm takes care of sparsity which is very effective.

2.2 Subspace Learning

According to the manifold hypothesis [19], instances $\boldsymbol{X} \in \mathbb{R}^{d \times n}$ usually exist on a submanifold or subspace $\mathbb{R}^{k \times n}$ having lower intrinsic dimensionality $k \leq d$. There have been different linear and nonlinear algorithms for approximating the subspace some of which are unsupervised and some are supervised. Well established basic unsupervised algorithms include *Principal Component Analysis (PCA)* [20], *Isomap* [21], and *Locally Linear Embedding* [22]. For supervised cases, *Supervised PCA (SPCA)* [23] and *Fisher Linear Discriminant Analysis (FLDA)* [20] are examples. Some subspace learning (dimensionality reduction) methods, such as PCA, SPCA, and FLDA, provide a projection matrix while others, such as Isomap and LLE have different approaches. We will focus on the former since projection-based subspace learning and matrix decomposition are tightly related.

Principal Component Analysis: PCA finds the directions where the data have the maximum variance. The columns of $\widetilde{U}$ in SVD are these directions in PCA if X is centered. If X is not centered, the columns of $\widetilde{U}$ are directions as in Latent Semantic Indexing (LSI) [24]. As PCA and LSI are equivalent to SVD, we do not use PCA in this work.

Supervised Principal Component Analysis: SPCA [23] is based on the Hilbert-Schmidt Independence Criterion (HSIC) [25]. The HSIC itself is based on the claim that the correlation of two random variables can be obtained by considering the Hilbert-Schmidt norm of their cross-covariance matrix. The HSIC is $(1/(n-1)^2)\,\mathbf{tr}(\boldsymbol{KHBH})$, where $\boldsymbol{K}$ and $\boldsymbol{B}$ are kernels over the two variables and $\boldsymbol{H} = \boldsymbol{I} - (1/n)\mathbf{11}^\top$ is the centering matrix. The higher the HSIC, the more dependency they have. SPCA tries to maximize the dependency of the projected instances $\boldsymbol{U}^\top\boldsymbol{X}$ and the labels $\boldsymbol{Y}$. It uses a linear kernel for the projected instances $\boldsymbol{K} = \boldsymbol{X}^\top\boldsymbol{UU}^\top\boldsymbol{X}$ and an arbitrary kernel $\boldsymbol{B}$ over $\boldsymbol{Y}$ (we use a linear kernel for $\boldsymbol{B} = \boldsymbol{Y}^\top\boldsymbol{Y}$ in this work). The columns of the projection matrix $\boldsymbol{U}$ in SPCA are the k eigenvectors of $\boldsymbol{XHBHX}^\top$ corresponding to the largest k eigenvalues, where $k = \min(d, n)$ in our work. An interesting thing about SPCA is that if we do not consider the similarities of labels (so that $\boldsymbol{B} = \boldsymbol{I}$), we have $\boldsymbol{XHHX}^\top$ which is the covariance matrix of $\boldsymbol{X}$ and thus it reduces to PCA.

Fisher Linear Discriminant Analysis: FLDA tries to find the projection direction $\boldsymbol{u}$ which maximizes the Fisher criterion $(\boldsymbol{u}^\top\boldsymbol{S_b}\boldsymbol{u})/(\boldsymbol{u}^\top\boldsymbol{S_w}\boldsymbol{u})$ where $\boldsymbol{S_b} = \sum_{c=1}^{|\mathcal{C}|} n_c(\bar{\boldsymbol{x}}_c - \bar{\boldsymbol{x}})(\bar{\boldsymbol{x}}_c - \bar{\boldsymbol{x}})^\top$ and $\boldsymbol{S_w} = \sum_{c=1}^{|\mathcal{C}|} \sum_{\boldsymbol{x}_i \in \text{class } c}(\boldsymbol{x}_i - \bar{\boldsymbol{x}}_c)(\boldsymbol{x}_i - \bar{\boldsymbol{x}}_c)^\top$ are between- and within-scatter matrices, respectively. The $\bar{\boldsymbol{x}}_j$ and $\bar{\boldsymbol{x}}$ denote the mean of the j-th class and the total mean, respectively. Maximizing the Fisher criterion results in a generalized eigenvalue problem $\boldsymbol{S_b}\boldsymbol{u} = \lambda\boldsymbol{S_w}\boldsymbol{u}$; hence, the columns of projection matrix $\boldsymbol{U}$ are the k leading eigenvectors of $\boldsymbol{S_w}^{-1}\boldsymbol{S_b}$ with largest eigenvalues. In this work, we take $k = \min(d, n, |\mathcal{C}| - 1)$ because the rank of $\boldsymbol{S_b}$ is at most $|\mathcal{C}| - 1$.

3   Instance Ranking and Numerosity Reduction

We propose several related algorithms for instance ranking and numerosity reduction based on matrix decomposition and subspace learning. Algorithms are proposed for both unsupervised and supervised cases. For supervised cases, both classification and regression are covered.

3.1   Instance Ranking Using Matrix Decomposition

As mentioned before, matrix decomposition factorizes a matrix into the product of two matrices $X = UV^\top$. The matrices $U \in \mathbb{R}^{d \times k}$ and $V \in \mathbb{R}^{n \times k}$ can be interpreted as kernel or similarity over columns (instances) and rows (features) of X, respectively. From another perspective, U and V can be considered as bases and coefficients of instances, respectively. Comparing the p-th columns in $X = UV^\top$, we have $x_p = \sum_{j=1}^{k} V(p,j)\, u_j \in \text{span}\{u_1, \ldots, u_k\}$; therefore, columns of U are the bases for column space or range of X, i.e., columns of U span the space of instances. We can present a similar argument for the row space of X and for V, regarding features rather than instances. In this work, we deal with the instances of data and thus U is of interest since its columns can be interpreted as basis vectors over instances. In SVD, for example, the columns of U are principal components of X showing its maximum variations.

Our primary contribution flows from the idea that the basis vectors capture the most informative directions in the data in terms of some type of information such as variation. Hence, it is expected that the more important instances are more similar to these basis vectors. Therefore, the instances can be ranked based on their similarity to the basis vectors. We have several basis vectors so we need to somehow fuse their information into one similarity score.

Measuring Similarity with Basis Vectors: Suppose that u is a basis vector and x is an instance. Different metrics can be used for measuring the similarity of these two vectors such as cosine, logistic, hyperbolic tangent, and exponential. The similarity measures can be in the range $[0, 1]$ with 1 for highest similarity. In this work, the cosine is used for the similarity metric because it can be written in simple closed-form matrix operations. If the two vectors x and u are already normalized and have unit length, the cosine is reduced to the inner product $\cos(x, u) = x^\top u$. Thus, we normalize the instances (columns of X), i.e., $\widetilde{X}(:,j) = X(:,j)/\|X(:,j)\|_2$, $\forall j = 1: n$, where $\widetilde{X}$ is the normalized dataset. We assume that basis vectors U are already orthonormal; otherwise, they should be normalized as well which is the case with the basis of NMF, DL, and PLU decomposition. Having k basis vectors as columns of $U \in \mathbb{R}^{d \times k}$ and n instances as columns of $X \in \mathbb{R}^{d \times n}$, the cosine of basis vectors and instances are $\mathbb{R}^{n \times k} \ni \cos(X, U) = \widetilde{X}^\top U$.

We would like these scores to be positive in order to be ready for fusion, so we use $|\widetilde{X}^\top U|_\varepsilon \in \mathbb{R}^{n \times k}$ in range $[\varepsilon, 1]$ where the safe absolute value $|A|_\varepsilon :=$ $\max(|A|, \varepsilon)$ prevents the elements of matrix A from being zero and ε is a small positive number (e.g., 0.001). The intuition of absolute value is that in measuring similarity with a basis vector, we should care only about the direction of the basis vector and not its sign of direction. The intuition of safe absolute value is that we want to fuse the k scores of every instance by multiplication so having a small score regarding one of the basis should not spoil all scores.

In order to fuse the scores of similarities of every instance with different basis vectors, we use weighted product of scores, $|s_1|_\varepsilon^{w_1} \times \cdots \times |s_k|_\varepsilon^{w_k}$. This can be written in logarithmic form, $-(w_1 \log |s_1|_\varepsilon + \cdots + w_k \log |s_k|_\varepsilon)$, which gives us a closed-form matrix where the negative signs cancel with those obtained from logarithms of scores in range $[\varepsilon, 1]$. Finally, the overall score of every instance with respect to the k basis vectors can be obtained as the entries of:

$$\mathbb{R}^n \ni \boldsymbol{s} = -\log\left(|\widetilde{\boldsymbol{X}}^\top \boldsymbol{U}|_\varepsilon\right)\boldsymbol{w}, \tag{1}$$

where $\mathbb{R}^k \ni \boldsymbol{w} = [w_1, \ldots, w_k]$ contains the weights regarding the k basis vectors. In case the eigenvalues are obtained (such as SVD, SPCA, and FLDA) or are not obtained (such as NMF, DL, PLU and QR decompositions) from decomposition or subspace learning, we use $w_i = \frac{1/\lambda_i}{\sum_j 1/\lambda_j}$ and $w_i = \frac{2i}{k(k+1)}$, $\forall i = 1, \ldots, k$, respectively. When having eigenvalues, the basis vectors and eigenvalues are sorted in descending order. Note that smaller weight gives more importance as the range of $|s_i|_\varepsilon$ is $[\varepsilon, 1]$. Moreover, as the weights should be positive, if we have any negative eigenvalue, we shift all values to become positive $\lambda_i \leftarrow \lambda_i - 2\lambda_k \mathbb{I}(\lambda_k < 0)$ where $\mathbb{I}(.)$ is the indicator function (1 if its condition is satisfied and 0 otherwise).

Unsupervised Cases: In unsupervised learning, we are given a dataset X without labels. The matrix $\boldsymbol{U}$ can be obtained from its decomposition, i.e., in SVD: $\boldsymbol{X} = (\widetilde{\boldsymbol{U}})(\boldsymbol{\Lambda}\widetilde{\boldsymbol{V}}^\top) = \boldsymbol{U}\boldsymbol{V}^\top$, in NMF: $\boldsymbol{X} = \boldsymbol{U}\boldsymbol{V}^\top$, in PLU: $\boldsymbol{X} = (\boldsymbol{P}\widetilde{\boldsymbol{L}})(\widetilde{\boldsymbol{U}}) = \boldsymbol{U}\boldsymbol{V}^\top$, in QR: $\boldsymbol{X} = (\boldsymbol{Q})(\boldsymbol{R}) = \boldsymbol{U}\boldsymbol{V}^\top$, and in DL: $\boldsymbol{X} = (\boldsymbol{D})(\boldsymbol{R}) = \boldsymbol{U}\boldsymbol{V}^\top$. To be more clear, for example for SVD, we take left singular matrix $\widetilde{\boldsymbol{U}}$ to be $\boldsymbol{U}$ and $\boldsymbol{\Lambda}\widetilde{\boldsymbol{V}}^\top$ to be $\boldsymbol{V}^\top$. The scores of instances are calculated using Eq. (1). The instances are sorted based on the scores $\boldsymbol{s}$ in descending order because the larger the score, the more important the instance. Let $m \in (0, n]$ be the desired number of instances to be selected for numerosity reduction. The reduced sorted dataset is denoted by $\mathbb{R}^{d \times m} \ni \hat{\boldsymbol{X}} = \boldsymbol{X}(:, 1:m)$ where $\boldsymbol{X}$ is already sorted.

Regression Cases: In regression, there exist some independent variables $\boldsymbol{X} \in \mathbb{R}^{d \times n}$ (we call them observations) and some dependant variables, or labels, $\boldsymbol{Y} \in \mathbb{R}^{\ell \times n}$. The goal of regression is to estimate (or predict) the $\boldsymbol{Y}$ from given $\boldsymbol{X}$. Instance ranking may help regression in terms of finding the most important instances for this prediction. Considering merely $\boldsymbol{X}$ takes into account the distribution and variation of data regardless of the labels. On the other hand, considering only $\boldsymbol{Y}$ ignores the effect of $\boldsymbol{X}$ and concentrates on the output labels to be predicted and the relation of instances in terms of labels. These two scenarios have their own merits and can be done separately because both distribution of data and prediction of labels are effective. However, if done, they should be fused in some way.

The explained methodology for unsupervised cases can be applied to $\boldsymbol{X}$ in order to find the important instances in terms of overall distribution of data ignoring the observations. Likewise, the same algorithm can be performed solely on $\boldsymbol{Y}$. In other words, the $\boldsymbol{X}$ and $\boldsymbol{Y}$ are considered as unsupervised cases separately. Let the scores obtained from processing $\boldsymbol{X}$ and $\boldsymbol{Y}$ be $\boldsymbol{s}_X$ and $\boldsymbol{s}_Y$, respectively. The fusion of these scores can be done by multiplying these two scores which are both positive: $\mathbb{R}^n \ni \boldsymbol{s} = \boldsymbol{s}_X \odot \boldsymbol{s}_Y$, where $\odot$ denotes Hadamard product.

After sorting $\boldsymbol{X}$ and $\boldsymbol{Y}$ using the obtained score $\boldsymbol{s}$, the numerosity is reduced as $\mathbb{R}^{d \times m} \ni \hat{\boldsymbol{X}} = \boldsymbol{X}(:, 1: m)$ and $\mathbb{R}^{\ell \times m} \ni \hat{\boldsymbol{Y}} = \boldsymbol{Y}(:, 1: m)$.

Classification Cases: In classification, the dataset $\boldsymbol{X} \in \mathbb{R}^{d \times n}$ and the corresponding possible labels $\boldsymbol{Y} \in \mathbb{R}^{\ell \times n}$ exist, while every column of $\boldsymbol{Y}$ encodes one of the $|\mathcal{C}|$ classes. The goal of classification is to identify the class of an instance. Data reduction may help classification in terms of better discrimination of classes.

Every important instance should be a satisfactory representative of its own class; therefore, the instances within a class, denoted by $\boldsymbol{X}_c$ for class c, can be processed using matrix decomposition for the sake of better representation of the class. If $\boldsymbol{U}_{X_c}$ denotes the basis matrix obtained from decomposition of $\boldsymbol{X}_c$ and $\widetilde{\boldsymbol{X}}_c$ is the normalized $\boldsymbol{X}_c$, the scores of instances in class c are obtained as $\mathbb{R}^{n_c} \ni \boldsymbol{s}_{X_c} = -\log\big(|\widetilde{\boldsymbol{X}}_c^\top \boldsymbol{U}_{X_c}|_\varepsilon\big)\boldsymbol{w}$. Then, $\boldsymbol{s}_X \in \mathbb{R}^n$ is obtained by putting together the scores of instances in all classes.

On the other hand, the instances should be important in terms of discrimination of classes. Here, as in the case of regression, we cannot find the scores for solely the labels $\boldsymbol{Y}$. The reason is that the labels of a class are all similar and the rank of $\boldsymbol{Y}$ will be only $|\mathcal{C}|$ with so many repetitive columns if it is one-hot encoded. Therefore, inspired by [26], we put the labels $\boldsymbol{Y}$ alongside $\boldsymbol{X}$ to yield a new matrix. However, in order to bias instances of every class to fall closer to each other in the space, it is best to choose $\boldsymbol{Y}$ to be encoded by one-hot encoding resulting in $\boldsymbol{E} \in \mathbb{R}^{|\mathcal{C}| \times n}$. Finally, concatenating $\boldsymbol{X}$ and $\boldsymbol{E}$ results in $\boldsymbol{D} := [\boldsymbol{X}^\top, \boldsymbol{E}^\top]^\top \in \mathbb{R}^{(d+|\mathcal{C}|) \times n}$. The same approach is used for finding the scores of instances based on $\boldsymbol{D}$: $\mathbb{R}^n \ni \boldsymbol{s}_D = -\log\big(|\widetilde{\boldsymbol{D}}^\top \boldsymbol{U}_D|_\varepsilon\big)\boldsymbol{w}$, where $\widetilde{\boldsymbol{D}}$ and $\boldsymbol{U}_D$ are the normalized $\boldsymbol{D}$ and basis vectors from decomposition of $\boldsymbol{D}$, respectively. Finally, the two scores $\boldsymbol{s}_X$ and $\boldsymbol{s}_D$ are fused similarly as before: $\mathbb{R}^n \ni \boldsymbol{s} = \boldsymbol{s}_X \odot \boldsymbol{s}_D$, and the reduced dataset ($\hat{\boldsymbol{X}}$ and $\hat{\boldsymbol{Y}}$) is obtained.

3.2   Instance Ranking Using Subspace Learning

The goal of subspace learning is to project data form the original d-dimensional space to a lower dimensional subspace with dimensionality k. The projection is formulated as $\boldsymbol{U}^\top \boldsymbol{X}$ where $\boldsymbol{U} \in \mathbb{R}^{d \times k}$ is the projection matrix. Interestingly, the projection matrix can be considered as the basis matrix in the matrix decomposition. The reason for this claim is that assuming $\mathbb{R}^{k \times n} \ni \boldsymbol{V}^\top := \boldsymbol{U}^\top \boldsymbol{X}$, the reconstruction of matrix $\boldsymbol{X}$ can be written as $\boldsymbol{X} \approx \boldsymbol{U}\boldsymbol{U}^\top \boldsymbol{X} = \boldsymbol{U}\boldsymbol{V}^\top$, which is the matrix decomposition of $\boldsymbol{X}$. Therefore, the columns of the projection matrix $\boldsymbol{U}$ can be interpreted as the basis vectors with which the instances can be compared. Recall that our focus in subspace learning is on methods which involve using a projection matrix.

Unsupervised Cases: For instance ranking and reduction using subspace learning in unsupervised cases, the unsupervised submanifold learning techniques, such as PCA, could also be used. Note that PCA is equivalent to SVD if $\boldsymbol{X}$ is centered.

Regression Cases: As introduced before, one of the recently developed supervised methods for subspace learning is SPCA [23]. An interesting fact about SPCA is that the labels Y in this method are not necessarily required to be a finite discrete set of values as in the classification case. Therefore, SPCA can also be utilized in regression problems. Having the dataset X and labels Y, the projection matrix U can be obtained by eigen-decomposition $XHBHX^\top = U\Lambda U^\top$, where columns of U are the eigenvectors of $XHBHX^\top$ and Λ is a diagonal matrix containing the corresponding eigenvalues. We take the first $k = \min(d, n)$ leading eigenvectors ($U \in \mathbb{R}^{d \times k}$) and $B = Y^\top Y$. After finding U, the scores of instances are calculated using Eq. (1) and the reduced dataset $\hat{X}$ and $\hat{Y}$ are obtained.

Classification Cases: For classification as a supervised scenario, any projection-based supervised subspace learning technique can be used. Here, we use two basic and effective supervised methods, namely SPCA and FLDA. Having dataset X and labels Y, SPCA is used exactly as explained for regression cases. For classification cases, FLDA is another possible approach to have. In FLDA, maximizing the between-scatter variance and minimizing the within-scatter variance yields to the generalized eigenvalue problem $S_b u = \lambda S_w u$, where S_b and S_w are calculated as in Sect. 2 and u and λ are the eigenvector and eigenvalue, respectively. If the S_w is not singular, the columns of projection matrix U are the leading eigenvectors of $S_w^{-1}S_b$. In order to guarantee the invertibility of S_w in case it is ill-posed, we slightly strengthen its main diagonal as $(S_w + \varepsilon I)^{-1}S_b$, where ε is a positive small number. Finally, the projection matrix U is found by the eigen-decomposition $(S_w + \varepsilon I)^{-1}S_b = U\Lambda U^\top$, where U and Λ contain eigenvectors and eigenvalues of $(S_w + \varepsilon I)^{-1}S_b$, respectively. The matrix $U \in \mathbb{R}^{d \times k}$ is the projection matrix. Note that, unlike SPCA, FLDA is only applicable to classification and not to regression.

4   Experiments

Datasets: Several datasets with various characteristics are used. For classification cases, we used five datasets, i.e., Page Blocks, Pima Indians Diabetes, Spambase, Image Segmentation, and Yeast datasets. The Facebook metrics dataset was used for regression experiments. For clustering cases, two datasets, Isolet and Iris, were utilized. The Pima dataset is available on Kaggle and the other datasets can be found in the UCI machine learning repository. As Table 1 shows, there is enough variety in the selected datasets in terms of number of instances, features, and classes. Note that in the Facebook Metrics dataset, we have $d = 7$ and $\ell = 12$. The Image Segment and Isolet datasets include negative values and thus NMF cannot be used for them. For all experiments, datasets were shuffled and the results are the average of values in 10-fold cross validation.

Classification Cases: Tables 2 and 3 report the results of experiments for classification. In all tables of this paper, underlined bold, bold, and underlined values show the best, the second best, and the third best values, respectively. The rates

Table 1. Characteristics of utilized datasets for experiments

	Page blocks	Pima	Spambase	Image segment	Yeast	Facebook	Isolet	Iris
#Instances	5473	768	4601	2310	1484	500	7797	150
#Features	10	8	57	19	8	19	617	4
$|\mathcal{C}|$	5	2	2	7	10	–	26	3

Table 2. Comparison of instance selection methods in classification (cont' in Table 3). Rates are accuracy and times are in seconds. Numbers in parentheses for ENN and DROP3 show the percentage of retained data.

Page Block Dataset

		SVD	NMF	PLU	QR	DL	SPCA	FLDA	SOS	SE	SDM	ENN	DROP3	NR
	Time:	5.74E-2	2.31E+1	5.96E-2	6.05E-2	3.28E+1	4.67E-1	**4.04E-2**	5.43E-1	1.51E-1	**2.95E-2**	3.90E-1	4.20E+3	×
1NN	20% data:	37.56%	**73.83%**	56.45%	**92.08%**	41.23%	**83.02%**	73.77%	1.93%	34.27%	33.12%	(95.61%)	(1.98%)	95.57%
	50% data:	64.59%	87.74%	71.97%	**93.51%**	67.02%	**89.34%**	83.55%	1.35%	**95.19%**	58.03%	95.66%	48.45%	
	70% data:	85.05%	91.81%	84.83%	**94.09%**	77.74%	**92.56%**	88.45%	1.20%	**95.28%**	75.05%			
LDA	20% data:	92.89%	**93.23%**	92.65%	90.17%	93.16%	**93.45%**	93.18%	0.36%	**93.71%**	92.34%	94.79%	80.26%	94.53%
	50% data:	93.62%	93.73%	93.53%	93.87%	93.66%	**94.04%**	**94.55%**	0.25%	**94.55%**	94.24%			
	70% data:	94.00%	94.04%	93.89%	94.06%	94.00%	94.18%	**94.51%**	0.25%	**94.48%**	94.40%			
SVM	20% data:	71.64%	65.87%	**82.52%**	**90.11%**	70.48%	64.18%	69.22%	2.52%	**89.07%**	71.09%	94.86%	87.33%	89.76%
	50% data:	71.09%	**86.48%**	83.17%	**91.85%**	82.07%	83.33%	82.21%	1.99%	**94.24%**	84.27%			
	70% data:	**91.10%**	90.31%	90.46%	**91.44%**	83.09%	88.08%	87.99%	1.53%	**94.37%**	89.20%			

Pima Dataset

		SVD	NMF	PLU	QR	DL	SPCA	FLDA	SOS	SE	SDM	ENN	DROP3	NR
	Time:	1.09E-2	2.39E+0	1.00E-2	9.52E-3	1.13E+1	1.40E-2	**6.31E-3**	2.13E-2	1.59E-2	**4.51E-3**	1.87E-2	5.00E+0	×
1NN	20% data:	**68.61%**	68.35%	68.09%	67.28%	67.96%	64.44%	**69.01%**	67.06%	44.40%	62.87%	(69.40%)	(13.44%)	68.48%
	50% data:	70.03%	69.78%	**70.05%**	66.64%	69.66%	69.26%	**70.04%**	67.04%	50.64%	66.39%	71.74%	64.06%	
	70% data:	**70.16%**	**69.91%**	67.96%	67.95%	69.14%	69.25%	69.78%	66.91%	65.62%	67.81%			
LDA	20% data:	**74.61%**	73.69%	72.39%	70.81%	**74.86%**	**74.61%**	74.08%	73.56%	54.55%	65.09%	77.07%	75.12%	77.73%
	50% data:	**76.30%**	75.39%	**76.16%**	73.81%	75.77%	75.64%	75.91%	**76.43%**	71.35%	73.68%			
	70% data:	**76.82%**	76.04%	76.69%	75.38%	**77.59%**	76.56%	**77.34%**	76.56%	76.55%	74.98%			
SVM	20% data:	**65.24%**	59.65%	60.17%	59.62%	58.19%	60.68%	60.63%	**63.52%**	42.04%	**63.78%**	67.18%	53.12%	64.57%
	50% data:	62.62%	57.39%	**65.10%**	61.57%	**63.02%**	60.04%	**65.23%**	62.25%	48.15%	60.01%			
	70% data:	**65.49%**	57.51%	**65.89%**	55.97%	59.36%	62.13%	58.70%	62.63%	55.23%	64.98%			

Spambase Dataset

		SVD	NMF	PLU	QR	DL	SPCA	FLDA	SOS	SE	SDM	ENN	DROP3	NR
	Time:	8.70E-2	2.81E+1	8.57E-2	7.84E-2	6.37E+1	3.81E-1	**5.92E-2**	3.25E-1	2.53E-1	**3.30E-2**	3.47E-1	1.57E+3	×
1NN	20% data:	68.15%	66.40%	67.76%	68.42%	67.94%	**69.44%**	64.61%	**73.74%**	72.48%	68.70%	(81.12%)	(12.25%)	83.09%
	50% data:	74.89%	74.07%	**75.28%**	75.04%	74.48%	74.41%	74.09%	**79.04%**	78.41%	73.70%	80.63%	63.29%	
	70% data:	79.59%	78.78%	79.48%	**79.65%**	78.78%	79.28%	79.28%	**81.26%**	81.24%	78.52%			
LDA	20% data:	85.72%	84.87%	85.33%	**87.26%**	83.65%	**86.17%**	84.43%	**87.06%**	72.48%	73.39%	87.72%	88.02%	88.76%
	50% data:	88.28%	88.41%	88.50%	**89.11%**	87.76%	**89.13%**	88.24%	87.71%	**89.58%**	81.52%			
	70% data:	89.21%	89.21%	89.35%	**89.56%**	89.34%	**89.63%**	89.39%	88.08%	**90.71%**	84.72%			
SVM	20% data:	67.76%	**74.87%**	67.31%	68.42%	70.65%	68.55%	64.76%	**83.32%**	68.61%	**71.91%**	84.35%	80.91%	81.63%
	50% data:	79.22%	**84.06%**	81.91%	**82.89%**	81.15%	79.39%	75.00%	**83.09%**	77.93%	78.52%			
	70% data:	83.24%	79.87%	**85.13%**	**84.50%**	82.39%	80.93%	79.30%	80.56%	**85.02%**	81.35%			

are accuracy and the reported times (in seconds) are the average times over folds for computing ranking/reduction. Our proposed methods are compared with our own implementations of SOS [2], SE [9], ascending Sorted By Distance from Mean (SDM), ENN [6], DROP3 [7], and No Reduction (NR) scenario. Three classifiers, 1-Nearest Neighbor (1NN), Linear Discriminant Analysis (LDA), and Support Vector Machine (SVM) are used with three different levels of reduction each, i.e., 20%, 50%, and 70%. Note that in ENN and DROP3, the amount of reduction cannot be selected and thus in highlighted comparisons for different reduction levels, we exclude them. In comparing time of ranking, SDM usually is

Table 3. Comparison of instance selection methods in classification (cont' of Table 2). Rates are accuracy and times are in seconds. Numbers in parentheses for ENN and DROP3 show the percentage of retained data.

		SVD	NMF	PLU	QR	DL	SPCA	FLDA	SOS	SE	SDM	ENN	DROP3	NR
						Image Segment dataset								
	Time:	3.04E−2	×	2.94E−2	3.64E−2	7.21E+1	9.97E−2	**1.83E−2**	5.86E−2	4.85E−2	**1.38E−2**	1.30E−1	3.13E+2	×
1NN	20% data:	81.21%	×	70.68%	78.00%	75.67%	76.75%	77.92%	**89.26%**	83.41%	80.99%	(95.25%)	(6.78%)	
	50% data:	**92.07%**	×	85.10%	86.83%	86.53%	86.75%	88.35%	**94.19%**	84.71%	87.05%	94.45%	60.34%	96.45%
	70% data:	**94.54%**	×	91.21%	91.03%	92.07%	90.77%	92.98%	**95.41%**	90.69%	90.00%			
LDA	20% data:	84.41%	×	76.01%	82.90%	75.93%	84.63%	**87.66%**	**90.51%**	85.23%	86.36%			
	50% data:	**90.43%**	×	83.89%	86.32%	87.31%	88.22%	90.21%	**91.16%**	86.27%	87.83%	67.57%	79.35%	91.55%
	70% data:	**91.25%**	×	89.56%	88.26%	90.25%	89.87%	91.08%	**91.34%**	90.30%	89.48%			
SVM	20% data:	72.12%	×	72.77%	71.42%	73.41%	**78.57%**	77.05%	**78.35%**	71.73%	71.86%			
	50% data:	**83.11%**	×	79.17%	80.60%	82.51%	82.33%	80.64%	**82.85%**	70.60%	81.34%	78.52%	55.10%	85.23%
	70% data:	83.72%	×	78.26%	78.87%	84.54%	83.03%	80.34%	**85.67%**	**86.79%**	84.32%			
						Yeast Dataset								
	Time:	1.93E−2	8.72E+0	1.93E−2	1.96E−2	6.44E−1	3.10E−2	1.20E−2	3.80E−2	3.03E−2	**9.72E−3**	5.77E−2	1.50E+1	×
1NN	20% data:	**53.63%**	34.49%	39.96%	47.51%	48.98%	44.47%	45.15%	50.00%	45.07%	**56.00%**	(50.44%)	(17.87%)	
	50% data:	**53.50%**	42.18%	46.09%	52.56%	51.48%	47.91%	50.20%	51.48%	45.95%	**55.33%**	56.00%	53.17%	52.28%
	70% data:	52.90%	47.64%	49.46%	53.63%	52.82%	50.47%	51.95%	52.62%	51.07%	**54.38%**			
LDA	20% data:	47.91%	44.86%	49.66%	54.72%	53.71%	49.32%	53.98%	**56.54%**	51.01%	**54.91%**			
	50% data:	54.92%	54.44%	55.53%	**57.75%**	56.33%	55.53%	**58.56%**	57.69%	51.21%	57.41%	56.88%	56.20%	58.90%
	70% data:	57.21%	56.40%	56.67%	58.02%	57.55%	57.28%	**59.23%**	58.76%	53.77%	57.95%			
SVM	20% data:	**56.81%**	44.66%	47.24%	52.63%	**55.80%**	49.73%	54.04%	54.58%	48.92%	55.46%			
	50% data:	**57.55%**	51.55%	52.63%	55.46%	**56.88%**	55.19%	56.47%	57.15%	49.19%	56.27%	57.42%	51.08%	57.35%
	70% data:	57.28%	54.99%	55.66%	56.47%	**57.55%**	**57.55%**	57.08%	57.35%	50.54%	57.35%			

Table 4. Comparison of instance selection methods in regression. The values are mean absolute error and times are in seconds.

		SVD	NMF	PLU	QR	DL	SPCA	SOS	SE	SDM	NR
						Facebook dataset					
	Time:	7.62E−3	1.63E+0	6.51E−3	7.01E−3	1.24E+1	**6.41E−3**	1.40E−2	8.40E−3	**4.81E−3**	×
LR	20% data:	8.09E+3	2.94E+4	1.62E+4	**5.28E+3**	1.57E+4	**5.71E+3**	6.07E+3	1.89E+4	7.50E+3	
	50% data:	6.45E+3	1.15E+4	8.95E+3	**4.85E+3**	8.61E+3	**5.20E+3**	6.25E+3	6.59E+3	5.69E+3	5.81E+3
	70% data:	6.08E+3	8.34E+3	6.90E+3	**4.95E+3**	6.76E+3	5.80E+3	5.90E+3	5.83E+3	**5.30E+3**	
RF	20% data:	8.64E+3	2.80E+4	1.54E+4	**5.53E+3**	1.49E+4	7.08E+3	**6.56E+3**	1.09E+4	6.58E+3	
	50% data:	7.03E+3	1.43E+4	1.04E+4	**5.02E+3**	9.69E+3	7.38E+3	6.76E+3	**6.10E+3**	7.19E+3	6.17E+3
	70% data:	6.76E+3	1.05E+4	7.66E+3	**5.03E+3**	7.52E+3	6.30E+3	6.26E+3	6.32E+3	**6.03E+3**	
MLP	20% data:	1.11E+4	5.37E+4	1.84E+4	**7.11E+3**	1.73E+4	1.66E+4	**7.46E+3**	4.70E+4	2.31E+4	
	50% data:	6.19E+3	1.45E+4	8.80E+3	**4.75E+3**	8.26E+3	**5.67E+3**	6.02E+3	5.62E+3	6.64E+3	5.72E+3
	70% data:	6.19E+3	1.15E+4	7.18E+3	**5.14E+3**	6.81E+3	**5.55E+3**	5.95E+3	5.86E+3	6.00E+3	

the best because its scoring is merely sorting distances. As reported in Tables 2 and 3, we always outperform SOS, SE, ENN, and DROP3 in terms of time. Also, in many cases, we hold the best or the second best positions in accuracy. In some cases, we even outperform no data reduction (NR). Not surprisingly, FLDA performs very well combined with the LDA classifier. Overall, SVD provides very good results which makes sense since it minimizes the reconstruction error. Also, the orthogonal bases of QR-decomposition seem to lead to good performance.

Regression Cases: The results of regression are reported in Table 4. Linear Regression (LR) as an example of linear methods and Random Forest (RF), and Multi-Layer Perceptron (MLP) as examples of non-linear methods are used. The number of trees and maximum depth in RF were both 100. In MLP, the ReLu

170   B. Ghojogh and M. Crowley

Table 5. Comparison of instance selection methods in clustering. Rates are adjusted rand index (best is 1) [27] and times are in seconds.

Isolet dataset

		SVD	NMF	PLU	QR	DL	SOS	SE	SDM	NR
	Time:	5.30E-1	×	**3.14E-1**	3.25E-1	3.62E+3	1.32E+0	3.39E-1	**2.10E-1**	×
K-means	20% data:	3.32E-1	×	3.24E-1	**5.31E-1**	3.67E-1	**4.80E-1**	3.02E-1	2.11E-1	
	50% data:	**4.40E-1**	×	3.95E-1	**4.95E-1**	4.19E-1	**4.95E-1**	4.15E-1	4.22E-1	4.86E-1
	70% data:	4.53E-1	×	4.57E-1	**4.82E-1**	4.66E-1	**4.88E-1**	4.51E-1	4.67E-1	
Birch	20% data:	**3.77E-1**	×	3.23E-1	**5.04E-1**	3.74E-1	**4.70E-1**	3.09E-1	2.29E-1	
	50% data:	4.58E-1	×	4.38E-1	**5.20E-1**	4.12E-1	**4.89E-1**	4.48E-1	3.92E-1	5.13E-1
	70% data:	4.76E-1	×	**4.84E-1**	**5.00E-1**	4.77E-1	4.82E-1	4.57E-1	4.79E-1	

Iris Dataset

		SVD	NMF	PLU	QR	DL	SOS	SE	SDM	NR
	Time:	**1.60E-3**	2.62E-1	**1.32E-3**	**1.60E-3**	9.23E-2	3.70E-3	3.00E-1	**1.90E-3**	×
K-means	20% data:	**6.84E-1**	2.41E-1	1.36E-1	1.32E-1	6.68E-2	**6.92E-1**	5.02E-1	4.89E-1	
	50% data:	**7.80E-1**	4.95E-1	6.21E-1	5.93E-1	**7.05E-1**	**7.38E-1**	4.85E-1	5.77E-1	7.03E-1
	70% data:	7.17E-1	7.18E-1	6.95E-1	6.53E-1	**7.32E-1**	**7.35E-1**	6.27E-1	7.12E-1	
Birch	20% data:	**6.89E-1**	3.77E-1	1.72E-1	1.07E-1	1.80E-1	**6.41E-1**	**5.10E-1**	2.90E-1	
	50% data:	**7.04E-1**	5.03E-1	**6.15E-1**	**6.35E-1**	6.05E-1	5.34E-1	5.11E-1	5.92E-1	5.93E-1
	70% data:	**6.85E-1**	6.16E-1	**6.77E-1**	6.61E-1	5.76E-1	5.79E-1	6.23E-1	6.04E-1	

activation function, ADAM optimizer, and two hidden layers with 100 and 50 neurons were used. Regarding existing methods for numerosity reduction, SOS and SE are not proposed for regression but we compare them here. However, ENN and DROP3 are only applicable to classification. Our method always outperforms SOS and SE in terms of both time and accuracy. QR-decomposition takes the best place probably because of orthogonality. SPCA also is performing well because it captures dependencies of labels and instances using HSIC. SVD also has acceptable result because of both orthogonality and minimum reconstruction error. In all cases, we even outperform NR interestingly.

Clustering Cases: Table 5 summarizes the results for clustering. Two clustering methods, i.e., K-means and Birch are used for experiments. We apply SOS and SE beyond their proposed use of classification. Again, our method always outperforms SOS and SE in ranking time and in most cases, we also have the best performance in comparisons. QR-decomposition is performing well in Isolet dataset. SVD shows promising results in both Isolet and Iris datasets. The possible reasons of good performances of SVD and QR were already explained. DL performs well enough in Iris dataset because it benefits from sparsity.

5 Conclusion

This paper proposed several related algorithms for instance ranking and numerosity reduction using the idea that comparison of instances with basis vectors, or projection directions, show the most informative directions of data and thus the greatest opportunity for data reduction. Our experiments on various datasets showed broad effectiveness of the proposed algorithms.

References

1. Arnaiz-González, Á., Díez-Pastor, J.F., Rodríguez, J.J., García-Osorio, C.: Instance selection of linear complexity for big data. Knowl. Based Syst. **107**, 83–95 (2016)
2. Kalegele, K., Takahashi, H., Sveholm, J., Sasai, K., Kitagata, G., Kinoshita, T.: On-demand data numerosity reduction for learning artifacts. In: 2012 IEEE 26th International Conference on Advanced Information Networking and Applications (AINA), pp. 152–159. IEEE (2012)
3. Garcia, S., Derrac, J., Cano, J., Herrera, F.: Prototype selection for nearest neighbor classification: taxonomy and empirical study. IEEE Trans. Pattern Anal. Mach. Intell. **34**(3), 417–435 (2012)
4. García, S., Luengo, J., Herrera, F.: Data Preprocessing in Data Mining. Springer, Cham (2015). https://doi.org/10.1007/978-3-319-10247-4
5. Carbonera, J.L., Abel, M.: Efficient prototype selection supported by subspace partitions. In: 2017 IEEE 29th International Conference on Tools with Artificial Intelligence (ICTAI), pp. 921–928. IEEE (2017)
6. Wilson, D.L.: Asymptotic properties of nearest neighbor rules using edited data. IEEE Trans. Syst. Man Cybern. **3**, 408–421 (1972)
7. Wilson, D.R., Martinez, T.R.: Reduction techniques for instance-based learning algorithms. Mach. Learn. **38**(3), 257–286 (2000)
8. Skalak, D.B.: Prototype and feature selection by sampling and random mutation hill climbing algorithms. In: Machine Learning Proceedings, pp. 293–301. Elsevier (1994)
9. Liu, C., Wang, W., Wang, M., Lv, F., Konan, M.: An efficient instance selection algorithm to reconstruct training set for support vector machine. Knowl. Based Syst. **116**, 58–73 (2017)
10. Ghojogh, B., Crowley, M.: Principal sample analysis for data reduction. In: 2018 IEEE International Conference on Big Knowledge (ICBK), pp. 350–357. IEEE (2018)
11. Golub, G.H., Reinsch, C.: Singular value decomposition and least squares solutions. Numerische mathematik **14**(5), 403–420 (1970)
12. Stewart, G.W.: On the early history of the singular value decomposition. SIAM Rev. **35**(4), 551–566 (1993)
13. Lee, D.D., Seung, H.S.: Algorithms for non-negative matrix factorization. In: Advances in Neural Information Processing Systems, pp. 556–562 (2001)
14. Biggs, M., Ghodsi, A., Vavasis, S.: Nonnegative matrix factorization via rank-one downdate. In: Proceedings of the 25th International Conference on Machine learning, pp. 64–71. ACM (2008)
15. Vavasis, S.A.: On the complexity of nonnegative matrix factorization. SIAM J. Optim. **20**(3), 1364–1377 (2009)
16. Golub, G.H., Van Loan, C.F.: Matrix Computations. Johns Hopkins University Press, Baltimore (1996)
17. Co, T.B.: Methods of Applied Mathematics for Engineers and Scientists. Cambridge University Press, New York (2013)
18. Mairal, J., Bach, F., Ponce, J., Sapiro, G.: Online dictionary learning for sparse coding. In: Proceedings of the 26th Annual International Conference on Machine Learning, pp. 689–696. ACM (2009)
19. Fefferman, C., Mitter, S., Narayanan, H.: Testing the manifold hypothesis. J. Am. Math. Soc. **29**(4), 983–1049 (2016)

20. Friedman, J., Hastie, T., Tibshirani, R.: The Elements of Statistical Learning. Springer Series in Statistics, vol. 1. Springer, New York (2001)
21. Tenenbaum, J.B., De Silva, V., Langford, J.C.: A global geometric framework for nonlinear dimensionality reduction. Science **290**(5500), 2319–2323 (2000)
22. Roweis, S.T., Saul, L.K.: Nonlinear dimensionality reduction by locally linear embedding. Science **290**(5500), 2323–2326 (2000)
23. Barshan, E., Ghodsi, A., Azimifar, Z., Jahromi, M.Z.: Supervised principal component analysis: visualization, classification and regression on subspaces and submanifolds. Pattern Recognit. **44**(7), 1357–1371 (2011)
24. Dumais, S.T.: Latent semantic analysis. Annu. Rev. Inf. Sci. Technol. **38**(1), 188–230 (2004)
25. Gretton, A., Bousquet, O., Smola, A., Schölkopf, B.: Measuring statistical dependence with Hilbert-Schmidt norms. In: Jain, S., Simon, H.U., Tomita, E. (eds.) ALT 2005. LNCS (LNAI), vol. 3734, pp. 63–77. Springer, Heidelberg (2005). https://doi.org/10.1007/11564089_7
26. Adel, T., Balduzzi, D., Ghodsi, A.: Learning the structure of sum-product networks via an SVD-based algorithm. In: Uncertainty in Artificial Intelligence (UAI), pp. 32–41 (2015)
27. Hubert, L., Arabie, P.: Comparing partitions. J. Classif. **2**(1), 193–218 (1985)

Hybrid Temporal Situation Calculus

Vitaliy Batusov[1][(✉)], Giuseppe De Giacomo[2], and Mikhail Soutchanski[3]

[1] York University, Toronto, Canada
vbatusov@cse.yorku.ca
[2] Sapienza Università di Roma, Rome, Italy
degiacomo@dis.uniroma1.it
[3] Ryerson University, Toronto, Canada
mes@scs.ryerson.ca

Abstract. We present a hybrid discrete-continuous extension of Reiter's temporal situation calculus, directly inspired by hybrid systems in control theory. While keeping to the foundations of Reiter's approach, we extend it by adding a time argument to all fluents that represent continuous change. Thereby, we ensure that change can happen not only because of actions, but also due to the passage of time. We present a systematic methodology to derive, from simple premises, a new group of axioms which specify how continuous fluents change over time within a situation. We study regression for our new hybrid action theories and demonstrate what reasoning problems can be solved. Finally, we show that our hybrid theories indeed capture hybrid automata.

Keywords: Situation calculus · Temporal reasoning · Hybrid systems

1 Introduction

Adding time and continuous change to situation calculus (SC) action theories has attracted a lot of interest over the years. A seminal book [16], refining the ideas of [13], extends situation calculus with continuous time. For each continuous process, there is an action that initiates the process at a moment of time, and there is an action that terminates it. A basic tenet of Reiter's temporal SC is that all changes in the world, including continuous processes such as a vehicle driving in a city or water flowing down a pipe, are the result of named discrete actions. Consequently, in his temporal extension of SC, fluents remain atemporal, while each instantaneous action acquires a time argument. As a side effect of this ontological commitment, continuously varying quantities do not attain values until the occurrence of a time-stamped action. For example, in Newtonian physics, suppose a player kicks a football, sending it on a ballistic trajectory. The question might be: given the vector of initial velocity, when will the ball be within 10% of the peak of its trajectory? In order to answer such questions either a natural, or an exogenous action, depending on the query, has to occur

Supported by the Natural Sciences and Engineering Research Council of Canada.

M.-J. Meurs and F. Rudzicz (Eds.): Canadian AI 2019, LNAI 11489, pp. 173–185, 2019.
https://doi.org/10.1007/978-3-030-18305-9_14

to deem the moment of interest for the query. Thus, before one can answer such questions, one needs the ability to formulate queries about the height of the ball at arbitrary time-points, which is not directly possible without an explicit action with a time argument, if the query is formed over atemporal fluents.

In Reiter's temporal SC, to query about the values of physical quantities in between the actions (agent's or natural), one could opt for an auxiliary exogenous action $watch(t)$ [18], whose purpose is to fix a time-point t to a situation when it occurs, and then pose an atemporal query in the situation which results from executing $watch(t)$. Similarly, one can introduce an exogenous action $waitFor(\phi)$ that is executed at a moment of time when the condition ϕ becomes true, where ϕ is composed of functional fluents that are interpreted as continuous functions of time. This approach has proved to be quite successful in cognitive robotics [8] and was used to provide a SC semantics for continuous time variants of the popular planning language PDDL [3].

In this paper we study a new variant of temporal SC in which we can *directly* query continuously changing quantities at arbitrary points in time without introducing any actions (natural or exogenous or auxiliary) that supply the moment of time. Our approach is query-independent. For doing so we take inspiration from the work on hybrid systems in control theory [4,12], which are based on discrete transitions between states that continuously evolve over time. Following this idea, the crux of our proposal is to add a new kind of axioms called *state evolution axioms* (SEA) to Reiter's successor state axioms (SSA). The SSA specify, as usual, how fluents change when actions are executed. Informally, they characterize transitions between different states due to actions. The state evolution axioms specify how the flow of time can bring changes in system parameters within a given situation while no actions are executed. Thus, we maintain the fundamental assumption of SC that all *discrete* change is due to actions, though situations now include a temporal evolution.

Reiter [16] shows how the SSA can be derived from the effect axioms in normal form by making the causal completeness assumption. We do similar work w.r.t. state evolution axioms, thus providing a precise methodology for axiomatization of continuous processes in SC in the spirit of hybrid systems. One of the key results of SC is the ability to reduce reasoning about a future situation to reasoning about the initial state by means of regression [16]. We show that a suitable notion of regression can be defined despite the continuous evolution within situations.

In hybrid automata, while continuous change is dealt with thoroughly, the discrete description is limited to finite state machines, i.e., it is based on a propositional representation of the state. SC, instead, is based on a relational representation. There are practical examples that call for an extension of hybrid systems where states have an internal relational structure and the continuous flow of time determines the evolution within the state [20]. Our proposal can readily capture these cases, by providing a relational extension to hybrid automata, which benefits from the representational richness of SC. Thus, our work can help to bring together KR and Hybrid Control, getting from the former the semantic richness of relational states and from the latter a convenient treatment of continuous time. The proofs of our theorems are available in [1].

2 Background

Situation Calculus. The situation calculus has three basic sorts (situation, action, object); formulas can be constructed over terms of these sorts. Reiter [16] shows that to solve many reasoning problems about actions, it is convenient to work with SC *basic action theories* (BATs) that have the following ingredients. For each action function $A(\bar{x})$, an *action precondition axiom* (APA) has the syntactic form $Poss(A(\bar{x}), s) \leftrightarrow \Pi_A(\bar{x}, s)$, meaning that the action $A(\bar{x})$ is possible in situation s if and only if $\Pi_A(\bar{x}, s)$ holds in s, where $\Pi_A(\bar{x}, s)$ is a formula with free variables among $\bar{x} = (x_1, \ldots, x_n)$ and s. A situation is a first-order (FO) term describing a unique sequence of actions. The constant S_0 denotes the *initial situation*, the function $do(\alpha, \sigma)$ denotes the situation that results from performing action α in situation σ, and $do([\alpha_1, \ldots, \alpha_n], S_0)$ denotes the situation obtained by consecutively performing $\alpha_1, \ldots, \alpha_n$ in S_0. The notation $\sigma' \sqsubseteq \sigma$ means that either situation σ' is a subsequence of σ or $\sigma = \sigma'$. The abbreviation $executable(\sigma)$ captures situations σ all of whose actions are consecutively possible. Objects are FO terms other than actions and situations that depend on the domain of application. Above, $\Pi_A(\bar{x}, s)$ is a formula *uniform* in situation argument s: it talks only about situation s and uses only domain-specific predicates (see [16]). For each relational fluent $F(\bar{x}, s)$ and each functional fluent $f(\bar{x}, s)$, respectively, a *successor state axiom* (SSA) has the form

$$F(\bar{x}, do(a, s)) \leftrightarrow \Phi_F(\bar{x}, a, s) \quad \text{or} \quad f(\bar{x}, do(a, s)) = y \leftrightarrow \phi_f(\bar{x}, y, a, s),$$

where $\Phi_F(\bar{x}, a, s)$ and $\phi_f(\bar{x}, y, a, s)$ are formulas uniform in s. A BAT $\mathcal{D}$ also contains the *initial theory*: a finite set $\mathcal{D}_{S_0}$ of FO formulas uniform in S_0. Finally, BATs include a set $\mathcal{D}_{una}$ of unique name axioms for actions (UNA). If a BAT has functional fluents, it is required to satisfy an explicit consistency property whereby each functional fluent is always interpreted as a function.

BATs enjoy the *relative satisfiability* property: a BAT $\mathcal{D}$ is satisfiable whenever $\mathcal{D}_{una} \cup \mathcal{D}_{S_0}$ is. This property allows one to disregard the problematic parts of a BAT, like the second order (SO) foundational axioms Σ for situations, when checking satisfiability. BATs benefit from *regression*, a reasoning mechanism for answering queries about the future (thereby solving the *projection* problem). The regression operator $\mathcal{R}$ is defined for sufficiently specific (*regressable*) queries about the future. $\mathcal{R}[\varphi]$ is obtained from a formula φ by a syntactic manipulation (see Defn. 4.7.4 in [16]). By a seminal result in [16], regression reduces SO entailment from a BAT $\mathcal{D}$ to FO entailment by compiling dynamic aspects of the theory into the query.

To accommodate time, Reiter adds a temporal argument to all actions and introduces two special functions: $time(a)$ refers to the time of occurrence of the action a, and $start(s)$ refers to the starting time of situation s, i.e., the time of the latest action of s. The points constituting the timeline with dense linear order are assumed to have the standard interpretation (along with $+$, $<$, etc [16]). To model exogenous events, Reiter develops a theory of *natural actions*—non-agent actions that occur spontaneously as soon as their precondition is satisfied. Such actions are marked using the symbol *natural*, and their semantics are encoded by a modification of *executable(s)*. We use natural actions to induce relational change based on the values of the continuous quantities.

Hybrid Systems. Hybrid automata are mathematical models used ubiquitously in control theory for analyzing dynamic systems which exhibit both discrete and continuous dynamics. [4] define a *basic hybrid automaton* (HA) as a system H consisting of: a finite set Q of *discrete states*; a *transition relation* $E \subseteq Q \times Q$; a *continuous state space* $X \subseteq \mathbb{R}^n$; for each $q \in Q$, a *flow function* $\varphi_q : X \times \mathbb{R} \mapsto X$ and a set $Inv_q \subseteq X$ called the *domain of permitted evolution*; for each $(q, q') \in E$, a *reset relation* $R_{q,q'} \subseteq X \times \mathcal{P}(X)$; a set $Init \subseteq \cup_{q \in Q}(\{q\} \times Inv_q)$ of *initial states*.

Like a discrete automaton, a HA has discrete states and a state transition graph, but within each discrete state its continuous state evolves according to a particular flow, e.g., it can be an implicit solution to a system of differential equations. The domain of permitted evolution delineates the boundaries which the continuous state X of the automaton cannot cross while in state q, i.e., $\varphi_q(X, t) \in Inv_q$. The reset relation helps to model discrete jumps in the value of the continuous state which accompany discrete state switching. A *trajectory* of a hybrid automaton H is a sequence $\eta = \langle \Delta_i, q_i, \nu_i \rangle_{i \in I}$, with $I = \{1, 2, \ldots\}$, where Δ_i is the *duration*, q_i is a state from Q, and $\nu_i : [0, \Delta_i] \mapsto X$ is a continuous curve along the flow φ_{q_i} (refer to [4] for details). A trajectory captures an instance of a legal evolution of a hybrid automaton over time. Duration Δ_i is the time spent by the automaton in the i-th discrete state it reaches while legally traversing the transition graph, obeying the reset relation. A trajectory is *finite* if it contains a finite number $|I|$ of steps and the sum $\Sigma_{i \in I} \Delta_i$ is finite.

3   Hybrid Temporal Situation Calculus

In our quest for a hybrid temporal SC, we reuse the temporal machinery introduced into BATs by Reiter, namely: all actions have a temporal argument and the functions *time* and *start* are axiomatized as before. We preserve atemporal fluents, but no longer use them to model continuously varying physical quantities. Rather, atemporal fluents serve to specify the context in which continuous processes operate. For example, the fluent $Falling(b, s)$ holds if a ball b is in the process of falling in situation s, indicating that, for the duration of s, the position of the ball should be changing as a function of time according to the equations of free fall. The fluent $Falling(b, s)$ may be directly affected by instantaneous actions $drop(b, t)$ (ball begins to fall at the moment t) and $catch(b, t)$ (ball stops at t), but the effect of these actions on the position of the ball comes about only indirectly, by changing the context of a continuous trajectory and thus switching the continuous trajectory that the ball can follow. Thus, a falling ball is one context, and a ball at rest is another. In general, there are finitely many parametrized context types which are pairwise mutually exclusive when their parameters are appropriately fixed, and each context type is characterized by its own continuous function that determines how a physical quantity changes.

To model continuously varying physical quantities, we introduce new functional fluents with a temporal argument. We imagine that these fluents can change with time, and not only as a direct effect of actions. For example, for the context where the ball is falling, the velocity of the ball at time t represented by fluent $vel(b, t, s)$ can be specified as $[Falling(b, s) \land y = vel(b, start(s), s) - g \cdot (t - start(s))] \rightarrow vel(b, t, s) = y$. Notice that this effect axiom does not mention actions and describes the evolution of vel within a single situation.

Deriving State Evolution Axioms. Our starting point is a *temporal change axiom* (TCA) which describes a single law governing the evolution of a particular temporal fluent due to the passage of time in a particular context of an arbitrary situation. An example of a TCA was given above for $vel(b, t, s)$. We assume that a TCA for a temporal functional fluent f has the general syntactic form

$$\gamma(\bar{x}, s) \land \delta(\bar{x}, y, t, s) \rightarrow f(\bar{x}, t, s) = y, \tag{1}$$

where t, s, $\bar{x}$, y are variables and $\gamma(\bar{x}, s)$, $\delta(\bar{x}, y, t, s)$ are formulas uniform in s whose free variables are among those explicitly shown. We call $\gamma(\bar{x}, s)$ the *context*, as it specifies the condition under which the formula $\delta(\bar{x}, y, t, s)$ is to be used to compute the value of fluent f at time t. Note that contexts are time-independent. The formula $\delta(\bar{x}, y, t, s)$ encodes a function (e.g., a solution to the initial value problem for a system of the ordinary differential equations [19]) which specifies y in terms of the values of other fluents at s, t. For each TCA (1) to be *well-defined*, we require that the background theory entails $\gamma(\bar{x}, s) \rightarrow \exists y \, \delta(\bar{x}, y, t, s)$. In other words, whatever the circumstance, the TCA must supply a value for the quantity modelled by f if its context is satisfied. A set of k well-defined temporal change axioms for some fluent f can be equivalently expressed as an axiom of the form (2) below, where $\Phi(\bar{x}, y, t, s)$ is $\bigvee_{1 \leq i \leq k}(\gamma_i(\bar{x}, s) \land \delta_i(\bar{x}, y, t, s))$. For each such axiom, we require that the background theory entails the condition (3).

$$\Phi(\bar{x}, y, t, s) \rightarrow f(\bar{x}, t, s) = y, \tag{2}$$

$$\Phi(\bar{x}, y, t, s) \land \Phi(\bar{x}, y', t, s) \rightarrow y = y'. \tag{3}$$

Condition (3) guarantees the consistency of the axiom (2) by preventing a continuous quantity from having more than one value at any moment of time. Provided (3), we can assume w.l.o.g. that all contexts in the given set of TCA are pairwise mutually exclusive w.r.t. the background theory $\mathcal{D}$.

Having combined all laws which govern the evolution of f with time into a single axiom (2), we can make a causal completeness assumption: *there are no other conditions under which the value of f can change in s from its initial value at $start(s)$ as a function of t.* We capture this assumption formally by the explanation closure axiom

$$f(\bar{x}, t, s) \neq f(\bar{x}, start(s), s) \rightarrow \exists y \, \Phi(\bar{x}, y, t, s). \tag{4}$$

Theorem 1. *Let* $\Psi(\bar{x}, s)$ *denote* $\bigvee_{1 \leq i \leq k} \gamma_i(\bar{x}, s)$. *The conjunction of axioms (2) and (4) in the models of (3) is logically equivalent to*

$$f(\bar{x}, t, s) = y \leftrightarrow [\Phi(\bar{x}, y, t, s) \vee y = f(\bar{x}, start(s), s) \wedge \neg\Psi(\bar{x}, s)]. \tag{5}$$

We call the formula (5) a *state evolution axiom* (SEA) for the fluent f. Note what the SEA says: f evolves with time during s according to some law whose context is realized in s or stays constant if no context is realized. The assumption (4) simply states that all reasons for change have been already accounted for in (2) and nothing is missed. It is important to realize that $\mathcal{D}_{se}$, a set of SEAs, complements the SSAs derived in [16] using a similar technique.

Hybrid Basic Action Theories. The SEA for a temporal fluent f does not completely specify the behaviour of f because it talks only about change within a single situation s. To complete the picture, we need a SSA describing how the value of f changes (or does not change) when an action is performed. A straightforward way to accomplish this would be by an axiom which would enforce continuity, e.g., $f(\bar{x}, time(a), do(a, s)) = f(\bar{x}, time(a), s)$. However, this choice would preclude the ability to model action-induced discontinuous jumps in the value of the continuously varying quantities, such as the sudden change of acceleration from 0 to -9.8m/s^2 when an object is dropped. To circumvent this, for each temporal functional fluent $f(\bar{x}, t, s)$, we introduce an auxiliary atemporal functional fluent $f_{init}(\bar{x}, s)$ whose value in s represents the value of the quantity modelled by f in s at the time instant $start(s)$. We axiomatize f_{init} using a SSA derived from an effect axiom for $f_{init}(\bar{x}, s)$ and a frame axiom of the form $\neg\exists y(e(\bar{x}, y, a, s)) \rightarrow f_{init}(\bar{x}, do(a, s)) = f(\bar{x}, time(a), s)$ stating that if no relevant effect is invoked by the action a, f_{init} assumes the most recent value of f. The SSA for f_{init} has standard syntax and describes how the initial value of f in $do(a, s)$ relates to its value at the same time instant in s (i.e., prior to a). To establish a consistent relationship between temporal fluents and their *init*-counterparts, we require that, in an arbitrary situation, the continuous evolution of each temporal fluent f starts with the value computed for f_{init} by its successor state axiom.

A *hybrid basic action theory* is a collection of axioms $\mathcal{D} = \Sigma \cup \mathcal{D}_{ss} \cup \mathcal{D}_{ap} \cup \mathcal{D}_{una} \cup \mathcal{D}_{S_0} \cup \mathcal{D}_{se}$ such that

1. Every action mentioned in $\mathcal{D}$ is temporal;
2. $\Sigma \cup \mathcal{D}_{ss} \cup \mathcal{D}_{ap} \cup \mathcal{D}_{una} \cup \mathcal{D}_{S_0}$ constitutes a BAT as per Definition 4.4.5 in [16];
3. $\mathcal{D}_{se}$ is a set of SEA of the form $f(\bar{x}, t, s) = y \leftrightarrow \psi_f(\bar{x}, t, y, s)$ where $\psi_f(\bar{x}, t, y, s)$ is uniform in s, such that $\mathcal{D}_{ss}$ contains an SSA for f_{init};
4. For each SEA of the form above, $\mathcal{D}_{una} \cup \mathcal{D}_{S_0}$ entails

$$\forall\bar{x}\forall t. \exists y(\psi_f(\bar{x}, t, y, s)) \wedge \forall y\forall y'(\psi_f(\bar{x}, t, y, s) \wedge \psi_f(\bar{x}, t, y', s) \rightarrow y = y'), \tag{6}$$

$$\exists y(f_{init}(\bar{x}, s) = y \wedge \psi_f(\bar{x}, start(s), y, s)); \tag{7}$$

A set $\mathcal{D}_{se}$ of SEA is *stratified* iff there are no temporal fluents $f_1, \ldots, f_n$ such that $f_1 \succ f_2 \succ \ldots \succ f_n \succ f_1$ where $f \succ f'$ holds iff there is a SEA in $\mathcal{D}_{se}$ where

f appears on the left-hand side and f' on the right-hand side. A hybrid BAT is *stratified* iff its $\mathcal{D}_{se}$ is.

Theorem 2. *A stratified hybrid BAT $\mathcal{D}$ is satisfiable iff $\mathcal{D}_{una} \cup \mathcal{D}_{S_0}$ is.*

Example 1. (See [1] for an illustartion and additional details). Consider a macroscopic urban traffic domain along the lines of [20]. For simplicity, we consider a single intersection of two 2-lane roads. Facing the intersection i are 4 incoming and 4 outgoing road segments. Depending on the traffic light, a car may turn left, turn right, or drive straight from an incoming lane to an outgoing lane. Each lane is denoted by a constant and each path through the intersection i is encoded using the predicates $st(i, r_1, r_2)$ (straight connection from lane r_1 to r_2 at intersection i), $lt(i, r_1, r_2)$ (left turn), and $rt(i, r_1, r_2)$ (right turn). The number of cars per unit of time that can pass through each connection is specified by the function $flow(i, r_1, r_2)$.

The outgoing lanes are of infinite capacity and are not modelled. The traffic lights are controlled by a simple looping automaton with the states $Green(i, r, s)$ (from lane r, go straight or turn right), followed by $RArr(i, r, s)$ (*right arrow*, i.e., only turn right), followed by $Red(i, r, s)$ (stop), and then $LArr(i, r, s)$ (only turn left), such that mutually orthogonal directions are in antiphase to each other. The switching between states for all r is triggered by the action $switch(i, t)$ with precondition $Poss(switch(i, t), s) \leftrightarrow start(s) \leq t$ via a set of simple SSA.

The continuous quantity we wish to model is the number of cars at intersection i queued up in lane r. For that, we use the temporal fluent $que(i, r, t, s)$ and its atemporal counterpart $que_{init}(i, r, s)$. Since the lane r may run dry, we call on the natural action $empty(i, r, t)$ to change the relational state:

$$Poss(empty(i, r, t), s) \leftrightarrow start(s) \leq t \land que(i, r, t, s) = 0,$$
$$a = empty(i, r, t) \land y = 0 \rightarrow que_{init}(i, r, do(a, s)) = y,$$
$$a \neq empty(i, r, t) \land y = que(i, r, time(a), s) \rightarrow que_{init}(i, r, do(a, s)) = y.$$

We can now formulate the TCA for que according to traffic rules. Cars do not move at a red light: $[Red(i, r, s) \land y = que_{init}(i, r, s)] \rightarrow que(i, r, t, s) = y$. When a non-empty lane r sees the left (or right) arrow, its queue decreases linearly with the rate associated with the left (resp., right) turn. For the signal $Green(i, r, s)$, the queue decreases with a combined rate of the straight connection and the right turn, i.e. $y = (que_{init}(i, r, s) - (flow(i, r, r') + flow(i, r, r''))) \cdot (t - start(s))$.

From these TCA, by Theorem 1, we obtain a SEA below (simplified for brevity). Notice that the last line comes not from the TCA but from the explanation closure (4) enforced by Theorem 1 and asserts the constancy of que in the context which the TCA did not cover (movement is allowed but the lane is empty). In general, the modeller only needs to supply the TCA for the contexts where the quantity changes with time.

$que(i, r, t, s) = y \leftrightarrow (\exists \tau \exists q_0 \exists r_L \exists r_S \exists r_R).$
$\quad \tau = (t - start(s)) \wedge q_0 = que_{init}(i, r, s) \wedge lt(i, r, r_L) \wedge st(i, r, r_S) \wedge rt(i, r, r_R) \wedge$
$\quad \big[LArr(i, r, s) \wedge q_0 \neq 0 \wedge y = (q_0 - flow(i, r, r_L) \cdot \tau) \vee$
$\quad Green(i, r, s) \wedge q_0 \neq 0 \wedge y = (q_0 - (flow(i, r, r_S) + flow(i, r, r_R)) \cdot \tau) \vee$
$\quad RArr(i, r, s) \wedge q_0 \neq 0 \wedge y = (q_0 - flow(i, r, r_R) \cdot \tau) \vee$
$\quad Red(i, r, s) \wedge y = q_0 \vee \neg Red(i, r, s) \wedge q_0 = 0 \wedge y = 0 \big].$

4   Regression

Projection is a ubiquitous computational problem concerned with establishing the truth value of a statement after executing a given sequence of actions. We solve it with the help of regression. The notions of uniform and regressable formulas trivially extend to hybrid BATs. The regression operator $\mathcal{R}$ as defined for atemporal BATs in Definition 4.7.4 of [16] can be extended to hybrid BATs in a straightforward way. When $\mathcal{R}$ is applied to a regressable formula W, $\mathcal{R}[W]$ is determined relative to a hybrid BAT. We extend $\mathcal{R}$ as follows.

Let $\mathcal{D}$ be a hybrid BAT, and let W be a regressable formula. If W is a non-fluent atom that mentions $start(do(\alpha, \sigma))$, then $\mathcal{R}[W] = \mathcal{R}[W|_{time(\alpha)}^{start(do(\alpha, \sigma))}]$. If W is a non-$Poss$ atom and mentions a functional fluent uniform in σ, then this term is either atemporal or temporal. The former case is covered by Reiter. In the latter case, the term is of the form $f(\bar{C}, \tau^\star, \sigma)$ and has a SEA $f(\bar{x}, t, s) = y \leftrightarrow \psi_f(\bar{x}, t, y, s)$, so we rename all quantified variables in $\psi_f(\bar{x}, t, y, s)$ to avoid conflicts with the free variables of $f(\bar{C}, \tau^\star, \sigma)$ and define $\mathcal{R}[W]$ to be $\mathcal{R}[\exists y. (\tau^\star = start(\sigma) \wedge y = f_{init}(\bar{x}, \sigma) \vee \tau^\star \neq start(\sigma) \wedge \psi_f(\bar{C}, \tau^\star, y, \sigma)) \wedge W|_y^{f(\bar{C}, \tau^\star, \sigma)}]$, where y is a new variable not occurring free in W, $\bar{C}$, $\tau^\star$, σ. Intuitively, this transformation replaces the temporal fluent f with either the value of f_{init} if f is evaluated at the time of the last action or, otherwise, with the value determined by the right-hand side of the SEA for f.

Theorem 3. *If W is a regressable sentence of SC and $\mathcal{D}$ is a stratified hybrid basic action theory, then $\mathcal{D} \models W$ iff $\mathcal{D}_{S_0} \cup \mathcal{D}_{una} \models \mathcal{R}[W]$.*

Example 2. Let the initial state in the previous example entail the following:

$\quad start(S_0) = 0, \; Red(I, in_1, S_0), \; que_{init}(I, in_1, S_0) = 100,$
$\quad flow(I, in_1, out_2) = 5, \; flow(I, in_1, out_3) = 15, \; flow(I, in_1, out_4) = 10.$

Let W be $que(I, in_1, 3, \sigma) < 95$, i.e., there are fewer than 95 cars in lane in_1 at time 3 in situation σ, where σ is $do([switch(I, 1), switch(I, 2)], S_0)$. In this narrative, the lane in_1 sees the red light, which at $t = 1$ switches to the left arrow, and at $t = 2$ to green. To check if $\mathcal{D} \models W$, we use Theorem 3 to reduce W to an equivalent statement about S_0:

$\mathcal{R}[que(I, in_1, 3, \sigma) < 95] = que_{init}(I, in_1, S_0) - 10(2 - 1) - (15 + 5)(3 - 2) < 95.$

The resulting query can be answered by FO means by plugging 100 for the initial number of cars at in_1: $100 - 10 - 20 = 70 < 95$, so the statement is true.

Regression can also be a powerful diagnostic tool. By analyzing the results of partial regression of a temporal query, one can attribute its validity to a particular action of the narrative. Let $\mathcal{R}^{\sigma'}$ be a variant of $\mathcal{R}$ which does not regress beyond σ'. We can establish whether $\mathcal{R}^{\sigma'}[W]$ is true for each $\sigma' \sqsubseteq \sigma$ as before. In our example, the query holds during and after $switch(I, 2)$ but is false before and at the instant of $switch(I, 1)$. We conclude that the action $switch(I, 1)$ as well as the time that has passed since $t = 1$ up to the time when $\mathcal{R}^{do(switch(I,1),S_0)}[W]$ became true are responsible for the fact that W holds at σ. Note that W can be an arbitrary regressable property of the dynamic system.

5   Comparison with Previous Approaches

Considering that discrete-continuous systems have been a hot topic for decades, it is impossible to fairly compare hybrid situation calculus to a representative subset of all work in that area. Hence, we draw comparisons only to approaches from the same paradigm.

A seminal work by Sandewall [17] points out that discarding information from a theory cannot lead to better inferences. He argues that differential calculus is the perfect language for modelling continuous change and that the essential task in describing physical systems is to provide a logical foundation for discrete state transitions. Pinto [13] presents initial proposals to introduce time into the situation calculus; these works focus on a so-called actual sequence of actions and introduced representation for occurrences of actions w.r.t. an external time-line. Ch. 6 of [13] discusses examples of continuous change and natural events following [17], but without using Sandewall's non-monotonic solution to the frame problem. It also introduces a class of objects called parameters that are used to name continuously varying properties such that each parameter behaves according to a unique function of time during a fixed situation. It is mentioned that parameters can be replaced with functional fluents of time, but this direction was not elaborated. Building on earlier work of [11,13,17] introduces time-independent fluents and situation-independent parameters that can be regarded as functions of time, but provides only an example, and no general methodology. [16] provides the modern axiomatization of time, concurrency, and natural actions in SC. However, [16] allows only atemporal fluents in contrast to [13]. For this reason, [18] proposes an auxiliary action $watch(t)$ (see below).

The example in Sect. 3 helps illustrate the differences with our approach. Consider Reiter's temporal SC [16]: since fluents are atemporal, the TCA above are replaced by effect axioms for the atemporal fluent $que(i, r, s)$, e.g.,

$$a = switch(i, t) \wedge \big[LArr(i, r, s) \wedge que(i, r, s) \neq 0 \wedge \exists r'[lt(i, r, r') \wedge$$
$$y = (que(i, r, s) - flow(i, r, r')\cdot(time(a) - start(s)))]\big] \rightarrow que(i, r, do(a, s)) = y.$$

Note that, in effect axioms, the change in que is associated with a named action. The modeller must replicate this axiom for each action which might affect the context $LArr(i, r, s) \land que(i, r, s) \neq 0$, and likewise for all other contexts and TCA. In our approach, the change in context is handled separately and does not complicate the axiomatization of continuous dynamics. The right-hand side of the resulting SSA, $\gamma_{que}(i, r, y, s) \lor que(i, r, s) = y \land \neg \exists y' \gamma_{que}(i, r, y', s)$, can be obtained from the right-hand side of the SEA above by replacing t with $time(a)$, $que_{init}(i, r, s)$ with $que(i, r, s)$, and the last line by $que(i, r, s) = y \land \neg \exists y' \gamma_{que}(i, r, y', s)$. Notice that the expression $\gamma_{que}(i, r, y, s)$ occurs twice—first due to the effect axiom (in a normal form) and then again due to explanation closure—see examples in Sect. 3.2.6 in [16]. In our approach, only the essential atemporal part of that expression appears. Furthermore, Reiter's version of the precondition axiom for $empty(i, r, t)$ is necessarily cumbersome because it mentions $que(i, r, t, s)$, whose evolution (and thus the value at t) depends on the current relational state of s. Therefore, the modeller must include the right-hand side of the SSA in the precondition, thereby increasing the size of the axioms by roughly the size of the SSA for the continuous fluent F for every occurrence of F in a precondition axiom while not adding any new information. Moreover, since fluents are atemporal, evaluating them at arbitrary moments of time t requires an auxiliary action.

The approach due to [18] introduces the special action $watch(t)$ to advance time to the time-point t. This allows one to access continuous fluents in between the agent actions, but at a cost: replacing $que(i, r, t, s)$ by $que(i, r, do(watch(t), s))$ in the precondition axiom makes the right-hand side non-uniform in s, violates Defn. 4.4.3 in [16], and thus steps outside of the well-studied realm of BATs. A later proposal due to [8] considers fluents whose values range over functions of time, but neither the fluents nor the actions have a temporal argument. Domain actions occur at the same instant as the preceding situation, and the mechanism for advancing time is the special action $waitFor(\phi)$ which simulates the passage of time until the earliest time-point where ϕ holds. Aimed specifically at robotic control, this approach relies on a cc-Golog program to trigger the $waitFor$ action.

Finzi and Pirri [6] introduce *temporal flexible situation calculus*, a dialect aimed to provide formal semantics and a Golog implementation for constraint-based interval planning which requires dealing with multiple alternating time-lines. To represent processes, they introduce fluents with a time argument. However, this time argument marks the instant of the process' creation and is not associated with a continuous evolution.

6   Modelling Hybrid Automata

Hybrid BATs introduced here are naturally suitable for capturing hybrid automata [12]. Given an arbitrary basic hybrid automaton H, c.f., Sect. 2, we proceed as follows. For every discrete state in the set Q, we introduce a constant q_i with $1 \leq i \leq |Q|$ and let $\mathcal{D}_{S_0}$ contain unique name axioms for all q_i. The transition relation E is encoded by a finite set of facts $E(q, q')$. Each flow φ_q is encoded by the function $flow$ such that $flow(q, x, t) = y$ iff $\varphi_q(x, t) = y$. Each set of invariant states Inv_q is encoded by the predicate $Inv(q, x)$ which holds iff $x \in Inv_q$. Each reset relation $R_{q,q'}$ is encoded by the predicate $R(q, q', x, y)$ which holds iff $y \in R_{q,q'}(x)$. The set of initial states $Init$ is encoded by the predicate $Init(q, x)$ which holds iff $(q, x) \in Init$.

Let $Q(s)$ denote the discrete and $X(t, s)$ the continuous state. Let $tr(q, q', y, t)$ be the action representing a transition from state q to q' at time t while resetting the continuous state to the value y. The automaton can be described as

$$Poss(tr(q, q', y, t), s) \leftrightarrow Q(s) = q \land E(q, q') \land R(q, q', X(t, s), y) \land Inv(q', y),$$

$$Q(do(a, s)) = q \leftrightarrow \exists q', y, t(a = tr(q', q, y, t)) \lor Q(s) = q \land \neg \exists q', y, t(a = tr(q, q', y, t)),$$

$$X_{init}(do(a, s)) = x \leftrightarrow \exists q \exists q' \exists t(a = tr(q, q', x, t)),$$

$$X(t, s) = x \leftrightarrow \bigvee_{i=1}^{k} [Q(s) = q_i \land x = flow(q_i, X_{init}(s), t)].$$

Theorem 4. *Let $\mathcal{D}$ be a satisfiable hybrid BAT axiomatizing a hybrid automaton H as above, let σ be an executable ground situation and let $\tau \geq start(\sigma)$. Then*

$$\mathcal{D} \models Init(Q(S_0), X_{init}(S_0)) \land (\forall a, s, t)[do(a, s) \sqsubseteq \sigma \land start(s) \leq t \leq time(a) \lor$$
$$s = \sigma \land start(\sigma) \leq t \leq \tau] \rightarrow Inv(Q(s), X(s, t))$$

if and only if a finite trajectory of H can be uniquely constructed from σ and τ.

Clearly, this axiomatization rules out non-trivial queries about the content of the states because its discrete states are a finite set without objects, relations, etc. A general hybrid BAT does not have this limitation. While classic HA are based on a finite representation of states and atomic state transitions, richer representations began to attract the interest of the hybrid system community. Of particular interest is the work by Platzer [15] based on FO dynamic logic extended to handle differential equations for describing continuous change. Our work contributes to this line of research by providing a very rich representation of the discrete states described relationally in FOL. Both [14] and our paper propose to go beyond finite-state HA. The key advantage of our work is in the availability of situation terms, and therefore, the regression operator. Thus, the usual SC-based reasoning tasks [16] can be solved in our hybrid BATs.

7   Conclusion

Inspired by hybrid systems, we have proposed a temporal extension of SC with a clear distinction between atemporal fluents, responsible for transitions between states, and temporal fluents, representing continuous change within a state. While this paper focuses on semantics, the connection with hybrid systems established here opens new perspectives for future work on automated reasoning as well. In hybrid systems, the practical need for robust specification and verification tools for HA resulted in the development of a multitude of logic-based approaches (see [4] for an overview). More recently, [7] show that certain classes of decision problems belong to reasonable complexity classes. These results provide foundations for verification of robustness in hybrid systems [9]. Platzer's

work offers some decidability results for verification based on quantifier elimi-
nations [14,15]. Note that the quantified differential dynamic logic [14], which
focuses on functions and does not allow for arbitrary relations on objects, cannot
encode SC action theories in an obvious way, i.e., it includes only one primitive
action (assignment), but BATs provide agent actions that can model a system
at a higher level of abstraction. Nevertheless, it may be interesting to study the
reductions of fragments of Golog [10] and BATs with or without continuous time
to such a dynamic logic, to exploit existing [14] and future decidability results.

On the other hand, while research in hybrid systems focuses on certain verification problems, the present paper, due to regression, proposes an approach
to solve other reasoning problems that cannot be formulated in hybrid systems.
Recent work on bounded theories [2,5] provides promising means to study decid-
able cases in this realm, which could be of interest to hybrid systems as well.

References

1. Batusov, V., De Giacomo, G., Soutchanski, M.: Hybrid temporal situation calculus. CoRR 1807.04861 (2018). https://export.arxiv.org/abs/1807.04861
2. Calvanese, D., De Giacomo, G., Montali, M., Patrizi, F.: First-order μ-calculus over generic transition systems and applications to the situation calculus. Inf. Comput. **259**(3), 328–347 (2018)
3. Claßen, J., Hu, Y., Lakemeyer, G.: A situation-calculus semantics for an expressive fragment of PDDL. In: Proceedings of AAAI-2007, Vancouver, British Columbia, Canada, 22–26 July 2007, pp. 956–961. AAAI Press (2007)
4. Davoren, J., Nerode, A.: Logics for hybrid systems (invited paper). Proc. IEEE **88**(7), 985–1010 (2000)
5. De Giacomo, G., Lespérance, Y., Patrizi, F.: Bounded situation calculus action theories. Artif. Intell. **237**, 172–203 (2016)
6. Finzi, A., Pirri, F.: Representing flexible temporal behaviors in the situation cal- culus. In: Proceedings of IJCAI-05, Edinburgh, UK, pp. 436–441 (2005)
7. Gao, S., Avigad, J., Clarke, E.M.: Delta-decidability over the reals. In: Lipovac, V., Scedrov, A. (eds.) Proceedings of LICS-2012, Dubrovnik, Croatia, 25–28 June 2012, pp. 305–314. IEEE Computer Society (2012)
8. Grosskreutz, H., Lakemeyer, G.: cc-Golog – a logical language dealing with con- tinuous change. Log. J. IGPL **11**(2), 179–221 (2003)
9. Kong, S., Gao, S., Chen, W., Clarke, E.: dReach: δ-reachability analysis for hybrid systems. In: Baier, C., Tinelli, C. (eds.) TACAS 2015. LNCS, vol. 9035, pp. 200– 205. Springer, Heidelberg (2015). https://doi.org/10.1007/978-3-662-46681-0_15
10. Levesque, H.J., Reiter, R., Lespérance, Y., Lin, F., Scherl, R.B.: GOLOG: a logic programming language for dynamic domains. J. Log. Program. **31**, 59–83 (1997)
11. Miller, R.: A case study in reasoning about actions and continuous change. In: Wahlster, W. (ed.) Proceedings of ECAI 1996, pp. 624–628 (1996)
12. Nerode, A.: Logic and control. In: Cooper, S.B., Löwe, B., Sorbi, A. (eds.) CiE 2007. LNCS, vol. 4497, pp. 585–597. Springer, Heidelberg (2007). https://doi.org/10.1007/978-3-540-73001-9_61
13. Pinto, J.: Temporal reasoning in the situation calculus. Ph.D. thesis, University of Toronto, Toronto, Canada (1994)

14. Platzer, A.: A complete axiomatization of quantified differential dynamic logic for distributed hybrid systems. Log. Methods Comput. Sci. **8**(4), 1–44 (2012)
15. Platzer, A.: A complete uniform substitution calculus for differential dynamic logic. J. Autom. Reason. **59**(2), 219–265 (2017)
16. Reiter, R.: Knowledge in Action: Logical Foundations for Specifying and Implementing Dynamical Systems. MIT press, Cambridge (2001)
17. Sandewall, E.: Combining logic and differential equations for describing real-world systems. In: Brachman, R.J., Levesque, H.J., Reiter, R. (eds.) Proceedings of KR 1989, Toronto, Canada, 15–18 May 1989, pp. 412–420. Morgan Kaufmann (1989)
18. Soutchanski, M.: Execution monitoring of high-level temporal programs. In: Beetz, M., Hertzberg, J. (eds.) Robot Action Planning, Proceedings of the IJCAI-99 Workshop, Stockholm, Sweden, pp. 47–54 (1999)
19. Teschl, G.: Ordinary differential equations and dynamical systems. AMS (2012). https://www.mat.univie.ac.at/~gerald/ftp/book-ode/ode.pdf
20. Vallati, M., Magazzeni, D., De Schutter, B., Chrpa, L., McCluskey, T.L.: Efficient macroscopic urban traffic models for reducing congestion: a PDDL+ planning approach. In: AAAI, pp. 3188–3194 (2016)

3D Depthwise Convolution: Reducing Model Parameters in 3D Vision Tasks

Rongtian Ye[1], Fangyu Liu[2], and Liqiang Zhang[1(✉)]

[1] Beijing Normal University, Beijing, China
carmoac@mail.bnu.edu.cn, zhanglq@bnu.edu.cn
[2] University of Waterloo, Waterloo, Canada
fangyu.liu@uwaterloo.ca

Abstract. Standard 3D convolution operations usually require larger amounts of memory and computation cost than 2D convolution operations. The fact increases the difficulty of the development of deep neural nets in many 3D vision tasks. In this paper, we investigate the possibility of applying depthwise separable convolutions in 3D scenario and introduce the use of 3D depthwise convolution. A 3D depthwise convolution splits a single standard 3D convolution into two separate steps, which would drastically reduce the number of parameters in 3D convolutions with more than one order of magnitude. We experiment with 3D depthwise convolution on popular CNN architectures and also compare it with a similar structure called pseudo-3D convolution. The results demonstrate that, with 3D depthwise convolutions, 3D vision tasks like classification and reconstruction can be carried out with more light-weighted neural networks while still delivering comparable performances.

Keywords: Depthwise convolution · Low latency models · Pseudo-3D convolution · 3D computer vision · 3D Reconstruction

1 Introduction

3D ConvNets have been widely used in almost all 3D computer vision tasks like 3D scene understanding (segmentation [1,2], classification [3,4], object detection [2,5]), human action/gesture recognition [6,7], video understanding [8,9], medical imaging [10,11], hyperspectral remote sensing images [12]. It is powerful tool to extract high dimensional features from 3D data.

In real world scenarios, such as robotics, self-driving cars, augmented reality, computer-aided medical diagnosis, etc., the tasks need to be carried out on a platform with rather limited computing resources, constraining neural nets to small sizes. This led us to think about the possibility of using light-weighted 3D ConvNets with small amount of parameters. Besides, we have noticed that conventional researches have only explored relatively shallow 3D ConvNets, whose reason may relate to the great time and memory consumption of deep 3D ConvNets experiments. Computation cost is the bottleneck here. The problem becomes

M.-J. Meurs and F. Rudzicz (Eds.): Canadian AI 2019, LNAI 11489, pp. 186–199, 2019.
https://doi.org/10.1007/978-3-030-18305-9_15

seeking a compromise (or balance) between 3D ConvNet's performance and its depth.

The number of parameters grows exponentially when convolution goes from 2D to 3D. For instance, a 2D convolution filter with shape 3×3 only has 9 parameters while a 3D one with the same side length, ie. a $3 \times 3 \times 3$ filter, would have 27 parameters. With even greater side length of filters, the problem becomes much more severe: there will be 125 parameters if the side length is 5. Inspired by MobileNets [13], which used the depthwise separable structure that factorized a standard 2D convolution into a depthwise convolution and a pointwise convolution (a 1×1 convolution), we factorize a standard 3D convolution into: 1. separate convolutions on separate channels; 2. pointwise convolutions on all the channels. In most cases, the number of parameters in convolutional layers is reduced more than 10 times. We do a formal analysis of this in Sect. 2.1 to compare the number of parameters in standard 3D convolution and 3D depthwise convolution.

In fact, there are already works in 3D vision community trying to reduce 3D convolution parameters with some success - to tackle the problem of using 3D ConvNets to learn video spatio-temporal representations (on a large scale), [9] proposed an architecture to reduce reasonably many number of parameters by splitting one $3 \times 3 \times 3$ convolution into a $3 \times 3 \times 1$ convolution and a $1 \times 1 \times 3$ convolution. It is referred as the pseudo-3D convolution. Our work shows a different approach to decompose a standard 3D convolution operation. We compare these two architectures in terms of number of parameters (in Sect. 2.2) and performance (in experiments). The results indicate that 3D depthwise convolution reaches comparable performance with even fewer parameters.

We also experiment 3D depthwise convolution on some off-the-shelf CNN architectures like VGG [14] and residual block [15]. In classification, it is shown that the number of parameters can be significantly reduced with little influence on neural nets' ability of extracting features. Besides classification, 3D depthwise convolution is open to be used in other tasks. We show an example by using 3D depthwise convolution in 3D reconstruction. Using 3D depthwise convolution in reconstruction networks brings several benefits: the original 3D vision tasks now can be carried out with fewer parameters (i.e. models of smaller size); more complex and deeper models with too many parameters are now applicable due to reduction of parameters by 3D depthwise convolution. In experiments, a deeper 3D depthwise ConvNet decoder with fewer parameters beat a relatively shallow standard 3D ConvNet decoder with more parameters.

To conclude, the major contributions of this paper are:

- proposes the use of 3D depthwise convolution to reduce parameters in 3D CNN models and analyzes the computing cost of 3D depthwise convolution in math, along with standard 3D convolution and pseudo-3D convolution;
- investigates the performance of 3D depthwise convolution on popular CNN models and does extensive experiments adopting 3D depthwise convolution in 3D reconstruction networks.

2   Proposed Method

In Sect. 2.1 we formally introduce the 3D depthwise operation. In Sect. 2.2, we state the difference of pseudo-3D and 3D depthwise, then analyze why 3D depthwise does have less parameters. In Sect. 2.3 we explain the use of 3D depthwise in reconstruction task and present the details of reconstruction networks we use.

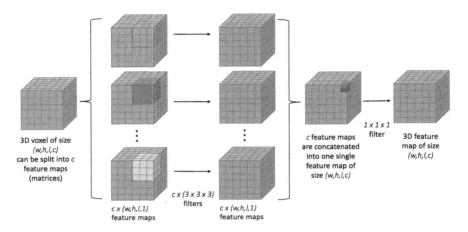

Fig. 1. We split the original 3D voxel grids (or 3D feature matrix) by the last dimension and get c feature maps. Notice that c is the number of channels. Then we use separate filters on separate feature maps. In the figure, it's c separate $3 \times 3 \times 3$ filters on c feature maps. After the filters, we stack all the results (ie. c feature maps) by the last dimension and get a feature map of the same size as the input (assuming paddings). And then we do a one-to-one convolution with filters of size $1 \times 1 \times 1$. It is essentially a 3D fully-connected layer that does an element-wise linear combination on channels at very single voxel. This step is analogous to the *pointwise* convolution in MobileNets.

2.1   3D Depthwise Convolution

Standard 3D Convolution. Given a 3D feature matrix with shape (l, w, h, c), where l, w, h represents length, width, height and c denotes channels, the natural way of doing convolution operation on it would be using a filter with size $k \times k \times k$ where k is the side length of filter, to go over the 3D matrix.

Formally, a standard convolution layer takes an input feature matrix F with shape (l_F, w_F, h_F, c_F) and outputs a feature matrix G of size (l_G, w_G, h_G, c_G). Notice that c_F, c_G are number of channels before and after the convolution. The kernel K here is of size $k \times k \times k \times c_F \times c_G$ where k is side length of the filter.

The output feature matrix for a standard 3D convolution is computed as

$$G_{x,y,z,n} = \sum_{i,j,l,m} K_{i,j,l,m,n} F_{x+i-1,y+j-1,z+l-1,m} \tag{1}$$

where x, y, z and i, j, l denotes voxel's spatial position and m denotes the input channel while n denotes the output channels.

And the computation cost is:

$$k \cdot k \cdot k \cdot c_F \cdot c_G \cdot l_F \cdot w_F \cdot h_F \qquad (2)$$

3D Depthwise Convolution. In 3D depthwise convolution, we decompose one 3D convolution operation into two steps, using two filters: first, apply separate filters for each individual channel; second, use $1 \times 1 \times 1$ filter to apply an pointwise linear combination on feature maps output by the first step. Notice that we apply Batchnorm [16] and ReLU [17] after both steps. The output feature matrix for a 3D depthwise convolution is computed as

$$\hat{G}_{x,y,z,m} = \sum_{i,j,k,m} \hat{K}_{i,j,k,m} F_{x+i-1,y+j-1,z+k-1,m} \qquad (3)$$

where x, y, z and i, j, k again denotes the spatial position of a voxel. $\hat{K}$ is a depthwise convolution kernel of size $k \times k \times k \times c$ (consisting of c filters). The m-th filter in $\hat{K}$ would be applied to the m-th channel in F. And the output of m-th filter becomes the m-th layer in $\hat{G}$. It is thus called depthwise/channelwise.

The computation cost is:

$$k \cdot k \cdot k \cdot c_F \cdot l_F \cdot w_F \cdot h_F \qquad (4)$$

After each channel is depthwisely filtered, it remains to combine them into a single new feature map. We adopt one more pointwise convolution, ie. a $1 \times 1 \times 1$ convolution at every position of feature maps, to carry out a linear combination of layers of all depths. It is essentially fusing the splitted channels back together and activating exchange of information adequately across channels.

The pointwise convolution has a cost of:

$$c_F \cdot c_G \cdot l_F \cdot w_F \cdot h_F \qquad (5)$$

And combining two steps together we get a cost of:

$$k \cdot k \cdot k \cdot c_F \cdot l_F \cdot w_F \cdot h_F + c_F \cdot c_G \cdot l_F \cdot w_F \cdot h_F \qquad (6)$$

Comparing Standard 3D Convolution and 3D Depthwise Convolution. We get a reduction by using (2) dividing (6):

$$\frac{k \cdot k \cdot k \cdot c_F \cdot l_F \cdot w_F \cdot h_F + c_F \cdot c_G \cdot l_F \cdot w_F \cdot h_F}{k \cdot k \cdot k \cdot c_F \cdot c_G \cdot l_F \cdot w_F \cdot h_F}$$

$$= \frac{k \cdot k \cdot k \cdot c_F + c_F \cdot c_G}{k \cdot k \cdot k \cdot c_F \cdot c_G} = \frac{1}{c_G} + \frac{1}{k^3} \qquad (7)$$

Channel size c_G is empirically speaking a large number (usually 32, 64, 128, etc.) which makes $\frac{1}{c_G}$ very small. $\frac{1}{k^3}$ is depending on the side length of kernel. Even when the kernel is small and of side length 2, $\frac{1}{k^3}$ is approaching to 0.1 already. Combining the two very small terms, (7) is easy to get to less than 0.1 which means achieving a more than 10 times parameters reduction.

2.2 Difference from Pseudo-3D Convolution

Using similar paradigm, we can analyze pseudo-3D convolution.

Both pseudo-3D and 3D depthwise convolutions are splitting one single standard 3D convolution into two separate convolutions. But the splitting philosophy is different as suggested in Figs. 1 and 2, which leads to very different number of parameters and behavior.

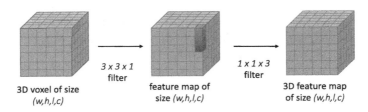

3 x 3 x 1
filter

1 x 1 x 3
filter

3D voxel of size
(w,h,l,c)

feature map of
size (w,h,l,c)

3D feature map
of size (w,h,l,c)

Fig. 2. The orange filter of size $1 \times 3 \times 3$ does a horizontal convolution (step one) and the red filter of size $3 \times 1 \times 1$ does a vertical convolution (step 2). Combining them together we have the pseudo-3D convolution. (Color figure online)

The first step of pseudo-3D convolution, which is basically a horizontal convolution, can be computed as

$$\hat{G}_{x,y,z,m} = \sum_{i,j,m} \hat{K}_{i,j,m,m} F_{x+i-1,y+j-1,z,m} \tag{8}$$

where x, y, z, i, j, m are all analogous to previous definitions.
It has a computation cost of:

$$k \cdot k \cdot c_F \cdot c_F \cdot l_F \cdot w_F \cdot h_F \tag{9}$$

The second step can be seen as a vertical convolution can be computed as:

$$G_{x,y,z,n} = \sum_{k,m} \hat{K}_{k,m,n} F_{x,y,z+k-1,m} \tag{10}$$

It has a computation cost of:

$$k \cdot c_F \cdot c_G \cdot l_F \cdot w_F \cdot h_F \tag{11}$$

The total computation cost of a pseudo-3D convolution is:

$$k \cdot k \cdot c_F \cdot c_F \cdot l_F \cdot w_F \cdot h_F + k \cdot c_F \cdot c_G \cdot l_F \cdot w_F \cdot h_F \tag{12}$$

To compare the parameters of 3D depthwise convolution and pseudo-3D convolution, we divide (6) by (12):

$$\frac{k \cdot k \cdot k \cdot c_F \cdot l_F \cdot w_F \cdot h_F + c_F \cdot c_G \cdot l_F \cdot w_F \cdot h_F}{k \cdot k \cdot c_F \cdot c_F \cdot l_F \cdot w_F \cdot h_F + k \cdot c_F \cdot c_G \cdot l_F \cdot w_F \cdot h_F}$$

$$= \frac{k^3 + c_G}{k^2 \cdot c_F + k \cdot c_G} \approx \frac{k}{c_F} \tag{13}$$

Comparing to channel size c_F, kernel side length k is usually really small (e.g. mostly likely to be 3 or 5 or 7). 3D depthwise, who is breaking the chain of multiplication by depth/channel, gains an edge over pseudo-3D, who is breaking the chain by kernel side length.

We also compare our proposed method with the pseudo 3D convolution idea in experiments, finding that the 3D depthwise achieves comparable performance with even fewer parameters.

2.3 Use 3D Depthwise Convolution in Reconstruction Networks

In this section, we state the use of 3D depthwise convolution in 3D reconstruction task.

3D reconstruction nets can be seen as autoencoders. For single view reconstruction, the input of neural net is an image, the output is a 3D voxel grid depicting the object. With (image, voxel grids) pairs as training data, we can do supervised learning on the autoencoder.

The autoencoder consists of an encoder, which is a 2D ConvNet to encode image into its vector representation, and a decoder, which is a upsampling 3D ConvNet to decode the encoded vector into 3D voxel grids. The general pipeline can be found in Fig. 3.

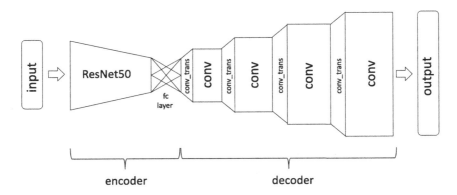

Fig. 3. A pipeline of the reconstruction networks.

In our setting, we use ResNet50 as the image encoder. Specifically, we extract the output of average pooling layer in ResNet50, which would be a 1-D vector with length 2048. Then, we use a fully-connected layer to map the vector into a 1024-length 1-D vector. We regard this vector as the output of encoder and input of decoder. Decoder is where we test the use of 3D depthwise convolution. The decoding steps are:

- use a transposed convolution kernel of size $4 \times 4 \times 4$ to map the 1-D vector to a feature matrix of size $(4, 4, 4, 256)$ where 256 is the number of channels;
- do several similar transposed 3D convolution to map the feature matrix to the size of $8 \times 8 \times 8 \times 128$, $16 \times 16 \times 16 \times 64$, then $32 \times 32 \times 32 \times 32$;
- use kernel of size $1 \times 1 \times 1$ to reduce number of channels back to 1 and get the $32 \times 32 \times 32 \times 1$ feature matrix we desired.

For the decoder, we designed convolution blocks with and without residual structure as suggested in Fig. 4. We tested their performance by using standard 3D convolution, 3D depthwise convolution, as well as pseudo-3D convolution. Experiment results of comparing the combinations of {residual block, regular block} × {standard 3D convolution, pseudo-3D convolution, 3D depthwise convolution} can be seen in Sect. 3.

(a) Regular convolution block.   (b) Residual convolution block.

Fig. 4. Two different blocks used in reconstruction networks. (a) is used in all Rec nets while (b) is used in all ResRec nets.

3   Experiments

3.1   Dataset Setup

We use ShapeNetCore [18] as our experiment dataset. To be consistent with some of our peers' works which we are comparing results with, we use the exact same dataset as [19]. The set contains $44,000$ 3D models and 13 categories with a train/val/test split of $[0.7, 0.1, 0.2]$. We use the $64 \times 64 \times 64$ resolution voxel grids for classification experiments and $32 \times 32 \times 32$ ones for reconstruction.

3.2   Classification

We start with the most common task - classification. We compare the performance of standard 3D convolution, pseudo-3D convolution and 3D depthwise convolution by applying them on VGGs [14] of different number of layers.

Training. We train every model for 20 epochs. The first 10 uses a learning rate of 10^{-5} and the second 10 uses 10^{-6}. Batch size is set to 8. Models take 3D voxel grids of size $(64, 64, 64)$ as input and output a vector of length 13 as prediction. Every element in the 13-d vector represents the probability of the input belongs to a corresponding class of object. And we use Cross Entropy as loss function.

Results. We found that standard 3D convolution, pseudo-3D convolution, 3D depthwise convolution's performances are comparable in this task. As suggested in Table 1, the parameters of convolution layers are reduced by more than 95% in all three VGGs. And the total numbers of parameters are reduced by 40%–60%. In these VGGs, the fully-connected layer takes 33, 579, 021 parameters and contributes to most of the parameters in the whole network. Depthwise convolutional layers are only using very minimum parameters comparing to it.

Table 1. Comparison of 3D depthwise convolution and standard 3D convolution on VGG in applications of classification task. "dw" is short for depthwise.

Method	Accuracy	# parameters in conv layers	Reduced by	# parameters in total	Reduced by
3D VGG13	95.11	28,261,824		61,840,845	
pseudo-3D VGG13	95.12	11,823,819	58.16%	45,402,840	26.58%
3D dw VGG13	95.10	1,174,237	**95.85%**	34,753,258	**43.80%**
3D VGG16	95.21	44,189,632		77,768,653	
pseudo-3D VGG16	94.93	18,906,827	57.21%	52,485,848	32.51%
3D dw VGG16	94.26	1,803,741	**95.92%**	35,382,762	**54.50%**
3D VGG19	94.95	60,117,440		93,696,461	
pseudo-3D VGG19	94.61	25,989,835	56.77%	59,568,856	36.42%
3D dw VGG19	94.71	2,433,245	**95.95%**	36,012,266	**61.56%**

3.3   3D Reconstruction

Previous works that focus on 3D reconstruction usually use several regular convolutions together with transpose convolutions in decoders to map encoded vectors back to 3D voxel [20–22]. The major constraint of applying more complex structure here is that 3D ConvNets have plenty of parameters, which makes deepening them very hard. For instance, to enhance the capability of decoder (the mapping from image encodings to 3D voxel), 3D-R2N2 [23] uses a residual block based architecture. However, due to the massive amount of parameters in the 3D neural nets, its depth (only 10 layers) is still not enough comparing to ResNets in 2D scenarios [15]. So, reducing number of parameters is the key to come up with a better (deeper) decoder here.

Figure 3 shows the pipeline of the reconstruction network we are experimenting on. We use identical encoders (ResNet50) for all experiments and test different decoder architectures, both with and without 3D depthwise, as suggested in Table 2.

Metrics. We use the mean Intersection-over-Union (mIoU) between ground truth and the models' predicted 3D voxel grids to evaluate performance of 3D reconstruction. Specifically, for an individual voxel grid, the mIoU is defined as

$$\text{mIoU} = \frac{\sum_{i,j,k}[I(p_{(i,j,k)}>t) \cdot I(y_{(i,j,k)})]}{\sum_{i,j,k}[I(I(p_{(i,j,k)}>t) + I(y_{(i,j,k)}))]} \tag{14}$$

where $I(\cdot)$ is an indicator function, $y_{(i,j,k)}$ is the ground truth at position (i,j,k) while $p_{(i,j,k)}$ is the prediction of it and t is a threshold. The higher the mIoU, the better the 3D reconstruction model is.

Training. We train every model for 120 epochs. The first 60 uses a learning rate of 10^{-6} and the second 60 uses 10^{-7}. Batch size is set to 32. Models take images of size $(224, 224)$ as input and output prediction (3D voxel) of size $(32, 32, 32)$. We use voxel-wise CrossEntropy as loss function. To evaluate a model, we set a threshold for binarizing every voxel, ie. choosing t in Eq. (14). We tested every one of $\{0.1, 0.3, 0.5, 0.7, 0.9\}$ and found that models generally have the highest mIoU when $t = 0.3$. For consistency, all experiments are using $t = 0.3$ as threshold.

Results. Table 3 shows that:

- With same number of layers, 3D depthwise loses ∼2.2% mIoU on average while pseudo-3D looses ∼1.7% on average. But 3D depthwise reduces significantly more parameters than pseudo-3D.
- A deeper 3D FCN with 3D depthwise convolution (ResRec-16 dw) achieves better accuracy with fewer parameters (# params in decoder: $17, 533, 792$) than a shallow standard FCN (Rec-6) with more parameters (# params in decoder: $21, 768, 928$) as suggested in Table 3.

Table 4 demonstrates the full quantitative results of different reconstruction networks, including two prior works and the combinations of {residual block, regular block} × {standard 3D convolution, pseudo-3D convolution, 3D depthwise convolution}. We notice that Rec-16, who has the largest depth and most parameters, also has the best performance.

We also show some qualitative results of using all 3 types of 3D convolutions in Fig. 5. Interestingly, although their quantitative performances are almost the same, different types of convolutions yields different styles of reconstructions. Pseudo-3D delivers smooth surfaces but tends to ignore some details. 3D depthwise generally produces rough surfaces however keeps the details better. Standard 3D convolution seems to be a trade-off solution for balancing smoothness and details.

Table 2. Architectures of reconstruction nets. Parameters written in the format of (kernel_size,output_channel_size); a pair of brackets presents a convolution block. "Res" means with residual block; "Rec" means reconstruction net; * means it's a residual block.

Layer	Output	Rec-6	ResRec-6	Rec-16	ResRec-16
Encoder	(1,1,1,2048)	ResNet50 (avgpool layer out)			
fc	(1,1,1,1024)	Fully connected layer			
convtrans1	(4,4,4,256)	conv_transpose $4 \times 4 \times 4, 256$, stride 1			
conv1		$[3 \times 3 \times 3, 256;$ $3 \times 3 \times 3, 256]$ $\times 1$	$[3 \times 3 \times 3, 256;$ $3 \times 3 \times 3, 256]$ $\times 1, *$	$[3 \times 3 \times 3, 256;$ $3 \times 3 \times 3, 256]$ $\times 2$	$[3 \times 3 \times 3, 256;$ $3 \times 3 \times 3, 256]$ $\times 2, *$
convtrans2	(8,8,8,128)	conv_transpose $2 \times 2 \times 2, 128$, stride 2			
conv2		$[3 \times 3 \times 3, 128;$ $3 \times 3 \times 3, 128]$ $\times 1$	$[3 \times 3 \times 3, 128;$ $3 \times 3 \times 3, 128]$ $\times 1, *$	$[3 \times 3 \times 3, 128;$ $3 \times 3 \times 3, 128]$ $\times 2$	$[3 \times 3 \times 3, 128;$ $3 \times 3 \times 3, 128]$ $\times 2, *$
convtrans3	(16,16,16,64)	conv_transpose $2 \times 2 \times 2, 64$, stride 2			
conv3		$[3 \times 3 \times 3, 64;$ $3 \times 3 \times 3, 64]$ $\times 1$	$[3 \times 3 \times 3, 64;$ $3 \times 3 \times 3, 64]$ $\times 1, *$	$[3 \times 3 \times 3, 64;$ $3 \times 3 \times 3, 64]$ $\times 2$	$[3 \times 3 \times 3, 64;$ $3 \times 3 \times 3, 64]$ $\times 2, *$
convtrans4	(32,32,32,32)	conv_transpose $2 \times 2 \times 2, 32$, stride 2			
conv4				$[3 \times 3 \times 3, 32;$ $3 \times 3 \times 3, 32]$ $\times 2$	$[3 \times 3 \times 3, 32;$ $3 \times 3 \times 3, 32]$ $\times 2, *$
conv5	(32,32,32,1)	$1 \times 1 \times 1, 1$, stride 1			

Table 3. Comparing reconstruction results of using different 3D convolutions in various decoder structures. "Res" means with residual block; "dw" is short for depthwise.

Method	mIoU	# param in 3D conv layers	Reduced by	# param in decoder	Reduced by
Rec-6	61.4	4,646,656		21,768,928	
Rec-6 pseudo	59.1	2,067,968	55.50%	19,190,240	11.85%
Rec-6 dw	58.1	201,600	**95.66%**	17,323,872	**20.42%**
ResRec-6	61.7	4,646,656		21,768,928	
ResRec-6 pseudo	61.0	2,067,968	55.50%	19,190,240	11.85%
ResRec-6 dw	60.4	201,600	**95.66%**	17,323,872	**20.42%**
Rec-16	63.4	9,404,160		26,526,432	
Rec-16 pseudo	61.2	4,185,600	55.49%	21,307,872	19.67%
Rec-16 dw	60.9	411,520	**95.62%**	17,533,792	**33.90%**
ResRec-16	63.1	9,404,160		26,526,432	
ResRec-16 pseudo	61.7	4,185,600	55.49%	21,307,872	19.67%
ResRec-16 dw	61.5	411,520	**95.62%**	17,533,792	**33.90%**

Table 4. Full results of all kinds of reconstruction networks. "Res" means with residual block; "dw" is short for depthwise; "ps" is short for pseudo.

Method	mIoU	Aero	Bench	Cabinet	Car	Chair	Display	Lamp	Speaker	Rifle	Sofa	Table	Phone	Vessel
3D-R2N2 [23]	55.1	56.7	43.2	61.8	77.6	50.9	44.0	40.0	56.7	56.5	58.9	51.6	65.6	53.1
V-lsm [19]	61.5	61.7	50.8	65.9	79.3	57.8	57.9	48.1	63.9	69.7	67.0	55.6	67.6	58.3
Rec-6	61.4	59.7	55.5	72.4	79.3	55.9	51.5	44.8	63.6	60.1	66.8	58.7	71.2	58.3
Rec-6 ps	59.1	56.9	52.6	69.5	76.4	54.2	49.7	44.5	63.4	58.1	65.6	56.2	65.4	55.4
Rec-6 dw	58.1	54.7	50.3	67.9	77.8	53.0	49.1	41.8	62.0	57.5	63.7	54.4	67.9	55.7
ResRec-6	61.7	60.7	57.3	71.9	79.2	56.8	52.5	44.9	62.1	60.2	68.2	59.1	70.9	58.4
ResRec-6 ps	61.0	59.2	54.5	71.8	79.5	55.5	50.6	44.9	63.8	59.8	67.2	57.9	69.9	57.9
ResRec-6 dw	60.4	57.8	52.5	71.1	78.9	55.1	51.5	43.5	63.3	59.7	66.1	56.9	71.7	57.7
Rec-16	**63.4**	63.3	59.2	74.2	80.0	57.7	54.4	46.7	64.7	62.0	69.5	60.0	74.1	58.1
Rec-16 ps	61.2	58.8	54.8	71.9	79.2	55.1	53.2	43.4	62.9	60.1	67.4	57.9	73.4	57.5
Rec-16 dw	60.9	60.5	52.9	71.2	79.6	55.5	50.6	40.6	65.8	61.3	68.4	57.2	70.5	58.3
ResRec-16	63.1	59.2	58.1	74.6	79.1	58.1	54.4	47.2	66.0	60.6	69.4	61.2	74.3	58.3
ResRec-16 ps	61.7	60.6	56.3	72.3	79.3	56.8	52.1	45.5	64.0	61.0	68.2	58.4	70.0	57.9
ResRec-16 dw	61.5	59.3	55.2	73.1	79.1	55.1	52.8	43.7	64.5	59.2	68.2	58.9	72.4	57.5

Table 5. Performances of 3D-R2N2s using different decoders. We also copy results for Rec-16 and ResRec-16 in last two lines for convenience of comparison.

Encoder	Decoder	mIoU
3D-R2N2 encoder	3D-R2N2 decoder	55.1
3D-R2N2 encoder	ResRec-16 decoder	59.5
ResNet50	Rec-16 decoder	63.4
ResNet50	ResRec-16 decoder	63.1

Depth Matters in Decoders. We notice that 3D-R2N2 [23], which was the state-of-the-art at that time, has a lower mIoU comparing to all other seemingly concise models. We show in experiments that model's depth has a major influence on decoder's performance. Both using residual blocks, the ResRec-16 decoder has more layers than 3D-R2N2 decoder. Keeping the encoder unchanged, by switching to the ResRec-16 decoder, the modified 3D-R2N2's mIoU raises more than 4% as suggested in Table 5.

Though we've seen significant improvement with the ResRec-16 and Rec-16 decoder in Table 5, the modified 3D-R2N2s' performances are still almost 4% behind ResRec-16 and Rec-16, leading to the conclusion that ResNet50 is also a better encoder than 3D-R2N2's encoder. Notice that the 3D-R2N2 encoder does intergrate residual block into it. But it's shallow comparing to ResNet50 and it's not pretrained on ImageNet.

This result confirms our intuition that, in current stage, we do need deeper 3D ConvNets to conduct 3D vision tasks. And 3D depthwise convolution can help us build deeper 3D ConvNets under when computing resources are constraining the model size.

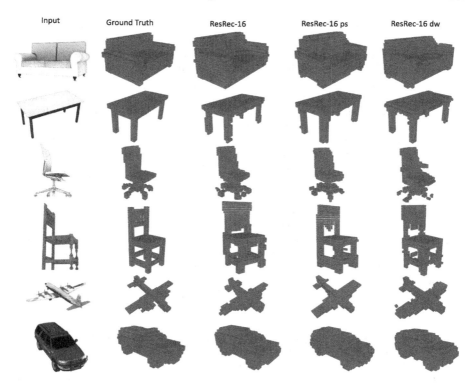

Fig. 5. Qualitative results of reconstruction using ResRec with standard, pseudo and depthwise 3D convolutions.

4   Conclusion

In this paper, we have proposed the use of 3D depthwise convolutions which improves 3D ConvNets' efficiency by reducing the number of parameters. We've shown that 3D depthwise convolution operation has comparable performance to standard 3D convolution operations in classification task with ∼95% fewer parameters. We've also demonstrated the potential of using it to improve 3D neural nets' performance in other 3D vision tasks like 3D Reconstruction and got decent results with minimum number of parameters in convolutional layers. And the experiment has further indicated that deeper 3D ConvNets are indeed needed in 3D vision tasks, where 3D depthwise convolution can help.

References

1. Tchapmi, L., Choy, C., Armeni, I., Gwak, J., Savarese, S.: SEGCloud: semantic segmentation of 3D point clouds. In: 2017 International Conference on 3D Vision (3DV), pp. 537–547. IEEE (2017)
2. Liu, F., et al.: 3DCNN-DQN-RNN: a deep reinforcement learning framework for semantic parsing of large-scale 3D point clouds. In: IEEE International Conference on Computer Vision (ICCV), pp. 5679–5688 (2017)
3. Maturana, D., Scherer, S.: Voxnet: A 3D convolutional neural network for real-time object recognition. In: 2015 IEEE/RSJ International Conference on Intelligent Robots and Systems (IROS), pp. 922–928. IEEE (2015)
4. Hegde, V., Zadeh, R.: FusionNet: 3D object classification using multiple data representations. arXiv preprint arXiv:1607.05695 (2016)
5. Song, S., Xiao, J.: Sliding shapes for 3D object detection in depth images. In: Fleet, D., Pajdla, T., Schiele, B., Tuytelaars, T. (eds.) ECCV 2014. LNCS, vol. 8694, pp. 634–651. Springer, Cham (2014). https://doi.org/10.1007/978-3-319-10599-4_41
6. Ji, S., Xu, W., Yang, M., Yu, K.: 3D convolutional neural networks for human action recognition. IEEE Trans. Pattern Anal. Mach. Intell. **35**(1), 221–231 (2013)
7. Molchanov, P., Gupta, S., Kim, K., Kautz, J.: Hand gesture recognition with 3D convolutional neural networks. In: Proceedings of the IEEE Conference on Computer Vision and Pattern Recognition Workshops, pp. 1–7 (2015)
8. Tran, D., Bourdev, L., Fergus, R., Torresani, L., Paluri, M.: Learning spatiotemporal features with 3D convolutional networks. In: Proceedings of the IEEE International Conference on Computer Vision, pp. 4489–4497 (2015)
9. Qiu, Z., Yao, T., Mei, T.: Learning spatio-temporal representation with pseudo-3D residual networks. In: 2017 IEEE International Conference on Computer Vision (ICCV), pp. 5534–5542. IEEE (2017)
10. Milletari, F., Navab, N., Ahmadi, S.A.: V-NET: Fully convolutional neural networks for volumetric medical image segmentation. In: 2016 Fourth International Conference on 3D Vision (3DV), pp. 565–571. IEEE (2016)
11. Çiçek, Ö., Abdulkadir, A., Lienkamp, S.S., Brox, T., Ronneberger, O.: 3D U-Net: learning dense volumetric segmentation from sparse annotation. In: Ourselin, S., Joskowicz, L., Sabuncu, M.R., Unal, G., Wells, W. (eds.) MICCAI 2016. LNCS, vol. 9901, pp. 424–432. Springer, Cham (2016). https://doi.org/10.1007/978-3-319-46723-8_49
12. Li, Y., Zhang, H., Shen, Q.: Spectral–spatial classification of hyperspectral imagery with 3D convolutional neural network. Remote. Sens. **9**(1), 67 (2017)
13. Howard, A.G., et al.: MobileNets: efficient convolutional neural networks for mobile vision applications. arXiv preprint arXiv:1704.04861 (2017)
14. Simonyan, K., Zisserman, A.: Very deep convolutional networks for large-scale image recognition. ICLR 2015 abs/1409.1556 (2014)
15. He, K., Zhang, X., Ren, S., Sun, J.: Deep residual learning for image recognition. In: Proceedings of the IEEE Conference on Computer Vision and Pattern Recognition, pp. 770–778 (2016)
16. Ioffe, S., Szegedy, C.: Batch normalization: accelerating deep network training by reducing internal covariate shift. In: International Conference on Machine Learning, pp. 448–456 (2015)
17. Le, Q.V., Jaitly, N., Hinton, G.E.: A simple way to initialize recurrent networks of rectified linear units. arXiv preprint arXiv:1504.00941 (2015)

18. Chang, A.X., et al.: ShapeNet: an information-rich 3D model repository. arXiv preprint arXiv:1512.03012 (2015)
19. Kar, A., Häne, C., Malik, J.: Learning a multi-view stereo machine. In: Advances in Neural Information Processing Systems, pp. 365–376 (2017)
20. Girdhar, R., Fouhey, D.F., Rodriguez, M., Gupta, A.: Learning a predictable and generative vector representation for objects. In: Leibe, B., Matas, J., Sebe, N., Welling, M. (eds.) ECCV 2016. LNCS, vol. 9910, pp. 484–499. Springer, Cham (2016). https://doi.org/10.1007/978-3-319-46466-4_29
21. Wu, J., Zhang, C., Xue, T., Freeman, B., Tenenbaum, J.: Learning a probabilistic latent space of object shapes via 3D generative-adversarial modeling. In: Advances in Neural Information Processing Systems, pp. 82–90 (2016)
22. Yan, X., Yang, J., Yumer, E., Guo, Y., Lee, H.: Perspective transformer nets: learning single-view 3D object reconstruction without 3D supervision. In: Advances in Neural Information Processing Systems, pp. 1696–1704 (2016)
23. Choy, C.B., Xu, D., Gwak, J.Y., Chen, K., Savarese, S.: 3D-R2N2: a unified approach for single and multi-view 3D object reconstruction. In: Leibe, B., Matas, J., Sebe, N., Welling, M. (eds.) ECCV 2016. LNCS, vol. 9912, pp. 628–644. Springer, Cham (2016). https://doi.org/10.1007/978-3-319-46484-8_38

Identifying Misaligned Spans in Parallel Corpora Using Change Point Detection

Andrea Pagotto[1], Patrick Littell[2], Yunli Wang[2], and Cyril Goutte[2(✉)]

[1] Carleton University, Ottawa, ON K1S 5B6, Canada
andreapagotto@cunet.carleton.ca
[2] National Research Council Canada, Ottawa, ON K1A 0R6, Canada
{patrick.littell,yunli.wang,cyril.goutte}@nrc.ca

Abstract. Parallel corpora are the basic resource for many multilingual natural language processing models. Recent advances in, e.g. neural machine translation have shown that the quality of the alignment in the corpus has a crucial impact on the quality of the resulting model, renewing interest in filtering automatically aligned corpora to increase their quality. In this contribution, we investigate the use of a fast change point detection method to detect possibly problematic parts of a parallel corpus. We demonstrate its performance on German-English corpora of 11k and 31k sentences, achieve a boundary identification performance above 80% and improve the detection of genuine parallel sentences up to 88%. To our knowledge this is the first application of change point detection to the problem of error detection in noisy corpora.

Keywords: Parallel corpus · Misalignement · Change point detection

1 Introduction

It has become increasingly clear that neural machine translation is highly sensitive to noisy data [3,5,13,16], compared to the relative robustness of statistical machine translation [7,12]. Thus, methods for identifying misaligned and otherwise inappropriate data in parallel corpora have become increasingly important to clean up large, automatically-acquired and automatically-aligned bilingual corpora. This need has elevated the task of *corpus filtering* to a research problem in its own right; for example, the shared task on Parallel Corpus Filtering [15] at the third conference on Machine Translation (WMT18) challenged teams to filter a very noisy web-crawled parallel German-English corpus [14]. It attracted 18 among the top machine translation (MT) teams worldwide and showed how proper filtering of the training data can dramatically impact the performance of the resulting MT systems.

Even the current-best filtering systems [11,18,19,21] consider each sentence pair as if its quality is independent of neighboring sentences. However, parallel data is often misaligned over a span of sentences [7,12], e.g. an off-by-one error

© Crown 2019
M.-J. Meurs and F. Rudzicz (Eds.): Canadian AI 2019, LNAI 11489, pp. 200–211, 2019.
https://doi.org/10.1007/978-3-030-18305-9_16

that corrupts several sentence pairs before being corrected, or spurious paragraphs (or even entire documents) that should not have been aligned to each other in the first place. The motivation for this work is that although sentence segmentation and alignment have reached a high level of performance [17,24], no amount of clever alignment can produce adequate sentence pairs from a pair of documents or parts of documents that are not translations of each other, a common situation when bilingual material is, for example, crawled from the web. Since corpus errors can occur over spans of sentence pairs, detecting them can be viewed as a time series problem, not merely a problem of considering sentence pairs as independent observations. Given a sequence of measurements of sentence-pair alignment in a parallel corpus, can we detect the points at which misaligned spans begin and end?

The investigation that follows concerns a narrow but novel intermediate question in the corpus-filtering pipeline: not how to measure the quality of sentence pairs (we use a classic measure of sentence alignment, Sect. 3.1), nor how best to clean up detected errors (we simply perform 30% thresholding according to the WMT18 challenge rules), but how to *interpret* the measurements. Do we consider each measurement to be directly indicative of sentence alignment quality, or do we consider each to be a noisy measurement of an underlying alignment quality that persists over a span of sentence pairs? This paper investigates the idea that using an estimation of the quality of underlying spans of unknown length, in place of the original measurements, will model corpus alignment errors better and thus can aid in the detection of misalignments for later cleaning.

Detecting changes in sequences of noisy measurements, also known as *change point detection*, is a well-studied topic in time series processing [2,6,10], but has not to our knowledge been applied to the problem of parallel data cleaning, or to machine translation problems in general. A change point is a point in a time series where the generative parameters underlying the time series data change; for example, a change in the fundamentals of a stock changing its behaviour [22], or a goal in a soccer game, suddenly changing the estimated sentiment scores of Twitter reactions to the game [8]. In our problem, the time series in question is a sequence of sentence pairs with corresponding alignment scores, and a change point represents when the pair of sentences either becomes misaligned or returns to well-aligned (e.g., when an off-by-one alignment error begins or ends, or when a completely misaligned document pair begins or ends).

In the following section, we introduce several change point detection methods and change point extraction approaches. In Sect. 3, we describe the two datasets we used in our experiments, while Sect. 4 reports the results we obtained on these datasets for two tasks: we identify the boundaries of text chunks artificially corrupted with noisy alignments with up to 80+% F-score, and we improve the detection of misaligned sentence pairs in a real, low-quality web-crawled corpus up to 88%.

2   Methods

We experimented with three different change point detection methods to detect significant changes in the alignment scores within a sentence-aligned bilingual corpus: bcp, a Bayesian retrospective method, ecp, a non-parametric retrospective method, and ocp, an online Bayesian change point detection method.

2.1   Bayesian Change Point Detection

The Bayesian change point detection of Barry and Hartigan [2] assumes that each block between two change points arises from a (multivariate) normal distribution. The algorithm estimates the posterior probability that a change occurred at each point in the time series. We use the implementation from the R package bcp [6]. The bcp algorithm runs fast: it is linear in the length of the time series, and handles multivariate time series. The biggest limitations are that it is designed to detect changes in the mean of independent Gaussian observations, and that it is retrospective, meaning that it works only when the entire time series has been collected. Post-processing is required to extract a list of change points from the posterior probabilities. We used the implementation in the R package bcp with default values, including the change point prior of $p_0 = 0.2$.

2.2   Non-parametric Change Point Detection

The non-parametric, hierarchical divisive algorithm of James and Matteson [10] uses recursive bisections, identifying change points using a non-parametric divergence measure proposed by Szekely [23]. We use the implementation of this algorithm from the R package ecp. As the divergence measure is non-parametric, this makes this method suitable to detect changes with minimal assumptions on the underlying distributions and no need for setting priors. The divisive approach by recursive bisections returns a number of consecutive *segments* between change points, without knowing the number of change points *a priori*. In addition, the implementation from the R package ecp handles multivariate time series. The main limitations of this method are that it is relatively slow compared to the other methods and that it works only retrospectively, once the entire batch of data is available.

2.3   Bayesian Online Change Point Detection

The Bayesian online change point detection algorithm of Adams and McKay [1] is designed to update the detection of change points sequentially as new data points are acquired, rather than wait until the entire time series is available. Denoting a time series up to time t as $\mathbf{x}_t = (x_1, \ldots x_t)$, it relies on the notion of *run*, which is the subset of the time series coming from the same underlying distribution. The *run length* r_t can range from $r_t = 1$, if x_t is the first data from a new run, to $r_t = t$, if the entire time series so far is a single run. In order to estimate the posterior run length distribution given the observed data $P(r_t|\mathbf{x}_t) = \frac{P(r_t,\mathbf{x}_t)}{P(\mathbf{x}_t)}$, the model relies on two components:

1. The underlying predictive model (UPM) $P(x_t|r_{t-1}, \mathbf{x}_{t-1})$ models the stochastic generation of new datum x_t, conditioned on the run length r_{t-1} and previously observed data. Note that the UPM in fact only depends on the observations between x_{t-1} and $x_{t-r_{t-1}}$, but we keep the conditioning on $\mathbf{x}_{t-1}$ for convenience.
2. The prior $P(r_t|r_{t-1})$ over the run length at time t, given the run length at the previous step.

The posterior run length distribution can then be estimated recursively as:

$$P(r_t, \mathbf{x}_t) = \sum_{r_{t-1}} P(r_t|r_{t-1})P(x_t|r_{t-1}, \mathbf{x}_{t-1})P(r_{t-1}, \mathbf{x}_{t-1}) \qquad (1)$$

While the UPM encodes the assumption about the underlying data distribution (e.g. a Gaussian assumption yields a Student t-distributed UPM), a crucial component is the conditional prior on the run length. As a run can only continue or restart, the prior has nonzero mass for only two outcomes:

$$P(r_t|r_{t-1}) = \begin{cases} 1/\lambda, & \text{if } r_t = 0, \\ 1 - 1/\lambda, & \text{if } r_t = r_{t-1} + 1, \\ 0, & \text{otherwise} \end{cases} \qquad (2)$$

where this constant *hazard function* implements a memoryless process with timescale λ [1]. Our implementation, using that prior and a multivariate Gaussian UPM, is available as R package ocp.[1]

The outcome of the recursive algorithm is the posterior $P(r_t|\mathbf{x}_{t-1})$ at each time t. Further post-processing is necessary to extract the best set of change points from the posteriors. We investigate three methods:

colmaxes: This method returns the list of change points corresponding to the run length with maximum probability at each t. Note that for most time steps t, the most probable run length is just one more than the most probable run length at the previous step, therefore corresponding to the same change point. However the method tends to be sensitive to local uncertainty: when several run lengths have similar probabilities, the maximally probable change point may change slightly from one time step to the next. This may result in many close change points in areas of uncertainty. This is good for recall, which may actually work well for some datasets, but hurts precision by overpredicting change points.

threshold: This method is similar to "colmaxes", except it only returns change points that have a probability higher than a set threshold. In our experiments, we use a threshold of 0.4. This method may still return change points that are close together, but it returns fewer change points from areas with a lot of uncertainty: the threshold forces the method to wait until a change point has a probability of at least 40% before returning it. This causes fewer change points to be returned overall.

---

[1] https://cran.r-project.org/web/packages/ocp/index.html.

maxCPs: This is the set of change points with highest combined overall probability, considering all the data, based on the optimal partition algorithm [9]. This method is less sensitive to fluctuations due to uncertainty and is less likely to return many change points close together, therefore returning fewer change points than the previous two methods in most cases. The output is impacted by the choice of the prior and its timescale parameter λ. As $\lambda = 12$ in our experiments, this results in the set of change points with the highest probability tending to have most runs around a length of 12, while the shorter and longer runs are less likely to be detected. Since the actual run lengths in our artificial dataset vary from 1 to 127 sentence pairs, this results in missing the detection of many smaller chunks (Sect. 4.1 and Fig. 2).

3   Data

We experiment with two datasets: one clean parallel corpus corrupted by artificial noise, and a real-world, web-crawled, noisy parallel corpus.

3.1   Clean Data with Synthetic Noise

For this dataset, we used an artificially corrupted corpus. We took a high-quality 11k-sentence German-English corpus, specifically the test corpus from the 2013 Workshop on Machine Translation (WMT13) shared task on news translation.[2] We added five kinds of errors, a.k.a. *mutations* below:

- Alignment to a random sentence.
- Partial or complete replacement of the sentence in one language with the corresponding portion in the other language. This simulates situations where sentences or parts of sentences remain untranslated.
- Offsetting a sentence in one language by a random number of words, so that the sentence contains material from the previous or next sentence.
- Random deletion of some percentage of a sentence's words.
- Replacement of the sentence in one language with a machine translation of the other sentence.

Since errors often impact a span of sentences, we corrupt chunks of sentences at a time. The size of the chunks vary from 1 sentence to 127 sentences. The same mutation is applied to all sentences in a chunk, with the same level of severity. For example, each sentence in a given chunk would have the same percentage of words deleted, although different chunks may have differing levels of severity.

It should be noted that these kinds of errors are not all equally detectable; for example, it is relatively easy to detect when a sentence has been replaced by a completely random sentence, and much harder to detect when it has been replaced with a machine translation. Some error boundaries are thus much harder to detect than others.

---

[2] http://www.statmt.org/wmt13/translation-task.html.

The sentence pairs were then scored using an IBM Model 4 [4] trained on past German-English corpora from the Workshop on Machine Translation. IBM Model 4 measures the alignment of sentences as the product of word-level alignment probabilities, given the previously aligned word and the word class of neighboring words. We chose IBM model 4 because it is well-understood and has been known for a long time to be good at alignment tasks [20].[3] The trained model was then run on the target sentence pairs to calculate alignment scores for each sentence pair.

To investigate the applicability of change point detection to this data, we analyzed the changes in the similarity scores before and after the start of a known new mutation type. If there is a significant difference in the mean of scores before and after a mutation boundary (a point where a new mutation type begins), then it is likely that the change point detection algorithm will be able to detect that as a change between two Gaussian distributions. The changes in mean around each mutation boundary were calculated and are shown in Fig. 1. It turns out that most boundaries have a mean separation greater than 1, but a large number have a smaller separation: 25% have a separation below 0.59. This is important, because the variance in scores within each chunk of sentences can be fairly large, as shown in Fig. 1: the median variance is 0.88. This corresponds to a situation where in some cases the difference in mean scores from one chunk to the next will be smaller than the variance in score within a chunk of sentences. As a consequence, we expect that some of the change points will be very difficult to detect by simply inspecting the scores. Based on Fig. 1, it also seems likely that a good portion of chunks have low variance in alignment scores while having high mean separations. In those situations, change point detection appears to be a promising approach for detecting mutation boundaries.

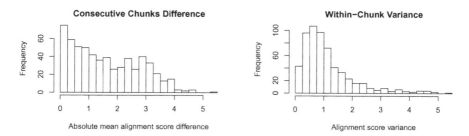

Fig. 1. Left: Distribution of mean differences in IBM4 alignment scores before and after boundaries of chunks with corrupted alignments (*mutation boundaries*). Right: Distribution of the variance in IBM4 alignment scores within sentence chunks with the same type of alignment corruption.

---

[3] Note that the scope of this paper is not to develop a better sentence alignment score, but to identify document boundaries to improve corpus filtering.

3.2   Real-World Noisy Data

In a second set of experiments, we use the German-English *ParaCrawl* corpus [14], a high-recall, low-precision web-crawled parallel corpus of around 1 billion words, in which roughly 70–75% of sentence pairs are estimated to be not genuinely parallel. This is the same corpus used in the WMT18 Parallel Corpus Filtering task [15].

We cannot directly measure the quality of change point detection here, since we do not know where misaligned chunks begin and end. Rather, we are testing the hypothesis that detecting these chunks using change point detection can allow us to take into account contextual information and ultimately lead to better detection of misaligned segments between change points.

In order to estimate the downstream performance on this corpus cleaning task, we manually annotated 318 randomly chosen sentence pairs for correct alignment (0 for non-parallel segments, 1 for well-aligned segments). Also, as running the change point detection algorithm on the entire *ParaCrawl* corpus would be expensive and largely uninformative, because sentence pairs very far from the annotated pairs will be largely irrelevant to the measurement, we selected 100 segment pairs around each of the annotated pairs (50 on each side). This results in 31,800 pairs on which we run the change point detection algorithm.

4   Results

4.1   Results on Synthetic Noise

In the dataset with synthetic noise, we know exactly the change points of interest, since we know the points at which the corruption algorithm switched between types of corruption. We can therefore compare our three change point detection algorithms according to their performance in identifying the true change points, something we cannot do on real-world noisy data where the change points are unknown.

Figure 2 shows the result of the online change point detection algorithm on a portion of the data (the first 500 sentence pairs). The detected change points (solid lines) clearly correspond to real boundaries (dotted lines) between chunks of different scores. This is mostly associated with sizable jumps in scores, and sometimes with more modest differences (e.g. around sentence #250 or #425). However it is clear as well that we are missing many of the shorter chunks, especially when the difference in score is not very apparent.

In order to quantify the performance more precisely, we compute the precision, recall and F-score of detected change points. We introduce some flexibility in evaluating a detected change point as correct, by counting a detected change point as correct (true positive) if it occurs within a certain window of a reference boundary. For example, a window size of five means that any detected change within five sentences from the true start of a new mutation chunk would be counted as a true positive detected change.

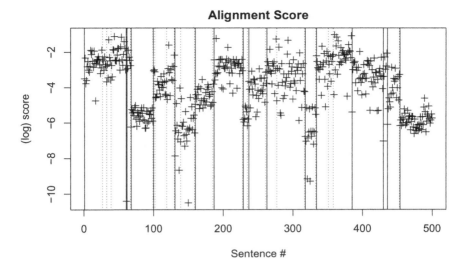

Fig. 2. Alignment scores for the artificial data (crosses), reference boundaries (dotted) and detected change points from online change point detection (ocp, solid), for the first 500 sentence pairs.

Figure 3 shows how the performance of different change point detection (and extraction) methods increases as the window size grows. The F-scores at the smaller window sizes show how well methods are able to detect changes very close to the actual true change point. Figure 3 suggests that the Bayesian online change point detection methods perform at much higher level even for window sizes as small as 2 or 3. The performance of all methods increases as the window size grows, as expected. The performance of bcp stays very low, mostly due to low precision caused by an over-detection of change points. The performance of ecp grows slowly and never reaches the level of the online method, because it can precisely detect the larger change points, but often misses the smaller change points and change points with smaller run lengths. The online change point detection methods perform best in this application because they are sensitive to small scale fluctuations and as a consequence can detect change points that are fairly close together, similarly to bcp, but without over-generating change points as much, so that they maintain a precision comparable to ecp.

Note that achieving an F-score of 1, even with extremely large window sizes, is only possible when the number of detected change points exactly matches the number of reference boundaries. In evaluation, only one detected event can be matched to a true event, so if there are more or less detected change points than true change points, then this would result in producing some false positives or false negatives, regardless of how big the window size is. Therefore, as the window size grows, the dominant effect being compared is picking the correct number of change points.

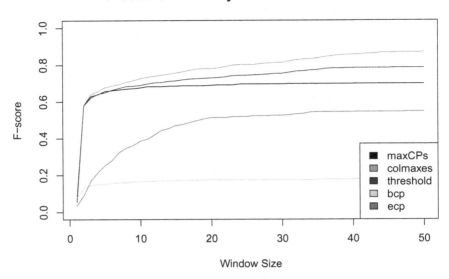

Fig. 3. F-score vs window size comparing change point detection methods on the artificial mutation data set.

4.2 Results on Real-World Noisy Data

The Bayesian online change point detection method was also applied to the real-world noisy data, in order to check whether detecting boundaries can improve our ability to predict which sentences pairs are correctly aligned vs. misaligned.

We compare the accuracy of three methods on the 318 human-labelled segment pairs:

Raw IBM scores: Using the IBM Model 4 score, as in the previous section, with no averaging.

Fixed-window averaging: The score of each pair of sentences t is a weighted average of the raw scores of t and its four surrounding sentences, with decreasing weights according to the distance from t : 0.25, 0.5, 1.0, 0.5, 0.25 for the sentences in positions $t-2$, $t-1$, t, $t+1$ and $t+2$, respectively.

Averaging between change points: This was obtained, as the name implies, by averaging IBM Model 4 scores between detected change points, resulting in a smoothed score that depends on the detected boundaries.

All three methods produce continuous scores. In order to predict misaligned sentence pairs, we label the top 30% of all scores as well-aligned (label 1) while the rest are predicted misaligned (label 0). The 30% threshold is motivated by the WMT18 corpus filtering task, where the largest task required to produce a filtered corpus of around 30% of the entire *ParaCrawl* corpus in order to train machine translation systems. This was also closely borne out in annotation as 29.25% of the annotated sentences were judged to be well aligned.

Performance is also compared to a majority-class baseline, which predicts the majority class (*misaligned*) for all sentence pairs, yielding an accuracy of 70.75% on our test set.

As seen in Table 1, averaging between change points improved classification by about 1.25% absolute over a strong baseline, and by more than 2.5% over raw scores, achieving an accuracy just above 88%. Only accuracy was reported, because the same number of sentences (30% of the test pairs) is returned in all cases, so that only the proportion of correctly identified sentences will be changing for each method.

Table 1. Accuracy in predicting aligned sentences

Method	Accuracy
Majority class	70.75%
Raw IBM scores	85.53%
Fixed-window averaging	86.79%
Averaging between change points	88.05%

This result shows that incorporating the change points could be a promising approach to improve machine translation training material by identifying segments of the document that have become corrupted or misaligned, so that they can be repaired or discarded.

5   Conclusion

We investigated the use of change point detection algorithms to detect boundaries of misaligned chunks in parallel corpora, in order to filter out misaligned segment pairs and improve alignment quality. We used a corpus with artificially corrupted alignments to evaluate the performance of various algorithms, and showed that Bayesian online change point detection performed significantly better than two alternatives, identifying up to above 80% of misaligned chunk boundaries. We then used that method on a subset of a real, low quality web-crawled parallel corpus and showed that the detection of misaligned chunks allowed us to improve the filtering performance up to 88%. This approach therefore shows promises in helping improve the quality of parallel data used for example to train machine translation models. Future directions include more extensive experiments estimating the impact of the detection of misaligned spans on the resulting machine translation quality.

References

1. Adams, R.P., MacKay, D.J.: Bayesian online changepoint detection. arXiv preprint arXiv:0710.3742 (2007)
2. Barry, D., Hartigan, J.A.: A Bayesian analysis for change point problems. J. Am. Stat. Assoc. **88**(421), 309–319 (1993)
3. Belinkov, Y., Bisk, Y.: Synthetic and natural noise both break neural machine translation. In: Proceedings of ICLR 2018 (2018)
4. Brown, P.F., Della Pietra, V.J., Della Pietra, S.A., Mercer, R.L.: The mathematics of statistical machine translation: parameter estimation. Comput. Linguist. **19**(2), 263–311 (1993)
5. Carpuat, M., Vyas, Y., Niu, X.: Detecting cross-lingual semantic divergence for neural machine translation. In: Proceedings of the First Workshop on Neural Machine Translation, pp. 69–79 (2017)
6. Erdman, C., Emerson, J.W., et al.: bcp: an R package for performing a Bayesian analysis of change point problems. J. Stat. Softw. **23**(3), 1–13 (2007)
7. Goutte, C., Carpuat, M., Foster, G.: The impact of sentence alignment errors on phrase-based machine translation performance. In: Proceedings of AMTA 2012 (2012)
8. Goutte, C., Wang, Y., Liao, F., Zanussi, Z., Larkin, S., Grinberg, Y.: Eurogames16: evaluating change detection in online conversation. In: Proceedings of the Eleventh International Conference on Language Resources and Evaluation (LREC-2018). European Language Resource Association (2018). http://aclweb.org/anthology/L18-1277
9. Jackson, B., et al.: An algorithm for optimal partitioning of data on an interval. IEEE Signal Process. Lett. **12**(2), 105–108 (2005)
10. James, N.A., Matteson, D.: ecp: an R package for nonparametric multiple change point analysis of multivariate data. J. Stat. Softw. **62**(1), 1–25 (2015). https://www.jstatsoft.org/index.php/jss/article/view/v062i07
11. Junczys-Dowmunt, M.: Dual conditional cross-entropy filtering of noisy parallel corpora. In: Proceedings of the Third Conference on Machine Translation: Shared Task Papers, pp. 888–895. Association for Computational Linguistics (2018). http://aclweb.org/anthology/W18-6478
12. Khadivi, S., Ney, H.: Automatic filtering of bilingual corpora for statistical machine translation. In: Montoyo, A., Muñoz, R., Métais, E. (eds.) NLDB 2005. LNCS, vol. 3513, pp. 263–274. Springer, Heidelberg (2005). https://doi.org/10.1007/11428817_24
13. Khayrallah, H., Koehn, P.: On the impact of various types of noise on neural machine translation. In: Proceedings of the 2nd Workshop on Neural Machine Translation and Generation, pp. 74–83 (2018)
14. Koehn, P., et al.: ParaCrawl corpus version 1.0 (2018). http://hdl.handle.net/11372/LRT-2610. LINDAT/CLARIN digital library at the Institute of Formal and Applied Linguistics (ÚFAL), Faculty of Mathematics and Physics, Charles University
15. Koehn, P., Khayrallah, H., Heafield, K., Forcada, M.: Findings of the WMT 2018 shared task on parallel corpus filtering. In: Proceedings of the Third Conference on Machine Translation, Volume 2: Shared Task Papers, Brussels, Belgium. Association for Computational Linguistics, October 2018
16. Koehn, P., Knowles, R.: Six challenges for neural machine translation. In: Proceedings of the First Workshop on Neural Machine Translation, pp. 28–39 (2017)

17. Lamraoui, F., Langlais, P.: Yet another fast, robust and open source sentence aligner. Time to reconsider sentence alignment? In: Machine Translation Summit XIV, Nice, France, September 2013
18. Lo, C.K., Simard, M., Stewart, D., Larkin, S., Goutte, C., Littell, P.: Accurate semantic textual similarity for cleaning noisy parallel corpora using semantic machine translation evaluation metric: the NRC supervised submissions to the parallel corpus filtering task. In: Proceedings of the Third Conference on Machine Translation (WMT 2018) (2018)
19. Lu, J., Lv, X., Shi, Y., Chen, B.: Alibaba submission to the WMT18 parallel corpus filtering task. In: Proceedings of the Third Conference on Machine Translation: Shared Task Papers, pp. 917–922. Association for Computational Linguistics (2018). http://aclweb.org/anthology/W18-6482
20. Mihalcea, R., Pedersen, T.: An evaluation exercise for word alignment. In: Mihalcea, R., Pedersen, T. (eds.) HLT-NAACL 2003 Workshop: Building and Using Parallel Texts: Data Driven Machine Translation and Beyond, Edmonton, Alberta, Canada, pp. 1–10. Association for Computational Linguistics, May 2003
21. Rossenbach, N., Rosendahl, J., Kim, Y., Graça, M., Gokrani, A., Ney, H.: The RWTH Aachen university filtering system for the WMT 2018 parallel corpus filtering task. In: Proceedings of the Third Conference on Machine Translation: Shared Task Papers, pp. 946–954. Association for Computational Linguistics (2018). http://aclweb.org/anthology/W18-6487
22. Saatçi, Y., Turner, R., Rasmussen, C.E.: Gaussian process change point models. In: Proceedings of the 27th International Conference on Machine Learning (2010)
23. Székely, G.J., Rizzo, M.L.: Hierarchical clustering via joint between-within distances: extending Ward's minimum variance method. J. Classif. **22**(2), 151–183 (2005). https://EconPapers.repec.org/RePEc:spr:jclass:v:22:y:2005:i:2:p:151-183
24. Xu, Y.: Confidence measures for alignment and for machine translation. Theses, Université Paris-Saclay, September 2016. https://tel.archives-ouvertes.fr/tel-01399222

In Vino Veritas: Estimating Vineyard Grape Yield from Images Using Deep Learning

Daniel L. Silver$^{(\boxtimes)}$ and Tanya Monga

Jodrey School of Computer Science, Acadia University,
Wolfville, NS B4P 2R6, Canada
danny.silver@acadiau.ca

Abstract. Agricultural harvest estimation is an important, yet challenging problem to which machine learning can be applied. There is value in having better methods of yield estimation based on data that can be captured with inexpensive technology in the field. This research investigates five approaches to using convolution neural networks (CNNs) to develop models that can estimate the weight of grapes on the vine from an image taken by a smartphone. The results indicate that a combination of image processing and deep CNN machine learning can produce models that are sufficiently accurate within a variety of grape for data captured at harvest time. The best approach involved transfer learning; where a CNN is developed starting from the weights of a pretrained density map model that learns to output the location of grapes in the image. The best model achieved a MAE of 157 g over a mean average weight of 1335 g, or a MAE% of 11.8.

Keywords: Convolution neural networks · Transfer learning ·
Counting objects in an image · Estimating a quantity in an image ·
Precision agriculture

1 Introduction

Estimating crop yield is important to commercial grape production because it informs many vineyard and winery decisions, such as the expected labour for harvest, the quantity of materials needed for wine production, and the blends of grapes to use [1]. However, yield estimation is a challenging and time consuming process. There is value in having better methods based on data that can be captured with inexpensive technology in the field, such as the combination of a smartphone and cloud-based computing. *In vino veritas* is a Latin phrase that means "in wine lies the truth", suggesting a person under the influence of wine is more likely to speak their true thoughts and desires. Let's see if neural networks can bring forth the truth about grape yield from images of fruit on the vine.

In this research, we investigate a method of estimating the weight of grapes on a vine, in grams, from a smart phone image of that vine prior to harvest.

© Springer Nature Switzerland AG 2019
M.-J. Meurs and F. Rudzicz (Eds.): Canadian AI 2019, LNAI 11489, pp. 212–224, 2019.
https://doi.org/10.1007/978-3-030-18305-9_17

Averaging the yields from a number of images, would allow the vineyard to better estimate the yield for a block. This is an interesting problem because the predictive model must learn to estimate the volume of grapes when many are occluded by other grapes. However, we know this is possible because viticulturists and experienced grape growers are able to make reasonable estimates based on having seen many harvests.

In this paper, we present an approach that is able to estimate Pinot Noir grape yield to nearly 90% mean accuracy on independent test sets of images [2]. We show that (1) a combination of image processing and deep learning methods can be used to accurately estimate crop yield while on the plant based on images taken by a hand-held smart phone, and that (2) transfer learning from a convolution-deconvolution network to a regression network can be used to improve regression accuracy for such problems.

The paper is organized as follows. Section 2 presents background on traditional approaches to grape yield estimation and more recent methods involving photographs and machine learning. Section 3 describes the dataset of 240 images of grape vines and associated harvest weights used. Section 4 presents the methods used to prepare the data and develop predictive models using machine learning. Section 5 compares a series of experiments that uses these methods. We conclude with a summary of the research, its contributions, and future work.

2   Background

2.1   Traditional Approaches to Estimating Grape Yield

Viticulturists traditionally estimate grape yield using a three step process [1]. They sample several clusters of grapes from a block of the vineyard, weigh each cluster and compute the average cluster weight (cw). They then count the number of clusters on several vines and compute the average number of clusters per vine (cv). They then estimate the total yield $= cw \times cv \times$ number of vines in the block. This can be done near harvest time or weeks prior to harvest, in which case the viticulturist must estimate the expected growth in the grapes and multiply the estimate by a factor (called the harvest multiplier) that reflects this growth. Experienced viticulturists will err in their estimates in the 10–15% range, whereas more junior grape growers will err by as much as 25%.

2.2   Recent Approaches to Estimating Crop Yield

A common approach to counting objects or estimating a quantity in an image is to use pixel-based segmentation methods, and then use image processing and pattern recognition software to determine the extent to which an object occurs in each segment. Various grape image segmentation techniques such as threshold segmentation, Mahalanobis distance segmentation, Bayesian classifier, direct three-dimensional histogram and linear color models have been investigated [3]. Grape berry detectors based on support vector machines (SVM) and histograms of oriented gradients (HOG) features have proven to be very efficient in the detection of white grape varieties [4]. The weight of grape clusters has been predicted using numerous metrics such as pixel area, volume, perimeter, berry number and berry size [5,6]. Specifically, a method of predicting the weight of grape clusters based on the number of berries pixels in the cluster has obtained an accuracy within 7.07%, but this method requires the grapes to be picked and images to be captured under ideal conditions with a white background [7].

An alternative approach is to use density estimation, which avoids the difficult task of learning to detect individual object instances. Instead, the problem is transformed into estimating a density function, or map, whose integral over any image region estimates the count or quantity of objects within that region [8]. Typically, one generates a variety of features from each pixel and its surrounding pixels using image processing methods and then trains a machine learning model to map these features to an image showing the location of the centre pixel or all associated pixels of the object(s) [9]. This is framed as a supervised learning problem that develops a model that maps from an $n \times m$ pixel space to an $n \times m$ density map. One needs only to add up the values for each pixel in the density map to obtain an estimate of the object count or quantity.

Deep convolutional neural networks, or CNNs, have proven to be very effective for image classification and regression problems [10]. In particular, CNNs have lead to significant improvements in the accuracy of regression based counting approaches [11]. Pretraining these deep networks using an unsupervised autoencoder approach is often used to overcome the problem of limited labeled images. An autoencoder network learns to reproduce the values presented at its input nodes at its output nodes. It consists of two parts; the *encoder* that maps the input data to a latent feature space, and the *decoder* that maps the latent features back to the data with as little error as possible. By training the autoencoder to copy the input to the output, the internal representation of the encoder takes on properties which are often useful for object classification [12]. The encoder portion of the network for an image is typically a CNN architecture. The decoder portion is a deconvolutional neural network that consists of one or more layers of deconvolution, upsampling, batch normalization and dropout.

Deep convolutional-deconvolutional neural networks have also been trained to accurately estimate a density map given an image containing objects of interest, as discussed above. Work by various groups has shown this approach to be fruitful for counting people [13], fish [14], palm trees [15], microscopy cells [16] and tomatoes [17].

3   The Grape Image Data

The data used in this research are images of Pinot Noir grapes on the vine and their associated harvest weight in grams per image. Images were captured at the Lightfoot & Wolfville Vineyards of Wolfville, Nova Scotia on harvest day, October 25, 2017, using the camera on a Samsung Galaxy S3 smartphone at a resolution of 3264×2448 pixels (see Fig. 1). Photographs of 40 vines were

(a) Original image.   (b) After Brightness normalization

Fig. 1. Brightness normalization.

taken from three different angles (left, centre, right) on one side of the plant, creating 120 images. The phone was held by hand, maintaining a similar height and distance so as capture the whole vine in each image. Prior to each photo a black and white calibration marker was place in a standard location next to the vine. The 40 images were then cropped into two parts, one for each half of a vine (called the left and right cordon), creating a total of 80 cordons. Each cordon of the vine was individually picked and weighed allowing us to create $3 \times 80 = 240$ examples, each composed of an image and its associated weight.

4   Methods and Approach

4.1   Image Preparation

The images were prepared to give the machine learning software a better chance of developing accurate models. Image processing methods were used to normalize the images in terms of scale, brightness, colour and contrast. This resulted in a more constrained set of images for the machine learning system.

Scale, Brightness, Colour and Contrast Normalization. The images of the grapes in the original photographs range in size depending upon the distance to the smart phone, thus making yield estimation more challenging. To normalize the size of the grapes in each image, the pre-processing software completes the following steps: find the calibration marker in the image (see Fig. 1); determine C_L, the width of the marker in pixels; calculate the scaling factor $S = D_L/C_L$, where D_L is the desired width of the marker in pixels; and scale the image by multiplying its height and width by S.

Variations in the brightness of the images can distort colors and hamper machine learning efforts. Therefore, it is important to transform all images to a common brightness level. Image brightness is calculated by taking the average value over all pixels to calculate the image's luminance, L. The image brightness is normalized out of 100 by multiplying each pixel value by 100/L (see Fig. 1).

Color and contrast are important for distinguishing objects in an image. A better distribution of these factors makes a photo look clearer and objects stand out from the background. Higher contrast provides better edges around objects, which assists the development of feature detectors in the machine learning software. To adjust colour and contrast, a technique called "histogram matching" is used in which the histogram of the pixels in each image is adjusted to match the histogram of a reference image that clearly distinguishes the grapes [18].

(a) Full image   (b) Left cordon   (c) Right cordon

Fig. 2. A full image of a vine and its associated cropped and resized images.

Cropping and Resizing of Images. After the images are adjusted for scale, brightness, color and contrast, the images are cropped to create one image of the left cordon of the vine and another of the right cordon. This was done manually, placing the cropped images on a white background of a fixed size. These fixed-size images are then resized to a resolution of 150 × 100 pixels (see Fig. 2).

Prior to machine learning, the image pixel values must be normalized and converted to an appropriate data type. The original images are in RGB format with individual pixels coded in Unit8 format with values ranging from 0–255. The neural network software required data to be in float32 format and normalized to a range of 0–1. Therefore, each prepared example consisted of 150 × 100 × 3 pixel values plus the associated grape yield (weight), all in float32 format.

4.2   Neural Network Approaches

The following describes five approaches for developing models to predict yield estimates using CNN deep learning. These will be tested and compared in the following section.

Basic CNN. A CNN architecture is comprised of a sequence of convolutional, max pooling, batch normalization, and dropout layers. Because we wish to predict a real-value weight at the network output, several dense layers are added to transform the convolutional feature maps into a grape yield estimate. Determining the optimal number of CNN layers and various learning parameters such as filter size and stride, dropout percentage, and learning rate required a lot preliminary testing. The final deep network architecture is shown in Fig. 5 and described in detail in Sect. 5.

Flipped Images CNN. We identified a variation in 50% of the 240 images used to train the basic CNN models, that if corrected, should improve model accuracy. Half of the images show a vine branching right (right cordon) and the other half show a vine branching left (left cordon). This variability is eliminated by flipping the left cordon images to look like they are right cordon images.

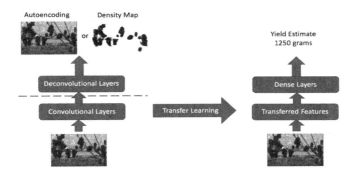

Fig. 3. Transfer learning from autoencoder or density map network representation to a yield model.

Transfer Learning from Autoencoder Network Representation. Figure 3 shows how we propose to use a CNN autoencoder to pretrain portion of a CNN based predictive model for our problem. The autoencoder network is trained to a high level of accuracy and the CNN encoder weights are transferred to the lower layers of a yield estimate model as the starting weights. The transfer learning approach provides an opportunity to pretrain the yield estimation network by learning useful internal representations from a large collection of unlabeled examples.

Transfer Learning from Density Map Network Representation. This approach takes advantage of transferring pretrained weights from an autoencoder while also isolating the location of the grapes in the image. An autoencoder-like density map network (see Fig. 3) is created to predict the location of the pixels of interest in the image; that is, the model outputs the probability of each pixel being part of a grape. This is a supervised learning approach that requires each training example to have a gray-scale target image that shows the location of the grape pixels. These density map target images were manually created using the quick mask feature of the GIMP programming tool. As with the autoencoder method proposed above, the density map model develops internal representation in the CNN encoder portion of network. This representation is then transferred to the CNN portion of a yield estimate network as a starting point for training.

Transfer Learning from Density Map Network Outputs. In this approach, density map images generated by the density map network are used directly as input to the yield estimate network (see Fig. 4). The density map

Fig. 4. Use of the density map network output as input to the yield model.

network is trained to a high level of accuracy and then its weights are fixed. This network creates density map images that are then used to train a CNN yield estimate network to predict the weight of the grapes in the original images.

5   Empirical Studies

This section summarizes experiments carried out to test and compare the five approaches described in Sect. 4.2 to develop deep neural network models that can estimate grape yield (weight in grams) from images.

5.1   Methods

All experiments used the 240 normalized Pinot Noir images of grapes on the vine ($150 \times 100 \times 3$ values) and associated harvest weights. A 5-fold cross-validation approach was taken with early stopping using a validation set. To maintain data independence, we placed all images of the same vine together within either the training, validation, or test set for each fold. The mean absolute error percentage or MAE% (and 95% confidence interval) is the main performance measure, calculated as the MAE in grams per image divided by the mean number of grams over all images.

The Tensorflow and Keras Python libraries were used to develop the deep neural networks. We used the parallel processing power of an NVIDIA GPU model GeForce GTX 530 6201 MB running on a PC equipped with 2.0 GB of RAM and an Intel Core i7-6700HQ CPU (2.60 GHz). Running on the Windows 10 operating system, it took 60–90 min to complete each cross-validation run.

Exp 1: Basic CNN. The best Basic CNN architecture consisted of 13 trainable layers organized into 4 convolutional blocks and 1 fully connected block (see Fig. 5). Each convolutional block contains at least one convolution layer having a stride of 1, followed by max pooling, batch normalization and dropout layers. Zero padding is applied to the images before passing them to the first block. Max-pooling is performed with a filter size of 2×2 and a stride of 2. The Rectified Linear Unit (ReLU) is used as the activation function throughout the network except the final two layers that use the sigmoid activation function to perform regression. The stochastic gradient descent optimization algorithm is used to train the network with a learning rate of 1×10^{-6}, a batch size of 32 images

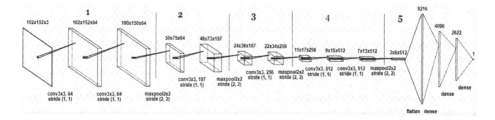

Fig. 5. Basic CNN architecture

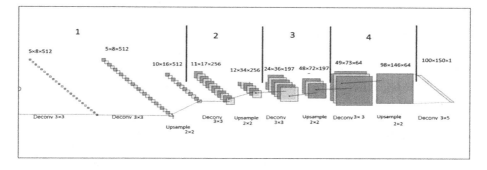

Fig. 6. Exp 3 and 4: CNN autoencoder deconvolution architecture

and a momentum value of 0.9. In addition, weight decay is set to 1×10^{-6}. The mean squared error is used as the cost function and the network was trained with early stopping for 150 epochs.

Exp 2: Flipped Images CNN. This experiment uses the flipped dataset of 240 Pinot Noir grape images and associated harvest weights as discussed in Sect. 4.1. The same CNN network architecture shown in Fig. 5 and the same learning parameters discussed in the prior experiment are used.

Exp 3: Transfer Learning from Autoencoder Network Representation. The experiment uses the same dataset of flipped images used in Experiment 2. The best architecture for the autoencoder was a deep convolution-deconvolution network consisting of 20 trainable layers organized into 9 blocks of layers; 4 convolutional blocks, 4 deconvolutional blocks and 1 final deconvolutional layer. The 4 convolutional blocks are the same as blocks 1–4 of the Basic CNN of Experiment 1 (see Fig. 5). Each deconvolutional block contains at least one deconvolution layer having a stride of 1, followed by upsampling, batch normalization and dropout layers (see Fig. 6). Upsampling is performed after each deconvolution with a filter size of 2×2 and a stride of 2. The ReLU activation function is used throughout the network except the last layer that uses a sigmoid function to produce density values in the 0–1 range. The Adadelta optimization algorithm is used with a learning rate of 1.0 and a batch size of 32 images. Weight decay is set to 1×10^{-6}. The mean squared error is used as the training cost function and the network was trained for 10,000 epochs. After the autoencoder network is trained, the CNN weights are transferred to the yield estimate model as pretrained starting weights, as shown in Fig. 3. The yield estimate model is then trained using the same network architecture and learning parameters as shown in Fig. 5 and discussed in the Experiment 1.

Exp 4: Transfer Learning from Density Map Network Representation. The experiment uses the same dataset as Experiment 2. The network architecture and the learning parameters discussed in the Experiment 3 for the autoencoder model (shown in Fig. 6) are used for the density map model, except the output image is $150 \times 100 \times 1$. The density map model was trained using 16 labelled examples and then its CNN encoder weights are transferred to the CNN layers of a yield estimate model as pretrained weights as shown in Fig. 3. The yield estimate model is then trained using the same network architecture and learning parameters as shown in Fig. 5 and discussed in the Experiment 1.

Exp 5: Transfer Learning from Density Map Network Outputs. This experiment uses the same dataset, density map network architecture, and learning parameters discussed in the Experiment 4. The density map images generated by the density map model described in Experiment 4 are used as input to a yield estimate model similar to the CNN in Fig. 5, only differing by the input being grey scale ($150 \times 100 \times 1$) versus RGB. Otherwise, the yield estimate model has the Basic CNN architecture and learning parameters discussed in Experiment 1.

5.2   Results and Discussion

Figure 7 shows the MAE% for each deep neural network approach taken. The Basic CNN models (Exp 1) produced an MAE of 236.192 g over all test images, with a 95% confidence interval of ± 9.534 g. Given 1335.111 as the average number of grams per image, this is equivalent to a MAE% of 17.68.

The best Flipped Images CNN models (Exp 2) produced an MAE of 206.01 g over all test images, with a 95% confidence interval of ± 12.95 g. This is equivalent to a MAE% of 15.43. Flipping all left cordon images improves the performance insignificantly over Exp 1 (difference of mean t-Test, p-value = 0.306), but is suggests keeping this modification for all remaining experiments.

The Transfer Learning from CNN Autoencoder models (Exp 3) produced an MAE of 240.49 g over all test images, with a 95% confidence interval of ± 11.59 g. This is equivalent to a MAE% of 18.01. Unfortunately, this suggests that the representation transferred from the autoencoder did not result in a better yield estimate model. This is likely because the latent features of the autoencoder do not focus on the grapes specifically. Examples of images generated by the autoencoder model and a scatter plot of actual versus predicted weights are shown in Fig. 8.

The Transfer Learning from Density Map Network Representation models (Exp 4) produced an MAE of 157.52 g over all test images, with a 95% confidence interval of ± 13.69 g. This is equivalent to a MAE% of 11.79. The generated images from the density map model are shown in Fig. 9. The representational

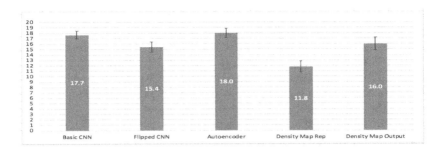

Fig. 7. The performance results of the five CNN-based models.

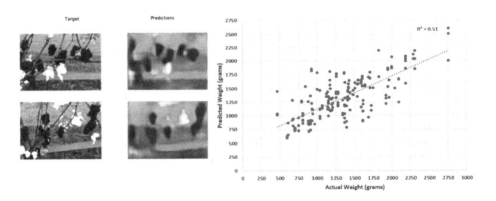

Fig. 8. Results of transfer learning from the CNN autoencoder representation.

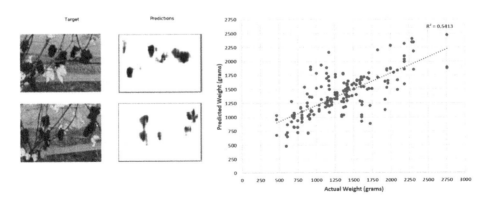

Fig. 9. Results of transfer learning from density map network representation.

knowledge transferred from the density map encoder as starting weights make a difference in the creation of a yield estimate model as compared to the results of Exp 2 (difference of mean t-Test, p-value = 0.118).

The Transfer Learning from Density Map Network Outputs models (Exp 5) produced an MAE of 213.56 g over all test images, with a 95% confidence interval of ±15.45 g. This is equivalent to a MAE% of 15.99. This approach did not do as well as transfer learning using the density map representation because the accuracy of the density map estimates were quite low. This is because we used only 16 labelled images to train the density map model for Exp 4 and 5. In the future we will develop this model using a larger dataset of labeled images to generate more accurate density maps for test images. These more accurate density map images should decrease the MAE of the yield estimate model.

6   Conclusion

The problem of estimating a quantity such as weight or density from an image is interesting and challenging. We have presented research that uses a combination of image processing and convolutional neural networks to develop models that can estimate the weight of grapes on the vine from a smartphone image. Empirical studies compare five approaches using deep neural network methods. The networks vary from a basic CNN regression model to those that employ transferred learning of representation from an autoencoder or density map network. The empirical results show that transfer learning from a network that estimates the location of grapes in the imagecomes close to reaching our goal of 90% accuracy with an MAE% as low as 11.8. This suggests that CNNs can be used for agricultural yield estimation as well as other image to quantity regressions.

There are two important contributions from this research. First, the prescribed combination of image processing, deep learning architectures and learning parameters allows the system to make accurate predictions despite having images capture by a handheld smartphone with significant variation in light level and image distance and angle to the grapes. Second, representational transfer of learned weights from a CNN density map network proved to be the most affective solution with an MAE% of 11.8. Transfer learning provides a method to pretrain the CNN network by starting from a representation developed for the related task of locating where grapes are in the image. This makes intuitive sense and can be used in similar image to quantity regression problems.

This past fall (2018), we collected images and harvest weights from twice as many vines as the prior year; and we did so each week for several weeks prior to harvest. These images will allow us to develop and test more robust predictive models and to attempt to predict grape yield from images taken several weeks prior to harvest. We plan to employ more advanced image processing methods prior to machine learning, including automated cropping of grapes from the original images. Most importantly, we will explore knowledge transfer using multiple task learning where one neural network is used to simultaneously predict both the density map (location of the grapes) and the weight of the grapes.

Acknowledgment. This research has been funded by Springboard Atlantic Inc. and Acadia University and made possible by Lightfoot & Wolfville Vineyard, NS, Canada.

References

1. Komm, B., Moyer, M.: Vineyard yield estimation. EM068e, Washington State University Extension Publishing, Pullman, Washington (2015)
2. Monga, T.: Image to grape yield estimation in the vineyard using deep learning. Acadia University (2018)
3. Font, D., Tresanchez, M., Martínez, D., Moreno, J., Clotet, E., Palacín, J.: Vineyard yield estimation based on the analysis of high resolution images obtained with artificial illumination at night. Sensors **15**(4), 8284–8301 (2015)
4. Škrabánek, P., Majerík, F.F.: Evaluation of performance of grape berry detectors on real-life images. In: Proceedings of the 22nd International Conference on Soft Computing, MENDEL 2016, pp. 217–224 (2016)
5. Nuske, S., Achar, S., Bates, T., Narasimhan, S., Singh, S.: Yield estimation in vineyards by visual grape detection. In: 2011 IEEE/RSJ International Conference on Intelligent Robots and Systems (IROS), pp. 2352–2358. IEEE (2011)
6. Diago, M.P., Correa, C., Millán, B., Barreiro, P., Valero, C., Tardaguila, J.: Grapevine yield and leaf area estimation using supervised classification methodology on RGB images taken under field conditions. Sensors **12**(12), 16988–17006 (2012)
7. Liu, S., Marden, S., Whitty, M.: Towards automated yield estimation in viticulture. In: Proceedings of the Australasian Conference on Robotics and Automation, Sydney, Australia, vol. 24, pp. 2–6 (2013). araa.asn.au
8. Lempitsky, V., Zisserman, A.: Learning to count objects in images. In: Lafferty, J.D., Williams, C.K.I., Shawe-Taylor, J., Zemel, R.S., Culotta, A. (eds.) Advances in Neural Information Processing Systems 23, pp. 1324–1332. Curran Associates, Inc. (2010). Accessed 15 July 2018
9. Saleh, S.A.M., Suandi, S.A., Ibrahim, H.: Recent survey on crowd density estimation and counting for visual surveillance. Eng. Appl. Artif. Intell. **41**, 103–114 (2015)
10. LeCun, Y., Bengio, Y., et al.: Convolutional networks for images, speech, and time series. In: The Handbook of Brain Theory and Neural Networks, vol. 3361, no. 10. The MIT Press (1995)
11. Cohen, J.P., Lo, H.Z., Bengio, Y.: Count-ception: counting by fully convolutional redundant counting. arXiv preprint arXiv:1703.08710 (2017)
12. Druzhkov, P., Kustikova, V.: A survey of deep learning methods and software tools for image classification and object detection. Pattern Recogn. Image Anal. **26**(1), 9–15 (2016)
13. Boominathan, L., Kruthiventi, S.S., Babu, R.V.: Crowdnet: a deep convolutional network for dense crowd counting. In: Proceedings of the 2016 ACM on Multimedia Conference, pp. 640–644. ACM (2016)
14. French, G., Fisher, M., Mackiewicz, M., Needle, C.: Convolutional neural networks for counting fish in fisheries surveillance video (2015)
15. Cheang, E.K., Cheang, T.K., Tay, Y.H.: Using convolutional neural networks to count palm trees in satellite images. arXiv preprint arXiv:1701.06462 (2017)
16. Xie, W., Noble, J.A., Zisserman, A.: Microscopy cell counting and detection with fully convolutional regression networks. Computer Methods in Biomechanics and Biomedical Engineering: Imaging & Visualization, pp. 1–10 (2016)

17. Rahnemoonfar, M., Sheppard, C.: Deep count: fruit counting based on deep simulated learning. Sensors **17**(4), 905 (2017)
18. Maini, R., Aggarwal, H.: A comprehensive review of image enhancement techniques. arXiv preprint arXiv:1003.4053 (2010)

Options in Multi-task Reinforcement Learning - Transfer via Reflection

Nicholas Denis[(✉)] and Maia Fraser[iD]

Department of Mathematics and Statistics, University of Ottawa, Ottawa, Canada
{ndeni032,mfrase8}@uottawa.ca

Abstract. Temporally extended actions such as *options* are known to lead to improvements in reinforcement learning (RL). At the same time, transfer learning across different RL tasks is an increasingly active area of research. Following Baxter's formalism for transfer, the corresponding RL question considers the benefit that an RL agent can achieve on new tasks based on experience from previous tasks in a common "learning environment". We address this in the specific context of goal-based multi-task RL, where the different tasks correspond to different goal states within a common state space, and we introduce Landmark Options Via Reflection (LOVR), a flexible framework that uses options to transfer domain knowledge. As an explicit analog of principles in transfer learning, we provide theoretical and empirical results demonstrating that when a set of *landmark* states covers the state space suitably, then a LOVR agent that learns optimal value functions for these in an initial phase and deploys the associated optimal policies as options in the main phase, can achieve a drastic reduction in cumulative regret compared to baseline approaches.

1 Introduction

The strength of reinforcement learning (RL) in modelling and solving sequential decision problems is apparent in the broad and increasing range of problem-types considered by RL researchers. One example of these are lifelong learning settings [19]. They assume an RL agent is faced with a sequence of MDPs that share certain properties and are thus viewed as coming from a common *environment*. As in transfer learning [3], rather than beginning fresh with each new task, an efficient lifelong agent should leverage past knowledge of its experience within the environment in order to solve new tasks as quickly as possible. We consider such transfer in the form of options.

Options, or "temporally extended actions" have been found to be useful in overcoming the curse of dimensionality in many settings [2]. They were originally motivated as a form of inductive transfer [22], appearing soon after the survey [23] where earlier approaches to "learning to learn" in RL had been addressed. The use of such actions inherently sets up a hierarchy in time-scales, with primitive actions at the bottom and longer-running options higher up.

© Springer Nature Switzerland AG 2019
M.-J. Meurs and F. Rudzicz (Eds.): Canadian AI 2019, LNAI 11489, pp. 225–237, 2019.
https://doi.org/10.1007/978-3-030-18305-9_18

How to structure control and learning within an RL hierarchy remain areas of active current research [5,7,12,18].

In this paper we introduce the Landmark Options Via Reflection (LOVR) framework. This is a general hierarchical framework for inductive transfer via options, that implements the shifting of control in the hierarchy through a dedicated action $a_{reflect}$ called "reflection". For simplicity we consider here only two levels. The framework assumes a set of landmark states, $\mathbb{L}$, that covers the state space in a manner to be made precise, analogous to coverings of a learning environment studied in [3]. In Phase I, the agent learns accurate Q-value functions associated to arriving at $l \in \mathbb{L}$, then based on these Q-value functions defines options for use in Phase II, where the agent can access options only through $a_{reflect}$. The action $a_{reflect}$ is subject to the same learning updates (e.g. Q-value) as other primitive actions, and thus provides an RL-native way to shift control between levels in the hierarchy. Further levels could be defined analogously by invoking further phases but this is out of the scope of the present paper. We focus here on the transfer-benefits achieved through LOVR.

The versions of LOVR described here work in lifelong-learning settings consisting of a sequence of episodic task MDPs, each with its own well-defined goal state, s_g, and the rest of the MDP common across tasks. We give theoretical results for LOVR in finite state spaces and also empirical results that show LOVR drastically reduces cumulative regret for a lifelong learning agent given a sequence of MDP tasks. This holds even with the number of landmark states being a small proportion of the state space.

2 Related Work

In the Separation of Concerns (SoC) framework [17,18] multiple agents estimate the value of a given state, and an aggregator function makes decisions by combining input of lower agents. Reflection via $a_{reflect}$ plays a similar role in our framework. Hierarchical RL approaches such as SoC or Feudal RL [6,18,24] allow for temporal abstraction at different levels; this is achieved at two levels by our use of landmark options.

Both MLSH [7] and LOVR are meta-algorithms for multi-task lifelong settings that allow various underlying RL agents. MLSH performs joint optimization over two sets of parameters, one for a higher-level agent which learns commonalities among tasks, and one for each task. Per task optimization is done in a reduced space of *sub*-policies which speeds discovery of a sequence leading to reward. Reducing learning problems is a motivation in our work as well, but the difference in our approaches is akin to learning in development vs. plasticity ([1] vs. [4]). Learning in mature nervous systems involves online training in both directions simultaneously [8] - similar to MLSH - while LOVR is a development-style agent, which first fine-tunes lower-level tools and then makes these available during learning by a higher-level agent.

Schaul et al. [16] introduced the Universal Value Function Approximator (UVFA) for learning goal-based tasks in the multi-task setting. Here, the state and goal representations are concatenated and used as inputs to the network, which learns to approximate the Q-values of the state with respect to the current goal, g, parameterized by weights θ. The UVFA network, $U_\theta(s; g)$, allows for a single network to learn several tasks.

Brunskill et al. [5] studied multi-task lifelong learning and provided PAC-MDP theoretical guarantees. Similar to LOVR, their framework involves two phases. Clustering of sub-policies provides the structure by which the higher-level agent accesses them. In contrast to LOVR, [5] allow for different tasks to have different transition kernels and reward functions. In their framework, during the second phase as the agent is experiencing the task MDP, the data acquired allows the agent to estimate which cluster the current MDP belongs to. In some sense, the decision to make such an estimation and (temporarily) commit to a policy built for the estimated MDP cluster could be seen as an instantiation of $a_{reflect}$, however only in spirit as $a_{reflect}$ is an action selected by a higher level policy (thus conditionally dependent on the current state), whereas in [5] this decision is dependent on a sequence of data.

Konidaris et al. [11] introduce the concept of an agent space, acting as an ego-centric view of the agent in a class of environments. Here, options learned in one MDP can be transferred to a new MDP coming from a similar class of environments (e.g. class of all grid worlds) via a mapping through the agent space. In this way, policies learned earlier can be transferred into the future as options (skills). A similar principle arises in [15].

3   Background

Multi-task and Lifelong Reinforcement Learning. We consider a multi-task RL setting where the agent is confined to a stationary environment, and throughout its lifetime is assigned a sequence of tasks. These tasks are represented as episodic MDPs $\mathcal{M}_i = \langle \mathcal{S}, \mathcal{A}, P, \mathcal{R}, s_{g_i}, \nu \rangle$, $i \in [T]$, with respective terminal states s_{g_i}. In summary: the state space $\mathcal{S}$, set of actions $\mathcal{A}$, transition probability kernel P, reward function $\mathcal{R}$ and initial state distribution ν all remain fixed across tasks and only the goal state varies. The goal state s_{g_i} thus encodes the i'th task, though when speaking of a single task this will be denoted simply s_g. In many settings it may be more applicable to consider goal sets $\mathcal{G}_i \subset \mathcal{S}$, rather than a single goal state. Our usage of terminal state follows the common formulation [21] for episodic tasks where one wants the agent, such as a robot, to perform a single specific function, and upon completion to become inactive and remain so until a new episode or task begins. In general one could let K_i be the number of episodes that $\mathcal{M}_i$ is run before the next task is assigned, but we will take $K_i \equiv K$ to be independent of i in the present paper. We consider the action-penalty reward structure [10] of -1 for all transitions, which reinforces the agent towards a policy of arriving at the current goal in as few steps as possible. The goal of the lifelong learning agent is to solve for a sequence of policies $\{\pi_i^*\}_{i=1}^T$ which minimize cumulative regret. It is clear that under this reward structure that $\pi^*(s)$ is the negative expected number of steps from state s to goal g, and since π^* is optimal, it encodes the (expected) most efficient path to the goal. For our empirical results we use two empirical measures of regret: the *empirical per-episode regret*: $\mathcal{R}^g_{episode}(\pi) := R_{\pi^*} - R_\pi$, which measures the difference in accumulated reward over that episode for the policy being considered vs. the optimal policy for the task (goal) g at hand, and the *empirical lifelong regret*: $\mathcal{R}_{life} := \sum_{t=1}^{T} \sum_{k=1}^{K} \mathcal{R}^t_{episode_k}$, which measures the cumulative empirical per-episode regret of an agent over a sequence of T tasks, each run for K episodes. Likewise, *per-episode regret* and *lifelong regret* are the expectations of the respective empirical regrets.

Options Framework. The options framework is a mathematically principled approach for temporally extended actions [22]. An option $o \in \mathcal{O}$ is a triple, $o = \langle \mathcal{I}_o, \pi_o, \beta_o \rangle$, where $\mathcal{I}_o \subseteq \mathcal{S}$ is the initiation set, π_o is the option policy that maps states to primitive actions, and β_o is the termination function which controls when to terminate π_o and return control back to π. The inclusion of options in an MDP results in a Semi-MDP (SMDP), where standard RL algorithms apply in the SMDP setting [22]. Landmark options were first introduced in [14,22] as policies leading towards "landmark" states. We retain this aspect but employ them in a different way.

4   Landmark Options via Reflection (LOVR) Framework

Landmark Value Functions, Landmark Coverings and Phase I. For all approaches considered here, we assume that during a preliminary phase the agent solves a set of tasks corresponding to a set of *landmark states* $\mathbb{L} \subset \mathcal{S}$, with associated Q-value functions. For $l \in \mathbb{L}$, we denote by $V^l(s)$ the value of state s with respect to arriving at the goal-state l. We allow this phase to be general, where the landmark states are picked by a researcher, or built *tabula rasa* online; in either case, we only assume the following covering conditions are achieved by the algorithm: (A1) $\exists \eta > 0$, $\forall s \in \mathcal{S}$ $\exists l \in \mathbb{L}$ such that $|V^l(s)| \leq \eta$, which we call η-reachability, and in addition (A2) $\forall l \in \mathbb{L}$, V^l is ϵ-optimal. A2 can be achieved with a PAC-MDP algorithm such as Delayed Q-learning [20] or through building a model (i.e. E^3 [9]), however as we demonstrate experimentally, even without this assumption we achieve impressive results. The LOVR framework assumes that upon completion of the first phase $A1 - A2$ are satisfied before the second phase begins.

Note that when $\mathbb{L} = \mathcal{S}$, A1 is automatically satisfied. Moreover, in this work we restrict ourselves to environments that satisfy what we call κ-*approximate reversibility*, namely $\exists \kappa \geq 1$, such that $V^{s*}(g) \leq V^{g*}(s)\kappa$, $\forall s, g \in \mathcal{S}$, where V^{s*} denotes the optimal value function associated with arriving at goal-state s. This assumption is trivially satisfied when the state space communicates with a finite diameter, and when $\kappa = 1$ is called *reversible* [13]. If $\mathcal{S}$ satisfies κ-approximate reversibility, then a natural metric on $\mathcal{S}$ can be defined that makes it possible to succinctly describe "good" choices of $\mathbb{L}$, namely, if this set is chosen such that η-balls around $\mathbb{L}$ cover $\mathcal{S}$ then applying any algorithm that achieves ϵ-optimality (e.g. PAC-MDP algorithms) automatically achieves conditions $A1 - A2$. We leave discussion of these details for the extended version of the paper and here simply assume that $A1$ and $A2$ hold at the end of Phase I.

Phase II. The LOVR framework uses the options framework to accomplish transfer in the goal-based multi-task RL setting by leveraging previously solved $\{Q^l\}_{l \in \mathbb{L}}$ from the first phase to define a set of landmark options $\mathcal{O}_l$ for the second phase. Options can only be accessed through a single action $a_{reflect}$, which acts as a gating mechanism to the set of landmark options. The effect of $a_{reflect}$ is then to execute a specific option $o_l \in \mathcal{O}_l$. Between initiation and termination, o_l follows the policy associated to the corresponding learned Q^l. LOVR's generality, however, allows flexibility to encode in a domain-specific manner both $a_{reflect}$'s choice of option, as well as the definition of *initiation and termination conditions* for every $o_l \in \mathcal{O}_l$. The latter determine respectively the states from which o_l can be initiated, and the state at which to stop following o_l and return control to the higher level policy. This flexibility allows the designer to control how options are invoked in a manner that promotes positive transfer and reduces negative transfer. To demonstrate this, we introduce two instantiations of the LOVR framework, termed LOVR1 and LOVR2, each used in different settings and with different biases. We provide a high level description of how they differ, and then a technical description of each.

Both LOVR1 and LOVR2 assume the same choice of $\mathbb{L}$ and the same Phase I to produce $\{Q^l\}_{l \in \mathbb{L}}$. We utilize LOVR1, in the setting where the agent is not provided knowledge of the goal state g and, hence, must first arrive at g to identify this goal, while LOVR2 is provided the goal state g at the start of each task. Subsequently, once g is known, both instantiations proceed similarly but differ in subtleties of initiation and termination. LOVR1 follows the landmark option policy o_l attempting to arrive all the way to the landmark state l, while LOVR2, on the other hand, does not attempt to arrive at the landmark state itself, but rather, terminates the landmark option once the agent is within η steps of the landmark, in expectation. The difference in these two termination conditions reflect two different biases. If the agent follows a landmark option to a given landmark l, the goal is expected to be at most $\eta \kappa$ steps away from l under some optimal policy, however en route to the landmark l, the agent may "pass through" a state s' from which g can be directly reached in $\lambda << \eta \kappa$ steps. LOVR1 will not be able to take advantage of this shortcut. It accepts this potential inefficiency, in exchange for being in expectation at most $\eta \kappa$ steps from the goal once it stops. LOVR2 on the other hand, terminates the landmark option sooner, at which point the agent is in expectation at most $\eta + \eta \kappa$ steps from the goal, but it might find a shorter path and so escapes the potential inefficiency just described for LOVR1. The two instantiations, LOVR1 and LOVR2 - aside from having access to g or not - thus differ primarily in how they handle the tradeoff between expected closeness at option's termination vs. ability to take a shortcut. In addition initiation is slightly different in the two, forcing LOVR2 to take options when it is far from the goal. We now describe the technicalities of both instantiations. Throughout, we assume the landmark option policy, $\pi_{o_l}(s)$

takes the action that maximizes the associated $Q^l(s,a)$. In describing LOVR1, for a given landmark option $o_l \in \mathcal{O}_l$, the initiation set $\mathcal{I}_{o_l}$ is defined as

$$
\mathcal{I}_{o_l} = \begin{cases} \mathcal{S} & if \ V^l(g) \geq V^{l'}(g), \ \forall l' \neq l \in \mathbb{L} \\ \emptyset & else. \end{cases} \tag{1}
$$

LOVR1's instantiation of $a_{reflect}$ chooses any landmark option o_l such that l satisfies 1. The initiation set is defined above, and the binary termination function, β_{o_l}, terminates o_l once the number of actions taken under this policy exceeds the value $|V^l(s)|$. Recall this value is an ϵ-optimal estimate of the expected number of steps from s to l under o_l. Thus, LOVR1 follows the optimal policy for the expected number of steps to reach the selected landmark. In describing LOVR2, we have,

$$
\mathcal{I}_{o_l} = \begin{cases} \{s \in \mathcal{S} | \eta < |V^l(s)|\} & if \ V^l(g) \geq V^{l'}(g), \ \forall l' \neq l \in \mathbb{L} \\ \emptyset & else. \end{cases}
$$

For LOVR2, we use the state dependant action set notation $\mathcal{A}(s)$ for the set of actions callable by π when in state s and we define: $\forall s \in \mathcal{S}$ s.t. $\eta < |V^l(s)|$, $\mathcal{A}(s) := \{a_{reflect}\}$, and $\forall s$ s.t. $\eta \geq |V^l(s)|$, $\mathcal{A}(s) := \mathcal{A}$, the set of primitive actions. In words, LOVR2 forces $a_{reflect}$ when the agent is at a state that is greater than η steps from the landmark state. Like in LOVR1, $a_{reflect}$ chooses any o_l such that l satisfies 1. LOVR2 then defines the initiation set as just stated, and defines the termination function, $\beta_{o_l}(s;g) = \mathbb{1}\{V^l(s) \geq V^l(g)\}$, i.e., while following the landmark option policy, β_{o_l} terminates the option at any state s such that $\eta \geq |V^l(s)|$. The idea behind LOVR2 is that the landmark option policy isn't attempting to arrive at the landmark state, rather to drive the agent to a pseudo-ball of states that can reach the landmark within η steps. While within that pseudo-ball, no landmark options can be taken, hence the agent must explore-exploit within that pseudo-ball. Rather than solving a large MDP defined over all of $\mathcal{S}$, the LOVR2 agent need only solve over a smaller constrained MDP defined over $\{s \in \mathcal{S} | \eta \geq |V^l(s)|\}$, which may be significantly smaller, and can also be controlled by the landmark covering parameter η.

5   Results

We test the two instantiations, LOVR1 and LOVR2, in two different settings. In the first, we use LOVR1 in a tabular gridworld where for each task the goal state is not provided to the agent. Hence, the agent has no access to $a_{reflect}$ nor the landmark options until the goal state has first been discovered. In the second setting, LOVR2 is tested in the function approximation setting in a visual gridworld. Here the goal state is provided to the agent. We provide some theoretical results for the first setting when $\mathbb{L} = \mathcal{S}$, and theoretical results for when the goal is provided to the agent, $\mathbb{L} \subset \mathcal{S}$, and under assumption A2.

Proposition 1. *Let $\mathcal{M}$ be an MDP as defined above with action-penalty reward structure, and P deterministic, $\mathcal{M}$ of finite diameter D and $\mathbb{L} = \mathcal{S}$. Let $D', d \in \mathbb{N}$ such that $d < D < D'$. Under A2, for $\epsilon < 1$, then given a sequence of episodic tasks in $\mathcal{M}$ parametrized by goal state, and with initialization of $Q(s, a) = -D'$, $\forall (s, a) \in \mathcal{S} \times \mathcal{A} - \{a_{reflect}\}$ and $Q(s, a_{reflect}) = -d$, $\forall s \in \mathcal{S}$, the lifelong cumulative regret of the LOVR1 agent after T tasks each lasting K episodes is $\leq T\big(2\mathcal{S}\mathcal{A}D + (K - 1)\big)$.*

The proof is omitted due to space restrictions, however follows easily from [10]. Next we provide a result that shows the asymptotic inefficiency of a LOVR2 agent, associated to forcing the agent to traverse towards a landmark first, before traversing towards the goal state g.

Proposition 2. *Consider LOVR2 and goal-based multi-task RL setting as described above. Given assumptions A1, A2 with $\eta \geq 1$, and environment satisfying κ-approximate reversibility for $\kappa \geq 1$. Let initial and goal states $s, g \in \mathcal{S}$ be arbitrary, then under the optimal LOVR2 policy the per-episode regret, $\mathcal{R}$, is no more than $\eta(1 + \kappa)$. Moreover, this proof is tight, in the sense that there exists an MDP such that under these conditions, the bounds are realized (Fig. 1).*

Proof. Denote $V^{g*}_{LOVR2}(s)$ the optimal value function associated to the optimal LOVR2 policy that first takes the optimal path landmark option from s, then upon termination of the landmark option, takes the optimal policy from some termination state s' to g. Recall that all value functions are non-positive. Then,

$$
\begin{aligned}
\mathcal{R} &= V^{g*}(s) - V^{g*}_{LOVR2}(s) && \text{definition of regret} \\
&\leq V^{g*}(s) - (V^{l*}(s) + V^{g*}(l)) && \text{see above} \\
&\leq V^{g*}(s) - (V^{l*}(s) + \kappa V^{l*}(g)) && \kappa\text{-approximate reversibility} \\
&\leq V^{g*}(s) - (V^{l*}(s) - \eta\kappa) && \eta\text{-reachability; } \eta, \kappa \geq 1 \\
&\leq V^{g*}(s) - (V^{l*}(g) + V^{g*}(s) - \eta\kappa) && \text{inefficiency of first arriving at } g \\
&\leq V^{g*}(s) - (V^{g*}(s) - \eta - \eta\kappa) && \eta\text{-reachability} \\
&= \eta(1 + \kappa).
\end{aligned}
$$

To see that the bounds are tight, we provide an example of an MDP where the bounds are realized (Fig. 1). It can be seen that this MDP satisfies κ-approximate reversibility, and condition A1. When the initial state is s_0, and the goal state g, the LOVR2 optimal policy will select landmark option o_{l_1}, until the termination condition is satisfied (in this case, $s = l_1$). From l_1, the agent will follow the optimal path to g. Thus, the regret of the optimal LOVR2 agent is $-\frac{1}{p_0} - (-\eta - \frac{1}{p_0} - \kappa\eta) = \eta(1 + \kappa)$.

LOVR1 Results. We first test LOVR1 experimentally using tabular Q-learning in a 15×15 gridworld domain. We assume that the agent is not provided the current goal state g, but only learns of g upon successfully arriving at $g \in \mathcal{S}$. Before g is known, the LOVR1 agent selects actions according to $\pi : \mathcal{S} \to \mathcal{A}$,

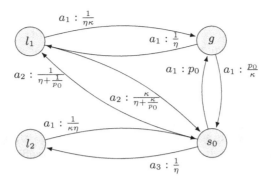

Fig. 1. Example of MDP that realizes bounds of proposition 2. Initial state s_0 with goal task g. The values listed represent the transition probabilities for the given action. For clarity we do not include self-loops in the figure, which occur for each action with probability $1 - p$, for p as listed. We allow $p_0 > 0$ to be arbitrary.

thus only has access to primitive actions. Upon arrival at the goal state, each subsequent episode the LOVR1 agent can now select actions according to $\pi : \mathcal{S} \to \mathcal{A}_+ := \mathcal{A} \cup \{a_{reflect}\}$, where $a_{reflect}$ selects the landmark option that maximizes $a_{reflect}$ as defined previously. In the first phase, we experiment with hand-selected landmarks $|\mathbb{L}| \in \{9, 25, 225\}$ (with respective pseudo-coverings of $\eta \in \{4, 2, 0\}$), where these associated Q-value functions are learned off-policy while the agent explores the environment taking actions randomly for a fixed number (100k) of steps. During the second phase of learning 100 tasks are sampled uniformly at random and each task is represented by a random goal and initial state, ensuring that $s_0 \neq g$. Each task is run for 1000 episodes where after every exploration episode (using ϵ-greedy, $\epsilon = 0.1$), an evaluation episode is carried out following the greedy policy. Greedy evaluation episodes terminate when the agent arrives at the goal or after for 5000 time steps have passed. For these experiments, as a proof of concept we aim to demonstrate the reduction in regret over a baseline Q-learning agent that does not use any form of transfer, as well as demonstrate the benefits of LOVR1, which constrains which options are available to the agent, over a version of the LOVR1 agent with access to $a_{reflect}$ but also can select any landmark options independently of $a_{reflect}$. We denote the agent with unrestricted access to the landmark options via its defined action set, $\mathcal{A}_{\mathcal{O}_l} := \mathcal{A}_+ \cup \mathcal{O}_l$. Note that the LOVR1 agent increases its set of actions by only a single action, while the naive $\mathcal{A}_{\mathcal{O}_l}$ agent increases its set of actions by the number of landmark options plus one. We consider both the deterministic setting and the stochastic setting where the agent successfully transitions along the intended cardinal direction with probability $p = 0.85$, and along any of the other cardinal directions uniformly with mass $1 - p$. Under these dynamics, it is clear that for the deterministic setting $\kappa = 1$, while it can be shown that under the stochastic setting, for all values of η considered, we have $\kappa < \frac{5}{2}$. Each set of experiments are repeated for n = 100 trials with a different random seed.

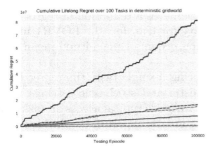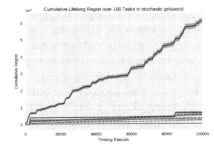

Fig. 2. Cumulative lifelong regret. Left: Deterministic, Right: Stochastic gridworld. Black- Baseline Q-learning agent, Blue-$|\mathbb{L}| = 9$, Green-$|\mathbb{L}| = 25$, Red-$|\mathbb{L}| = 225$ landmarks. Dashed: $\mathcal{A}_{\mathcal{O}_l}$, solid: $\mathcal{A}_+$. Mean $\pm$ 1 s.d. is shown. (Color figure online)

Figure 2 demonstrates that the LOVR1 implementations drastically reduce the regret over the lifetime of the agent in both the stochastic and deterministic settings, as compared to the baseline Q-learning agent with no form of transfer. We observe also that implementations of LOVR1 using $\mathcal{A}_+$ consistently outperform those using $\mathcal{A}_{\mathcal{O}_l}$. These empirical findings are a nice proof of concept, since using $\mathcal{A}_+$ only augments the set of actions by a single element, $a_{reflect}$, while adding the landmark options as well would significantly increase the number of actions, and hence sample complexity and regret, as many landmark options are not useful for a given task, and hence can result in negative transfer. Table 1

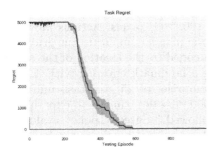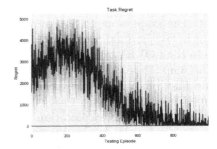

Fig. 3. Single Task Regret (Task 1). Left: Deterministic, Right: Stochastic gridworld. Black- Baseline Q-learning agent, Red-$|\mathbb{L}| = 225$ landmarks with $\mathcal{A}_+$. Mean $\pm$ 1 s.d. is shown. (Color figure online)

Table 1. Mean fold reduction in cumulative lifelong regret over 100 tasks as compared to baseline.

$|\mathbb{L}|$, $\mathcal{A}$	9, $\mathcal{A}_+$	9, $\mathcal{A}_{\mathcal{O}_l}$	25, $\mathcal{A}_+$	25, $\mathcal{A}_{\mathcal{O}_l}$	225, $\mathcal{A}_+$	225, $\mathcal{A}_{\mathcal{O}_l}$
Deterministic	10.6	4.9	24.3	5.2	1148.3	159.9
Stochastic	56.9	12.0	186.9	7.2	1491.4	6.8

demonstrates that though the $\mathcal{A}_{\mathcal{O}_l}$ agent still achieves a reduction in regret over a baseline agent with no options and no ability for transfer, the LOVR1 implementation drastically outperforms it, achieving orders of magnitude mean fold reduction in cumulative lifelong regret.

To demonstrate the regret experienced during a single task, the regret from the first task was plotted for both deterministic and stochastic settings with $L = 225$, using $\mathcal{A}_+$ (Fig. 3). Note these results are quite typical across all tasks. In the deterministic setting, the agent experiences greater regret in the first episode, but upon selecting $a_{reflect}$, follows a landmark option that has extremely small regret for the remaining 999 episodes. In comparison, the flat Q-learning agent with no landmarks must solve each task separately, requiring a few hundred episodes before finding the optimal policy. Similar results can be seen in the stochastic setting. Though the stochastic plot appears to receive no regret for the LOVR agent, very small (but non-zero) regret is accumulated.

Taken together, the results from LOVR1 experiments demonstrate a strong proof of concept for using landmark value functions for transfer learning on future goal-based tasks. For the following set of experiments we test the utility of landmark value functions in transfer learning using function approximation (neural networks).

LOVR2 Results. We considered next a set of experiments to test the utility of landmark coverings using function approximation (convolutional neural networks), to experiment with a different instantiation of LOVR, specifically LOVR2 (described previously), and the setting where the agent is provided the current goal state g.

For LOVR2 experiments we constructed an MDP environment as seen in Fig. 4. The state is an RGB image of size (100, 100) pixels. Actions move the avatar in the cardinal directions in a stochastic manner similar to the gridworld used in the LOVR1 experiments. A goal is encoded by the location of the avatar on the map. Here, the LOVR2 agent learns a landmark covering with $|\mathbb{L}| = 8$ in the first phase, and the second phase comprises of 25 goal task uniformly sampled from $\mathcal{S} \setminus \mathbb{L}$. Each task is run for 1000 episodes, similar to experiments for LOVR1 (ϵ-greedy exploration episode followed by greedy policy evaluation episode). Each episode terminates when either the goal has been reached, or a maximum of 250 time steps have elapsed. The start of each episode the avatar start state is sampled uniformly at random amongst the four corner positions of the map. Each set of experiments is repeated 5 times. We instantiate the LOVR2 agent using a Deep Q-network (DQN), and compare its performance to a baseline DQN agent with no transfer, and to a baseline transfer agent using a UVFA network trained with a similar phase I using the same landmark goal states as the LOVR2 agent. Note that UVFA encodes the goal state into the state representation, while our LOVR2 implementation does not, but accesses the goal state information via the options framework encoding.

As the main goal of this work is to assess mechanisms for positive transfer learning *given the agent has previously solved* a set of prior tasks, we used supervised learning to train the LOVR2 and UVFA agent networks on learning the optimal Q-value functions for each of the landmark value functions, for both LOVR2 and UVFA agents. We leave a more thorough exploration and study of how to efficiently construct landmark coverings and learning landmark value functions for future work. Hence, beginning the second phase the agent has access to approximately optimal Q-value functions, Q_θ^l, $\forall l \in \mathbb{L}$, parameterized by the network weights θ, and the UVFA agent has access to an approximately optimal UVFA, $U_\theta(s; g)$, $\forall g \in \mathbb{L}$.

Figure 4 (right) shows the mean cumulative lifelong regret of each agent. We see that the baseline DQN agent solves each task, eventually, however without any transfer mechanism is forced to solve each task from scratch, and hence has the highest lifelong regret. As expected, the UVFA agent outperforms the baseline DQN agent, however the LOVR2 drastically outperforms both baseline agents, receiving an almost 4-fold mean reduction in lifelong regret over the baseline DQN agent and an over 3-fold mean reduction in lifelong regret over the UVFA agent (Table 2). Taken together, the two sets of experiments demonstrate empirically that the LOVR framework can provide a principled approach towards transfer learning through landmark options, and can result in a significant reduction in lifelong regret in the goal-based multi-task RL setting.

Table 2. Statistical summary of lifetime of agent implementations. Mean (1 s.d.) $\times 10^5$ reported. Fold reduction reported with respect to baseline DQN agent regret. S.C. - sample complexity

Agent	Lifelong S.C.	Lifelong regret	Fold reduction in regret
DQN	36.88 (0.72)	31.07 (0.72)	1.00
UVFA	31.65 (0.12)	25.84 (0.12)	1.20
LOVR2	13.76 (0.22)	7.95 (0.22)	3.91
Oracle	0.58	0	∞

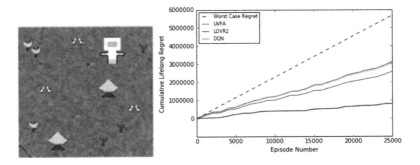

Fig. 4. LOVR2 experimental results. Left: Image of the MDP state space. Right: Cumulative lifelong regret over 25 tasks. Mean across 5 runs $\pm$ s.d. is plotted for various implementations.

Discussion. We presented the landmark options via reflection (LOVR) framework as a general approach to solving goal-based multi-task RL problems where the use of options can be shown to reduce regret over the lifetime of the agent. We show that landmark states that suitably cover $\mathcal{S}$ allow for an attractive mechanism for transfer learning through encoding properties of $\{Q^l\}_{l \in \mathbb{L}}$ into the options framework. In doing so, access of a high level controller to a possibly large set of options is restricted in a manner that promotes positive transfer and limits negative transfer across future tasks. For the setup considered in this study, theoretical results with respect to regret bounds as well as empirical results in both tabular and function approximation settings, both demonstrate the strength of LOVR as a principled approach for transfer learning in RL.

One attractive aspect, which we leave for future work, is that by using landmark options, the LOVR agent must only solve the optimal path from l to g that is constrained to lie in an $\eta\kappa$-pseudo ball around l. Given any base learning algorithm with sample complexity and regret dependant on $|\mathcal{S}|$, this $\eta\kappa$-pseudo ball may have a considerably smaller set of states to search over, and thus solving for the optimal policy from l to g requires significantly less interactions with the environment than solving from scratch the optimal policy from s to g, using the same base learning algorithm. Hence, up until some number of episodes K, LOVR will achieve a provable reduction in regret.

As η is a hyper-parameter, this allows for a clear mechanism to trade off what an acceptable level of regret may be for tasks given in the second phase, and the amount of training time that can be allocated to the first phase. For example, given a consumer product such as a floor cleaning agent that can be given commands such as "clean under the table", the commercial success of such an agent would depend on how quickly it can learn a sequence of optimal policies. If an agent had to learn completely from scratch, such a product would fail. However, a LOVR agent trained in a virtual environment on a set of landmark tasks before leaving the factory (phase I) could allow for more efficient transfer to tasks given in the home. The framework presented here is a first step towards a general framework for transfer in goal-based RL, in which the set $\mathbb{L}$ of landmarks plays a key role and will be the subject of further study in follow-up work.

Acknowledgments. The authors thank Doina Precup for helpful discussions.

References

1. Ahissar, M., Hochstein, S.: The reverse hierarchy theory of visual perceptual learning. Trends Cogn. Sci. **8**, 457–464 (2004)
2. Barto, A., Mahedevan, S.: Recent advances in hierarchical reinforcement learning. Discrete Event Dyn. Syst. **13**, 341–379 (2003)
3. Baxter, J.: A model of inductive bias learning. J. Artif. Intell. Res. **12**, 149–198 (2000)
4. Bourne, J., Rosa, M.: Hierarchical development of the primate visual cortex, as revealed by neurofilament immunireactivity: early maturation of the middle temporal area (MT). Cereb. Cortex **16**, 405–514 (2006)

5. Brunskill, E., Li, L.: Sample complexity of multi-task reinforcement learning. In: Conference on Uncertainty in Artificial Intelligence (UAI) (2013)
6. Dayan, P., Hinton, G.: Feudal reinforcement learning. In: NIPS, pp. 271–278 (1998)
7. Frans, K., Ho, J., Abbeel, P., Schulman, J.: Meta learning shared hierarchies. Technical report (2017). arxiv:1710.09767 [cs.LG]
8. Guergiuev, J., Lillicrap, T., Richards, B.: Towards deep learning with segregated dendrites. Technical report (2016). arxiv:1610.00161 [cs.LG]
9. Kearns, M., Singh, S.: Near-optimal reinforcement learning in polynomial time. Mach. Learn. **49**, 209–232 (2002)
10. Koenig, S., Simmons, R.: Complexity analysis of real-time reinforcement learning. In: AAAI, pp. 99–105 (1993)
11. Konidaris, G., Barto, A.: Building portable options: skill transfer in reinforcement learning. In: IJCAI, pp. 895–900 (2007)
12. Laroche, R., Fatemi, M., Romoff, J., van Seijen, H.: Multi-advisor reinforcement learning. Technical report (2017). arxiv:1704.00756 [cs.LG]
13. Liu, Y., Brunskill, E.: When simple exploration is sample efficient: identifying sufficient conditions for random exploration to yield PAC RL algorithms. In: European Workshop on Reinforcement Learning (2018)
14. Mann, T., Mannor, S., Precup, D.: Approximate value iteration with temporally extended actions. J. Artif. Intell. Res. **53**, 375–438 (2015)
15. Perkins, T., Precup, D.: Using options for knowledge transfer in reinforcement learning. Technical report UM-CS-99-34 (1999)
16. Schaul, T., Horgan, D., Gregor, K., Silver, D.: Universal value function approximators. In: ICML (2015)
17. van Seijen, H., Fatemi, M., Romoff, J., Larcohe, R., Barnes, T., Tsang, J.: Hybrid reward architecture for reinforcement learning. Technical report (2017). arxiv:1706.04208 [cs.LG]
18. van Seijen, H., Fatemi, M., Romoff, J., Laroche, R.: Separation of concerns in reinforcement learning. Technical report (2017). arxiv:1612.05159 [cs.LG]
19. Silver, D., Yang, Q., Li, L.: Lifelong machine learning systems: beyond learning algorithms. In: AAAI Spring Symposium: Lifelong Machine Learning, pp. 49–55 (2013)
20. Strehl, A., Li, L., Wiewiora, E., Langford, J., Littman, M.: PAC model-free reinforcement learning. In: ICML, pp. 881–888 (2006)
21. Sutton, R., Barto, A.: Reinforcement Learning: An Introduction. MIT Press, Cambridge (2016)
22. Sutton, S., Precup, D., Singh, S.: Between mdps and semi-MDPs: a framework for temporal abstraction in reinforcement learning. Artif. Intell. **112**, 181–211 (1999)
23. Thrun, S., Pratt, L.: Learning to Learn. Kluwer Academic Publishers, Norwell (1998)
24. Vezhnevets, A., et al.: Feudal networks for hierarchical reinforcement learning. Technical report (2017). arxiv:1703.01161 [cs.LG]

The Invisible Power of Fairness. How Machine Learning Shapes Democracy

Elena Beretta[1,3]([⊠]) [ID], Antonio Santangelo[1] [ID], Bruno Lepri[3] [ID],
Antonio Vetrò[1,2] [ID], and Juan Carlos De Martin[1] [ID]

[1] Nexa Center for Internet & Society, DAUIN, Politecnico di Torino, Turin, Italy
{elena.beretta,antonio.santangelo,antonio.vetro,demartin}@polito.it
[2] Future Urban Legacy Lab, Politecnico di Torino, Turin, Italy
[3] Fondazione Bruno Kessler, Trento, Italy
lepri@fbk.eu

Abstract. Many machine learning systems make extensive use of large amounts of data regarding human behaviors. Several researchers have found various discriminatory practices related to the use of human-related machine learning systems, for example in the field of criminal justice, credit scoring and advertising. Fair machine learning is therefore emerging as a new field of study to mitigate biases that are inadvertently incorporated into algorithms. Data scientists and computer engineers are making various efforts to provide definitions of fairness. In this paper, we provide an overview of the most widespread definitions of fairness in the field of machine learning, arguing that the ideas highlighting each formalization are closely related to different ideas of justice and to different interpretations of democracy embedded in our culture. This work intends to analyze the definitions of fairness that have been proposed to date to interpret the underlying criteria and to relate them to different ideas of democracy.

Keywords: Machine learning · Fairness · Equity · Discrimination

1 Introduction

Nowadays, machine learning systems make an extensive use of large amounts of data on human behaviors, collected through various channels (e.g. social media, apps, mobile phone usage, credit card transactions, etc.). The widespread resort to these data for disparate purposes is rapidly transforming several domains of our daily life. However, the increasing use of automated-software and machine learning systems, for example in the field of criminal justice [2], advertising [25], and credit scoring [15], is raising a wide range of legal and ethical issues. It is obviously impossible to provide an exhaustive mapping of all the problems, but through an easy expedient a useful logical scheme summarizing the various ones can be supplied. As a matter of fact, if we move from one of the many elementary definitions of "algorithm" - an encoded procedure to transform *input* (data) in

© Springer Nature Switzerland AG 2019
M.-J. Meurs and F. Rudzicz (Eds.): Canadian AI 2019, LNAI 11489, pp. 238–250, 2019.
https://doi.org/10.1007/978-3-030-18305-9_19

output (expected result) through a series of calculations - we understand that we are confronted with three constitutive elements: (1) *input*; (2) *procedure*; (3) *output*. A major concern in the first phase is the presence of biases in the input dataset (this is not the only one: for example, personal data protection is also a relevant issue). In the second phase, a largely debated issue is the transparency and accessibility of the procedure - the problem of the *black box* [21]; while the main problematic aspect of the third phase is the possible discriminatory effects of the algorithmic decision. A given ethical and/or legal problem must be studied with reference to: (a) the phase in which it is mainly located, and (b) the possible propagation of the issue throughout the algorithm's elaboration path. Since many application domains of machine learning algorithms are not protected by law against discrimination, the attention of actors belonging to different sectors is increasingly focusing on the way algorithms encode prejudices and lead to disproportionate results [20]. As a further development of the emergence of adverse outcomes, many researchers are involved in finding solutions to overcome the problem of discrimination in automated software systems by embedding the idea of *fairness* in the algorithm's structure. In this paper, we contribute to the debate on fairness in machine learning by discussing the impact that its recent mathematical formalizations may have on societies. In particular, we reflect on the meaning of each definition of fairness in relation to the different ideas of democracy and equality they take with them. Our hypothesis is twofold: on one side, we suppose different democratic cultures may foster the diffusion of different definitions of fairness; on the other side, the different ways of designing machine learning tools fit in different ways in the various societies and may fit better than others, especially because in a same social contexts various ideas of democracy may coexist. We describe in Sect. 2 the role of fairness in the field of machine learning; several recent mathematical implementations of fairness and various democracy typologies are exposed and analyzed in Subsects. 2.1 and 2.2, respectively. In Sect. 3 we discuss commonalities among fairness definitions and democracy typologies, pointing out challenges and open issues for the fair machine learning research domain. Finally, Sect. 4 draws some conclusions and discuss potential future works on this topic.

2 Why Fairness Matters?

The way machine learning systems act is a crucial matter for our societies [20]. Algorithms are designed to recognize situations leading to satisfactory outcomes, they are modeled to look for patterns and characteristics in individuals that have historically brought to success, thereby not making things fair if randomly employed, but replicating models and past practices. A relevant example is the recent case, exposed by the investigative website ProPublica[1], about the COMPAS recidivism tool, an algorithm used to inform criminal sentencing decisions

---

[1] https://www.propublica.org/article/machine-bias-risk-assessments-in-criminal-sentencing.

by predicting recidivism. In particular, the study found that COMPAS is evaluating useful risk factors for classification, such as socio-economic status and type of crime, which share high covariance with some sensitive attributes, in this case the race. Although the reason for this high covariance may be the result of a preexisting prejudice present in a wider system than one considered for the analysis of recidivism, it should be underlined that the study found that COMPAS was significantly more likely to falsely label as high-risk of recidivism black defendants because of the above mentioned interaction, and to underestimate the risk of white defendants by less accurately detecting the false negative rate. Although research is increasingly concentrating in finding solutions to mitigate the segregating effect of some algorithms, many efforts will have to be done to consider when an algorithm fails, for whom it fails and what are the social costs of the failure [9]. Hence, designing a fair algorithm entails two aspects, closely related to each other: firstly, to evaluate the meaning of choosing one kind of fairness instead of another in a certain society; and secondly, to assess the degree of social acceptability subordinated to the context and to the selected fairness criterion. Considering that more than one definition of fairness cannot be achieved simultaneously [18], design choices have a relevant impact on the effect that the algorithm's outcomes will have on society. It is therefore quite relevant starting to figure out which kind of societal values and democratic concepts are tied to the current mathematical formalizations of fairness. In the following sections, we analyze the most relevant mathematical definitions of fairness provided in the machine learning research area, assuming that a decision on how to design a fair machine learning system could be acceptable with more probability as much as it corresponds to the ideas of democracy shaped by individuals in the social context in which it will be used.

2.1 Fairness in Machine Learning Domain

In the recent years, several formal definitions of fairness have been suggested by the machine learning community. In Table 1, we report the most widespread ones grouped by similar characteristics: in particular, the first column indicates the name of the partitioning, while the second one the extended name of the fairness definition. Finally, the third column contains scientific references which are then further specified in Table 2.

First of all, we provide some mathematical notations that compose a typical setup in a machine learning domain: X denotes the features of an individual; Y denotes the target variable; A denotes a sensitive attribute (i.e. gender, race, etc.); C denotes a classifier; S denotes a score function or a conditional expectation. For example, the frequency of an event given certain observed characteristics can be written as $S = E[Y|X]$; t is a threshold. In case of binary classifiers, the score value causes the acceptance of classifier outputs when it is above t, otherwise causes the rejection.

Then, we introduce and briefly describe the fairness definitions listed in Table 1, supplied with examples regarding risk assessment in the criminal justice domain. Individuals rated high risk of re-offending are classified by 0, otherwise

Table 1. Fairness in machine learning literature

Partition	Definition	Reference
Group fairness	Statistical parity	[1,3,9]
	Accuracy parity	[10]
	False positive parity	[5,7]
	Positive rate parity	[4,11,12]
	Predictive parity	[6,7]
	Predictive value parity	[3]
	Equal opportunity	[4,7,9]
	Equal threshold	[4,7]
	Well-calibration	[2]
	Balance for positive class	[2]
	Balance for negative class	[2]
Individual fairness		[1]
Counterfactual fairness		[9]
Preference-based fairness	Preferred treatment	[8]
	Preferred impact	[8]
Fairness through unawareness		[1,4,9]

1 - that means low risk of recidivism. The variable *race* has been considered as a sensitive and protected attribute.

Group Fairness. Below, we introduce several formal definitions of *group fairness*.

Statistical parity. Classifier C satisfies *statistical parity* if $P_a(C = 1) = P_b(C = 1)$ for all groups a, b - i.e. $a = black, b = white$. This means that both black and white people should have equal probability to be classified as low risk.

Accuracy parity. Classifier C satisfies *accuracy parity* if $P_a(C = Y) = P_b(C = Y)$ for all groups a, b. This means that both black and white people should have equal probability to be correctly classified as low risk, if belonging to actual low risk rate, and correctly classified as high risk, if belonging to actual high risk rate.

False positive parity. Classifier C satisfies *false positive parity* if $P_a(C = 1 \mid Y = 0) = P_b(C = 1 \mid Y = 0)$ for all groups a, b. This means that both black and white people with actual high risk rate should have equal probability to be incorrectly classified as low risk (False Positive Rate).

Positive rate parity. Classifier C satisfies *positive rate parity* if $P_a(C = 1 \mid Y = i) = P_b(C = 1 \mid Y = i)$, $i \in 0, 1$, for all groups a, b. This means that both black and white people should have equal probability to be incorrectly classified as low risk - False Positive Rate - and to be correctly classified as low risk (True Positive Rate).

Table 2. References

Reference	Paper number
[12]	[1]
[18]	[2]
[2]	[3]
[15]	[4]
[9]	[5]
[24]	[6]
[7]	[7]
[28]	[8]
[19]	[9]
[10]	[10]
[27]	[11]
[4]	[12]

Predictive parity. Classifier C satisfies *predictive parity* if $P_a(Y = 1 \mid C = 1) = P_b(Y = 1 \mid C = 1)$, for all groups a, b. This means that both black and white people with low risk predicted score (Positive Predictive Value) should have equal probability to really belong to the low risk class.

Predictive value parity. Classifier C satisfies *predictive value parity* if $(P_a(Y = 1 \mid C = 1) = P_b(Y = 1 \mid C = 1)) \wedge (P_a(Y = 0 \mid C = 0) = P_b(Y = 0 \mid C = 0))$ for all groups a, b. This means that both black and white people with low risk predicted score (Positive Predicted Value) should have equal probability to really belong to low risk class, and both black and white people with high risk predicted score (Negative Predictive Value) should have equal probability to really belong to high risk class.

Equal opportunity. Classifier C satisfies *equal opportunity* if $P_a(C = 1 \mid Y = 1) = P_b(C = 1 \mid Y = 1)$ for all groups a, b. This means that both black and white people with actual low risk rate should have equal probability to be incorrectly classified as high risk (False Negative Rate). Since mathematically a classifier that satisfies False Negative Rate equity satisfies at the same time True Positive Rate equity, the definition also implies that both black and white people with actual low risk rate should have equal probability to be correctly classified as low risk.

Equal threshold. Classifier C satisfies *equal threshold* if $P_a(Y = 1 \mid S = s) = P_b(Y = 1 \mid S = s)$, $s \in [0, 1]$, for all groups a, b. This means that both black and white people should have equal score threshold t under which they are classified at low risk, and above which they are classified at high risk.

Well-calibration. Classifier C satisfies *well-calibration* if $P_a(Y = 1 \mid S = s) = P_b(Y = 1 \mid S = s) = s$, $s \in [0, 1]$, for all groups a, b. This means that both black and white people with the same score should be treated comparably *"with respect to the outcome, rather than treating black and white people with the same score differently based on the race group they belong to"* [18].

Balance for positive class. Classifier C satisfies *balance for positive class* if $E_a(S \mid Y = 1) = E_b(S \mid Y = 1)$, for all groups a, b. This means that both black and white people with an actual low risk rate should have the same expected value assigned by the classifier C (a classifier uses the characteristics of individuals to identify which class - or group - they belong to). That is to say, it should not happen that the scoring process is "systematically more inaccurate for negative cases - high risk score - in one group than the other" [18].

Balance for negative class. Classifier C satisfies *balance for negative class* if $E_a(S \mid Y = 0) = E_b(S \mid Y = 0)$, for all groups a, b. This means that both black and white people with an actual high risk rate should have the same expected value assigned by the classifier C. That is to say, it should not be that the scoring process is "systematically more inaccurate for positive cases - low risk score - in one group than the other" [18].

Individual Fairness. Given a set of individuals V and a set of outcomes $A = \{0, 1\}$, and considering a metric on individuals d: $V \, x \, V \rightarrow R$ and randomized mappings $M : V \rightarrow \Delta A$, *individuals fairness* is achieved if a randomized classifier, mapping individuals to distributions over outcomes, minimizes expected loss subject to the (D, d)-Lipschitz condition of $D(Mx, My) \leq d(x, Y)$ [12].

This means that two individuals are similarly classified if they are considered similar with respect to a particular task, such as to pay off a debt with a bank.

Counterfactual Fairness. Classifier C satisfies *counterfactual fairness* if $P(C_{A \leftarrow a} (U^2) = y \mid X = x, A = a) = P(C_{A \leftarrow a'}(U) = y \mid X = x, A = a)$. That is, given a set of attributes (*education level, type of crime, drugs problems* and protected attribute $A = race$) and an outcome $\hat{Y}$ to be predicted (*recidivism*), a graph is counterfactually fair if *race* is not directly linked to $\hat{Y}$ through any other attributes.

Intuitively, this means that a decision is fair towards an individual if it is the same in (i) the actual world and (ii) a counterfactual world where the individual belonged to a different demographic group (i.e. white instead of black).

Preference-Based Fairness. Here, we introduce new formalization of fairness [28] that are inspired by the concepts of fair division in economics and game theory [3, 26].

---

[2] "*U is a set of latent background variables, which are factors not caused by any variable in the set V of observable variables*" [19].

Preferred treatment. Classifier C satisfies *preferred treatment* if $B_a(C_a) \geq B_{a'}(C_{a'})$, for all a, a' $\in A^3$. This means that the preferred condition is preserved if each group obtains more benefit from their own classifier then it would be assigned from any other classifier. In other words, both black and white people should prefer *"the set of decisions they receive over the set of decisions they would have received had they collectively presented themselves to the system as members of a different sensitive group"* [28].

Preferred impact. Classifier C satisfies *preferred impact* if $B_a(C) \geq B_a(C')$, for all a $\in$ A. This means that the preferred condition is preserved if a classifier C, with respect to any other classifier, assigns at least the same benefit for all groups. In other words, both black and white people should prefer *"the set of decisions they receive over the set of decisions they would have received under the criterion of impact parity"* [28].

Fairness Through Unawareness. Classifier C satisfies *fairness through unawareness* if X: $X_a = X_b \rightarrow C_a = C_b$ for both individuals a, b. This means that for example the attribute *race* should not be used to train the classifier and thus to take a decision (i.e. granting or not a loan).

2.2   Fairness and Democracy

In order to understand which idea of fairness can be affirmed in a certain society, it may be useful to relate it to the type of democracy in force within the latter. In this way, we may reflect on the concept of equality that this idea of democratic life brings with it. As the philosopher Christiano writes [8], democracy is a procedure for taking collective decisions characterized by the fundamental equality of those who participate in it. According to Bozdag and Van Den Hoven [6], there are at least four different ways of conceiving democracy, which must be seen as some sort of ideal types, that interact and confront each other in reality, giving rise to the different democratic regimes in force in the world. The first is the model of *liberal democracy*, commonly centred on the defence of fundamental rights and freedoms (i.e. self-determination, choice, private property, expression, etc.) from all forms of coercive power [11,16]. It is an individualistic vision of democratic life, according to which democracy must be the place that preserves the freedoms of the individual, characterizing itself as a procedure which, through electoral competition and voting, represents and aggregates the will of individuals. In this case, the criterion of justice on which the concept of equality between persons is based denies the maxim of egalitarianism, according to which all men must be (at the very least) equal in everything, but admits the equality of all only in something, that is, in the so-called fundamental rights [5, p. 95]. The second model is that of *deliberative democracy* [6], seen as a way

---

[3] B_a *is the fraction of beneficial outcomes received by users sharing a certain value of the sensitive attribute a* [28].

of taking collective decisions, based on a confrontation between free and equal citizens. This confrontation must be aimed at finding logical solutions to satisfy the good of all. Here, the emphasis is less on the individual and more on her/his participation in common life. Democracy is not seen as the mere aggregation of the will of individuals in a vote, but as a place where people can talk to each other, make their points of view count and try to understand, together, which is the best one. The fulcrum of all this is the exercise of a rationality, in a certain sense, devoid of the will to affirm partisan interests, which is very similar to the rationality theorized by Habermas [14] and Rawls [22,23]. In this case, equality is above all in the possibility of having access to the deliberative process and in ensuring that one can confront others, even if s/he is part of a minority. The third model of democracy is what Bozdag and Van Den Hoven [6] call *republican* or *protesting*. It is based on the idea that all citizens must be free from the arbitrary exercise of power by someone. For this reason, they must be able to control the activity of those who govern them, being at the same time able to challenge it. More than the consensus of deliberative democracy, therefore, in this case it is fundamental the power to challenge choices and points of view that are not shared. It is very important the transparency and publicity of the actions of the rulers, and it is essential that anyone can question them, in a way that can produce some effect. This is the kind of equality that the supporters of this model of democracy are aiming for. The fourth model is that of *competitive democracy* [17]. It logically contrasts with that of deliberative democracy, because its proponents do not believe that people, by confronting and using pure rationality, can reach an agreement on the common good. Often, the positions of each one are based on passions, partisan interests, instances that one does not want or cannot negotiate. Democracy, then, must consist of a set of procedures that allow the supporters of the various positions to express themselves and to confront each other, in order to prevail. This must be done knowing that the opponents, having lost a confrontation, will have the opportunity to assert their demands in future occasions. Equality, in this case, is similar to the one existing in a competitive context, in which the participants have, at the beginning, the same opportunities to assert themselves. It is defined, in fact, as *equality of opportunity* [5, pp. 24–26]. To these four models, however, it is necessary to add a fifth that we could define as *egalitarian democracy*, following Bobbio's reasoning [5, pp. 26–38]. It is based on the ideal of guaranteeing a *de facto* equality which, as Bobbio himself writes (ibidem, p. 27), is economic equality. In practice, egalitarian democracy tries to reduce the distance between those who have less and those who have more, with equalizing mechanisms such as progressive taxes, subsidies to the poor, and so on. Hence, the principle on which it is based is the opposite of that of competitive democracy, in the sense that its proponents recognize the differences between people rather than what they have in common, and try to favor those who are disadvantaged.

3   Results

In the previous section we exposed the most widespread fairness criteria formalized in the machine learning domain, observing that the achievement of fairness is subordinated to various probabilistic constraints, each of them related to one or more parts which an algorithm is composed of (see Sect. 1). The connection among machine learning approaches and democratic structures similarly exists: as well as fair parameters can be encoded both in procedure and in algorithms' outcome, even democratic values can be accomplished both in procedural and in outcome part of a decision. Therefore, as we show in Fig. 1, selecting a fairness criterion instead of another is a trade-off process where the decision-maker is called upon to determine if the achievement of the desired goal entails preserving parity (all individuals are equal on the basis of some protected attributes) or satisfying preferences (individuals may have different preferences independent from protected attributes), and if it has to be reached by means of procedure constraints or impact factors. In other words, should the algorithm be fair in the way it takes a decision or should the decision be fair itself? The first column refers to the practice of preserving a kind of equality among people, while the second one refers to the practice of satisfying preferences; rows indicate in which part of an algorithm the fairness criteria are embedded and if inequality is mitigated from the beginning or in a second phase. The following partitioning (Fig. 1), inspired by Gajane and Pechenizkiy [13] is the result of our reasoning, which is explained in details for each democracy type in the following subsections.

Fig. 1. Trade-off in fairness selection process related to democracy typologies. Legend: *Competitive* (green), *Liberal* (yellow), *Egalitarian* (orange) (Color figure online)

Liberal Democracy. Liberal democracy is a model based on self-determination and preservation of individual freedom. Analyzing the mathematical formulation of *individual fairness*, we believe that such a notion of fairness underlies an idea of liberal democracy; in fact, even though attention is paid to protect minorities' rights, individual diversity is chiefly preserved in preferential way, considering individuals not because of an intrinsic differentiation but rather because of some different tasks for which they are similarly treated.

Competitive Democracy. This kind of democracy expresses the idea that actors cannot reach an agreement among themselves on a specific result by reason of utility; this reasoning implies the condition occurring when allocation of goods is such that it is not possible to improve the condition of one actor without worsening the condition of another one. So the best that can be done is guaranteeing equality of opportunity among groups through preserving and protecting minorities' rights. Therefore, from the point of view of competitive democracy, the mechanism of satisfying preferences could be seen as considering like a game which each individual should have the same opportunity to play. In this sense, almost all the fairness definitions that we have reported reflect the competitive democracy criteria; in fact, the majority of them encodes the idea that all people should have the same possibility of being correctly classified (i.e. each individual should have the same opportunity to receive a loan if s/he deserves it). From a statistically point of view, fairness definitions clustered in the *group fairness* partition are not substantially very different, the disparity mainly consists on the observer's point of view, or rather the stakeholder's; i.e., in *statistical parity* we assume the society's point of view that wonder *"Is the selected set demographically balanced?"*; in *false positive parity* we assume the defendant's point of view that wonder *"What is the probability that I'm incorrectly classified as high-risk?"*; in *predictive value parity* we assume the decision-maker point of view that wonder *"Of those I have labeled at high risk, how many will be recidivists?"*.

Egalitarian Democracy. This type of democracy is inspired to Bobbio's notion of *distributive egalitarianism*, intended as a way of redistributing material resources between advantaged and disadvantaged people [5, pp. 30–38]. But if we extend the idea of redistribution also to political resources, knowledge, etc., then we can apply it to the fairness criteria we have reported. In fact, no definition of fairness completely fulfills egalitarian democracy criteria, but preference-based models are the closest ones; in fact, both definitions of preference-fairness refer to the concept of *benefit*, which as expressed in the definition of Zafar *et al.* [28] is very similar to the concept of favoring individuals belonging to minorities. We exclude republican and deliberative democracy from the general classification reported in Fig. 1. The problems that are faced, when reflecting on the republican model of democracy, are usually touched, within the ethics of data and algorithms, when referring to the issues of transparency and accountability of computer tools. Since the attention mainly focuses on monitoring who exercises the power and on controlling the decisions, we believe actually the whole general ground of fair machine learning has been inspired by *republican democracy* values. But none of the definitions of fairness we have reported have directly to do with transparency and accountability, which have to be considered as the results of the whole fairness process.

Instead, *deliberative democracy* is a model based on agreement and on the idea that all individuals are free and equal to each others, when having the possibility of seeing their instances represented and taken into consideration in the contexts where decisions about them and their lives are taken. Its ethical premises are very similar to the ones of *competitive democracy*. In fact, strong attention is given to preserve minorities' rights and to give everyone equal possibilities to participate to the game of democratic debate. But once again, we think none of the fairness definitions we have reported has directly to do with guaranteeing that democratic decisions are taken acknowledging the instances of everybody. Moreover, we believe at the present the field of fair machine learning still lacks fairness definitions built on egalitarian democratic principles. In fact, if almost all fairness definitions are to some extent based on the principle that everyone should have equality of opportunities, egalitarian democratic definitions can be considered in contrast to this principle, because they rely on a sort of an *inequality principle*, according to which a disadvantaged individual must be favored over others. Criteria relating to most of the fairness definitions are linked to competitive democracy which may be a following stage of the egalitarian democracy, because competitive democracy rules, according to which everyone should have the opportunity to fairly participate at the competition, can be better implemented if inequalities are previously resolved through a distributive logic.

4   Discussion and Conclusions

Over the last decade machine learning has radically changed the way we take a decision. Many researchers studied the performances of statistical models and human judgments, demonstrating how in some circumstances models based on machine learning have surpassed those based on human judgments [1]. Moreover, by now machine learning algorithms are well known to use a large-amount of data to induce a particular rule starting from a generalized datum; in this article we have addressed problems related to the correctness of these rules. The examples enhancing machine learning models have in fact shown how often the results reflect bias and human prejudices in different contexts - especially those in which we try to reproduce and analyze human behavior. Hence, several efforts have been devoted to finding solutions that re-calibrate balances in outputs of machine learning systems. However, the debate around *fairness* the in machine learning domain displays a profound and relatively worrying lack: although researchers are acting and reacting in a positive and proactive way, data scientists and computer engineers are increasingly involved in taking decisions that affect individuals, operating as judges in decision-making processes and constituting a sort of *invisible power* - although in a positive way - whereas society at large should play this role. In this article, we enrich these efforts and the related debate, highlighting that meanings of fairness, underlying to statistical constraints, and the democratic values are strictly connected; in fact, we carry on the idea that the spread of one fairness definition instead of another is highly justified by the ideas of justice and democracy shared among the society in which the above-mentioned fairness criterion is selected. We consider this qualitative analysis as a prelude of quantitative researches. For example, our study may pave the way for the implementation of an agent-based model that mimics the decision-making, where agents are informed by different kind of democratic values. The aim of the model would be to analyze which statistical parameters lead to specific fairness criteria and under which probabilities. Moreover, we plan to further develop our work by proposing a new formalization of fairness based on egalitarian democratic principles.

References

1. Barocas, S., Hardt, M., Narayanan, A.: Fairness and Machine Learning. fairml-book.org (2018). http://www.fairmlbook.org
2. Berk, R., Heidari, H., Jabbari, S., Kearns, M., Roth, A.: Fairness in criminal justice risk assessments: the state of the art. Sociol. Methods Res. (2018)
3. Berliant, M., Thomson, W.: On the fair division of a heterogeneous commodity. J. Math. Econ. **21**, 201–216 (1992)
4. Binns, R.: Fairness in machine learning: lessons from political philosophy. In: Friedler, S.A., Wilsonf, C. (eds.) Proceedings of Machine Learning Research, vol. 81, pp. 149–159 (2018)
5. Bobbio, N.: Eguaglianza e libertá. Einaudi, Torino, Italy (1995)
6. Bozdag, E., van den Hoven, J.: Breaking the filter bubble: democracy and design. Ethics Inf. Technol. **17**, 249–265 (2015)
7. Chouldechova, A.: Fair prediction with disparate impact: a study of bias in recidivism prediction instruments. Big Data (2017)
8. Christiano, T.: Democracy. Stanford Encyclopedia of Philosophy (2006)
9. Corbett-Davies, S., Pierson, E., Feller, A., Goel, S., Huq, A.: Algorithmic decision making and the cost of fairness. In: Proceedings of the 23rd ACM SIGKDD International Conference on Knowledge Discovery and Data Mining, KDD 2017 (2017)
10. Dieterich, W., Mendoza, C., Brennan, T.: Compas risk scales: demonstrating accuracy equity and predictive parity. Technical report, Northpointe Inc. (2016)
11. Dunn, J.: Western Political Theory in the Face of the Future, vol. 3. Cambridge University Press, Cambridge (1979)
12. Dwork, C., Hardt, M., Pitassi, T., Reingold, O., Zemeln, R.: Fairness through awareness. In: Proceedings of the 3rd Innovations in Theoretical Computer Science Conference, pp. 214–226. ACM (2012)
13. Gajane, P., Pechenizkiy, M.: On formalizing fairness in prediction with machine learning arXiv:1710.03184 (2018)
14. Habermas, J.: Between Facts and Norms: Contributions to a Discourse Theory of Law and Democracy. MIT Press, Cambridge (1998)
15. Hardt, M., Price, E., Srebro, N.: Equality of opportunity in supervised learning. In: Advances in Neural Information Processing Systems (2016)
16. Held, D.: Models of Democracy. Stanford University Press, Palo Alto (2006)
17. Held, D.: The Democratic Paradox. Verso, London (2009)
18. Kleinberg, J., Mullainathan, S., Raghavan, M.: Inherent trade-offs in the fair determination of risk scores. In: Proceedings of Innovations in Theoretical Computer Science, ITCS 2017 (2017)
19. Kusner, M.J., Loftus, J.R., Russell, C., Silva, R.: Counterfactual fairness. In: Proceedings of 31st Neural Information Processing Systems, NIPS 2017 (2017)
20. O'Neil, C.: Weapons of Math Destruction: How Big Data Increases Inequality and Threatens Democracy. Crown Publishing Group, New York (2016)
21. Pasquale, F.: The Black Box Society: The Secret Algorithms That Control Money and Information. Harvard University Press, Cambridge (2015)

22. Rawls, J.: A Theory of Justice. Harvard University Press, Harvard (1971)
23. Rawls, J.: The Idea of Public Reason. The MIT Press, Cambridge (1997)
24. Simoiu, C., Corbett-Davies, S., Goel, S.: The problem of infra-marginality in outcome tests for discrimination. Ann. Appl. Stat. **11**, 1193–1216 (2017)
25. Sweeney, L.: Discrimination in online ad delivery. Queue Storage **11**(3), pages 10 (2013)
26. Varian, H.: Equity, envy, and efficiency. J. Econ. Theory **9**, 63–91 (1974)
27. Zafar, M.B., Valera, I., Rodriguez, M.G., Gummadi, K.P.: Fairness beyond disparate treatment and disparate impact: learning classification without disparate mistreatment. In: Proceedings of the 26th International Conference on World Wide Web, WWW 2017 (2017)
28. Zafar, M.B., Valera, I., Rodriguez, M.G., Gummadi, K.P., Weller, A.: From parity to preference-based notions of fairness in classification. In: Proceedings of the 31st Conference on Neural Information Processing Systems, NIPS 2017 (2017)

Maize Insects Classification Through Endoscopic Video Analysis

André R. de Geus[1(✉)], Marcos A. Batista[2], Marcos N. Rabelo[2],
Celia Z. Barcelos[1], and Sérgio F. da Silva[2]

[1] Federal University of Uberlândia, Uberlândia, Brazil
`{geus.andre,celiazb}@ufu.br`
[2] Federal University of Goiás, Catalão, Brazil
`marcos.batista@pq.cnpq.br, rabelo@dmat.ufpe.br, sergio@ufg.br`

Abstract. An early identification of insects in grains is of paramount importance to avoid losses. Instead of sampling and visual/laboratory analysis of grains, we propose carrying out the insect identification task automatically, using endoscopic video analysis methods. As the classification process of moving objects in video relies heavily on precise segmentation of moving objects, we propose a new background subtraction method and comparing their results with the main methods of the literature according to a comprehensive review. The background subtraction method relies on a binarization process that uses two thresholds: a global and a local threshold. The binarized results are combined by adding details of the object obtained by the local threshold in the result to the global threshold. Experimental results performed through visual analysis of the segmentation results and using a SVM classifier suggest that the proposed segmentation method produces more accurate results than the state-of-art background subtraction methods.

Keywords: Insects classification · Background subtraction ·
Feature extraction · SVM

1 Introduction

The growing need for food to meet the increasing global demand requires harvesting of grains to be held with minimal losses for final consumption. However, stored grains are highly susceptible to insect infestation. Currently there are several types of grain storages available, however most of them are susceptible to insect infestation. As intensive use of insecticides are discouraged due to their harmful effects on individuals both managing and consuming the grain, a timely identification of insects in the grain is of great importance for taking measures that avoid losses.

Undoubtedly, periodic sampling of grains is one of the most effective methods for identifying insects in grain. However, sampling incurs high labor costs. Also, considering a real metal silo, sampling is a difficult process once it is necessary to take distributed samples throughout the storage as a whole.

© Springer Nature Switzerland AG 2019
M.-J. Meurs and F. Rudzicz (Eds.): Canadian AI 2019, LNAI 11489, pp. 251–262, 2019.
https://doi.org/10.1007/978-3-030-18305-9_20

The process of insects identification can be greatly simplified by installing endoscopic cameras in silos and then performing computer vision analysis to determinate if there are insects in the grains and classify them. There are two main approaches for such computer vision methods: automatic feature learning methods and those that rely on pre-designed feature extraction.

Feature learning methods aim at learning automatically an adequate description of the data by optimizing some computer vision-based criteria. There are many variations of such methods including supervised [1], unsupervised [2] and reinforcement [3] learning methods. Considering the supervised methods which are more related to the scope of this research we have methods that perform as of the simple object classification by considering static images, up to methods which recognize complex activities in videos taking into account spatio-temporal information. For example, [4–6] have proposed feature learning methods based on convolutional deep learning for event recognition. According to papers [4–6] the proposed deep-learning based methods have achieved impressive results in both accuracy and recognition speed allowing real-time applications. In the range of supervised methods there are those named semantic segmentation that can be defined as the task of grouping together image parts that belong to the same object class. Semantic segmentation methods when applied to videos produces the same kind of results aimed at in this research, which is the classification of objects in videos. In the literature there are several important researches on semantic segmentation. The collaborators Girshick et al. [7] proposed convolutional neural networks (CNNs) to localize and segment objects, and a paradigm for training large CNNs when labeled training data is scarce. The study of Long et al. [8] defined a deep CNN architecture that combines semantic information from a deep and coarse layer with appearance information from a shallow and fine layer to produce an accurate and detailed semantic segmentation. Proposed in Chen et al. [1] is an elaborated methodology based on deep convolutional neural networks and reached the new state-of-art for the PASCAL VOC-2012 semantic image segmentation task and improved the results on three other datasets.

Pre-designed feature extraction methods when applied to object recognition in videos use a pre-processing step to detect the moving objects. The main approaches for detecting moving objects are optical flow [9] and background subtraction [10]. Briefly, optical flow techniques calculate velocity vectors of moving objects. These velocity vectors, once computed, can be used for a wide variety of tasks ranging from navigation to exploration of autonomous agents in their operating environments. Methods of background subtraction estimate and maintain a background model that is subtracted from the current frame to determine the foreground (segmentation of moving objects).

This research focuses on computer vision methods based on pre-designed feature extraction which relies on background subtraction (BS) methods. This choice is due to the fact that BS methods allow to delimitation of the object boundaries, which enable highly relevant shape descriptions for maze insect classification, since it is known that different species of insects causing the damage have different shapes.

As known from many research studies, object segmentation plays a key role in computer vision systems aimed at classify objects. In this research we propose a segmentation method of moving objects, and its results were compared with the most promising methods of literature according to [10]. The main innovation of the proposed background subtraction method is in the binarization process, where we use two thresholds: a global and a local. The global threshold is adjusted to capture only pixels that have a high probability of belonging to a moving object. For the local threshold we used the Sauvola method [11]. The local thresholding captures details of the object, which are added to the result of the global threshold. We consider that regions of the local threshold result are added to the final result if they are 8-connected to some region of the global threshold result. For object description we used the elliptic Fourier descriptor introduced in [12], which is robust to object size, translation and rotation. Although several studies compare background subtraction methods, as far as we are aware, this is the first study to compare these methods in the context of insect classification based on endoscopic videos. Experiments with real videos, obtained with an endoscopic camera, show that the proposed method produces more precise segmentation results than state-of-art methods.

The remainder of this article is organized as follows: Section 2 describes the proposed methodology. Section 3 describes and discusses the results. Finally, Sect. 4 summarizes the main contributions of this investigation.

2 Methodology

The first step of the methodology proposed in this paper consists of a background subtraction composed of four main steps: bootstrapping, background updating, subtraction of the background, and binarization. The bootstrapping step, that initializes the background model, uses the same approach of [13]. Background updating is performed by using a learning rate according to Eq. 1,

$$BG^{(t)} = (\alpha * I^{(t)}) + ((1 - \alpha) * BG^{(t-1)})$$ (1)

where α is the learning-rate parameter that determines how fast the background incorporates the moving objects. The subtraction of the background is performed according to Eq. 2,

$$M^{(t)}(i,j) = \frac{(I^{(t)}(i,j) - BG^{(t)}(i,j)) * v^T}{3}$$ (2)

where t is the index of the current frame, v^T is the identity vector, $I^{(t)}(i,j)$ and $BG^{(t)}(i,j)$ are the vectors of color components (R, G, B) of the pixel (i,j). Therefore, $M^{(t)}$ is a two-dimensional array at time t that represents the grayscale intensity level of each pixel.

The foreground mask, which is an image in gray scale corresponding the difference between the background model and the current frame, is binarized using two different thresholds: a user-defined global threshold to identify pixels where a large intensity variation between the current frame and the background model occurs, having a high probability of belonging to a real object in the scene; and a local threshold to identify pixels belonging to the object with a small variation compared to the local background model. The global threshold is represented by T_{global} and the local threshold employed is the one introduced in [11] and therefore called $T_{sauvola}(i,j)$ due to the first author.

The computation of the binary masks using global and local thresholds are given by the Eqs. 3 and 4, respectively. To compute the local thresholds we use a window size of 55 × 55 pixels. This investigation introduces the use of local threshold and proposed the combination of the binarized masks produced by the threshold methods, bringing contribution on the previous research of [14] that used only a global threshold for binarization.

$$B_{global}^{(t)}(i,j) = \begin{cases} 1, & \text{if } M^{(t)}(i,j) > T_{global} \\ 0, & \text{otherwise} \end{cases} \tag{3}$$

$$B_{local}^{(t)}(i,j) = \begin{cases} 1, & \text{if } M^{(t)}(i,j) > T_{sauvola}^{(t)}(i,j) \\ 0, & \text{otherwise} \end{cases} \tag{4}$$

The final binarized mask is calculated by a composition of the two binarized masks, B_{global} and B_{local}. The composition operation is given by Eq. 5. The idea behind Eq. 5 is to take pixels with large intensity variation that have high probability of belonging to a moving object and aggregate them to spatially connected regions, which have a meaningful local variation regarding the local background.

$$B_{final}^{(t)} = B_{global}^{(t)} \cup \{R \in (B_{local}^{(t)})\} \tag{5}$$

where R are regions 8-connected to $B_{global}^{(t)}$.

Figure 1 illustrates the binarization process. Figures 1d and e, respectively show the results of the global and local thresholds considering the mask in Fig. 1c. Figure 1f shows the composition result, where we can clearly see that the object details obtained by the local threshold are added to object segmentation obtained by the global threshold.

The methodology used to evaluate the proposed background subtraction method consists of a visual analysis of the segmentation results and analysis of the classification accuracy. In the object classification experiments we used elliptical Fourier descriptor [12] for feature extraction and the Support Vector Machine [15] for classification. In order to demonstrate how the Elliptical Fourier descriptor captures the shape of the objects closely, we illustrate the reconstruction of the contour of maize insects from the coefficients calculated by the descriptor.

Figure 2 shows the elliptical Fourier approximation to the shape of insects using their set of coefficients generated using 30, 20, 10 and 5 harmonics. As we can see 30 harmonics are enough for a reasonable approximation of the objects contour. In our experiments we calculated the coefficients with 30 harmonics producing overall 120 coefficients that represent each object contour.

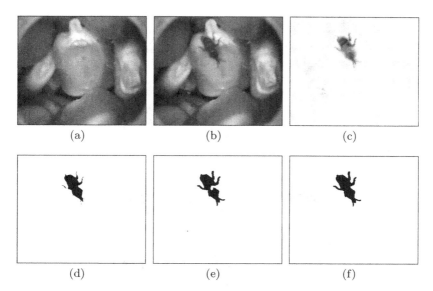

Fig. 1. (a) Background updated model; (b) Input frame; (c) mask; (d) global threshold binarized; (e) local threshold binarized image; (f) final binarized image.

For classification we use an SVM classifier that is well-known for producing high classification rates mainly for problems with a small number of classes, which is the case of this research. Given that SVM relies on kernel functions to map the feature space in a higher dimensional feature space, before the estimation of support vectors, we experiment two different kernel functions: a polynomial and a radial basis function.

3 Results

3.1 Data Set

Our video data set consists of short videos recorded within a container with bulk maize, using an endoscopic camera with a 640×426 pixel resolution. The use of a low resolution camera is based on its feasibility in a real-world implementation in metal silos due to its low cost, where a large number of endoscopic cameras are required.

In most tropical countries, *Sitophilus zeamais*, *Tribolium castaneum* and *Sitotroga cerealella* are the main insects that cause damage to stored maize. However, *Sitotroga cerealella* is a surface insect and is not recorded in our videos.

3.2 Baselines Methods

To compare the proposed method we chose the best background subtraction methods reported in the comprehensive review by Sobral and Vacavant [10] and

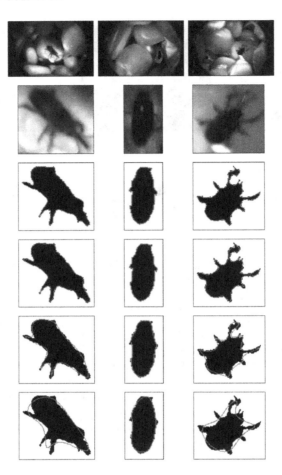

Fig. 2. Contour reconstruction using the elliptical Fourier coefficients calculated using different numbers of harmonics. First row – original frames; Second row – cropped image of insects; Third until sixth row – binarized insect image with contour approximation on red-line using 30, 20, 10 and 5 harmonics, respectively.

implemented by the same authors in the BGS library[1]. Table 1 lists the methods of the BGS Library used and the settings for each of these, which are the same as those used by Sobral and Vacavant at [10]. For more details concerning each method, the reader is referred to reference [10].

3.3 Experiments

We compare the proposed background subtraction method that uses a combination of well-known formulations for bootstraping, background updating and

---

[1] https://github.com/andrewssobral/bgslibrary.

Table 1. Background subtraction algorithms used and the parameters settings of each algorithm. The parameter settings used are the same as those in the Sobral and Vacavant investigation [10].

Method ID	Method name	Reference	Settings
Basic method: mean and variance over time			
AdaptiveBackground-Learning	Adaptive Background Learning	[10]	$T = 15$, $\alpha = 0.5$
Fuzzy based method			
FuzzyChoquetIntegral	Fuzzy Choquet Integral	[16]	$T = 0.67$, $LF = 10$, $\alpha_{learn} = 0.5$, $\alpha_{update} = 0.05$, $RGB + LBP$
Statistical method using one Gaussian			
DPWrenGABGS	Gaussian Average	[17]	$T = 12.15$, $LF = 30$, $\alpha = 0.05$
Statistical method using multiple gaussians			
MixtureOfGaussian-V1BGS	Gaussian Mixture Model	[18]	$T = 10$, $\alpha = 0.01$
Type-2 Fuzzy based method			
T2FGMM_UM	Type-2 Fuzzy GMM-UM	[19–21]	$T = 1$, $k_m = 2.5$, $n = 3$, $\alpha = 0.01$
Statistical method using color and texture features			
MultiLayerBGS	Multi-Layer BGS	[22]	Original default parameters from [22]
Method based on eigenvalues and eigenvectors			
DPEigenbackground-BGS	Eigenbackground/SL-PCA	[23]	$T = 255$, $HS = 10$, $ED = 10$
Neural method			
LBAdaptiveSOM	Adaptive SOM	[24]	$LR = 180$, $LR_{training} = 255$, $\sigma = 100$, $\sigma_{training} = 240$, $TS = 40$

subtraction and has as its innovation concerning the binarization process that produces a precise boundary delineation for a classification process relying on shape feature description with state-of-art BGS methods from the literature.

Regarding Eq. 1, in the experiments we used a relatively low learning parameter, namely 0.01, due to how quickly the insects move. The main contribution of the paper is the binarization process that is designed to capture small details regarding shape of the insects presents in the maize that apply a fundamental role to differentiate them.

3.4 Results and Discussion

Figure 3 shows the results of applying the proposed method and the selected state-of-art methods in three different frames extracted from our data set. For better visualization the negative of each binary mask is shown. In a visual analysis one can see that our approach delineates more precisely the insects shape. Among the experimented methods, those that achieved more accurate results are our approach, MixtureOfGaussianV1BGS, LBAdaptiveSOM and MultiLayer-BGS, respectively.

To perform classification using SVM, we selected 602 samples of frames containing maize weevil (Sitophilus zeamais) and 606 containing brown beetles (Tribolium castaneum). As the elliptical Fourier descriptor is applied to a single closed contour we took the largest object in each binary mask; this operation is required due to some methods producing small disconnected ghost regions. In the classification experiments we compared the proposed method only with MixtureOfGaussianV1BGS, LBAdaptiveSOM and MultiLayerBGS methods, since the other methods of background subtraction did not produce satisfactory segmentation results; this is noted through the segmentation results in Fig. 3. The selected kernels were polynomial and RBF with parameters determined by Grid-Search. The correct classification rates using k-fold cross-validation with $k = 10$ are shown on Table 2. One observes that the proposed segmentation method achieves a higher classification rate accuracy than the state-of-art methods.

Table 3 shows the confusion matrices for each combination of background subtraction method and classifier. In many cases a *Sitophilus zeamais* is incorrectly classified as *Tribolium castaneum* due to poorly segmented details of the insect, such as legs and antenna.

Table 2. Classification accuracy of the proposed method compared with state-of-art background subtraction methods.

Method	SVM polynomial	SVM RBF
Proposed method	97.10%	91.63%
MixtureOfGaussianV1BGS	91.72%	88.32%
LBAdaptiveSOM	75.16%	75.16%
MultiLayerBGS	90.96%	87.66%

The training complexity are the same when comparing the proposed method with baseline methods, considering that we already have segmented the images for training. All segmented objects are described by the same algorithm

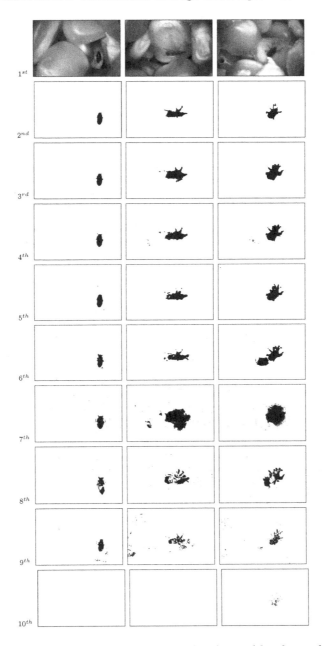

Fig. 3. Original images in the first row and results obtained by the methods analyzed in the following rows: second row – proposed method; third row – MixtureOfGaussianV1BGS; fourth row – LBAdaptiveSOM; fifth row – MultiLayerBGS; sixth row – DPWrenGABGS; seventh row – DPEigenbackgroundBGS; eighth row – AdaptiveBackgroundLearning; ninth row – FuzzyChoquetIntegral; tenth row – T2FGMM_UM. The settings of each algorithm are given in the Table 1.

Table 3. Confusion matrices regarding the classification experiments. A corresponds to the class *Tribolium castaneum*, B to the class *Sitophilus zeamais* and Acc. to the accuracy for each class. The matrices on the left are SVM polynomial kernels and the matrices on the right are RBF kernels.

Proposed	A	B	Acc.
A	594	12	98.02%
B	23	579	96.18%

(a)

Proposed	A	B	Acc.
A	593	13	97.85%
B	88	514	85.38%

(b)

MOG	A	B	Acc.
A	593	13	97.85%
B	87	515	85.55%

(c)

MOG	A	B	Acc.
A	591	15	97.52%
B	126	476	79.07%

(d)

LBA	A	B	Acc.
A	591	15	97.52%
B	285	317	52.66%

(e)

LBA	A	B	Acc.
A	591	15	97.52%
B	285	317	52.66%

(f)

M.Layer	A	B	Acc.
A	593	13	97.85%
B	107	495	82.23%

(g)

M.Layer	A	B	Acc.
A	591	15	97.52%
B	134	468	77.74%

(h)

(the elliptical fourier descriptor), which produces the same number of features, using the same set parameters. Moreover, we used the same classifier for each insects classification approach. However, when considering a real time application, is important to take into account the time for preprocessing the video, i.e., the time taken to perform background subtraction, which produces a binary image. We estimated the complexity by executing all methods from one of our videos on the same computer and recorded the time taken to segment all frames. All methods were implemented in C++ and executed on a machine Intel Core i7, 2.3 Ghz, 16 GB RAM running macOS. Table 4 shows the number of frames per seconds (FPS) taken to segment a video with 1150 frames comparing the proposed method and state-of-art background subtraction methods.

Table 4. Performance results for video segmentation in frames per second.

Method	FPS
Proposed method	23.31
MixtureOfGaussianV1BGS	20.39
LBAdaptiveSOM	7.79
MultiLayerBGS	4.44

4 Conclusions

As known from many research studies, object segmentation plays a crucial role in computer vision systems aimed at object classification. In this study we present an object segmentation method based on the background subtraction technique. Our experiments through use of a video data set containing insects shows that the proposed method achieves more accurate segmentation results than state-of-art background subtraction methods listed in a comprehensive research review. We also confirm the superiority of our background subtraction method in terms of classification, where our results were up to 5.3% superior in terms of classification accuracy. As object segmentation plays a key role in object classification, our results indicate that the proposed method can be successfully applied in other areas of applications such as surveillance systems. Furthermore, as far as we are aware, this is the first study carried out that compare background subtraction methods applied to insect segmentation from endoscopic videos. As a future investigation, we intend to experiment the proposed methods to identify people in video surveillance systems.

Acknowledgment. The authors thank the Brazilian agencies CNPq, CAPES, FAPEG and FAPEMIG for the financial support.

References

1. Chen, L.C., Papandreou, G., Kokkinos, I., Murphy, K., Yuille, A.L.: DeepLab: semantic image segmentation with deep convolutional nets, atrous convolution, and fully connected CRFs. IEEE Trans. Pattern Anal. Mach. Intell. **40**(4), 834–848 (2018)
2. Le, Q.V.: Building high-level features using large scale unsupervised learning. In: 2013 IEEE International Conference on Acoustics, Speech and Signal Processing, pp. 8595–8598 (2013)
3. Mnih, V., et al.: Human-level control through deep reinforcement learning. Nature **518**(7540), 529–533 (2015)
4. Luo, S., Yang, H., Wang, C., Che, X., Meinel, C.: Action recognition in surveillance video using ConvNets and motion history image. In: Villa, A.E.P., Masulli, P., Pons Rivero, A.J. (eds.) ICANN 2016. LNCS, vol. 9887, pp. 187–195. Springer, Cham (2016). https://doi.org/10.1007/978-3-319-44781-0_23
5. Wang, C., Yang, H., Meinel, C.: Exploring multimodal video representation for action recognition. In: 2016 International Joint Conference on Neural Networks (IJCNN), pp. 1924–1931 (2016)
6. Luo, S., Yang, H., Wang, C., Che, X., Meinel, C.: Real-time action recognition in surveillance videos using ConvNets. In: Hirose, A., Ozawa, S., Doya, K., Ikeda, K., Lee, M., Liu, D. (eds.) ICONIP 2016. LNCS, vol. 9949, pp. 529–537. Springer, Cham (2016). https://doi.org/10.1007/978-3-319-46675-0_58
7. Girshick, R., Donahue, J., Darrell, T., Malik, J.: Rich feature hierarchies for accurate object detection and semantic segmentation. In: 2014 IEEE Conference on Computer Vision and Pattern Recognition, pp. 580–587 (2014)

8. Long, J., Shelhamer, E., Darrell, T.: Fully convolutional networks for semantic segmentation. In: 2015 IEEE Conference on Computer Vision and Pattern Recognition (CVPR), pp. 3431–3440 (2015)
9. Fortun, D., Bouthemy, P., Kervrann, C.: Optical flow modeling and computation: a survey. Comput. Vis. Image Underst. **134**, 1–21 (2015)
10. Sobral, A., Vacavant, A.: A comprehensive review of background subtraction algorithms evaluated with synthetic and real videos. Comput. Vis. Image Underst. **122**, 4–21 (2014)
11. Sauvola, J., Pietikainen, M.: Adaptive document image binarization. Pattern Recogn. **33**(2), 225–236 (2000)
12. Kuhl, F.P.: Elliptic fourier features of a closed contour. Comput. Graph. Image Process. **18**, 1982 (1982)
13. Calderara, S., Melli, R., Prati, A., Cucchiara, R.: Reliable background suppression for complex scenes. In: Proceedings of the 4th ACM International Workshop on Video Surveillance and Sensor Networks, Santa Barbara, CA, USA, pp. 211–214 (2006)
14. de Geus, A.R., Batista, M.A., dos Santos Filho, T.A., da Silva, S.F.: Segmentação de classifição de insetos em milho à granel por meio de análise de vídeo. In: Simpósio Brasileiro de Automação Inteligente, Natal, RN, BRA, pp. 1–6 (2015)
15. Cortes, C., Vapnik, V.: Support-vector networks. Mach. Learn. **20**(3), 273–297 (1995)
16. El Baf, F., Bouwmans, T., Vachon, B.: Fuzzy integral for moving object detection. In: IEEE International Conference on Fuzzy Systems, Hong Kong, pp. 1729–1736 (2008)
17. Wren, C., Azarbayejani, A., Darrell, T., Pentland, A.: Pfinder: real-time tracking of the human body. IEEE Trans. Pattern Anal. Mach. Intell. **19**(7), 780–785 (1997)
18. KaewTraKulPong, P., Bowden, R.: An improved adaptive background mixture model for real-time tracking with shadow detection. In: Remagnino, P., Jones, G.A., Paragios, N., Regazzoni, C.S. (eds.) European Workshop on Advanced Video Based Surveillance Systems, pp. 135–144. Springer, Boston (2002). https://doi.org/10.1007/978-1-4615-0913-4_11
19. El Baf, F., Bouwmans, T., Vachon, B.: Type-2 fuzzy mixture of Gaussians model: application to background modeling. In: Bebis, G., et al. (eds.) ISVC 2008. LNCS, vol. 5358, pp. 772–781. Springer, Heidelberg (2008). https://doi.org/10.1007/978-3-540-89639-5_74
20. El Baf, F., Bouwmans, T., Vachon, B.: Fuzzy statistical modeling of dynamic backgrounds for moving object detection in infrared videos. In: IEEE Conference on Computer Vision and Pattern Recognition, Miami, FL, USA, pp. 60–65 (2009)
21. Bouwmans, T., El Baf, F.: Modeling of dynamic backgrounds by type-2 fuzzy gaussians mixture models. MASAUM J. Basic Appl. Sci. **1**(2), 265–276 (2010)
22. Yao, J., Odobez, J.: Multi-layer background subtraction based on color and texture. In: IEEE Conference on Computer Vision and Pattern Recognition, Minneapolis, MN, USA, pp. 1–8 (2007)
23. Oliver, N., Rosario, B., Pentland, A.: A bayesian computer vision system for modeling human interactions. IEEE Trans. Pattern Anal. Mach. Intell. **22**(8), 831–843 (2000)
24. Maddalena, L., Petrosino, A.: A self-organizing approach to background subtraction for visual surveillance applications. IEEE Trans. Image Process. **17**(07), 1168–1177 (2008)

Collaborative Clustering Approach Based on Dempster-Shafer Theory for Bag-of-Visual-Words Codebook Generation

Sabrine Hafdhellaoui, Yaakoub Boualleg$^{(\boxtimes)}$ ⓘ, and Mohamed Farah ⓘ

Univ. Manouba, ENSI, RIADI, LR99ES26 (SIIVT Team), Manouba, Tunisia
hafdellaouisabrine@gmail.com, yaakoub.boualleg@ensi-uma.tn,
mohamed.farah@riadi.rnu.tn

Abstract. Feature encoding methods play an important role in the performance of the recognition tasks. The Bag-of-Visual-Words (BoVW) paradigm aims to assign the feature vectors to the codebook visual words. However, in the codebook generation phase, different clustering algorithms can be used, each giving a different set of visual words. Thus, the choice of the discriminative visual words set is a challenging task. In this work, we propose an enhanced bag-of-visual-words codebook generation approach using a collaborative clustering method based on the Dempster-Shafer Theory (DST). First, we built three codebooks using the k-means, the Fuzzy C-Means (FCM), and the Gaussian Mixture Model (GMM) clustering algorithms. Then, we computed the Agreement Degrees Vector (ADV) between the clusters of the pairs (k-means, GMM) and (k-means, FCM). We merged the obtained ADVs using the DST in order to generate the clusters weights. We evaluated the proposed approach for Remote Sensing Image Scene Classification (RSISC). The results proved the effectiveness of our proposed approach and showed that it can be applied for different recognition tasks in various domains.

Keywords: Bag-of-Visual-Words · Codebook generation ·
Collaborative clustering · Dempster-Shafer Theory ·
Remote sensing image scene classification

1 Introduction

With the rapid advance of imaging technologies, a huge amount of visual information is becoming more and more available in different digital archives, whether they are publicly available on the Web or used for specialized applications such as in the Remote Sensing (RS) for which many freely high resolution images such as SPOT and SENTINEL are available. All these available satellite data are very useful for a wide range of critical applications in agriculture, deforestation, urban planning, etc.

© Springer Nature Switzerland AG 2019
M.-J. Meurs and F. Rudzicz (Eds.): Canadian AI 2019, LNAI 11489, pp. 263–273, 2019.
https://doi.org/10.1007/978-3-030-18305-9_21

The Remote Sensing Image Scene Classification (RSISC) aims to label and identify each image scene with its corresponding class. Since the classification performance is strongly affected by the effectiveness of the features vector, considerable efforts have been made to develop powerful feature representations. Early, image classification methods have been intensively using handcrafted global low-level visual features that are extracted from the whole image such as color, texture, and shapes [4]. Next, researches have focused more on local low-level features that are extracted from the interest points within the image such as Scale-Invariant Feature Transform (SIFT) [14], Speed Up Robust Feature (SURF) [2], Local Binary Pattern (LBP) [16], Histogram of Oriented Gradient (HOG) [6], and Pyramid Histogram of Oriented Gradient (PHOG) [3].

However, due to the high dimensionality of these features, as well as the time they need to be computed and processed, researches tend to map low-level image visual features into mid-level image representations through feature encoding methods such as Bag-of-Visual-Words (BoVW) [20], Vector of Locally Aggregated Descriptors (VLAD) [11], and Improved Fisher Kernel (IFK) [17].

Recently, Convolutional Neural Networks (CNNs), which are Deep-Learning (DL) architectures, showed significant progress in computer vision tasks. However, they have many limitations. First, learning CNN models from scratch requires a huge amount of labelled data. In addition, parameters tuning is an uninterpretable process and requires high computational power. Moreover, they are highly prone to overfitting. Transfer-Learning strategies have been proposed to alleviate the cited limits through fine tuning pretrained models or by using them as features extractors. More recently and motivated by the results of the use of CNNs for extracting deep features, new feature representations are proposed. In [5], the authors proposed to use the extracted deep convolutional feature maps from the pretrained CNN instead of using dense SIFT features. Also, in [13] multi-scale convolutional feature maps are aggregated using IFK to generate a better image representation.

Over the years, mid-level image representations, especially the BoVW method, have received increasing interest from the image classification community. This is because they have proven to be efficient for discriminating feature representations. The BoVW method was firstly proposed by Szelinski [22] based on the work of Sivic and Zisserman [20]. The main idea is to obtain image descriptions from a training set in order to generate a codebook or book of visual words by clustering the image features and using clustering centres as the words of the codebook. Then, the image is represented by the histogram of the visual words.

Most of the existing BoVW-based methods for RSISC have extensively explored various features as well as combinations of various strategies to generate the codebook of visual words. Sujatha et al. [21] proposed a multi-dictionaries model by combining the dictionaries resulting from the FCM clustering algorithm with different subsets of SIFT descriptors. In fact, during the feature-grouping step, n subsets of SIFT descriptors are randomly selected, and n dictionaries are generated using FCM. n histograms are therefore generated for each image and the final result is obtained from the concatenation of these n histograms. Jonathan et al. [15] investigated the use of a Dual BoVW model (Dual-BoVW) in a relatively conventional framework to perform image classification. They showed the superiority of a BoVW with the combination of two local-feature descriptors by creating a dual codebook which contains both local features (Dual BoVW) compared to the conventional BoVW methods (BoVW and HOG-BoVW) with a single codebook. Zurita et al. [23] proposed a hybrid classification in the BoVW Model. Firstly, SIFT descriptor was used in the feature extraction phase. Then a dictionary of words was created through a clustering process using k-means, Expectation Maximisation algorithm (EM) and k-means in combination with EM. Finally, for the classification, they compared the algorithms of Support Vector Machine (SVM), Gaussian Naïve Bayes (GNB), k-Nearest Neighbours (kNN), Decision Tree (DT), Random Forest (RF), Neural Network (NN) and AdaBoost in order to determine the performance and accuracy of every method.

In the codebook generation phase, Sivic and Zisserman. [20] used the k-means classifier to construct the vocabulary of the BoVW model. Farquhar et al. [8] have proposed an extended grouping based on a generative model called the Gaussian Mixture Model (GMM). Avrithis and Kalantidis. [1] have proposed an approximate version of GMM, called Approximate Gaussian Mixture (AGM), Sujatha et al. [21] have proposed to use the Fuzzy C-Means classifier (FCM). Each of these classification models produces different visual words for the same sample of images, which poses the problem of choosing the best set and therefore the best classifier.

Besides, in order to improve the classification results, several authors have proposed techniques for fusing multiple unsupervised classifiers. Gançarski and Wemmert. [10] proposed a multiple-view voting method to combine unsupervised classifications. Forestier et al. [9] proposed the collaborative-clustering method to fuse data. This method allows the user to exploit different heterogeneous images in a global system. The process consists of three stages: initial and parallel execution of the classifiers, result refinement and result unification. In the second step, different classifications need to converge through an assessment and a resolution of the existing conflicts. The search is done two by two. For two results, the correspondence between classes is recommended via a similarity measure. Once the refinement is complete, the results are unified using a voting algorithm.

In order to overcome the problem caused by the conflict generated by different clustering algorithms in the codebook generation and to enhance the BoVW model performance, we propose a new codebook generation approach that uses at the same time the different results produced by several classifiers in a collaborative clustering method based on the Dempster-Shafer Theory (DST). Firstly, the extracted image features are clustered into k clusters using three classifiers that are k-means, GMM, and FCM. We then apply a collaborative clustering of these three clustering results, taking the results of k-means as reference clusters since k-means is usually used for this step. The collaboration is done between the results of k-means and GMM on the one side, then between those of k-means and FCM on the other side. Indeed, for each cluster resulting from k-means, we associate a mass function. These masses measure the degree of agreement between k-means results and those of GMM as well as those of FCM. Then we use the DST to fuse the results. Finally, we obtain for each k-means cluster (visual word) a weight which indicates the cluster's confidence. The Fusion-Agreement-Degree vector (visual words weights) is used for reweighting the final image representation.

The rest of this paper is structured as follows. In Sect. 2, we describe the DST for information fusion. In Sect. 3, we focus on the proposed approach. Section 4 describes the experimental implementation and evaluation of the obtained results. We will end with a conclusion summarizing our proposal.

2   The Dempster-Shafer Theory

Information fusion is a multilevel process which serves to combine information from multiple sources to improve decision-making. We report in this section necessary theoretical elements of the DST. This theory comes from the work of Dempster [7] which was resumed by Shafer [18]. It allows modelling information imperfections, particularly the conflicts. The formalism can be described as follows. Let θ be the framework of discernment, which describes all the possible hypotheses $\theta = \{H_1, H_2, H_3, \cdots, H_k\}$. The set 2^θ of all the partitions of θ is given by:

$$2^\theta = \{A, A \subseteq \theta\} = \{\{H_1\}, \{H_2\}, \cdots, \{H_k\}, \{H_1 \cup H_2\}, \cdots, \theta\} \qquad (1)$$

A first magnitude called 'mass of belief' can be constructed. This magnitude characterizes the veracity of a proposition A for an information source S. The mass m associated with this source is defined over all the partitions of the framework of discernment θ, i.e. 2^θ, as follows:

$$m : \begin{cases} 2^\theta \longrightarrow [0,1] \\ A \longrightarrow m(A) \end{cases} \qquad (2)$$

where $\sum_{A \in 2^\theta} m(A) = 1$ and $m(\emptyset) = 0$.

Each subset $A \subset \theta$ where $m(A) > 0$ is called a focal element of m. The union of the focal elements is called the nucleus. The complete ignorance of the hypotheses set corresponds to $m(\theta) = 1$.

The DST defines precisely a mass combination rule when there are different sources. Let $A, B \in 2^\theta$ and two sources S_1 and S_2 expressing the masses of belief $m_1(*)$ and $m_2(*)$ on the elements of 2^θ. The mass of the hypothesis A resulting from the fusion of masses $m_1(*)$ and $m_2(*)$ by the application of the Dempster rule is given by:

$$\begin{cases} (m_1 \bigoplus m_2)(A) = \frac{\sum_{B_1 \cap B_2 = A} m_1(B_1)m_2(B_2)}{1-K} \\ (m_1 \bigoplus m_2)(\emptyset) = 0 \end{cases} \qquad (3)$$

where K is defined by:

$$K = \sum_{B_1 \cap B_2 = \emptyset} m_1(B_1)m_2(B_2) \qquad (4)$$

As part of the Dempster combination rule, the mass on the empty set $\emptyset$ must be zero and the sum of the masses on 2^θ must equal 1. Therefore, it is necessary to redistribute the mass assigned to the empty set on all the other masses. For this, the final mass distribution must be renormalized with the renormalization coefficient K. The rule of Dempster is a rule of a conjunctive consensus normalized by the conflict K. This global conflict is the sum of the partial conflicts resulting from the empty intersections of the focal elements of the different mass functions.

3   Proposed Approach

In this work, we propose an enhanced BoVW codebook generation based on a collaborative clustering approach. The proposed codebook uses the clustering results of several unsupervised classifiers at the same time. We use DST to reduce the conflict between the obtained clustering results and to generate the visual words' weights. The proposed weighted cookbook is tested within an image classification framework for RSISC.

As we can see from Fig. 1, the proposed approach takes as input the image descriptors and outputs a new feature representation based on a weighted codebook with a visual words weights vector, through four steps.

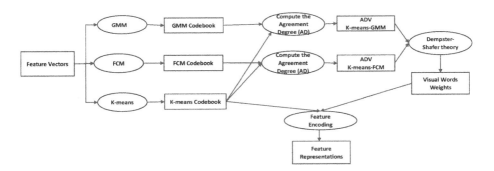

Fig. 1. The overall scheme of the proposed Bag-of-Visual-Words codebook generation.

3.1   Codebook Generation

We apply different unsupervised clustering algorithms separately on the input image feature set. We use three clustering algorithms namely k-means, FCM, and GMM.

Let R_1, R_2, R_3 denote the set of clustering results (codebooks) that are given by k-means, GMM, and FCM respectively, where $||R_1|| = N_1$, $||R_2|| = N_2$, and $||R_3|| = N_3$. And let C denote the obtained clusters as follow: $C_i^1 \in R_1$, $C_j^2 \in R_2$, and $C_l^3 \in R_3$ where $i \in [1, N_1]$, $j \in [1, N_2]$, and $l \in [1, N_3]$.

3.2   Modelling

According to the traditional BoVW-based methods and inspired by [12], we consider that the clustering result provided by k-means presents the reference classes. To merge the obtained clustering results, we build the mass functions of the unsupervised classifiers GMM and FCM, by measuring the degree of agreement between the reference classes (k-means results) and the obtained clusters using GMM and then using FCM. In order to assign the masses, we look for the proportions of the reference classes in each cluster using the intersection matrix ($intrscM$) that is obtained using intercluster correspondence function [9] (see Eq. 5). Next, the Agreement Degree Vector (ADV) is generated using the Maximum function on the $intrscM$ as described by Eq. 6 that represents the correspondence between the most similar clusters.

Define Mass Functions by Collaborating k-Means with GMM:

$$intrscM(R_1, R_c) = \begin{bmatrix} a_{1,1}^{1,c} & \cdots & a_{1,N_c}^{1,c} \\ \vdots & & \vdots \\ a_{N_1,1}^{1,c} & \cdots & a_{N_1,N_c}^{1,c} \end{bmatrix} \tag{5}$$

where $a_{i,j}^{1,c} = \frac{|C_i^1 \cap C_j^c|}{|C_i^1|}$.

$$m_1 = ADV(R_1, R_c) = \begin{pmatrix} max(a_{1,1}^{1,c} & \cdots & a_{1,N_c}^{1,c}) \\ \vdots \\ max(a_{N_1,1}^{1,c} & \cdots & a_{N_1,N_c}^{1,c}) \end{pmatrix} \tag{6}$$

where $c = 2$ to define mass functions by collaborating k-means with GMM and $c = 3$ to Define mass functions by collaborating k-means with FCM.

3.3   Fused Decision

In our method, there are two distinct and independent sources of information: the GMM and the FCM algorithms. Each source has its own mass vector (ADV) that is defined on the clustering result of k-means ($ADV(R_1, R_2) \neq ADV(R_1, R_3)$).

In order to benefit from both information sources, we fuse the normalized ADVs using the DST. We use the orthogonal Sum of DST (Eq. 3) where m_1 and m_2 are the mass functions corresponding to each cluster in R_1 with GMM clusters R_2 ($ADV(R_1, R_2)$) and with FCM clusters R_3 ($ADV(R_1, R_3)$), respectively. Finally, we get a new vector FADV (Fused-Agreement Degree Vector) where each value represents a weight associated with a k-means cluster (visual word).

3.4   Feature Encoding

Based on the clustering results of the k-means algorithm, with N_1 visual words (codebook size), we encode an input image using a global histogram representation which is determined by the frequency of each codebook visual word within the image. Next, the obtained N_1-dimensional image feature representation is reweighted using the visual words weights (FADV) based on a simple pairwise multiplication function.

4 Experiments and Results

4.1 Dataset

In order to evaluate our proposed approach, our experiments were conducted on the "NWPU-RESISC45" dataset [1] [5] which was proposed for REmote Sensing Image Scenes Classification (RESISC) by the Northwestern Polytechnical University (NWPU). The dataset is the largest publicly available aerial image dataset with 31500 remote sensing RGB images. It consists of 45 land-use classes, with 700 images per class. The aerial scene images of this dataset are acquired from Google Earth (Google Inc.) covering more than 100 countries and regions around the world. The image size is 256×256 pixel with a different special resolution that varies from 30 to 0.2 m. Figure 2 shows some sample images from this dataset.

4.2 Experimental Setup

To compare our proposed approach with the RSISC methods that are based on the BoVW feature encoding method, we selected two baseline methods; the traditional BoVW that uses the SIFT features, and BoCF which uses the convolutional feature maps from the VGG16 pretrained model [19]. In order to compare the classification performances of the proposed method with BoCF obtained results [5], we use the same training/test rate, the dataset was randomly split into 10% for the training and 90% for the test.

In the first experiment, we uses the $128D$ dense SIFT feature vector to describe the dataset's images and we compared the results with the BoVW results. Secondly we used the $13 \times 13 \times 256$ deep-feature maps extracted from the pooling layer from the last convolutional block of the pretrained VGG16 CNN model in order to compare the obtained results with the BoCF results. Similarly to the BoVW and BoCF methods, in the codebook generation phase, for the k-means, GMM and FCM clustering algorithms, we set the number of clusters C to 500, 1000, 2000, 5000, and 10000 and we investigated the impact of the codebook size on the classification performance of the proposed approach.

The classification results were obtained using a linear SVM classifier. We evaluated the classification performance using the same evaluations metrics that used in [5] (i.e. Overall Accuracy (OA) and the confusion matrix metrics). All the experiments were performed with a t2.2xlarge machine that is available on the Amazon Web Service EC2 instance[2]. The proposed approach was implemented in python 2.7.

---

[1] http://www.escience.cn/people/JunweiHan/NWPU-RESISC45.html.
[2] https://aws.amazon.com/ec2/instance-types/.

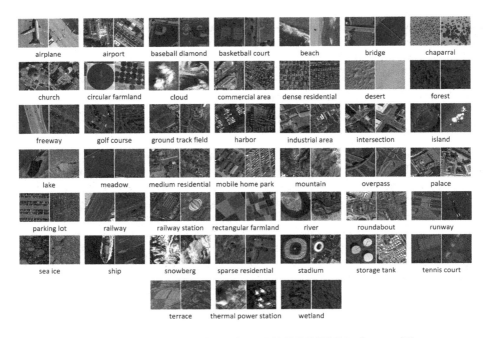

Fig. 2. Sample images from the NWPU-RESISC45 dataset [5]

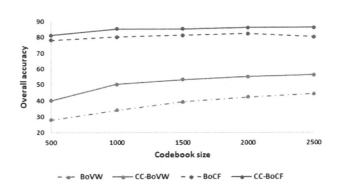

Fig. 3. Overall accuracies on the NWPU-RESISC45 dataset of BoVW with SIFT, BoCF with VGG16 deep features, and their enhanced versions using the proposed collaborative clustering method CC-BoVW and CC-BoCF respectively.

4.3 Results and Discussion

Figure 3 shows the Overall Accuracies (OA) that are obtained by using the baseline methods, the BoVW using SIFT features and the BoCF using the VGG16 deep features, and their enhanced versions using the proposed collaborative clustering method CC-BoVW and CC-BoCF in terms of the codebook size.

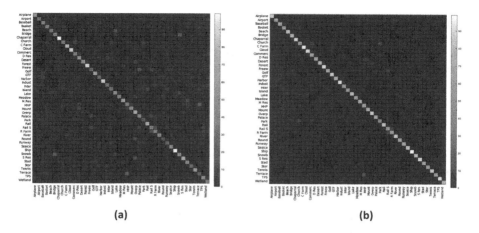

Fig. 4. Confusion matrices showing classification performance on the NWPU-RESISC45 dataset for BoVW with SIFT (a) and the enhanced version using the proposed collaborative clustering method CC-BoVW (b).

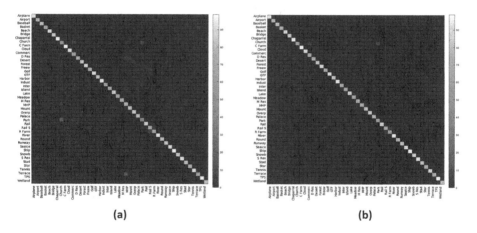

Fig. 5. Confusion matrices showing classification performance on the NWPU-RESISC45 dataset for BoCF with VGG16 deep features (a) and the enhanced version using the proposed collaborative clustering method CC-BoCF (b).

From Fig. 3, we can observe that the overall accuracy is influenced by the codebook size; for the BoCF, the codebook size 5000 gives the best OA. However, for the other methods, the largest codebook size gives the highest OA. The proposed approach achieves a better performance compared to the baseline methods on all the codebook size variation. For coodebook size 10000, our proposed approach CC-BoVW achieves 54.2% OA, which is better than the traditional BoVW by 10.2%. Also, for the proposed CC-BoCV, the OA is boosted by 5.8% compared to the traditional BoCF. For more details, the classification performance for each class is presented on the confusion matrices where the rows and columns of the matrix represent actual and predicted classes, respectively. In Fig. 4, the BoVW is compared to the proposed CC-BoCV, and in Fig. 5, the BoCF is compared with the proposed CC-BoCF.

5 Conclusion

In this paper, we propose a new Bag-of-Visual-Words codebook generation approach, based on a collaborative clustering method using the DST. In the codebook generation, each clustering algorithm gives a different codebook. In order to reduce the conflict between these sources of information, the collaborative clustering method was used to associate a weight value for each visual word of the k-means codebook. This weight represents the fused agreement degree of the GMM and the FCM codebooks with the k-means clusters. To fuse the obtained agreement degree vectors, we used the orthogonal sum of the DST. The proposed approach was evaluated on a remote sensing image scene classification framework, which achieves encouraging results compared to the RSISC baselines, and this showed that it can be applied for different recognition tasks in various domains.

References

1. Avrithis, Y., Kalantidis, Y.: Approximate gaussian mixtures for large scale vocabularies. In: Fitzgibbon, A., Lazebnik, S., Perona, P., Sato, Y., Schmid, C. (eds.) ECCV 2012. LNCS, vol. 7574, pp. 15–28. Springer, Heidelberg (2012). https://doi.org/10.1007/978-3-642-33712-3_2
2. Bay, H., Tuytelaars, T., Van Gool, L.: SURF: speeded up robust features. In: Leonardis, A., Bischof, H., Pinz, A. (eds.) ECCV 2006. LNCS, vol. 3951, pp. 404–417. Springer, Heidelberg (2006). https://doi.org/10.1007/11744023_32
3. Bosch, A., Zisserman, A., Munoz, X.: Representing shape with a spatial pyramid kernel. In: Proceedings of the 6th ACM International Conference on Image and Video Retrieval, pp. 401–408. ACM (2007)
4. Cheng, G., Han, J.: A survey on object detection in optical remote sensing images. ISPRS J. Photogramm. Remote Sens. **117**, 11–28 (2016)
5. Cheng, G., Li, Z., Yao, X., Guo, L., Wei, Z.: Remote sensing image scene classification using bag of convolutional features. IEEE Geosci. Remote Sens. Lett. **14**(10), 1735–1739 (2017)
6. Dalal, N., Triggs, B.: Histograms of oriented gradients for human detection. In: IEEE Computer Society Conference on Computer Vision and Pattern Recognition, 2005, CVPR 2005, vol. 1, pp. 886–893. IEEE (2005)
7. Dempster, A.P.: Upper and lower probabilities induced by a multivalued mapping. Ann. Math. Statist. **38**(2), 325–339 (1967). https://doi.org/10.1214/aoms/1177698950
8. Farquhar, J., Szedmak, S., Meng, H., Shawe-Taylor, J.: Improving 'bag-of-keypoints' image categorisation: Generative models and pdf-kernels (2005)

9. Forestier, G., Wemmert, C., Gançarski, P.: Multisource images analysis using collaborative clustering. EURASIP J. Adv. Sig. Process. **2008**(1), 11 (2008)
10. Gançarski, P., Wemmert, C.: Collaborative multi-strategy classification: application to per-pixel analysis of images. In: Proceedings of the 6th International Workshop on Multimedia Data Mining: Mining Integrated Media and Complex Data, pp. 15–22. ACM (2005)
11. Jégou, H., Douze, M., Schmid, C., Pérez, P.: Aggregating local descriptors into a compact image representation. In: 2010 IEEE Conference on Computer Vision and Pattern Recognition (CVPR), pp. 3304–3311. IEEE (2010)
12. Karem, F., Dhibi, M., Martin, A., Bouhlel, M.S.: Credal fusion of classifications for noisy and uncertain data. Int. J. Electr. Comput. Eng. (IJECE) **7**(2), 1071–1087 (2017)
13. Li, E., Xia, J., Du, P., Lin, C., Samat, A.: Integrating multilayer features of convolutional neural networks for remote sensing scene classification. IEEE Trans. Geosci. Remote Sens. **55**(10), 5653–5665 (2017)
14. Lowe, D.G.: Distinctive image features from scale-invariant keypoints. Int. J. Comput. Vis. **60**(2), 91–110 (2004)
15. Maas, J.L., Okafor, E., Wiering, M.A.: The dual codebook: combining bags of visual words in image classification. In: Proceedings of the 28th Benelux Artificial Intelligence Conference (BNAIC), pp. 46–71 (2016)
16. Ojala, T., Pietikainen, M., Maenpaa, T.: Multiresolution gray-scale and rotation invariant texture classification with local binary patterns. IEEE Trans. Pattern Anal. Mach. Intell. **24**(7), 971–987 (2002)
17. Perronnin, F., Sánchez, J., Mensink, T.: Improving the fisher kernel for large-scale image classification. In: Daniilidis, K., Maragos, P., Paragios, N. (eds.) ECCV 2010. LNCS, vol. 6314, pp. 143–156. Springer, Heidelberg (2010). https://doi.org/10.1007/978-3-642-15561-1_11
18. Shafer, G.: A Mathematical Theory of Evidence, vol. 42. Princeton University Press, Princeton (1976). ISBN 9780691100425
19. Simonyan, K., Zisserman, A.: Very deep convolutional networks for large-scale image recognition. arXiv preprint arXiv:1409.1556 (2014)
20. Sivic, J., Zisserman, A.: Video google: a text retrieval approach to object matching in videos. In: null, p. 1470. IEEE (2003)
21. Sujatha, K., Keerthana, P., Priya, S.S., Kaavya, E., Vinod, B.: Fuzzy based multiple dictionary bag of words for image classification. Procedia Eng. **38**, 2196–2206 (2012)
22. Szeliski, R.: Computer Vision: Algorithms and Applications. Springer Science & Business Media, Heidelberg (2010). https://doi.org/10.1007/978-1-84882-935-0
23. Zurita, B., Luna, L., Hernandez, J., Ramírez, J.: Hybrid classification in bag of visual words model. Circ. Comput. Sci. **3**(4), 10–15 (2018)

Memory-Efficient Backpropagation for Recurrent Neural Networks

Issa Ayoub and Hussein Al Osman[(✉)]

University of Ottawa, Ottawa, Canada
{Iayou005,Hussein.AlOsman}@uottawa.ca

Abstract. Recurrent Neural Networks (RNN) process sequential data to capture the time-dependency in the input signal. Training a deep RNN conventionally involves segmenting the data sequence to fit the model into memory. Increasing the segment size permits the model to better capture long-term dependencies at the expense of creating larger models that may not fit in memory. Therefore, we introduce a technique to allow designers to train a segmented RNN and obtain the same model parameters as if the entire data sequence was applied regardless of the segment size. This enables an optimal capturing of long-term dependencies. This technique can increase the computational complexity during training. Hence, the proposed technique grants designers the flexibility of balancing memory and runtime requirements. To evaluate the proposed method, we compared the total loss achieved on the testing dataset after every epoch while varying the size of the segments. The results we achieved show matching loss graphs irrespective of the segment size.

Keywords: Backpropagation through time · Gradient calculation ·
Recurrent Neural Networks

1 Introduction

Artificial Neural Networks (ANNs) belong to a class of machine learning algorithms that can be applied to supervised and unsupervised learning problems. They are composed of connected nodes called artificial neurons which are organized into an input layer, intermediate layers, and an output layer. The neurons are connected through unidirectional weighted edges (which carry the output of a neuron to the next layer in the case of the input and intermediate layers' neurons). To calculate the output of a neuron in the intermediate or output layers, we compute the weighted sum of values it collects through its incident edges and feed the result into a non-linear activation function. Training the ANN involves varying the weights and biases at the input of each neuron by calculating gradients of the error with respect to each variable in the model's parameter through backpropagation [1] and optimizing the weights and biases using the gradient descent algorithm.

Commonly, ANNs, such as Convolutional Neural Networks (CNNs), consider their input data to be independent and identically distributed [1]. However, RNNs break

© Springer Nature Switzerland AG 2019
M.-J. Meurs and F. Rudzicz (Eds.): Canadian AI 2019, LNAI 11489, pp. 274–283, 2019.
https://doi.org/10.1007/978-3-030-18305-9_22

from this convention by processing input data sequentially [1], which involves the application of the recurrence formula over multiple time steps:

$$h_t = \sigma(W_{hh}h_{t-1} + W_{xh}x_t) \tag{1}$$

Where σ denotes a non-linear function, h_t is the current state or the hidden vector at time t, which is a function of the hidden state at a previous time step h_{t-1} multiplied by the transition matrix W_{hh} added to the current input x_t multiplied by the weight matrix W_{xh}.

Theoretically, RNNs can be extended to infinitely long sequences. An RNN is similar to a state machine where the output at each time step is a function of the current input and the output of the previous step [1]:

$$y_t = W_{hy}h_t \tag{2}$$

Where y_t is the output of the RNN cell at time step t, h_t is the current hidden state, and W_{hy} is a weight matrix.

To train an RNN, we apply backpropagation over time [2]. By unraveling the network, we consider the RNN to be a feed forward network with multiple layers. Hence, we can train the RNN through the conventional backpropagation algorithm. However, in this case, all time steps share the same set of weights.

Theoretically, gradients at each time step are back propagated until time 0 and updating the model parameters is done by summing the gradients followed by optimizing using gradient descent.

As the number of time steps increases, the depth of the RNN, once unraveled, increases as well. Hence, we may encounter the vanishing or exploding gradients problems during model training [3]. We can address the exploding gradients problem by clipping the gradients. However, to solve the vanishing gradients problem, researchers have proposed new RNN architectures such as the Long Short-Term Memory (LSTM) unit [4] and Gated Recurrent Unit (GRU) [3, 5]. These architectures solve the vanishing gradient problem by applying an identity function to the input of the LSTM or GRU cells. The identity function has a derivative of 1, hence gradients are back propagated without vanishing [3]. In this paper, we will use the term RNN to refer to both LSTM and GRU cells as the proposed method applies to both architectures.

Given the requirement needed by the standard back propagation through time (BPTT) algorithm, a memory constraint arises when we unroll the RNN for a large number of time steps [6, 7] as the internal states and activations of all cells are stored in memory to compute the gradients. As a result, researchers commonly employ truncated-BPTT by dividing the network into a fixed number of segments of length S [2, 8]. The last hidden state from a segment is fed as an initial hidden state to the next segment. However, dividing the network into a fixed number of segments prohibits the RNN cells from capturing long-term dependencies [9]. This problem arises since RNN gradients are summed up during training. Summing S gradients from a single segment is not equivalent to summing all gradients in the network. Therefore, when the RNN is truncated, the gradients only back propagate within each segment as opposed to across the entire network.

We propose a backpropagation technique for a segmented RNN that enables the same training behavior as the non-truncated RNN. The proposed method benefits from the advantage of truncated backpropagation in terms of minimizing the training memory footprint to that of a single segment of length S while maintaining the same result as the non-truncated backpropagation. Furthermore, the method permits the application of any loss function during training. The latter can be calculated using the output from all the time steps as opposed to being limited to an RNN segment.

We organize the rest of this paper as follows: we survey the related work in Sect. 2, detail our method in Sect. 3, present and discuss our results in Sect. 4, and provide concluding remarks in Sect. 5.

2   Background and Related Work

To perform backpropagation in a typical neural network, we calculate and store in memory all intermediate results in a forward pass and compute the gradients in a backward pass. This process requires only one forward pass. Figure 1 describes the order through which node related computations are executed in the forward and backward passes. In Fig. 1, the nodes are evaluated in the same order as they are labeled, assuming that all neurons in the same layer are activated simultaneously. Therefore, after evaluating the first layer, its activations are stored in memory as they are needed for the evaluation of the second layer. The same scenario is repeated to evaluate the third layer.

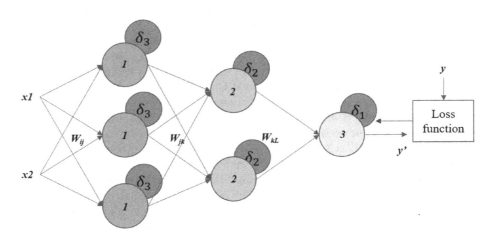

Fig. 1. Sequence of node generation during forward and backward pass

After calculating the output, the loss is computed and the error is propagated from the third to the first layer. Computing gradients δ_1 requires the activations of the third layer. To compute gradients δ_2, the activations of the second layer and δ_1 are needed.

Hence, performing back propagation requires the maintenance of all the activations in memory, which entails a memory consumption of $O(n)$ where n is the number of layers.

Thus, the memory requirement increases as the number of layers, n, of a network grows since all intermediate results are maintained in memory until they are no longer needed for the backpropagation process.

In the case where memory resources are severely limited, only few intermediate results need to be stored in memory. When the outputs of a layer are computed, the corresponding inputs can be removed from memory. According to this scheme, and using the example of Fig. 1, after calculating the activations for the second layer, those of the first layer are removed. Likewise, after computing the activations for the third layer, those of the second layer are cleared. When we calculate the error δ_1, to back propagate the layer, we need to evaluate the activations of the second layer again as they were not maintained. As a result, the first and second layers' activations are re-evaluated. These activations are cleared once gradients δ_2 are calculated. This procedure continues until all gradients are computed and results in $O(n^2)$ evaluations where n is the number of layers. Thus, this method becomes impractical for considerably deep networks due to the quadratic relationship between the computational complexity and number of layers.

In situations where memory resources are less restricted, we can truncate the neural graph into several segments. The output of the last node and the gradient calculated with respect to the initial node for each segment is maintained in memory. Hence, we only need to re-compute the output of nodes within the segment being back propagated to compute its gradients. Chen et al. [10] proved that any network of n layers can be trained using $O(\sqrt{n})$ memory at the cost of an additional forward operation per mini-batch.

When we apply the Chen et al. [10] scheme to RNNs, after dividing the sequence of length T into $\sqrt{T}$ segments, the output intermediate hidden state of a segment is stored in memory and subsequently used to compute intermediate activations in the next segment during the backpropagation of the latter. Thus, the Chen et al. [10] technique reduces the cost of memory allocation to $O(\sqrt{T})$ for RNNs while requiring an additional feed forward pass on average.

Jaderberg et al. [6] attempt to decouple segments in RNNs using synthetic gradients. They unroll the RNN and add an additional time step at the end of each segment. The last cell in each segment is responsible for predicting future gradients. The aim of the synthetic gradients is to approximate future gradients of the losses with respect to the hidden state activations at any time step $t < T$ where T is the length of the data sequence. The future gradients are the outputs of function (3):

$$f_g = Wx + b \tag{3}$$

Where x is the last hidden state, from the additional unrolled RNN cell, W is the weights of the linear function, b is the bias and f_g is the future gradient. This method does not support the application of a loss function that depends on the entire data sequence. For example, for the cross-cultural sub-challenge of AVEC2018 [11], the organizers propose the use of Concordance Correlation Coefficient (CCC) as a loss function to optimize the network's parameters which depend on the entire data sequence, rather than a segment of the network.

3   Proposed Method

An RNN can be truncated into several, typically equally sized, segments to relieve the training memory requirements [2, 8]. The segment size can affect the calculation of the loss and gradients if they are calculated from each data segment independently as opposed to the entire sequence. Consequently, a small segment size can prevent the model from capturing long-term dependencies [6]. Conversely, a large segment size would better learn long-term dependencies at the expense of increased memory resource requirements. The segment size can be increased until the number of segments is reduced to 1, in which case the RNN is no longer truncated and the memory resource requirements are the most stringent. The calculation of some loss functions, such as the CCC, require using the outputs for the entire data sequence. Hence, calculating the loss based on each segment separately is not feasible in this case.

The proposed method uses a truncated RNN training strategy. However, we introduce a mechanism to ensure that the calculation of the loss matches that of non-truncated RNNs. Furthermore, when we back propagate gradients from the final time step to time step 0, the sum of gradients with respect to the model parameters at time step 0 is equal to the sum of gradients we would obtain if the RNN was not truncated. Hence, the modeler can benefit from the less stringent memory requirement that a truncated RNN enables while achieving training results that reproduce those of a non-truncated RNN.

We summarize the proposed approach as follows:

- Step 1: Truncate the RNN into m segments of size S.
- Step 2: Feed the data points into the RNN and record predictions and the last hidden state for each segment. The last recorded hidden state in a segment is fed to the next segment. Concatenate predictions from all segments (See Fig. 2).
- Step 3: Given the ground truth, compute the loss and the gradients of the loss with respect to the predictions (L_t and $\frac{\partial E_t}{\partial o_t}$ in Fig. 2).
- Step 4: Feed the data points into the RNN starting from segment m and traversing backwards towards segment 0 while at each segment (See Fig. 3):
  - A - Back propagate the recorded gradients of predictions with respect to the model parameters which we refer to as local parameters. Then, sum the local gradients according to Eq. (4):

$$\frac{\partial E}{\partial w} = \sum_{t=1}^{S} \frac{\partial E_t}{\partial w} = \sum_{t=1}^{S} \left(\frac{\partial E_t}{\partial o_t} * \frac{\partial o_t}{\partial h_t} * \left(\prod_{k=t+1}^{S} \frac{\partial h_k}{\partial h_{k-1}} \right) * \frac{\partial h_t}{\partial w} \right) \quad (4)$$

Where E is the error, w are the weights, o and h are the predictions and hidden states at time step t.

- B - Back propagate the gradients of the last hidden state with respect to the model parameters using Eq. (5). We refer to these as the future gradients.

$$\frac{\partial h_S}{\partial w} = \sum_{t=0}^{S} \frac{\partial h_t}{\partial w} * \prod_{k=t+1}^{k=S} \frac{\partial h_k}{\partial h_{k-1}} \tag{5}$$

- C - Back propagate the gradients of the predictions with respect to the initial hidden states according to Eq. (6).

$$\frac{\partial E}{\partial h_0} = \sum_{t=0}^{S} \frac{\partial E_t}{\partial o_t} * \frac{\partial o_t}{\partial h_t} * \prod_{k=1}^{k=t} \frac{\partial h_k}{\partial h_{k-1}} \tag{6}$$

- D - Add local and future gradients to obtain the total gradients.
- E - Back propagate the gradients from the last hidden state, if it exists, to the initial hidden state and add them to the gradients collected at C.
- F - Repeat step 4 until we reach segment 0.
- Step 5: Apply the gradients to the model parameters.

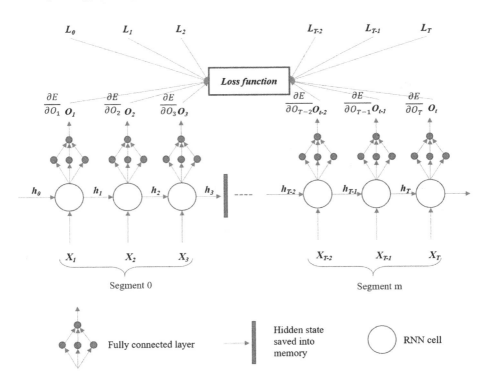

Fig. 2. This figure demonstrates what happens during the initial forward pass for the RNN. All inputs are fed sequentially starting from segment 0 to segment m, where the length of a segment is 3 time steps in this case. At the boundaries between segments, the hidden states are saved and used as the initial hidden state in the next iteration. Predictions are recorded and saved in memory. At the end, predictions, O_t, are concatenated and the partial derivative of the error E is computed with respect to the predictions given the Labels L_t.

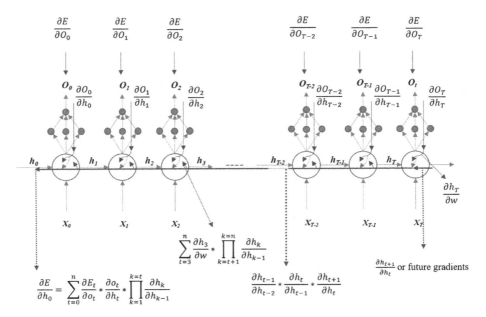

Fig. 3. This figure illustrates the gradient propagation within one segment

4 Evaluation

To evaluate the proposed scheme, we will train several truncated RNN models with various segment sizes and compare them to a non-truncated RNN in terms of the accuracy, execution time, and memory consumption during training. Hence, the objective of the evaluation is to:

1. Illustrate the relationship between memory consumption and execution time during training; and
2. Demonstrate that all compared models achieve the same accuracy at the end of each epoch and thus exhibit identical training behavior.

As a case study, we will employ an affective computing problem to demonstrate the performance of the proposed method. However, our training technique can be used to train RNN models corresponding to any other domain.

4.1 Database

We employed part of the SEWA, sentiment analysis in the wild, database [12] which was used for the AVEC2018 competition [11]. The cross-cultural emotion sub-challenge of AVEC2018 aims at detecting emotions from audiovisual recordings of spontaneous human-human interactions. The recordings consist of one participant using a personal device to record a video conversation about commercial products with another subject. The measured emotions are divided into 3 different dimensions, valence, arousal and liking. The first describes the negativity and positivity of the underlying emotions. The second corresponds to the level of activation of the underlying emotion. The third represents the user's preference of the commercial product. All three dimensions range from −1 to +1. The details of the data distribution are shown in Table 1.

Table 1. SEWA database subjects' distribution

Set	Subject	Culture
Training	34	German
Development	14	German

4.2   Experimental Setup

We built the RNN model deployed with AVEC2018 cross cultural sub-challenge using the tensorflow framework [13] as a regressor composed of 2 LSTM layers with 64 and 34 number of units for the first and second layer's cell respectively, followed by a fully connected layer to predict all three dimensions of emotions. The model is trained using the Adam optimizer for 100 epochs. The metric used for evaluating the model is the CCC whose inverse is employed to calculate the loss during training. The formula for CCC is described in Eq. (7),

$$\rho_c = \frac{2\rho\sigma_x\sigma_y}{\sigma_x^2 + \sigma_y^2 \left(\mu_x - \mu_y\right)^2} \tag{7}$$

Where ρ is the Pearson correlation coefficient between vectors x and y; σ_x and σ_y are the standard deviation of x and y respectively and μ_x and μ_y are the means of x and y respectively.

4.3   Results

In Fig. 4, we depict the CCC accuracy plotted against the number of epochs for three models that estimate arousal (Fig. 4a), valence (Fig. 4b), and liking (Fig. 4c). The first model is a truncated RNN with a segment size of 2 time steps. The second model is a truncated RNN with a segment size of 100 time steps. We apply the proposed method for training these truncated models. The third model is a non-truncated RNN that considers the entire data sequence (which is composed of 1500 time steps). The results in Fig. 4 are rendered using tensorboard. Although we are training three models, the three CCC curves are fully overlapping. This demonstrates that the three models are behaving identically during training.

Figure 5 illustrates the relationship between the time and memory requirement as the number of segment size decreases from 1500 to 2. Note that a segment size of 1500 corresponds to a non-truncated RNN as a single segment covers the entire data sequence. The inverse relationship between the memory consumption and execution time during training is evident. This tradeoff gives modelers the flexibility of reducing memory consumption at the cost of increasing the training time without limiting their ability to capture long term dependencies.

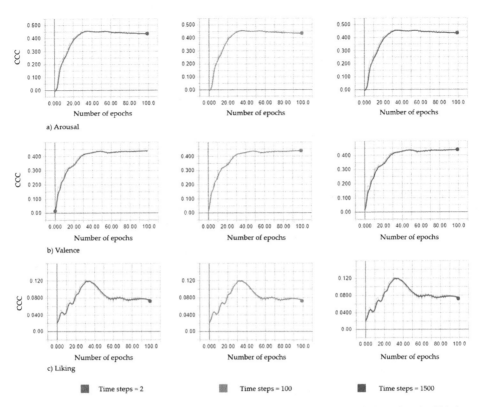

a) Arousal

b) Valence

c) Liking

Time steps = 2   Time steps = 100   Time steps = 1500

Fig. 4. The accuracy of Arousal, Valence and Liking when segment size is 2, 100 and 1500 time steps; given the same random initialization and random seed for all models.

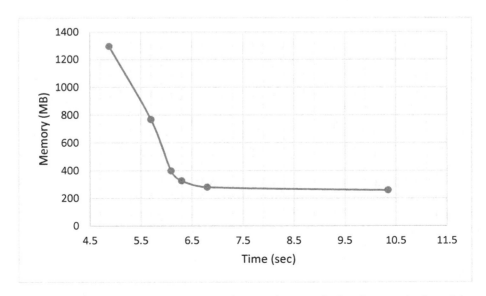

Fig. 5. This figure shows the memory vs. time requirement, displayed on y-axis, for training different RNN models unrolled into different segment sizes, displayed on x-axis

5 Conclusion

To capture long-term dependencies, it is imperative to unroll the RNN over the entire data sequence. This goal is difficult to achieve as it requires fitting a large model in memory. Hence, this obliges model designers to truncate the RNN into segments and optimize the model parameter within each segment. Therefore, we modified the truncated-BPTT procedure to achieve the same results we would obtain if the model was trained without truncation. We evaluated the proposed scheme using the AVEC2018 dataset and showed that we can obtain matching loss graphs regardless of the number and size of the segments.

Finally, our work can be extended by applying any neural network architecture, such as introducing a fully connected layer or convolutional neural network before the RNN layers, while using any dataset depending on the research problem.

References

1. Goodfellow, I., Bengio, Y., Courville, A.: Deep Learning. MIT Press, Cambridge (2016)
2. Werbos, P.J.: Backpropagation through time: what it does and how to do it. Proc. IEEE **78** (10), 1550–1560 (1990)
3. Pascanu, R., Mikolov, T., Bengio, Y.: On the difficulty of training recurrent neural networks (2013)
4. Hochreiter, S., Urgen Schmidhuber, J.: Long short-term memory. Neural Comput. **9**(8), 1735–1780 (1997)
5. Cho, K., et al.: Learning Phrase Representations using RNN Encoder-Decoder for Statistical Machine Translation (2014)
6. Jaderberg, M., et al.: Decoupled Neural Interfaces using Synthetic Gradients (2017)
7. Gruslys, A., et al.: Memory-Efficient Backpropagation Through Time (2016)
8. Williams, R.J., Peng, J.: An efficient gradient-based algorithm for on-line training of recurrent network trajectories. Neural Comput. **2**(4), 490–501 (1990)
9. Jaeger, H., Jaeger, H.: A tutorial on training recurrent neural networks, covering BPPT, RTRL, EKF and the "echo state network" approach (2002)
10. Chen, T., Xu, B., Zhang, C., Guestrin, C.: Training Deep Nets with Sublinear Memory Cost (2016)
11. Ringeval, F., et al.: AVEC 2018 workshop and challenge. In: Proceedings of 2018 Audio/Visual Emot. Chall. Work. – AVEC 2018, pp. 3–13 (2018)
12. SEWA database. https://db.sewaproject.eu/. Accessed 12 Jan 2019
13. Abadi, M., et al.: TensorFlow: Large-Scale Machine Learning on Heterogeneous Distributed Systems (2016)

A Behavior-Based Proactive User Authentication Model Utilizing Mobile Application Usage Patterns

Yosef Ashibani$^{(\boxtimes)}$ ⓘ and Qusay H. Mahmoud ⓘ

Department of Electrical, Computer and Software Engineering,
University of Ontario Institute of Technology, Oshawa, ON L1G 0C5, Canada
{yosef.ashibani, qusay.mahmoud}@uoit.net

Abstract. Access to smart home networks is mostly achieved through end-user devices, especially mobile phones, but such devices are susceptible to theft or loss. In this paper, we present a user authentication model based on application access events, using only a small amount of information, thus reducing the computation time. To validate our model, we utilize a public real-world dataset collected from real users over a long period of time, in an uncontrolled manner. The model is evaluated by differentiating between users who utilize shared apps at the same daily intervals. In addition, we evaluate various classification approaches regarding legitimate user classification in compliance with the history of app usage events. The results demonstrate the capacity of the presented model to authenticate users with high true positive and true negative rates.

Keywords: Behavior-based authentication · Continuous authentication · Classification · App access events

1 Introduction

The growth in the Internet of Things (IoT) has brought an increased number of security threats. User authentication, which is crucial, keeps unauthorized users from gaining access to services. Hence, an effective authentication mechanism that prevents unauthorized access is essential, either at the entry point or during the session. As a result, it is important to provide continuous authentication techniques that periodically examine the access situation and the legitimacy of the communicated party. Many of the end-user devices, including smartphones and tablets, from which access to IoT devices is achieved, are currently equipped with advanced authentication approaches, such as fingerprint readers, to access the devices or their applications (apps). Since these methods are only employed at the time of access, many ensuing issues should be taken into consideration. Firstly, many users avoid setting short access sessions after initial authentication; as a result, after the entry point, devices are vulnerable to use by others. Secondly, as the devices are susceptible to loss or theft, traditional authentication credentials could be observed by unauthorized users. Hence, it is important to present a method that overcomes these issues and can be implicitly and continuously employed in the background.

© Springer Nature Switzerland AG 2019
M.-J. Meurs and F. Rudzicz (Eds.): Canadian AI 2019, LNAI 11489, pp. 284–295, 2019.
https://doi.org/10.1007/978-3-030-18305-9_23

The number of apps employed on these devices is increasing and, as an example, in addition to pre-installed apps, over two million mobile apps are already available in major app stores, with more added daily [1]. Moreover, the enhanced functionalities of home hubs, for instance the Samsung Connect Home hub and the Wink hub, in smart home networks, provide sophisticated authentication mechanisms. Thus, these devices can provide behavior-based user authentication for continuously monitoring and authenticating remote user access. Therefore, it is important to consider continuous authentication techniques that periodically examine the access situation and the legitimacy of the communicating party. A behavior-based authentication method is built on the assumption that users present a unique behavior, like browsing particular pages and using specific apps, while interacting with their mobile devices. The advantage of considering behavior-based authentication is that much information, such as apps, device resources and network information generated from apps, can be retrieved and utilized for user authentication in the background. This paper aims to present a remote continuous user authentication approach based on app access events of the most utilized apps by users on mobile phones. This approach continuously authenticates users to smart home hubs in the background without requiring further action. In addition, it employs the unique user app usage events utilized by all users on end-devices.

To this end, the contribution of this paper is an authentication approach based on a machine learning model that authenticates users based on the most used apps. This model utilizes only a small amount of information, reducing computation time and requiring less memory. The remainder of this paper is organized as follows: Sect. 2 presents a related work summary, while the proposed model is described in Sect. 3. The evaluation of the model and results are presented in Sect. 4, and Sect. 5 concludes the paper and suggests future research directions.

2   Related Work

Many behavior-based authentication methods have been presented in the related literature. As demonstrated in [2], user behavior relies on the power consumption of accessed apps. Utilizing device resources, the authors in [3] noted that working on a particular device results in varying power consumption. Another study in [4], which presents a power estimation approach based on user usage patterns, finds that there is a high correlation between these patterns and device power consumption. Furthermore, using built-in battery voltage sensors, the authors in [5] present a power model construction approach that monitors power consumption per app on user devices. This approach obtained an absolute Average Error Rate of less than 10%. However, these studies do not employ app access patterns for user authentication. Moreover, it is difficult to model power consumption only for specific apps as they are running in the background.

The authors in [6] show that users can be identified and anomalies can be detected according to interactions with their mobile apps, use of text messages and calling behavior. Based on the assumption that mobile phone users tend to use apps in different locations at different times, host-based behavior profiling is described by the same authors. The authors in [7] and [8] report on building user behavior profiles, including calling and migration patterns, over service provider networks. Considering the assumption that users perform similar tasks at certain times during weekdays, a user profile approach that gathers behavioral information, such as text messages, calls and geographic location, is proposed in [9]. For observed actions, such as habitual or good events, this approach assigns a score. The study in [10], which proposes an anomaly-based detection approach based on monitoring users' actions, such as sending text messages or calls, used four different machine learning classifiers for evaluating the performance of the presented model.

As seen in the literature, several studies have considered smartphone app usage patterns for user authentication. A further study [11] utilizes two derived features, namely the time of the last viewed email and the user's GPS location, from the used mobile device. In this approach, an overall score is provided for authentication decisions. The work in [12] presents a continuous authentication model on smartphones based on app usage data. However, the authors do not limit their study to apps that are used by all users; rather, it considers all apps, including those that are only used by individual users. This is due to the selected datasets that include alternative apps. A behavior profiling technique, which makes use of historical app usage that continuously verifies mobile users, is proposed in [13], with an achieved Equal Error Rate (EER) of 9.8%. However, the selected datasets are dated and the number of mobile apps since then has dramatically increased. Furthermore, the time frame is limited to a maximum of 22 days. Additional studies, [6–9] and [10], consider text messages and calls behavior for unauthorized usage detection. As the number of mobile apps increases, traditional text messages and calls are being replaced with apps that achieve the same purpose. In addition, unauthorized users who access mobile devices will, most probably, tend to operate inconspicuously. Hence these traditional text messages and calls provide insufficient evidence of the intended user. The results in our previous work [14, 15], show that users can be identified based on app access time and the generated traffic while accessing these apps during the selected period. In this paper, our model focuses on differentiating between users who utilize shared apps at the same daily intervals.

3 Proposed Model

This section presents an authentication model that builds a classification model from learning users' app access events on their end-devices and authenticating them, based on the produced pattern. In the system model, which has been implemented in previous work [16, 17], we consider that users access to smart home devices by their end-devices through home hubs to access and control these devices. In addition, we assume that the smart home hub is trusted, by which all information related to app access history is collected to training and authentication stages, and have the ability to provide the required processing capabilities. Panda's library is used to build the proposed model. The presented machine learning classification model, as shown in Fig. 1, the model monitors the app access logs on end-user devices during the access sessions, in order to detect deviations from built normal patterns.

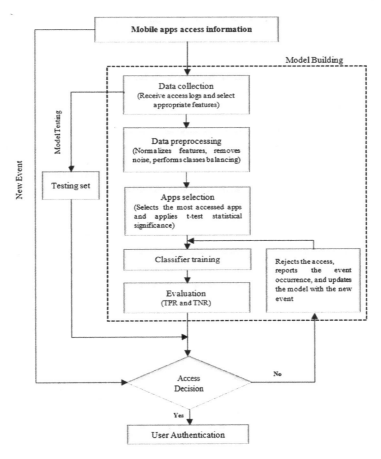

Fig. 1. The proposed model

3.1 Dataset Analysis

For evaluating the proposed approach, a public real-world dataset, collected from real users in an uncontrolled manner, are utilized. The dataset is the UbiqLog4UCI [18], which was collected from 35 users over a period of around one year, and contains 581,829 instances (events) and 1418 unique used apps. The dataset was collected in the form of records, each of which represents a sample of the observations in the dataset. Although the dataset contains 35 users, only the data of 30 users were used, while the data of 5 users were removed during this study due to errors and missing app usage information. Although this dataset contains other features such as calling information, SMS and location, these features were discarded because the aim of this paper is solely on app usage history during access to host-based apps. The meaning and the fields of the dataset utilized in this paper are listed in Table 1. There are many advantages of the dataset utilized in this paper, one of which is that the dataset collection procedure runs in the background and does not require root privileges.

Table 1. Utilized features from the dataset

	Feature	Meaning	Type
1	User ID	The user's identifier	Text
2	Timestamp	The date of this data entry	Date
3	Package (process) name	The app's identifier	Text

The interaction time in this study is considered as the total access time to the app in the foreground, meaning that the user is continuously interacting with the app during the session. As an overview, Fig. 2 shows the average daily interaction time of four randomly selected users from the dataset, over a period of 90 consecutive days. The average daily access time ranged from approximately 76 min to 242 min. Moreover, 30% of the interaction time of the selected apps had an average range between 30 and 180 min in all the dataset. Even though there are changes in the average of the daily access time over a period of 90 days, as Fig. 2 shows, the difference among users is clear. Furthermore, the average access per user can be distinguished in contrast with other users. Hence, this result indicates that the total access time can be utilized for user authentication.

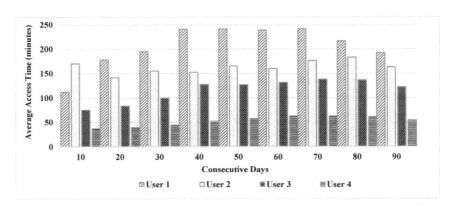

Fig. 2. Average access time to apps per day in the Dataset

3.2   Data Preprocessing

Next to receiving the apps access information, the model then normalizes features, while considering the imbalanced representation of classes among class observations in the dataset vary from vary from 0.17% to 8.6%. Since classes' observations are variant, considering such difference is necessary to avoid disregarding minority classes over majority classes. To deal with this issue, sampling-based methods [19] were applied. However, this does not ensure that the minority classes will be part of the sampling as the validation and training sets are from containing the same samples, leading to overfitting. To avoid such a problem, oversampling is achieved after performing cross-validation on the training data. This step avoids the validation and training sets from having the same samples, thus eliminating overfitting.

3.3   Apps Selection

Prior to the model evaluation, we need to examine if there is a significant difference between the two data samples, all apps and most used apps, in the dataset. For the most used apps, we consider ones that are accessed twice or more by all users. In order to determine the difference between the two data samples in the dataset, student t-test statistical significance [20] is applied, based on the following formula, with an assumption of unequal variances:

$$t = \frac{\bar{g}_1 - \bar{g}_2}{\sqrt{(p_1^2/a_1) + ((p_2^2/a_2)}} \tag{1}$$

$$df = \frac{[(p_1^2/a_1) + (p_2^2/a_2)]^2}{\frac{(p_1^2/a_1)^2}{a_1 - 1} + \frac{(p_2^2/a_2)^2}{a_2 - 1}} \tag{2}$$

$$p_1^2 = \frac{\sum_{i=1}^{n_1}(g_i - \bar{g}_1)^2}{a_1 - 1} \tag{3}$$

$$p_2^2 = \frac{\sum_{j=1}^{n_2}(g_j - \bar{g}_2)^2}{a_2 - 1} \tag{4}$$

Where $\bar{g}_1$ and $\bar{g}_2$ are the sample means, p^2 is the sample variance, a_1 and a_2 are the sample sizes with a degree of freedom df, using Satterthwaite's approximation. The standard deviation of the dataset that results from including all apps is 8.347, and 7.1806 for including the most used apps. The calculated t value is 0.891, which is less than 2.018, with a degree of freedom of 42 at p = 0.05. Consequently, there is no significant difference between the two data samples, and it is clear that both data samples come from the same distribution.

3.4   Classifier Training

Since the focus here is on classifying multi-users rather than one user, in order to select the most suitable classifiers, different classification strategies are applied, including inherently multiclass algorithms, such as: Gaussian NB (NB) [21], K-Nearest Neighbor KNN [22], MLP Classifier (MLP), Nearest Centroid(NC) [23], Linear Discriminant Analysis (LDA), Ridge Classifier (RC) [24], Quadratic Discriminant Analysis (QDA), Random Forest Classifier (RF) [25]; a One-Vs-One Multiclass classification strategy such as SVC, Gaussian Process Classifier (GPC) [26]; and a One-Vs-All Multiclass classification such as Gradient Boosting Classifier (GBC) [27] and AdaBoost Classifier (ABC).

We evaluate the mentioned algorithms on the training set, utilizing the listed classification algorithms as Fig. 3 shows. The accuracy measures for each algorithm are evaluated 5 times. As seen from the same Figure, six classifiers, KNN, MLP, NC, RF, GBC and AdaB, provide the best accuracy compared with other algorithms.

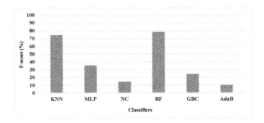

Fig. 3. Performance evaluation of the proposed model on the training set

3.5 User Authentication

After the classification model is built, it monitors the access events of users to detect abnormalities from built access patterns based on the access events. The proposed model classifies every access point received from the users' devices and identifies related users for this event. When detecting a abnormality from the user's built pattern or in case of misclassifying a non-user event, the authentication procedure rejects the next access request and requires other authentication factors, such as a password, from the user while reporting the occurrence to the home network administrator, and updating the model with the new observation. Anomalies are detected immediately as the detection depends on the previous access patterns on the end-user device.

4 Evaluation and Results

For evaluating the proposed approach, the dataset is divided into 70% for training the model and 30% for validation and the oversampling is performed after applying cross validation on the training set, and then we apply a 5-fold cross validation [28] utilizing the six selected classification algorithms. For the sake of comparison, the evaluation is performed based on two classification methods: one-vs-all and all-vs-all.

4.1 One-vs-All

The one-vs-all classification method trains a single classifier per class, including the samples of that class, as positive and all others as negative. We apply the selected classification algorithms to the UbiqLog4UCI dataset. In this approach, each user is classified against the remaining users to train a single classifier for each user. In this scenario, observations of the chosen class are considered positive while the remaining observations of the other users are regarded as negative. The class label for a targeted user is assigned to 0 and the class for all other users combined is set to 1. The resulting highest accuracies from the best classification algorithms are shown in Fig. 4.

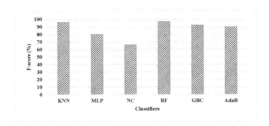

Fig. 4. One-vs-all performance of the proposed model

4.2 All-vs-All

The all-vs-all, also known as the multi-class classification, is a classification task with two or more classes. This type of classification is based on the assumption that each observation is assigned to only one label and builds one classifier for each class.

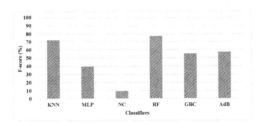

Fig. 5. All-vs-all performance of the proposed model

We apply the selected classification algorithms to the dataset and Fig. 5 shows the resulting F-score from the best classification algorithms.

Figures 4 and 5 show that RF classifier attains the best F-score in both approaches. RF classification algorithms have advantages compared to other classification algorithms such as KNN, NB, AdaB, GBC and MLP Classifiers. These advantages include: overcoming the problem of overfitting; providing accurate predictions of many types of applications [29]; rapidly performing out-of-sample predictions; and being less sensitive to outliers in the training data; measuring the importance of each feature with model training.

4.3 Performance Evaluation Based on the TPR and TNR

This section presents the obtained results of testing the performance of the proposed model with regard to the true positive rate (TPR) and true negative rate (TNR). The average TPR and TNR based on the one-vs-all classification approach of the proposed model is presented in Fig. 6. Our main objective is to increase the TPR and TNR as possible, followed by reducing the FPR and FNR. The model is tested based on the

TPR and TNR on the dataset utilizing the selected classification algorithms, as shown in Fig. 6, where we see that the RF classification algorithm performs the best TPR and TNR followed by the KNN.

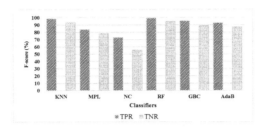

Fig. 6. Performance evaluation of the proposed model based on TPR and TNR

4.4   Performance Evaluation Based on the Number of Days

The dataset is examined for user authentication in an incremental manner in terms of the number of days. For consistency, this test is applied for a period of 90 consecutive days. The presented model presents high TPR and high TNR values, and the achieved TPR and TNR is shown in Table 2. As observed from the same Table, the RF algorithm provides the best TPR, followed by the KNN and GBC. In addition, it is noticeable that the TPR and TNR do not significantly change for the period 10 to 90 days.

Table 2. Performance evaluation based on the number of days

Period (days)	Classification algorithms											
	KNN		MPL		NC		RF		GBC		AdaB	
	TPR	TNR	TPR	TNR	TPR	TNR	TPR	TNR	TPR	TNR	TPR	TNR
10	98.55	93.20	86.30	79.70	74.20	56.20	99.30	94.40	96.00	90.20	94.20	88.60
20	98.60	93.30	86.80	77.50	72.80	56.10	99.30	94.40	92.60	89.80	93.80	88.00
30	98.48	93.50	84.59	77.76	72.87	55.07	99.32	94.46	96.03	89.35	93.96	87.36
40	98.57	93.27	85.53	77.14	71.53	55.75	99.35	94.32	95.87	88.91	93.16	87.10
50	98.58	92.97	85.07	79.24	71.53	55.76	99.36	94.34	95.87	88.91	94.80	87.10
60	98.57	93.27	83.16	79.29	71.53	55.76	99.35	94.34	95.87	88.91	93.16	87.08
70	98.57	93.27	83.89	78.43	71.53	55.76	99.30	94.33	95.87	88.91	93.16	87.07
80	98.67	93.27	86.40	76.51	71.53	55.76	99.34	94.32	95.87	88.91	93.16	91.90
90	98.57	93.27	83.60	79.36	71.53	55.76	99.34	94.33	95.87	88.91	93.16	87.081

5   Conclusion and Future Work

In this paper, we present a user authentication model based on app access events, utilizing a minimal amount of information, thus reducing computation time and using less memory. For validation of the proposed model, a dataset collected from real users in an uncontrolled manner, over a period of almost one year, is utilized. The results demonstrate the capacity of the presented model to authenticate users with both high true positive and true negative rates. Even though there are some changes in access time to apps, these changes do not considerably affect accuracy since we utilized the most regularly used apps. Moreover, the apps access patterns are different for each user even when considering the most used apps by users and users can be discriminated by the proposed model. In conclusion, the increasing number of mobile apps as well, as the opportunity to constantly install additional apps on end-devices, will provide the possibility to discriminate and authenticate users with an appropriate level of accuracy.

While we considered only 30 users in evaluating our model, future work will examine a higher number of users with potentially similar behavior. We also assume in this work that the smart home hub, by which all information related to app access history is collected to training and authentication stages, is trusted. However, if the hub is attacked, this approach offers limited privacy. This issue is beyond the scope of this paper and will be considered in future work. Future work will also include a consideration of all apps for user authentication in addition to presenting a window of accessed apps and applying it to obtain continuous user authentication. In order to continuously authenticate the user, previous access instances will be considered for future access requests. Running time, which is an important factor for IoT devices, especially those with limited computing resources, will also be considered.

References

1. ICTC: The Application of Everything: Canada's Apps Economy Value Chain (2014)
2. Murmuria, R., Stavrou, A., Barbará, D., Fleck, D.: Continuous authentication on mobile devices using power consumption, touch gestures and physical movement of users. In: International Workshop on Recent Advances in Intrusion Detection, pp. 405–424 (2015)
3. Zhou, K., Medsger, J., Stavrou, A., Voas, J.M.: Mobile application and device power usage measurements. In: Sixth International Conference on Software Security and Reliability (SERE), pp. 147–156. IEEE (2012)
4. Shye, A., Scholbrock, B., Memik, G.: Into the wild: studying real user activity patterns to guide power optimizations for mobile architectures. In: Proceedings of the 42nd Annual IEEE/ACM International Symposium on Microarchitecture, pp. 168–178 (2009)
5. Zhang, L., et al.: Accurate online power estimation and automatic battery behavior based power model generation for smartphones. In: Proceedings of the Eighth IEEE/ACM/IFIP International Conference on Hardware/Software Codesign and System Synthesis, pp. 105–114 (2010)
6. Li, F., Clarke, N., Papadaki, M., Dowland, P.: Behaviour profiling for transparent authentication for mobile devices. In: European Conference on Cyber Warfare and Security, pp. 307–315. Academic Conferences International Limited (2011)
7. Gosset, P.: Fraud Detection Concepts : Final Report. CiteSeer, Doc Ref. AC095/VOD/W22/DS/P/18/1, pp. 1–27 (1998)
8. Hall, J., Barbeau, M., Kranakis, E.: Anomaly-based intrusion detection using mobility profiles of public transportation users. In: International Conference on Wireless and Mobile Computing, Networking and Communications (WiMob'2005), vol. 2, pp. 17–24. IEEE (2005)

9. Shi, E., Niu, Y., Jakobsson, M., Chow, R.: Implicit authentication through learning user behavior. In: Burmester, M., Tsudik, G., Magliveras, S., Ilić, I. (eds.) ISC 2010. LNCS, vol. 6531, pp. 99–113. Springer, Heidelberg (2011). https://doi.org/10.1007/978-3-642-18178-8_9

10. Damopoulos, D., Menesidou, S.A., Kambourakis, G., Papadaki, M., Clarke, N., Gritzalis, S.: Evaluation of anomaly-based IDS for mobile devices using machine learning classifiers. Secur. Commun. Netw. 5(1), 3–14 (2012)

11. Kalamandeen, A., De Lara, E., Lamarca, A.: Ensemble: cooperative proximity-based authentication, pp. 331–343 (2010)

12. Mahbub, U., Komulainen, J., Ferreira, D., Chellappa, R.: Continuous Authentication of Smartphones Based on Application Usage, arXiv preprint arXiv:1808.03319 (2018)

13. Li, F., Clarke, N., Papadaki, M., Dowland, P.: Active authentication for mobile devices utilising behaviour profiling. Int. J. Inf. Secur. 13(3), 229–244 (2014)

14. Ashibani, Y., Mahmoud, Q.H.: A behavior profiling model for user authentication in IoT networks based on app usage patterns. In: 44th Annual Conference of the IEEE Industrial Electronics Society (IECON), pp. 2841–2846 (2018)

15. Ashibani, Y., Mahmoud, Q.H.: A user authentication model for IoT networks based on app traffic patterns. In: 9th Annual IEEE Information Technology; Electronics and Mobile Communication Conference (IEEE IEMCON), pp. 632–638 (2018)

16. Ashibani, Y., Kauling, D., Mahmoud, Q.H.: A context-aware authentication framework for smart homes. In: Canadian Conference on Electrical and Computer Engineering (CCECE), pp. 1–5 (2017)

17. Ashibani, Y., Kauling, D., Mahmoud, Q.H.: Design and implementation of a contextual-based continuous authentication framework for smart homes. Appl. Syst. Innov. 2(1), 1–20 (2019)

18. Rawassizadeh, R., Momeni, E., Dobbins, C., Mirza-babaei, P.: Lesson learned from collecting quantified self information via mobile and wearable devices. J. Sens. Actuator Netw. 4(4), 315–335 (2015)

19. López, V., Fernández, A., Moreno-Torres, J.G., Herrera, F.: Analysis of preprocessing vs. cost-sensitive learning for imbalanced classification. Open problems on intrinsic data characteristics. Expert Syst. Appl. 39(7), 6585–6608 (2012)

20. Massey, A., Miller, S.J.: Tests of Hypotheses Using Statistics. Mathematics Department, Brown University, Providence, RI 2912, pp. 1–32 (2006)

21. Almeida, T.A., Yamakami, A.: Compression-based spam filter. Secur. Commun. Netw. 9(4), 1327–1335 (2016)

22. Kim, K.S., Choi, H.H., Moon, C.S., Mun, C.W.: Comparison of k-nearest neighbor, quadratic discriminant and linear discriminant analysis in classification of electromyogram signals based on the wrist-motion directions. Curr. Appl. Phys. 11(3), 740–745 (2011)

23. Li, M., et al.: Coupled k-nearest centroid classification for non-iid data. In: Nguyen, N.T., Kowalczyk, R., Corchado, J.M., Bajo, J. (eds.) Transactions on Computational Collective Intelligence XV. LNCS, vol. 8670, pp. 89–100. Springer, Heidelberg (2014). https://doi.org/10.1007/978-3-662-44750-5_5

24. Singh, A., Prakash B.S., Chandrasekaran, K.: A comparison of linear discriminant analysis and ridge classifier on twitter data. In: International Conference on Computing, Communication and Automation (ICCCA), pp. 133–138 (2016)

25. Amasyali, M.F., Ersoy, O.K.: Classifier ensembles with the extended space forest. IEEE Trans. Knowl. Data Eng. 26(3), 549–562 (2014)

26. Hensman, J., Matthews, A., Ghahramani, Z.: Scalable Variational Gaussian Process Classification (2015)

27. Rawat, M., Goyal, N., Singh, S.: Advancement of recommender system based on clickstream data using gradient boosting and random forest classifiers. In: 8th International Conference on Computing, Communications and Networking Technologies, ICCCNT, pp. 1–6 (2017)
28. Abdi, L., Hashemi, S.: To combat multi-class imbalanced problems by means of over-sampling techniques. IEEE Trans. Knowl. Data Eng. **28**(1), 238–251 (2016)
29. Qi, Y.: Random forest for bioinformatics. In: Zhang, C., Ma, Y. (eds.) Ensemble Machine Learning: Methods and Applications, pp. 307–323. Springer, Boston (2012). https://doi.org/10.1007/978-1-4419-9326-7_11

Sparseout: Controlling Sparsity in Deep Networks

Najeeb Khan$^{(\boxtimes)}$ and Ian Stavness

Department of Computer Science, University of Saskatchewan, Saskatoon, Canada
{najeeb.khan,ian.stavness}@usask.ca

Abstract. Dropout is commonly used to help reduce overfitting in deep neural networks. Sparsity is a potentially important property of neural networks, but is not explicitly controlled by Dropout-based regularization. In this work, we propose Sparseout a simple and efficient variant of Dropout that can be used to control the sparsity of the activations in a neural network. We theoretically prove that Sparseout is equivalent to an L_q penalty on the features of a generalized linear model and that Dropout is a special case of Sparseout for neural networks. We empirically demonstrate that Sparseout is computationally inexpensive and is able to control the desired level of sparsity in the activations. We evaluated Sparseout on image classification and language modelling tasks to see the effect of sparsity on these tasks. We found that sparsity of the activations is favorable for language modelling performance while image classification benefits from denser activations. Sparseout provides a way to investigate sparsity in state-of-the-art deep learning models. Source code for Sparseout could be found at https://github.com/najeebkhan/sparseout.

1 Introduction

Sparsity is often thought to be a desirable property for artificial neural networks. This is likely rooted in early neuroscience studies that discovered sparse coding in the visual cortex [1] hypothesizing that at any given time, only a small number of neurons are used to encode sensory information. Sparsity has been observed both in connectivity [2] and representation [1]. To mimic sparse coding from brain studies, researchers have devised approaches to encourage sparsity when training ANNs.

Sparsity has been used to regularize models by imposing a *sparsity* constraint on the activations of the neural network [3]. Many useful properties are ascribed to sparsity in the literature. It has been hypothesized that neurons that are rarely active are more interpretable than those that are active most of the time [4]; images of natural world objects can be described in terms of sparse statistically independent events; neural networks with sparsity constraints learn filters that resemble the mammalian visual cortex area V1 [1] and area V2 [5]; and sparsity allows faster learning [6].

© Springer Nature Switzerland AG 2019
M.-J. Meurs and F. Rudzicz (Eds.): Canadian AI 2019, LNAI 11489, pp. 296–307, 2019.
https://doi.org/10.1007/978-3-030-18305-9_24

One of the main motivations behind sparsity based training methods is the biological plausibility of these algorithms. However, recent studies have questioned the pervasiveness of neural sparsity. The biological studies that provide evidence for sparsity are performed when the subject is passive and in reality sparsity might not be the mechanism that the brain uses in active tasks [7]. Empirically for DNNs, new methods that may discourage sparsity, such as Maxout [8] and DARC1 [9], have achieved better performance than sparse methods in certain domains such as computer vision. Therefore it is not clear whether or not sparsity is a generally desirable property for DNNs. We hypothesize that sparsity will benefit some learning tasks and hinder others. Therefore, new DNN training approaches that include the flexibility to either encourage sparsity, where necessary, and discourage sparsity otherwise could provide improved task performance.

There are many approaches for affecting the sparsity of a DNN during training including the use of certain activation functions such as rectified linear units [10]. One of the main ways is through regularization and several deterministic regularization algorithms have been proposed to train deep neural networks with sparse weights [11–13] and sparse activations [10, 14–17].

Training deep neural networks with deterministic regularization and backpropagation results in correlated activities of the neurons. To prevent such co-adaptations as well as regularize the models, stochastic regularization methods are used. Stochastic methods such as Dropout, Bridgeout and Shakeout have been shown to be equivalent to ridge, bridge and elastic-net penalties on the model weights. Previous stochastic regularization methods that explicitly encourage sparsity, i.e., Shakeout and Bridgeout require a new set of masked weights per training example in a mini-batch making them computationally expensive. Therefore, these existing methods cannot be applied to large fully connected architectures. Likewise, Shakeout and Bridgeout cannot be easily applied to other convolutional architectures such as DenseNet and Wide-ResNet that provide current state-of-the-art performance for image classification, because they cannot be used with highly optimized black-box implementations such as cuDNN [18].

In this paper, we propose Sparseout, a stochastic regularization method that is capable of either encouraging or discouraging sparsity in deep neural networks. It provides an L_q-norm penalty on the network's activations and therefore can vary activation sparsity by its q parameter. The computational cost of Sparseout is comparable to Dropout and it can be applied to existing optimized CNN and LSTM blocks, making it applicable to state-of-the-art architectures. We provide theoretical and empirical results demonstrating the bridge-regularization capability of Sparseout. We use Sparseout to evaluate whether or not sparsity is beneficial for two distinct learning tasks: image classification and language modeling.

2   Related Work

Due to the over-parameterization of deep neural networks, they suffer from large generalization error, specifically, when the dataset size is relatively small. This phenomenon is known as over-fitting. Generalization error is upper bounded by the model complexity [19] thus overfitting could be reduced by controlling the complexity of the model.

One way to control the complexity of a model is to impose constraints on the parameters of the model such as the weights in the neural networks. Such model regularization methods can be classified into deterministic and stochastic methods. Deterministic methods either remove redundant weights [13] or penalize large magnitude weights. Weight penalties are imposed by adding a regularization term to the loss function consisting of a norm of the weight matrix [20].

Stochastic methods randomly perturb the weights so as to achieve minimal co-dependency between neurons [21–23] as well as regularizing the model at the same time. Stochastic regularization has become the standard practice in training deep learning models and have outperformed deterministic regularization methods on many tasks. Stochastic regularization techniques have a Bayesian model averaging interpretation as well as they posses an equivalence to weight penalties for linear models. In terms of Bayesian estimation, weight penalties are equivalent to imposing a prior distribution on the model weights.

Beside the weight penalty interpretation, a reason for the effectiveness of stochastic regularization methods could be the prevention of correlated activations. It has been shown that high correlation between activations of the neurons results in overfitting. DeCov [24], reduces overfitting by adding a penalty term to the cost function consisting of the co-variances among the activations of the neurons over a mini-batch.

Another approach to control model complexity, inspired by sparse coding [1], is to impose a *sparsity* constraint on the activations of the neural network [3]. To encourage sparsity of the activations, an L_1 norm of the activations is added to the cost function [5]. Another form of penalty is to add the KL-divergence of the expected activations and a preset target sparsity value ρ [4]. Liao et al. have used a clustering approach to obtain sparse representation by encouraging activations to form clusters [17].

Another related technique that normalizes activations in the network so as to have zero mean and unit variance is Batchnorm [25]. Although, the primary purpose of Batchnorm is accelerating training/optimization of the neural network rather than regularization, Batchnorm has reduced the need for stochastic regularization in certain domains. The above mentioned sparsity-inducing methods are deterministic and thus may result in correlated activations. In this paper we propose Sparseout that implicitly imposes an L_q penalty on the activations thus allowing us to choose the level of sparsity in the activations as well as the stochasticity preventing correlated neural activities. Sparseout is different than Bridgeout [23] in that it is applied to activations rather than the weights. Therefore, Sparseout is orders of magnitude faster and practical than Bridgeout. We believe that Sparseout is the first theoretically-motivated technique that is capable of simultaneously controlling sparsity in activations and reducing correlations between them, besides being equivalent to Dropout for $q = 2$.

3   Sparseout

Consider a feedforward neural network layer l, the output of l-th is given by

$$a^l = \sigma(W^l a^{l-1} + b^l), \tag{1}$$

where W^l and b^l are the weight matrix and bias vector for the l-th layer, σ is a non-linear activation function and a^{l-1} is the output of the previous layer.

The Sparseout perturbed output of the l-th layer $\widetilde{a}^l$ is given by

$$\widetilde{a}_i^l = \begin{cases} a_i^l - |a_i^l|^{q/2} & r_i = 0 \\ a_i^l + |a_i^l|^{q/2}\left(\frac{1-p}{p}\right) & r_i = \frac{1}{p} \end{cases} \tag{2}$$

where r is a random mask vector randomly sampled from a Bernoulli distribution with probability p and scaled by $1/p$ and q specifies the normed space to which the activations are restricted. Since the random mask is scaled by $1/p$ during training, no changes to the neural network are needed during testing. Since the training of the neural networks is performed using the back-propagated gradients of the error, the gradient of the Sparseout perturbed output is given by

$$\frac{\partial \widetilde{a}_i^l}{\partial a_i^l} = 1 + \frac{q}{2}|a_i^l|^{\frac{q}{2}-1}(r_i - 1)\operatorname{sgn}\left(a_i^l\right), \tag{3}$$

where sgn is the sign function.

Since Sparseout operates on the activations of the neural networks similar to Dropout, Sparseout can be implemented with minimal changes to the existing Dropout implementation. Sparseout can be used with the highly optimized black-box implementations such as cuDNN [18]. The above Sparseout formulation is applicable to any feedforward network layer such as convolutional or fully connected layers as well as layers in recurrent neural networks.

Theorem 1. *Sparseout is equivalent to an L_q penalty on the features of a generalized linear model.*

For a generalized linear model with parameters β, log partition function A and the perturbed design matrix $\widetilde{X}$ of dimension $n \times d$, the negative log likelihood function could be split into a mean squared error term and a penalty term [26, Eq. 6] given by

$$R(\beta) \approx \sum_{i=1}^{n} \frac{A''(X_i \cdot \beta)}{2} Var[\widetilde{X}_i \cdot \beta] \tag{4}$$

For the Sparseout perturbation $\widetilde{X}_{i,j} = X_{i,j}[1 + |X_{i,j}|^{\frac{q-2}{2}}(r_j - 1)]$, the variance of $\widetilde{X}_i \cdot \beta$ is given by

$$Var[\widetilde{X}_i \cdot \beta] = \sum_{j=1}^{d} \frac{1-p}{p}|X_{i,j}|^q \beta_j^2 \tag{5}$$

Substituting in Eq. 4 we have

$$\hat{R}(\boldsymbol{\beta}) = \frac{1-p}{2p}||\boldsymbol{\Gamma X}||_q^q, \tag{6}$$

where $\Gamma = [\boldsymbol{\beta}^T V(\boldsymbol{\beta})\boldsymbol{\beta}]^{\frac{2}{q}}$ and $V(\boldsymbol{\beta}) = Diag(A)$.   ∎

Theorem 2. *For non-negative activation functions, Dropout is equivalent to Sparseout when $q = 2$.*

Setting $q = 2$ in Eq. 2 and considering the fact that $\boldsymbol{a}$ is non-negative, we have $\tilde{a}_i^l = r_i a_i^l$, which is identical to the Dropout perturbation [21].   ∎

L_q-normed spaces with different values of L_q exhibit different sparsity charactersitics. For $q < 2$ the norm space is sparse while for $q > 2$ the norm space is dense [27]. With Sparseout we can select the norm space of the activations by choosing the value of the hyper-parameter q. Thus, Sparseout allows us to control the level of sparsity in the activations of the neural networks.

4   Experimental Results

4.1   Sparsity Characterization

To verify that Sparseout is capable of controlling sparsity of a neural network's activations, we train an autoencoder with a hidden layer of 512 rectified linear units on the MNIST dataset. Dropout and Sparseout are applied to the hidden layer activations with $p = 0.5$ and different values of q for Sparseout. We measure sparsity of the hidden layer activations during testing (when no perturbations are applied to the activations). To measure sparsity we use the Hoyer's measure H [28]:

$$H(\mathbf{x}) = \frac{\sqrt{d} - \frac{|\mathbf{x}|_1}{|\mathbf{x}|_2}}{\sqrt{d} - 1} \tag{7}$$

where $\mathbf{x}$ is a d-dimensional vector, $|\mathbf{x}|_1$ is the L_1-norm and $|\mathbf{x}|_2$ is the L_2-norm of $\mathbf{x}$. A vector consisting of equal non-zero values has $H = 0$ while vectors only having one non-zero element has $H = 1$. Figure 1 shows the Hoyer's sparsity measure on the test set as the training progresses for different values of q. As the value of q decreases below 2, we see an increase in the sparsity of the activations, whereas for q values greater than 2 the sparsity is reduced. For $q = 2$, Sparseout results in the same sparsity as Dropout. These results confirm our theoretical analysis that Sparseout can be used to control sparsity of the activations in the neural networks.

4.2   Computational Cost

Sparseout is computationally efficient and incurs similar training cost as Dropout. We train an autoencoder with two hidden layer sizes on MNIST with a

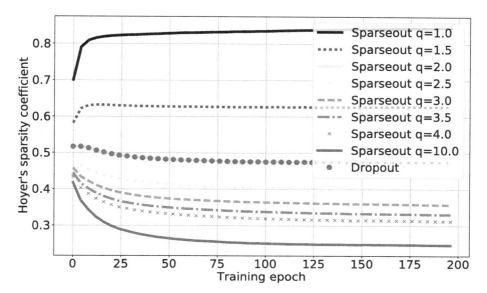

Fig. 1. Average Hoyer's sparsity measure of the hidden layer activations calculated over the MNIST test set for an autoencoder trained on MNIST with Dropout versus Sparseout with different values of q.

batch size of 128 both on Nvidia GTX 1080 GPU. As shown in Table 1, Sparseout is only fractionally more expansive than Dropout while Bridgeout is an order of magnitude more expensive even for this simple model. Also doubling the hidden layer size results in a doubling of the execution time for Brigdeout while Sparseout and Dropout have almost constant execution time due to utilization of GPU parallelism.

Table 1. Average execution time in seconds per epoch for different types of stochastic regularization for an autoencoder with different hidden layer sizes.

Hidden layer size	Backprop	Dropout	Sparseout	Bridgeout
1024	5.2	5.3	5.8	31.6
2048	5.6	5.6	6.0	57.2

4.3   Image Classification

Image classification is one of the key areas where deep neural networks have been highly successful achieving state-of-the-art results. The CIFAR datasets [29] are a standard benchmark for image classification. The CIFAR-10 dataset consists of color images of size 32×32 each belonging to one of the ten classes of objects. The dataset is divided into a training set of 50000 images and a test set of 10000 images. The CIFAR-100 dataset is similar to CIFAR-10 except that the images are divided into 100 classes of objects, thus making the classification task more harder than CIFAR-10. We used the standard pre-processing of mean and standard deviation normalization. Random cropping and random horizontal flips were used for data augmentation.

We use the wide residual network (WRN) architecture to evaluate the effect of sparsity on classification accuracy using Sparseout. WRNs achieved state-of-the-art accuracy on several image classification tasks including CIFAR-10 and CIFAR-100. WRNs are based on deep residual networks [30] that use identity links between the input and output of each layer known as the residual connections, but they employ fewer and wider layers. The residual connections helps in training very deep neural networks consisting of upto a thousand layers.

We employ a WRN with the basic building block shown in Fig. 2. The stochastic regularization is applied between the convolutional layers. Each convolutional layer is preceded by batch normalization and rectified linear unit activation function. In our experiments we use the WRN architecture WRN-28-10 with depth 28 and a widening factor of 10. A Dropout probability of $p = 0.3$ was used. Stochastic gradient descent with a mini-batch of 64 was used to train the networks. The learning rate was annealed from 0.1 at 60, 120 and 160 epochs by a factor of 0.2 as in the original WRN paper [31].

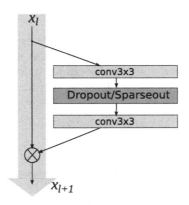

Fig. 2. The basic building block of wide residual network architecture with either Dropout or Sparseout stochastic regularization. Figure adapted from [31]

Image Classification Results. For image classification we found that Sparseout with $q > 2$ resulted in better performance compared to values of $q < 2$ as shown in Fig. 3. For $q < 2$ the accuracy drops as the training progresses beyond around 100 epochs indicating overfitting. As shown in Table 2, error rate for $q = 2.5$ is about 1% lower than Dropout for CIFAR-10 and 2.5% lower for CIFAR-100. Our baseline results are comparable to the baselines reported in the literature for CIFAR-100 and better for CIFAR-10 [31,32].

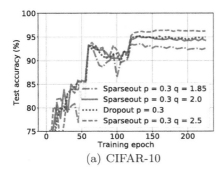

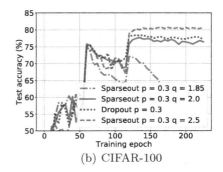

(a) CIFAR-10   (b) CIFAR-100

Fig. 3. Test accuracy during training of WRN on CIFAR-10 and CIFAR-100 for Dropout versus Sparseout with different values of q.

Table 2. Mean test errors for the same WRN network trained with Dropout and Sparseout with different values of q. Lower sparsity (high q) results in lower test error.

Model	p	q	CIFAR10	CIFAR100
WRN-28-10-Sparseout	0.3	1.5	7.58	25.87
WRN-28-10-Sparseout	0.3	1.85	5.92	24.65
WRN-28-10-Sparseout	0.3	2.0	4.72	21.91
WRN-28-10-Dropout	0.3	-	4.59	21.66
WRN-28-10-Sparseout	0.3	2.5	3.63	19.07
WRN-28-10-Dropout [31]	$p = 0.3$		3.89	18.85
WRN-28-10-$L_{0_{hc}}$ [32]	-		3.83	18.75

4.4 Language Modelling

Another task for which deep learning has been widely used is natural language processing (NLP). The dimensionalty of NLP tasks is very high and sparse; therefore, sparsity is likely to play an important role in such tasks. Language modelling (LM) assigns a probability to a sequence of words. LM is an important component of several NLP tasks such as speech recognition, information retrieval and machine translation among others. Since LM is a sequential task recurrent neural networks are used for it. Vanilla RNN are difficult to train due to vanishing and exploding gradients problem. To overcome these limitations, long short term memory (LSTM) models are used instead [33].

The LSTM model is a type of recurrent neural network with layers consisting of *memory cells*. The weights of the nodes in a memory cell learn the long term information while a node with a self-connected edge retains short term information. The input gate, forgetting gate and output gate help in controling the flow of information in the LSTM. For a detailed review of the LSTM formulation see Lipton et al. [34].

We adapt the baseline LSTM architecture for word level language modelling from Merity et al. [35]. The model consists of 3 layers of 1150 units. To train the baseline model we used the same hyper-parameters used by Merity et al.[1] except that we used only stochastic gradient descent for training.

We replace variants of Dropout with variants of Sparseout in the LSTM model. Variational Dropout [36] is replaced with variational Sparseout where a single random mask is used within a forward and backward pass. Embedding Dropout applied to the word embedding layer is similarly replaced with embedding Sparseout.

We evaluate the model on two standard word-level language modelling datasets where the task is to predict the next word and the performance is evaluated on perplexity which is the negative log likelihood raised to the exponent. The first dataset is the Penn Treebank dataset [37] that contains 1 million words and a vocabulary size of 10,000. The second dataset is the WikiText-2 dataset [38] which contains over 100 million words and a vocabulary of size 30,000.

Language Modeling Results. Applying Sparseout with $q > 2$ resulted in significant overfitting as shown in Fig. 4. For $q < 2$ we found that Sparseout resulted

Table 3. Single model test perplexities on PTB and Wiki-2 datasets for LSTM models trained with Dropout, Sparseout with $q = 2.25$ to reduce sparsity, and Sparseout with $q = 1.75$ to increase sparsity. Higher sparsity results in lower (better) relative perplexity.

Model	Penn Tree Bank	WikiText-2
LSTM-Sparseout ($q = 2.25$)	62.7	70.18
LSTM-Dropout	62.13	68.34
LSTM-Sparseout ($q = 1.75$)	60.57	67.17
AWD-LSTM-Dropout [35]	57.3	65.8

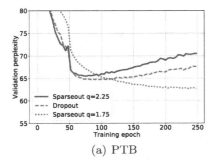

(a) PTB

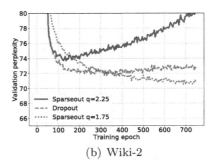

(b) Wiki-2

Fig. 4. Validation perplexity for language modeling on PTB and Wiki-2 datasets for LSTM models trained with Dropout, Sparseout with $q = 2.25$ to reduce sparsity, and Sparseout with $q = 1.75$ to increase sparsity.

---

[1] https://github.com/salesforce/awd-lstm-lm.

in better prediction performance than Dropout. For PTB dataset Sparseout results in 2.5% reduction in relative perplexity. For Wiki-2 dataset the reduction in relative perplexity is 1.25% as shown in Table 3.

5   Discussion

Existing literature is contradictory on whether sparsity is a good [10, 14–17, 32] or bad [7–9, 39, 40] property for deep neural networks. No previous study has evaluated sparse vs. non-sparse networks in a controlled fashion with stochastic regularization. In this study, we propose a new bridge-regularization scheme, Sparseout, which has the flexibility to control sparsity and the efficiency to be applied to large networks.

We evaluated Sparseout with two distinct network architectures and machine learning tasks: CNNs for image classification and LSTMs for language modeling. Our empirical results show that lower sparsity improves image classification performance, whereas higher sparsity improves performance on language modeling. These results align with the fundamental differences between data types: relatively tiny densely-featured images vs. sparsely-featured high-dimensional language data.

In this study, we chose the most suitable architecture for each task: CNNs for IID image classification and RNNs for sequential language modelling. Therefore, we evaluated task-architecture in a coupled manner. For each task, image classification or language modelling, we tested two datasets (CIFAR10/CIFAR100 or PTB/WikiText-2) and obtained consistent results regarding the benefit or lack thereof of sparse activations. It is possible, however, that the inherent sparse nature of convolutional layers requires spreading of the activations over all the neurons while enforced parsimony of representation is helpful for the fully connected gates in an LSTM. Therefore, decoupling the effect of data type from that of architecture is an important consideration we plan to investigate as future work.

Acknowledgments. This work was supported by the Natural Sciences and Engineering Research Council of Canada (NSERC).

References

1. Olshausen, B.A., Field, D.J.: Sparse coding with an overcomplete basis set: a strategy employed by V1? Vis. Res. **37**(23), 3311–3325 (1997)
2. Morris, G., Nevet, A., Bergman, H.: Anatomical funneling, sparse connectivity and redundancy reduction in the neural networks of the basal ganglia. J. Physiol.-Paris **97**(4), 581–589 (2003)
3. Thom, M., Palm, G.: Sparse activity and sparse connectivity in supervised learning. J. Mach. Learn. Res. **14**(Apr), 1091–1143 (2013)
4. Hinton, G.: A practical guide to training restricted Boltzmann machines. Momentum **9**(1), 926 (2010)

5. Lee, H., Ekanadham, C., Ng, A.Y.: Sparse deep belief net model for visual area V2. In: Advances in Neural Information Processing Systems, pp. 873–880 (2008)
6. Schweighofer, N., Doya, K., Lay, F.: Unsupervised learning of granule cell sparse codes enhances cerebellar adaptive control. Neuroscience **103**(1), 35–50 (2001)
7. Spanne, A., Jörntell, H.: Questioning the role of sparse coding in the brain. Trends in Neurosci. **38**(7), 417–427 (2015)
8. Goodfellow, I., Warde-farley, D., Mirza, M., Courville, A., Bengio, Y.: Maxout networks. In: Proceedings of the 30th International Conference on Machine Learning (ICML-13), pp. 1319–1327 (2013)
9. Kawaguchi, K., Kaelbling, L.P., Bengio, Y.: Generalization in deep learning. arXiv preprint arXiv:1710.05468 (2017)
10. Glorot, X., Bordes, A., Bengio, Y.: Deep sparse rectifier neural networks. In: Proceedings of the Fourteenth International Conference on Artificial Intelligence and Statistics, pp. 315–323 (2011)
11. Hanson, S.J., Pratt, L.Y.: Comparing biases for minimal network construction with back-propagation. In: Advances in Neural Information Processing Systems, pp. 177–185 (1989)
12. LeCun, Y., Denker, J.S., Solla, S.A., Howard, R.E., Jackel, L.D.: Optimal brain damage. In: NIPS, vol. 2, pp. 598–605 (1989)
13. Han, S., Pool, J., Tran, J., Dally, W.: Learning both weights and connections for efficient neural network. In: Advances in Neural Information Processing Systems, pp. 1135–1143 (2015)
14. Chauvin, Y.: A back-propagation algorithm with optimal use of hidden units. In: Advances in Neural Information Processing Systems, pp. 519–526 (1989)
15. Mrázová, I., Wang, D.: Improved generalization of neural classifiers with enforced internal representation. Neurocomputing **70**(16), 2940–2952 (2007)
16. Wan, W., Mabu, S., Shimada, K., Hirasawa, K., Hu, J.: Enhancing the generalization ability of neural networks through controlling the hidden layers. Appl. Soft Comput. **9**(1), 404–414 (2009)
17. Liao, R., Schwing, A., Zemel, R., Urtasun, R.: Learning deep parsimonious representations. In: Advances in Neural Information Processing Systems, pp. 5076–5084 (2016)
18. Chetlur, S., et al.: cudnn: Efficient primitives for deep learning. arXiv preprint arXiv:1410.0759 (2014)
19. Shalev-Shwartz, S., Ben-David, S.: Understanding Machine Learning: From Theory to Algorithms. Cambridge University Press, New York (2014)
20. Neyshabur, B., Tomioka, R., Srebro, N.: Norm-based capacity control in neural networks. In: Conference on Learning Theory, pp. 1376–1401 (2015)
21. Srivastava, N., Hinton, G.E., Krizhevsky, A., Sutskever, I., Salakhutdinov, R.: Dropout: a simple way to prevent neural networks from overfitting. J. Mach. Learn. Res. **15**(1), 1929–1958 (2014)
22. Kang, G., Li, J., Tao, D.: Shakeout: a new regularized deep neural network training scheme. In: Thirtieth AAAI Conference on Artificial Intelligence (2016)
23. Khan, N., Shah, J., Stavness, I.: Bridgeout: stochastic bridge regularization for deep neural networks. arXiv preprint arXiv:1804.08042 (2018)
24. Cogswell, M., Ahmed, F., Girshick, R., Zitnick, L., Batra, D.: Reducing overfitting in deep networks by decorrelating representations. arXiv preprint arXiv:1511.06068 (2015)
25. Ioffe, S., Szegedy, C.: Batch normalization: accelerating deep network training by reducing internal covariate shift. In: International Conference on Machine Learning, pp. 448–456 (2015)

26. Wager, S., Wang, S., Liang, P.S.: Dropout training as adaptive regularization. In: Advances in Neural Information Processing Systems, pp. 351–359 (2013)
27. Park, C., Yoon, Y.J.: Bridge regression: adaptivity and group selection. J. Stat. Plann. Inference **141**(11), 3506–3519 (2011)
28. Hoyer, P.O.: Non-negative matrix factorization with sparseness constraints. J. Mach. Learn. Res. **5**(Nov), 1457–1469 (2004)
29. Krizhevsky, A., Hinton, G.: Learning multiple layers of features from tiny images. Technical report, Citeseer (2009)
30. He, K., Zhang, X., Ren, S., Sun, J.: Deep residual learning for image recognition. In: Proceedings of the IEEE Conference on Computer Vision and Pattern Recognition, pp. 770–778 (2016)
31. Zagoruyko, S., Komodakis, N.: Wide residual networks. arXiv preprint arXiv:1605.07146 (2016)
32. Louizos, C., Welling, M., Kingma, D.P.: Learning sparse neural networks through l_0 regularization. In: International Conference on Learning Representations (2018)
33. Sundermeyer, M., Schlüter, R., Ney, H.: LSTM neural networks for language modeling. In: Thirteenth Annual Conference of the International Speech Communication Association (2012)
34. Lipton, Z.C., Berkowitz, J., Elkan, C.: A critical review of recurrent neural networks for sequence learning. arXiv preprint arXiv:1506.00019 (2015)
35. Merity, S., Keskar, N.S., Socher, R.: Regularizing and optimizing LSTM language models. arXiv preprint arXiv:1708.02182 (2017)
36. Gal, Y., Ghahramani, Z.: Dropout as a Bayesian approximation: insights and applications. In: Deep Learning Workshop, ICML (2015)
37. Marcus, M.P., Marcinkiewicz, M.A., Santorini, B.: Building a large annotated corpus of english: the penn treebank. Comput. Linguist. **19**(2), 313–330 (1993)
38. Merity, S., Xiong, C., Bradbury, J., Socher, R.: Pointer sentinel mixture models. arXiv preprint arXiv:1609.07843 (2016)
39. Rigamonti, R., Brown, M.A., Lepetit, V.: Are sparse representations really relevant for image classification? In: 2011 IEEE Conference on Computer Vision and Pattern Recognition (CVPR), pp. 1545–1552. IEEE (2011)
40. Gulcehre, C., Cho, K., Pascanu, R., Bengio, Y.: Learned-norm pooling for deep feedforward and recurrent neural networks. In: Calders, T., Esposito, F., Hüllermeier, E., Meo, R. (eds.) ECML PKDD 2014. LNCS (LNAI), vol. 8724, pp. 530–546. Springer, Heidelberg (2014). https://doi.org/10.1007/978-3-662-44848-9_34

Crowd Prediction Under Uncertainty

Luis Da Costa[⊠] and Jean-François Rajotte

Centre de Recherche Informatique de Montréal (CRIM),
Montréal, QC H3N 1M3, Canada
{luis.dacosta,Jean-Francois.Rajotte}@crim.ca
http://www.crim.ca/

Abstract. On this paper we use a newly-published method, **DFFT**, to estimate counts of crowds on unseen environments. Our main objective is to explore the relationship between noise in the input snapshots (of crowds) that a DFFT-based pipeline requires with the errors made on the predictions. If such a relationship exists we could apply our pipeline to the understanding of crowds; this application is extremely important to our industrial partners, that see utility in predicting crowds for objectives such as security of spatial planning. Our explorations indicate the possibility of such a characterization, but it depends on features of the actual environment being studied. Here we present 2 simulated environments of different difficulty, and we show how the predictions DFFT issues are of varying quality. We discuss the reasons we hypothesize are behind these performances and we set the ground for further experiments.

Keywords: Multi-agent systems · AI applications · Crowd behaviour

1 Introduction

The objective of this paper is to explore the application of a recent set of results in the field of multi-agent systems to an industrial R&D problem that we've been working on. Under the initiative of the city of Montreal (Canada) we are exploring the possibility of counting people on crowds using a network of cameras strategically installed in the city's cultural heart. Specific examples of events of interest are music concerts or political rallies, which raise needs on the city officials from the point of view of security, proper scheduling of services (*e.g.*, food and transportation), and localisation of the material needed for the event (*e.g.*, a stage).

In addition of the counting of the crowds we also want to analyse the behaviour of the groups of people: mean visit time on a place, trajectory taken by individuals or small groups moving at once, and punctual studies about entries and exits from/to important landmarks (subway stations, *e.g.*).

This project is well underway, as we will show on Sect. 2. The current focus on this project is on the **prediction** of crowds, specially for events that are *similar* to what we have observed, but not exactly the same; we are starting by

© Springer Nature Switzerland AG 2019
M.-J. Meurs and F. Rudzicz (Eds.): Canadian AI 2019, LNAI 11489, pp. 308–319, 2019.
https://doi.org/10.1007/978-3-030-18305-9_25

considering the case where we want to predict crowds on a different geographical configuration of the environment (for example, a different shape of a stage for a concert, or the same stages moved closer to points of interests). A precise prediction would allow us to give recommendations geared towards specific goals in crowd-management.

On a recent article ([4]) the authors use classical density functional theory to quantify micro-behaviour of crowds (rules of behaviour for individuals) from their macro-behaviour (local counts of density). A series of measurements of the population density at different points of the crowd is all that is required for this derivation; moreover, the different components of this characterization are functionally combinable: a component that is precisely derived under certain circumstances can be used as a proxy for another set of settings. This point is crucial for us as we expect to be able to build a precise understanding of the dynamics for certain kind of crowds but we must predict behaviours on slightly different conditions.

The ideas on [4] are certainly very promising our specific project; there is however an important difference: the input of this method assumes we have an exact idea of the counting in the crowd. In our case this is an unrealistic expectation because of different environmental conditions (*e.g.*, lights, distance of crowd from cameras; see Sect. 2 for an extended explanation). We have however the intuition that this method could be useful for us if we characterized the error on our prediction as a function of the error on the input measurements. On this paper we will explore this question. For doing so we build an artificial environment where we can run different instances of crowd problems; this artificial environment was built with the NetLogo modelling language [7], and it's described on Sect. 3. The results we obtain are shown and discussed on Sect. 4, and we delineate our next steps on Sect. 5.

2   Motivation and Background

In this section we present a summary of our current industrial problem, with a hint of the performances obtained (Sect. 2.1); we also present a recently published method (DFFT) upon which we will rely to build our prediction pipeline (Sect. 2.2).

2.1   Industrial Problem

Our main practical problem (motivated by a project on population understanding, spearheaded by the city of Montreal, Canada) revolves around the generation of analytics concerning crowds. The most basic of these analytics is the simple counting of the crowds, for which a proof of concept already exists. Let's show some examples of input images that we have used, along with the results we obtain. The images come from a set of cameras installed along the city, with locations, angles, and distances optimized for different purposes.

310   L. Da Costa and J.-F. Rajotte

Fig. 1. Sparse crowd, at day: ground truth = 217 people, estimation = 207 ± 19

Figures 1 and 2 show a sparse crowd under a natural sunlight, from a distance; our method (an extension of [6], re-trained on several publicly available datasets) focuses on counting the crowd under a region of interest, visible as the clearer zone on the images (we don't try to count the people that are on the shadowy zone of the images). Our method estimates densities (shown on the right side of the picture) and integrates over these to have estimated counts.

Fig. 2. Sparse crowd, at day: ground truth = 206 people, estimation = 211 ± 26

Figure 3 shows a dense(r) crowd under a slightly dimmer light than on Figs. 1 and 2. The results obtained are not as good as before; we attribute this on the changing light (which is paramount in image analysis) but also on the difference of density, which affects the interactions between individuals in the image.

Fig. 3. Dense crowd, afternoon: ground truth = 260 people, estimation = 200 ± 21

Figures 4 and 5 show the effort in estimating the crowds under different artificial lightning, at night. As we can observe, the counts are really off the mark.

Fig. 4. Dense crowd, at night: ground truth = 597 people, estimation = 207 ± 15

Fig. 5. Dense crowd, at night: ground truth = 628 people, estimation = 287 ± 16

We are currently studying the interesting points of the project, grouped under the name *analysis of the behaviour of groups of people*: mean time of visitors on a place, the trajectory taken by small groups moving at once, how much time each visitor stops at different specific points (exhibitions) and punctual studies about entries and exits from/to important landmarks (subway stations, bathrooms, food stands).

2.2 Density-Functional Fluctuation Theory, DFFT

When studying crowds of agents (people, animals, particles in general) we are confronted with 2 "typical" ways of working: either we can describe certain high-level phenomena we observe in the crowd, or we can postulate certain rules governing the behaviour of the agents. We would call these approaches "top-down" for the former and "bottom-up" for the latter.

Bottom-up approaches are very popular as there is a great amount of literature describing how simple individual agent rules can derive in interesting, "unexpected" behaviour at system's level, and there is fascinating research concerning the characterization of these emergent phenomena (see, *e.g.*, [2] or [8] for further details). A recent question that has been raised concerns the role of social interactions and environmental influences on living systems [4]; or, expressed another way, can we fully describe the micro-interactions of a system by dividing them into only 2 classes: (1) social interactions (individual-to-individual, or agent-agent, interactions), and (2) environmental influences (how does the environment affect the dynamics and patters of a macro-picture of the system). If this idea is correct the hypothesis is that on any crowd the individuals on it will (a) try to position themselves at the best possible spots, while (b) respecting some kind of individual-to-individual restriction (*e.g.*, avoiding overcrowded areas when there is insufficient "personal space" [4]).

There has been progress and studies on this line of enquiry recently; one of them is referenced as [4], where researchers applied classical density functional theory to quantify, directly from observations of local crowd-density, the rules that predict mass behaviours under new circumstances.

Mathematically this conjecture can be represented with 2 functions:

1. a *"vexation"* function $V(x)$ that assigns a desirability value to each location x, and
2. a *"frustration"* function $f(n(x))$ ($n(x)$ being a local-density function for locations x).

As their names indicate both these functions' values are better if they are smaller. The relative preferability of location x is the sum of vexation and frustration [4],

$$V(x) + f(n(x)). \tag{1}$$

This model is then completed with a rule that expresses the tendency for an agent to look for better locations: when an agent moves from location x to location x' we can compute a dissatisfaction

$$\Delta H = (V(x') + f(n(x'))) - (V(x) + f(n(x))) \tag{2}$$

that we try to minimize. Starting with this formulation we can obtain an expression for a probability of seeing a certain density N on a certain bin

$$P_b(N) = z_b^{-1} \frac{1}{N!} (e^{-v_b})^N e^{-f_N} \tag{3}$$

where z_b is a bin-dependent normalization constant, v_b is the average value of vexation V over bin b, and $f_N = f(\frac{N_b}{A}) * A$ approximates the total frustration contribution of bin b of area A [4].

The authors examine at this point Eq. 3 to extract functions f and V, using different numerical optimizations assumptions: for example, when $f = 0$ (meaning that the individual-individual interactions are neutral) the predictions take a form a single-parameter Poisson distribution, for which we can then reason about the structure of V (or v_b, under the form of a vexation averaged over a bin). Such derivations and an interesting example of real-life prediction (walking *Drosophila melanogaster* on a controlled setting) are all very well explained in [4].

The setting makes explicit the physical independence between f and V, which means that it should be possible to combine these functions extracted from diverse environments in order to produce new predictions; in [4] this is shown by predicting populations of flies under different environmental conditions.

The same way of working is used here; we use 2 environments, one of which we have the possibility of observing in detail (**env-1**) and one from whom we want to make predictions (**env-2**). The method would consist of 3 steps:

1. **Characterize dense environment (env-1)**, using DFFT. For example, use the snapshot shown at Fig. 3, and the temporal sequence around it, to derive precise functions for this, "environment 1", f_1, V_1
2. **Study a different environment (env-2) at low density.** Now we make the hypothesis (because of our knowledge of the problem at hand) that the individual-to-individual interactions are the same between the 2 environments (*i.e.*, the f function is, asymptotically, the same). In order to have an idea of the preferred locations on this "environment 2" we apply DFFT with low-density crowds on it. This analysis would yield f_2, V_2, where we hypothesize that (asymptotically, when working with dense crowds) $f_2 \approx f_1$
3. **Predict on the second environment.** Since we have an idea of the vexation of **env-2** (V_2), and a precise measure of the frustration part (f_1, since $f_2 \approx f_1$), we can then make predictions for high density crowds on **env-2**.

3 DFFT Under Noisy Crowd Counting: Experimental Setting

Our objective is to build a pipeline that would allow the prediction of crowds on environments similar (but not identical) to what we have observed. We would be interested in issuing results for environments geographically different: for example, if we apply our method to a music concert where the stage is far from the exit doors of a concert hall, can we predict the crowd on a concert where the stage is closer to the door, and the stage the shape of an "**L**"?

The usage of DFFT for our purposes seems like a promising avenue. However, as we showed in Sect. 2.1 our images are noisy, and so are our counts. This poses both a challenge and an opportunity, as we would need some ground truth in order to characterize the error incurred. If we can construct a pipeline with ground truth included, we could try and derive a function characterizing the prediction error in function of the input error.

For building this function we need essentially any environment for which the overall behaviour can reasonably be reduced to 2 components (as presented in page 5): (1) **social interactions** (individual-to-individual, or agent-agent, interactions), and (2) **environmental influences** (how does the environment affect the dynamics and patters of a macro-picture of the system). The family of such system is quite large, and includes the kind of multi-agent simulation we can build by specifying local rules of behaviour for the agents. So we decided to build, using NetLogo [7], 2 kinds of worlds for such crowd simulations; the first one ("*Gradient-Descent World*") has simpler dynamics than the second one ("*Bar El Farol*").

3.1   Gradient-Descent World

This world is a non-toroidal space where every cell has an associated "weight", directly associated with its "desirability": the lower the weight, the higher the desirability. A specific cell of the environment has weight 0 and all other cells have a weight equal to the Euclidean distance to that point. The agents in such a world implement a "weighted random motion": each one of them will move randomly while trying to navigate towards lower weights. They remember their last move and probabilistically backtrack their steps if they moved towards a region with higher weights.

This strategy is illustrated on Fig. 6, where an agent randomly chooses to move to a cell that is less advantageous than the original, and so the probabilities of movement are adjusted to allow backtracking. This creates a randomized gradient-descent walk.

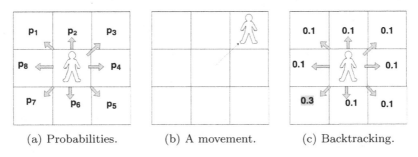

(a) Probabilities.   (b) A movement.   (c) Backtracking.

Fig. 6. Probabilities of movement, centered at agent's location (a). By default all $p_i = 1/8 = 0.125$ If weight of agent's cell at beginning of movement (a) is lower than weight of agent's cell at the end (b; the agent only sees the weight of a cell when it moves in the cell) then the probabilities of backtracking become better than random (fixed a priori; on c that probability is 0.3)

3.2   (Extended) Bar El Farol

We now consider a modified version of the classic problem based on a bar called **El Farol** in Santa Fe, New Mexico. The problem was first introduced on [1] as an example of how one might model economic systems of boundedly rational agents who use inductive reasoning. *The bar is popular—especially on Thursday nights when they offer Irish music—but sometimes becomes overcrowded and unpleasant. In fact, if the patrons of the bar think it will* *be overcrowded they stay home; otherwise they go enjoy themselves at El Farol. This model explores what happens to the overall attendance at the bar on these popular Thursday evenings, as the patrons use different strategies for determining how crowded they think the bar will be* [3,5]. An agent will go to the bar on Thursday night if they think that there will not be more than a certain number (parametrizable) of people there. Each agent has access to a set of prediction strategies and the actual attendance figures of the bar from previous Thursdays.

We implemented specific changes for the crowd to behave as such; from the geographic point of view, the bar exists on the right-hand side of the world, and has "doors", which are places where the agents can enter and exit the bar (Fig. 7a). This renders the movement of the agents more directed. A second change corresponds to the movement of the agents: in our case, the agents decide which way they will go, and then they move on that direction (if possible) on the step. This causes a fluid movement from one side of the world to another.

We also implemented the same local rules as described above but with a different position of the bar: instead of it being on a corner, this bar is located on the center of the world (Fig. 7b). We call this setting **Center is Better**.

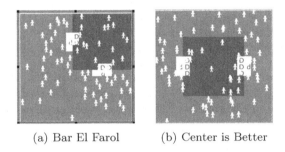

(a) Bar El Farol   (b) Center is Better

Fig. 7. Artificial Environments. Bar is in blue and doors are in white. (Color figure online)

4 Results and Discussion

4.1 Experiments and Measurements

Experiments. We apply the prediction pipeline shown on page 6 (consisting on 3 steps: (1) characterize dense environment **env-1**, (2) study a different environment, **env-2**, at low density, then (3) predict populations at high density on **env-2**) to the 2 experiments described below; all worlds physically consist of 55 spaces (numbered from -27 to $+27$), in both directions (vertical and horizontal), for a total of 3025 positions. There can be only 1 agent per square of the environment, so that is also the maximum number of agents in these worlds. The size of each zone is $5 * 5$ spaces, so in total we have $11 * 11 = 121$ zones (see Fig. 8). Each simulation is ran for 5000 iterations.

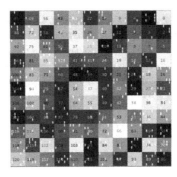

Fig. 8. Division of physical environment in 121 zones, size $5 * 5$

1. **Random World Experiment** Described on Sect. 3.1 These environments
   have a "special" square that has weight 0, and all other squares have a weight
   corresponding to the Euclidean distance to this square.
   (a) **env-1: Up-Right.** Special square is at coordinates $(13, 13)$. 1458 agents
       live in this environment.
   (b) **env-2: Center.** Special square is square at coordinates $(0, 0)$. 1458 agents
       live in the dense version of this environment, 100 in the sparse version.
2. **Bar El Farol Experiment** Described on Sect. 3.2
   (a) **env-1: Bar El Farol.** 1458 agents live in this environment, and the
       threshold for the bar is set at 500 agents.
   (b) **env-2: Center is Better.** The **dense version** has 1458 agents, and
       the threshold for the bar is set at 500 agents. The **sparse version** has
       120 agents, and the threshold for the bar is set at 42 agents.

Measure. For each experiment we compare the density predicted with the real
(measured) density. We calculate a relative measure, on a quite straightforward
way: if we measure $\hat{v}$, an approximation of v, then the error e is $e = abs(\hat{v} - v)/\hat{v}$.

Noise on Input. For each cell in the input snapshots we apply a noise η to the
actual count we make; when the real value is v, its noised value $\tilde{v}$ is a random
value in the interval $[(1 - \eta/100) * v, (1 + \eta/100) * v]$ In other words, η is a
percentage of noise on the actual value.

4.2   Random World Experiment

Let's start by running the predictions without noising the input measures; the
plot of real density vs. density predicted is shown on Fig. 9. The predictions'
fit (Fig. 9a) looks reasonable, although the relative error is big at low densities.
This explains the large errors at the left of Fig. 9b We observe that the error
bottoms up at density ≈ 0.009, which corresponds with $1458 * 0.009 \approx 13$ agents
on a square. Our hypothesis is that at that density and higher the cells are not
spatially decorrelated, which is one of the pre-requisites for the method in [4] to
work. Further investigations are needed to confirm this hypothesis (Fig. 12).

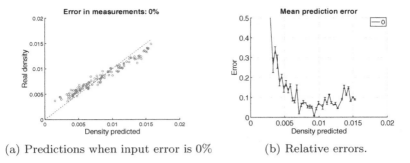

(a) Predictions when input error is 0%   (b) Relative errors.

Fig. 9. Predictions on clean input. For very low densities (<0.003) the errors are quite high ($\in [0.8, 1.5]$)

Figure 10 shows the fit when input error is 25% and Fig. 11 when input error is 50% It's interesting to see how the prediction error on noisy inputs is lower than the input errors themselves. We also observe the same inflexion point on the predictions' fit, where it seems that the values are better on densities around 0.01 Our hypothesis is that at higher densities the pre-conditions for DFFT break - but, again, this has to be empirically tested and theoretically demonstrated.

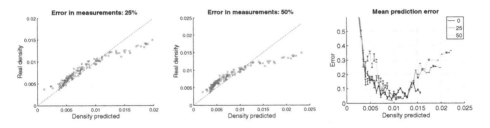

Fig. 10. Input error: 25%   **Fig. 11.** Input error: 50%   **Fig. 12.** All errors.

4.3   Bar El Farol Experiment

Let's start by running the predictions without noising the input measures; the plot of real density vs. density predicted is shown on Fig. 13a. The fit we obtain is not as good as the one obtained on the previous experiment (Fig. 9a), an observation corroborated by the errors obtained on Fig. 13b. This corresponds with the intuition that this problem is harder than the first one: the agents are more sophisticated (they make predictions about their environment), and the physical environment has some particularities. (*e.g.*, it has doors were the agents have to go by, which become overcrowded when the agents change opinions and want to go to/move from the bar. If this is the case there is the possibility that we are violating the pre-suppositions of the DFFT method, which causes the predictions to be low-quality.)

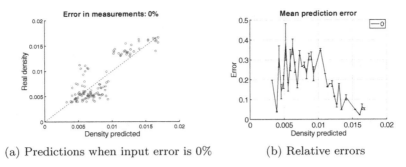

(a) Predictions when input error is 0%   (b) Relative errors

Fig. 13. Predictions on clean input.

Let's see what happens when the input is noisy (Figs. 14 and 15) The fit is noisier than in the case of clean input, but the prediction does not degrade dramatically. The comparative errors on predictions are shown in Fig. 16.

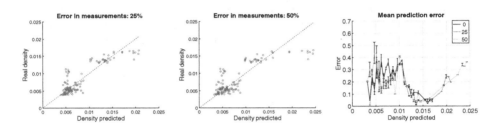

Fig. 14. Input error: 25%   **Fig. 15.** Input error: 50%   **Fig. 16.** All errors.

5   Conclusions and Future Work

In this paper we explored the possibility of using a recently published method (DFFT [4]) to predict crowds at unseen events; the particularity of our (industrial) use-case is that the input images (that we use to numerically train the model) are noisy. We have shown how there is a wide variability on the performance of our current crowd-counting system, and for such reason it would be important to be able to characterize the prediction error a method would incur into.

Our first explorations show promise, but also indicate the need for careful analysis of our results. On simple simulated environments the prediction of the counts are somehow accurate (*i.e.*, accurate on certain ranges); when we use more complicated simulations (agents that take decisions based on internal rules, physical environments containing funnels where agents can get "stuck") the predictions are much less accurate. Our first research effort for the near future will be to try and characterize different kinds of environments where we can reasonable expect good performance of our method.

At the same time we have been collecting snapshots of real events (music and other kinds of shows) so now we have a good dataset, clean and annotated on which we will be able to apply the method described here. Doing so could have important ramifications for our industrial partners, who would benefit from any insight concerning security and landscaping of physical environments for city events.

Acknowledgements. We thank our colleagues from the CRIM who provided insight and expertise that greatly assisted the research, and we appreciate the comments made by 3 anonymous reviewers on an earlier version of the manuscript. This work was partially supported by the MEI (*Ministère de l'Économie et Innovation*) of the Government of Québec.

References

1. Arthur, W.B.: Inductive reasoning and bounded rationality. Am. Econ. Rev. **84**(2), 406–411 (1994)
2. Axelrod, R.: The Complexity of Cooperation: Agent-Based Models of Competition and Collaboration. Princeton University Press, Princeton (1997)
3. Fogel, D., Chellapilla, K., Angeline, P.: Inductive reasoning and bounded rationality reconsidered. IEEE Trans. Evol. Comput. **3**(2), 142–146 (1999)
4. Méndez-Valderrama, J.F., Kinkhabwala, Y.A., Silver, J., Cohen, I., Arias, T.: Density-functional fluctuation theory of crowds. Nat. Commun. **9**(3538) (2018). https://doi.org/10.1038/s41467-018-05750-z
5. Rand, W., Wilensky, U.: NetLogo El Farol model. Center for Connected Learning and Computer-Based Modeling, Northwestern University, Evanston, IL (1997). http://ccl.northwestern.edu/netlogo/models/ElFarol
6. Vishwanath Sindagi, V.M.P.: CNN-based cascaded multi-task learning of high-level prior and density estimation for crowd counting. CoRR abs/1707.09605 (2017). http://arxiv.org/abs/1707.09605
7. Wilensky, U.: NetLogo. Center for Connected Learning and Computer-Based Modeling, Northwestern University, Evanston, IL (1999). http://ccl.northwestern.edu/netlogo/
8. Wilensky, U.: An Introduction to Agent-Based Modeling: Modeling Natural, Social, and Engineered Complex Systems with NetLogo. The MIT Press, Cambridge (2015)

Optimized Random Walk with Restart for Recommendation Systems

Seyyed Mohammadreza Rahimi[1], Rodrigo Augusto de Oliveira e Silva[1],
Behrouz Far[2], and Xin Wang[1(✉)]

[1] Department of Geomatics Engineering, University of Calgary, Calgary, AB, Canada
{smrahimi,radeoliv,xcwang}@ucalgary.ca
[2] Department of Electrical and Computer Engineering,
University of Calgary, Calgary, AB, Canada
far@ucalgary.ca

Abstract. Many sophisticated recommendation methods have been developed to produce recommendations to the users. Among them, Random Walk with Restart (RWR) is one of the most widely used techniques. However, RWR has a large time complexity of $O(k(n+m)^3)$ and memory complexity of $(O(n+m)^2)$. The change reducing the computational complexity is of great practical importance. In this paper, we propose an optimized version of random walk with restart, called the Optimized Random Walk with Restart (ORWR) and conduct theoretical and empirical studies on its performance. Mathematical analysis shows that using this technique the time complexity reduces to $O(nm^2)$ and the memory complexity to $O(nm)$. Experiments on three different recommendation problems using real-world datasets confirms the proposed ORWR method improves both time and memory cost of the recommendation.

1 Introduction

Recommendation systems have been extensively applied in various applications across different domains. Many e-commerce services such as Amazon and Netflix mainly depend on recommendation systems to increase their profits by selling what consumers are interested in against overloaded information of products.

Random walk with restart (RWR)—which is widely used in many applications in biology, computer science, etc. [1–3]—is commonly used as a graph-based model for recommendation systems. RWR recommends items to a user with the user's personalized ordering of node-to-node proximities which are measured by a random surfer in a user-item bipartite graph.

Researchers have proposed random-walk-with-restart-based methods for recommendations in different applications [4,5]. However, the focus of these studies have been on the semantics of the application. Thus the methods are not optimized to the specific requirements of the recommendation systems. On the other hand, researchers who focused on the optimization of random walk models, tried to optimize the random walk model in general, so that it can be utilized in a

© Springer Nature Switzerland AG 2019
M.-J. Meurs and F. Rudzicz (Eds.): Canadian AI 2019, LNAI 11489, pp. 320–332, 2019.
https://doi.org/10.1007/978-3-030-18305-9_26

wider set of problems [6,7]. With all these advancements, the time and memory complexity of Random Walk with Restart is still of $O(k(n+m)^3)$ where k is the number of steps before convergence, n is the number of users and m is the number of items. Given the large number of users and items of a typical recommendation, the computational cost can be critical for the system. In this study, we focus on optimizing the RWR for recommendation systems based on their specific characteristics. Additionally, we are looking to theoretically study the performance improvements of the proposed methods. Lastly, we want to empirically study the proposed method in different recommendation scenarios and compare it with existing popular recommendation methods.

To summarize, our contributions in this study are:

1. A novel Optimized Random Walk with Restart (ORWR) method is proposed to improve the performance of random walk with restart in recommendation systems. ORWR improves RWR in terms of both time and memory complexity. More specifically, RWR's memory and time complexity are of $O(n+m)^2$ and $O(k(n+m)^3)$ respectively, where n is the number of users and m is the number of items and k is the number of steps before convergenve. The ORWR reduces the time complexity to $O(knm^2)$ and memory complexity to $O(nm)$.
2. A mathematical proof of the proposed method is provided to prove the compatibility of the method to the Random walk with Restart. Sufficient analysis of the time and memory complexity improvements is also provided.
3. The proposed method is evaluated on three different applications namely, movie recommendation, route recommendation and location recommendation. A comparison between the original RWR and Matrix Factorization (MF) in terms of precision, recall, runtime and used memory is conducted.

The rest of the paper is organized as follows. First, a review of the literature on random walk with restart and recommendation systems is given in Sect. 2. Section 3 discusses the proposed Optimized Random Walk with Restart (ORWR) and analysis of the proposed method. Experimental results, comparing the performance of ORWR, RWR and MF are discussed in Sect. 4 followed by final conclusions and future studies in Sect. 5.

2   Related Works

Random walk—first introduced as a problem by Pearson in 1905 [8]—is a very common technique in the graph analysis. It has many applications for solving various problems. It can be used in the image processing for the image segmentation [1,2], e.g. to predict the microRNA-disease associations [3]. Some scholars have utilized random walk as recommendation models. For example, Yin et al. [4] utilized random walk with restart to make link recommendations on social network. Gori and Pucci [5] proposed ItemRank based on random walk, which is a scoring algorithm for ranking items based on the expected user preferences. The top ranked items are then recommended to the interested users.

Tong *et al.* [6] utilize linear correlations and block-wise community structures to propose a Fast Random Walk with Restart method using low-rank matrix approximation combined with graph partitioning, and discuss the applications of the proposed method. Yu *et al.* [7] proposed an iterative version of the random walk with restart that focuses on quickly updating the random walk matrix if some entries of the adjacency matrix are changed.

Some researchers tried to limit the number of steps a random walker takes in order to estimate the random walk results in a faster way. For example, Fouss *et al.* [9] proposed that running random walk for just 3 or 5 steps is enough. This method results in improved time complexity of the random walk sacrifices the precision of the recommendation. Also, such method can cause recommendation lists dominated by the high degree items [10].

Christoffel *et al.* [10] proposed an improvement over this method by introducing a reweighting function to remove the effect of Blockbusters. However, they did not propose suggestions to improve the estimation error caused by simulation sampling and does not improve the memory usage of the model. Also, the base 3-step random walk model would result in the same similarity values to that of the user-based collaborative filtering and will in turn produce the same predictions.

Some researchers tried to optimize PageRank which is a very similar method to Random Walk with Restart. For example, Fouss *et al.* [11] studied various scoring algorithms for collaborative-recommendation tasks, which showed a better performance in methods that utilize Laplacian matrix. Nikolakopoulos *et al.* [12] proposed a block teleportation rank method that allows faster convergence compared to the same method, without exploring lumpability, and PageRank.

These studies usually utilize the general random walk method. Similarly, the proposed optimizations for random walk with restart (RWR) consider the general random walk scenario for optimizations. However, as we will discuss in the following sections, recommendation systems require a specific type of graph. The special shape of the graph can be utilized to reduce both time and memory complexity of RWR. More specifically, if the users and items are sorted in a special order, the adjacency matrix will be partitioned into 4 sub-matrices, which calculations on two of them will be identical to running RWR on the whole matrix. That way we can improve RWR's time and memory complexity.

3   Methodology

RWR is one of the most commonly used methods for graph-based collaborative filtering in recommendation systems. In a recommendation system, we usually use a special graph that is the user-item graph.

Definition 1. *User-item graph:* *A user-item graph is a bipartite graph defined as:*

$$G = (V, E)$$

where V is the set of nodes, given a set of users U and a set of items I, $V = U \cup I$. And E is the set of edges. Each member of the edges set, defined as $e = (u, i, r_{ui}) \in E$ shows a that the user u has given the item i a rating of $r_u i$. Figure 1 (A) is an example of such graph.

Given G and a query user u, RWR computes a personalized ranking of items w.r.t the user u. Node set of the graph (V) has a number of users U and items I, i.e., $V = U \cup I$. A sample user-item graph is shown in Fig. 1 (A). RWR exploits a random surfer to produce the personalized ranking of items for a user u by letting her move around the graph G. Suppose the random surfer started from the user node u and she is at the node v currently. Then the surfer takes one of the following actions: random walk or restart. Restarting probability of RWR indicates that the surfer moves to a neighbor node of the current node with probability $1 - p$, or restarts the surfer back to the starting node u with probability p. If the random surfer visits node v many times, then node v is highly related to node u; thus, node v is ranked high in the personalized ranking for u. The random surfer is likely to frequently visit items that are rated highly by users who give ratings similarly to the query user u, which is consistent with the intuition of collaborative filtering, i.e., if a user has similar taste with the query user u, and the similar user likes an item i, then u is likely prefer the item i. We consider the probability of random walker visiting an item as the ranking score of that item. the items with higher scores, get higher ranks and are recommended to the user.

In this study, we are going to optimize the runtime and memory consumption of RWR for recommendation systems. Before we explain the optimizations we provide a definition of the notation used in this paper in Table 1.

3.1 RWR Probability Formula

In a graph G, let A be the transition matrix (i.e., column normalized adjacency matrix) of G, whose entry $A_{u,v} = 1/d_v$ if $(u, v) \in E$, and 0 otherwise. Here, d_v denotes the in-degree of v. Given query node $q \in V$, and restart probability

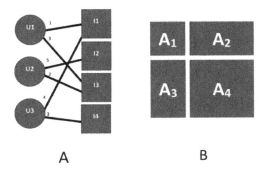

A   B

Fig. 1. (A) A sample user-item bipartite graph. (B) General outline of a user-item adjacency matrix

---

Algorithm 1. Random Walk with Restart(G,p)

1: $A \leftarrow AdjacencyMatrix(G)$ $\triangleright$ G is the input graph
2: $Q^0 \leftarrow queryMatrix(A)$ $\triangleright$ Q^0 is a diagonal unity matrix with the same size as A
3: $i \leftarrow 0$
4: $\epsilon \leftarrow \infty$ $\triangleright$ ϵ is the estimation error
5: **while** $\epsilon > \theta$ **do** $\triangleright$ θ is the convergence threshold
6:     $i \leftarrow i + 1$
7:     $Q^i \leftarrow (1-p)AQ^{i-1} + pQ^0$ $\triangleright$ p is the input restarting probability
8:     $\epsilon \leftarrow |Q^i - Q^{i-1}|$
9: **end while**
10: **return** Q^i

$p \in (0,1)$, the proximity of node u w.r.t. q, denoted by $Q_{u,q}$, is recursively defined as follows [6]:

$$Q^i = (1-p)AQ^{i-1} + pQ^0$$

$$Q = lim_{i \to \infty} Q^i$$

This pseudo-code of the RWR method is given in Algorithm 1. Assuming k steps before convergence, the runtime of this algorithm is of $O(k(n+m)^3)$, that is k steps of multiplying a $(n+m) * (n+m)$ matrix to another. The memory usage is of $O((n+m)^2)$, that is the some matrices of the size $(n+m) * (n+m)$.

3.2 Optimization of Random Walk with Restart

The rows and columns of the adjacency matrix can be rearranged to have the following order $(u_1, u_2, ..., u_n, i_1, i_2, ..., i_m)$. This creates a matrix with an outline similar to that of the Fig. 1(B). The four parts of the matrix are explained below:

- A_1 is a $n \times n$ matrix showing the user-user connections in the graph. For example, for users of a social network, this part can show the friendship links between them. In general, this part of the adjacency matrix shows the similarity of the users.
- A_2 is a $n \times m$ matrix capturing the information about the user-item interactions, i.e. the ratings. It can either be explicit or the implicit ratings of the items from each user.

Table 1. Notations used in the paper

Notation	Definition
n	Number of users
m	Number of items
A	Adjacency matrix
Q^k	Random walker state at the k-th step
$x_{i,j}$	Entry on row i and column j of matrix X

- A_3 is a $m \times n$ matrix showing the item-user interactions, i.e. the ratings. It can either be explicit or the implicit ratings.
- A_4 is a $m \times m$ matrix storing the item-item connections in the graph. In general, this part of the adjacency matrix shows the similarity of the items.

Given that the user-behavior graph in our system is an undirected bipartite graph (Fig. 1(A)) we can make the following conclusions:

1. $A_3 = A_2^T$
2. A_1 and A_4 are 0 as there are not user-user or item-item information stored in this system.

From an entry perspective the matrix can be seen as:

$$a_{i,j} = \begin{cases} 0 & \text{where } i \leq n \text{ and } j \leq n \\ 0 & \text{where } i > n \text{ and } j > n \\ r_{i,j} & \text{otherwise} \end{cases}$$

With these assumptions, we can keep updating these four segments instead of the whole matrix. With this strategy, the first step of the RWR calculations will result in a Q^1 matrix of that is the row-normalized version of the adjacency matrix. Similar to the adjacency matrix, we can divide Q^1 into four quadrants. Q_1^1 and Q_4^1 are zeros, and Q_2^1 and Q_3^1 contain the normalized values of the user-item ratings. The second step will results in the four segments of generated by the following equations.

$$Q_1^2 = Q_1^1 \cdot Q_1^1 + Q_2^1 \cdot Q_3^1 = Q_2^1 \cdot Q_3^1$$

$$Q_2^2 = Q_1^1 \cdot Q_2^1 + Q_2^1 \cdot Q_4^1 = 0$$

$$Q_3^2 = Q_3^1 \cdot Q_1^1 + Q_4^1 \cdot Q_3^1 = 0$$

$$Q_4^2 = Q_3^1 \cdot Q_2^1 + Q_4^1 \cdot Q_4^1 = Q_3^1 \cdot Q_2^1$$

From an entry point of view, the entry on the ith row and j^{th} column of the matrix Q^2 is calculated as:

$$q_{i,j}^2 = \begin{cases} 0 & \text{where } i \leq n \text{ and } j > n \\ 0 & \text{where } i > n \text{ and } j \leq n \\ \sum_{k=1}^{n+m} q_{i,k}^1 q_{k,j}^1 & \text{otherwise} \end{cases}$$

As seen in the equations the second step of the random walker results in the Q_2^2 and Q_3^2 matrices to be zeros and Q_1^2 and Q_4^2 to contain values. In general all even number of steps will have the same outline. Calculating one more step of the random walker we get the following four segments for the third step.

$$Q_1^3 = Q_1^2 \cdot Q_1^1 + Q_2^2 \cdot Q_3^1 = 0$$

$$Q_2^3 = Q_1^2 \cdot Q_2^1 + Q_2^2 \cdot Q_4^1 = Q_1^2 \cdot Q_2^1$$

$$Q_3^3 = Q_3^2 \cdot Q_1^1 + Q_4^2 \cdot Q_3^1 = Q_4^2 \cdot Q_3^1$$
$$Q_4^3 = Q_3^2 \cdot Q_2^1 + Q_4^2 \cdot Q_4^1 = 0$$

Similar to Q^2, the entry on the ith row and j^{th} column of the matrix Q^3 is calculated as:

$$q_{i,j}^3 = \begin{cases} 0 & \text{where } i \leq n \text{ and } j \leq n \\ 0 & \text{where } i > n \text{ and } j > n \\ \sum_{k=1}^{n+m} q_{i,k}^2 q_{k,j}^1 & \text{otherwise} \end{cases}$$

We can see that at steps 1 and 3 Sections 1 and 4 of the matrix are zeros. Similarly, at steps 2 and 4 Sections 2 and 3 of the matrix are zeros. To be more general. Q_1^n and Q_4^n are zeros if n is an odd number, Q_2^n and Q_3^n are zeros if n is an even number. In general RWR, the random agent can start from any node and land on any node. However, in recommendation systems, we are only interested in predicting the rating of each user for each item. Thus, the random walker agent is limited to starting from a user node and landing on an item node. More specifically, the random walker can only take an odd number of steps, so that it can land on an item node. Thus, we propose the following optimization routine.

In ORWR, we:

1. Only update the second quadrant of the matrix.
2. Only calculate it for odd number of steps.

Therefore we can define the updating equation of the optimized RWR for recommendation systems as:

$$Q_2^{2k+1} = (1-p)Q_2^1 Q_3^1 Q_2^{2k-1} + pQ_2^1$$

where p is the restarting probability. Then the objective of the ORWR can be defined as

$$Q_2^\infty = lim_{k \to \infty} Q_2^{2k+1}$$

The pseudo-code of the ORWR method is given in Algorithm 2.

3.3   Analysis

The ORWR has a more significant impact if the size of A_1 and A_4 matrices are higher than A_2 and A_3 matrices. Given that these two matrices have a size of $n \times n$ and $m \times m$ and the size of A_2 and A_3 is $n \times m$ and $m \times n$ respectively, the relative number of calculations of ORWR to RWR is:

$$\frac{\text{Time complexity of ORWR}}{\text{Time Complexity of RWR}} = \frac{knm^2}{k(n+m)^3} = \frac{nm^2}{(n+m)^3}$$

where k is the number of steps it takes the method to converge. The most performance improvement happens when n is much larger that m or m is much

---

Algorithm 2. Optimized Random Walk with Restart(G,p)

1: $A \leftarrow R$ ▷ R is the user-item-rating matrix
2: $Q_2^1 \leftarrow normalize(A)$ ▷ normalize by dividing each item by the sum of all items in that row
3: $Q_3^1 \leftarrow normalize(transpose(A))$
4: $k \leftarrow 0$
5: $\epsilon \leftarrow \infty$ ▷ estimation error
6: **while** $\epsilon \geq \theta$ **do**   ▷ θ is the convergence threshold
7:      $k \leftarrow k + 1$
8:      $Q_2^{2k+1} \leftarrow (1-p)Q_2^1 Q_3^1 Q_2^{2k-1} + pQ_2^1$ ▷ p is the input restarting probability
9:      $\epsilon \leftarrow |Q_2^{2k+1} - Q_2^{2k-1}|$
10: **end while**
11: **return** Q_2^{2k+1}

---

larger than n. The least performance improvement is when $n = m/2$. In that case the number of calculations of the RWR is 6.75 times higher than the ORWR.

Similar analysis can be done for the memory complexity.

$$\frac{\text{Memory complexity of ORWR}}{\text{Memory Complexity of RWR}} = \frac{nm}{(n+m)^2}$$

If no sparse matrix data structure is used for matrix storage, the memory required for ORWR is at least a quarter of the memory used by RWR (which happens when n and m are the same value.)

4 Experiments

In this section, we evaluate ORWR and compare it to other existing recommendation systems on three real-world datasets. The summary of the datasets are given in Table 2. MovieLens has 1 million ratings of 6000 users on 4000 movies. It is used to compare the performance of these models on a dataset with explicit ratings. Gowalla dataset contains check-ins of users to a set of locations. This dataset is used for a location recommendation problem. Check-ins are implicit ratings and the collaborative filtering method is used to predict the behavior of the user as in the Behavior-based Location Recommendation (BLR) [16]. The Geolife [15,17–19] data set is used for route recommendation using MAP2R method [20]. The collaborative filtering is used to predict the route behavior of the user.

4.1 Experimental Setup

ORWR, RWR, MF [21], P_β^3 [9] and P_β^5 [10] were developed in Matlab and testing was done on a Core i7 MacBook Pro with 16 GB of Ram and an SSD Drive. The spatial codes for route recommendation and location recommendation were developed in C# and Java respectively, this code is shared between all the methods and it uses the output of the Matlab code to generate recommendations.

Table 2. Datasets used in this study

Dataset name	Users	Items	Ratings
MovieLens [13]	6000	4000	1000000
Gowalla (Check-ins) [14]	5462	5999	104851
GeoLife [15]	22	21124	29759

4.2 Recommendation Performance

In this experiment, we measure the quality of the recommendations. Before discussing the results, we need to define measures to quantify the quality of the recommendations.

Performance Measures. To quantify and compare the quality of the recommendations, we use precision and recall as defined below.

For movie recommendation and location recommendation precision and recall are defined as follows.

$$Precision = \frac{|\text{Recommended Items} \cap \text{Correct Items}|}{|\text{Recommended Items}|}$$

$$Recall = \frac{|\text{Recommended Items} \cap \text{Correct Items}|}{|\text{Correct Items}|}$$

Precision is the ratio of the recommended items that match the chosen item by the user to all recommended items. Recall, on the other hand, is the ratio of actual chosen items predicted by the recommendation model to all the actual chosen items.

For route recommendation we redefined the precision and recall as follows.

$$Precision_R = \frac{|\text{Recommended Route Segments} \cap \text{User Route Segments}|}{|\text{Recommended Route Segments}|}$$

$$Recall_R = \frac{|\text{Recommended Route Segments} \cap \text{User Route Segments}|}{|\text{User Route Segments}|}$$

$Rrecision_R$ measures the portion of the recommended route segments that are taken by the user. Whereas, $Recall_R$ measures the portion of the user route correctly predicted by the route recommendation.

Results. As shown in Fig. 2, RWR and ORWR perform similarly in terms of recommendation results. And they are outperforming MF in location recommendation and route recommendation. However, MF is outperforming both in movie recommendation. That is inline with previous studies that claimed MF is better with explicit ratings and RWR is better with implicit ratings [22]. P_β^3 and P_β^5, on the other hand, are performing similarly and result in precision and recall values smaller than that of the ORWR and RWR.

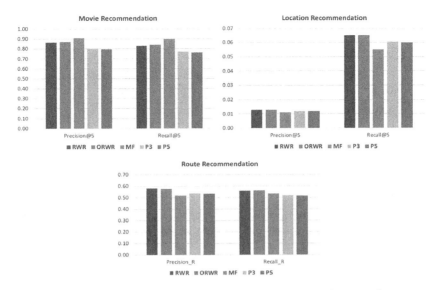

Fig. 2. Precision and recall values of ORWR, RWR, MF, P^3_β and P^5_β on different recommendation tasks

4.3   Runtime Analysis

One of the main contributions of this study is the improved computational performance of the RWR. We designed this experiment to compare the runtime of the proposed method to the original RWR and the matrix factorization.

Performance Measure. The performance measure in this case is the time it takes to learn a model based on the training dataset i.e the training time. The reported run-time is the average of 5 runs on different training datasets of the same size. The reported run-time values are in seconds and smaller values indicate better performance. For MF, RWR and ORWR the training is completed once the model has converged. To be more specific, when the change in the learned matrices is less than the given threshold.

Results. The runtime of RWR, ORWR and MF for learning a model for each of the three tasks is given in Table 3. As shown in Table 3, the ORWR is the best performing model in terms of runtime and the second-best model is MF with RWR being the worst performing model. It can also be observed that the performance improvement of ORWR over RWR is much larger for route recommendation compared to the movie and location recommendation. That is because for route recommendation the number of items is much larger than the number of users. As predicted in the analysis section, the performance improvement is more noticeable. Similarly, P^3_β and P^5_β are better performing than the

Table 3. Model training time (in seconds) of RWR, ORWR and MF for three recommendation problems

	Movie recommendation	Location recommendation	Route recommendation
RWR	8.14	4.32	6460
ORWR	0.54	0.16	17
MF	3.2	2.12	38
P_β^3	5.45	3.12	1293
P_β^5	6.32	4.58	2585

RWR, however, since their optimization targets the number of steps not the matrix size, the runtime improvements is not as impressive as the ORWR.

5 Conclusions and Future Works

In this study, an Optimized Random Walk with Restart method (ORWR) is proposed for recommendation systems. This optimization utilizes the special shape of the user-item graph in a recommendation system to reduce the number of calculations needed by the Random walk with Restart. The time and memory complexity analysis shows that ORWR method improves on RWR in terms of both time and memory by a factor of at least 4. These recommendation performance and runtime comparison of the method has been performed on real-world datasets for three types of recommendation systems, which shows time superiority of the proposed method as well as better recommendation quality for implicit feedback scenarios.

For future studies, we will focus on the application of the ORWR on other recommendation scenarios. Also, we will study the application of proposed model to fast random walk with restart techniques to provide an even faster ORWR method for recommendation systems. Also, as the user-item matrix of a recommendation system is constantly updating, an iterative version of ORWR can also be implemented to reduce the number of updates required if a single value is changed.

Acknowledgments. The research is supported by the Natural Sciences and Engineering Research Council of Canada Discovery Grant to Xin Wang and Behrouz Far, and National Natural Science Foundation of China (No. 61772420).

References

1. Dong, X., Shen, J., Shao, L., Gool, L.V.: Sub-markov random walk for image segmentation. IEEE Tran. Image Process. **25**(2), 516–527 (2016)
2. Grady, L.: Random walks for image segmentation. IEEE Trans. Pattern Anal. Mach. Intell. **28**(11), 1768–1783 (2006)

3. Liu, Y., Zeng, X., He, Z., Zou, Q.: Inferring microrna-disease associations by random walk on a heterogeneous network with multiple data sources. IEEE/ACM Trans. Comput. Biol. Bioinf. **14**(4), 905–915 (2017)
4. Yin, Z., Gupta, M., Weninger, T., Han, J.: A unified framework for link recommendation using random walks. In: 2010 International Conference on Advances in Social Networks Analysis and Mining (ASONAM), pp. 152–159, 08 2010
5. Gori, M., Pucci, A.: Itemrank: a random-walk based scoring algorithm for recommender engines. In: Proceedings of the 20th International Joint Conference on Artifical Intelligence. IJCAI 2007, pp. pp. 2766–2771. Morgan Kaufmann Publishers Inc., San Francisco (2007)
6. Tong, H., Faloutsos, C., Pan, J.: Fast random walk with restart and its applications. In: Sixth International Conference on Data Mining (ICDM 2006), pp. 613–622, December 2006
7. Yu, W., Lin, X.: IRWR: incremental random walk with restart. In: Proceedings of the 36th International ACM SIGIR Conference on Research and Development in Information Retrieval. SIGIR 2013, pp. 1017–1020. ACM, New York (2013)
8. Pearson, K.: The problem of the random walk. Nature **72**(1865), 294 (1905)
9. Cooper, C.S., Lee, S.H., Radzik, T., Siantos, Y.: Random walks in recommender systems: exact computation and simulations. In: WWW (2014)
10. Christoffel, F., Paudel, B., Newell, C., Bernstein, A.: Blockbusters and wallflowers: accurate, diverse, and scalable recommendations with random walks. In: RecSys (2015)
11. Fouss, F., Pirotte, A., Renders, J., Saerens, M.: Random-walk computation of similarities between nodes of a graph with application to collaborative recommendation. IEEE Trans Knowl. Data Eng. **19**(3), 355–369 (2007)
12. Nikolakopoulos, A.N., Korba, A., Garofalakis, J.D.: Random surfing on multipartite graphs. In: 2016 IEEE International Conference on Big Data (Big Data), pp. 736–745, December 2016
13. Movielens | grouplens. https://grouplens.org/datasets/movielens/. Accessed 17 Oct 2018
14. Zhou, D., Wang, B., Rahimi, S.M., Wang, X.: A study of recommending locations on location-based social network by collaborative filtering. Proc. Can. AI **2012**, 255–266 (2012)
15. Download geolife gps trajectories. https://www.microsoft.com/en-us/download/details.aspx?id=52367 Accessed 17 Oct 2018
16. Rahimi, S.M., Wang, X., Far, B.: Behavior-based location recommendation on location-based social networks. In: Kim, J., Shim, K., Cao, L., Lee, J.-G., Lin, X., Moon, Y.-S. (eds.) PAKDD 2017. LNCS (LNAI), vol. 10235, pp. 273–285. Springer, Cham (2017). https://doi.org/10.1007/978-3-319-57529-2_22
17. Zheng, Y., Zhang, L., Xie, X., Ma, W.Y.: Mining interesting locations and travel sequences from GPS trajectories. In: Proceedings of the 18th International Conference on World Wide Web. WWW 2009, pp. 791–800. ACM, New York (2009)
18. Zheng, Y., Li, Q., Chen, Y., Xie, X., Ma, W.Y.: Understanding mobility based on GPS data. In: Proceedings of the UbiComp 2008. UbiComp 2008, pp. 312–321. ACM, New York (2008)
19. Zheng, Y., Xie, X., Ma, W.Y.: Geolife: a collaborative social networking service among user, location and trajectory. IEEE Data(base) Engineering Bulletin, June 2010

20. Cui, G., Wang, X.: MaP2R: a personalized maximum probability route recommendation method using GPS trajectories. In: Kim, J., Shim, K., Cao, L., Lee, J.-G., Lin, X., Moon, Y.-S. (eds.) PAKDD 2017. LNCS (LNAI), vol. 10235, pp. 168–180. Springer, Cham (2017). https://doi.org/10.1007/978-3-319-57529-2_14
21. Koren, Y., Bell, R., Volinsky, C.: Matrix factorization techniques for recommender systems. Computer **42**(8), 30–37 (2009)
22. Park, H., Jung, J., Kang, U.: A comparative study of matrix factorization and random walk with restart in recommender systems. CoRR abs/1708.09088 (2017)

Inter and Intra Document Attention for Depression Risk Assessment

Diego Maupomé$^{(\boxtimes)}$, Marc Queudot, and Marie-Jean Meurs

Université du Québec à Montréal, Montréal, QC, Canada
maupome.diego@courrier.uqam.ca, meurs.marie-jean@uqam.ca

Abstract. We take interest in the early assessment of risk for depression in social media users. We focus on the eRisk 2018 dataset, which represents users as a sequence of their written online contributions. We implement four RNN-based systems to classify the users. We explore several aggregations methods to combine predictions on individual posts. Our best model reads through all writings of a user in parallel but uses an attention mechanism to prioritize the most important ones at each timestep.

1 Introduction

In 2015, 4.9 million Canadians aged 15 and over experienced a need for mental health care; 1.6 million felt their needs were partially met or unmet [7]. In 2017, over a third of Ontario students, grades 7 to 12, reported having wanted to talk to someone about their mental health concerns but did not know who to turn to [6]. These numbers highlight a concerning but all too familiar notion: although highly prevalent, mental health concerns often go unheard. Nonetheless, mental disorders can shorten life expectancy by 7–24 years [9].

In particular, depression is a major cause of morbidity worldwide. Although prevalence varies widely, in most countries, the number of persons that would suffer from depression in their lifetime falls between 8 and 12% [15]. Access to proper diagnosis and care is overall lacking because of a variety of reasons, from the stigma surrounding seeking treatment [23] to a high rate of misdiagnosis [25]. These obstacles could be mitigated in some way among social media users by analyzing their output on these platforms to assess their risk of depression or other mental health afflictions. The analysis of user-generated content could give valuable insights into the users mental health, identify risks, and help provide them with better support [3,11]. To promote such analyses that could lead to the development of tools supporting mental health practitioners and forum moderators, the research community has put forward shared tasks like CLPsych [2] and the CLEF eRisk pilot task [1,18]. Participants must identify users at risk of mental health issues, such as eminent risk of depression, post traumatic stress disorder, or anorexia. These tasks provide participants with annotated data and a framework for testing the performance of their approaches.

M.-J. Meurs and F. Rudzicz (Eds.): Canadian AI 2019, LNAI 11489, pp. 333–341, 2019.
https://doi.org/10.1007/978-3-030-18305-9_27

In this paper, we present a neural approach to identify social media users at risk of depression from their writings in a subreddit forum, in the context of the eRisk 2018 pilot task. From a technical standpoint, the principal interest of this investigation is the use of different aggregation methods for predictions on groups of documents. Using the power of Recurrent Neural Networks (RNNs) for the sequential treatment of documents, we explore several manners in which to combine predictions on documents to make a prediction on its author.

2 Dataset

The dataset from the eRisk2018 shared task [18] consists of the written production of `reddit` [22] English-speaking users.

The dataset was built using the writings of 887 users, and was provided in whole at the beginning of the task. Users in the RISK class have admitted to having been diagnosed with depression; NO_RISK users have not. It should be noted that the users' writings, or posts, may originate from different separate discussions on the website. The individual writings, however, are not labelled. Only the user as a whole is labelled as RISK or NO_RISK. The two classes of users are highly imbalanced in the training set with the positive class only counting 135 users to 752 in the negative class. Table 1 presents some statistics on the task dataset.

We use this dataset but consider a simple classification task, as opposed to the early-risk detection that was the object of the shared task.

Table 1. Statistics on the eRisk 2018 task dataset

	Training		Test	
	Risk	Control	Risk	Control
# users	135	752	79	741
# writings	49,557	481,837	40,665	504,523
Submissions/subject	367.1	640.7	514.7	680.9
Words/submission	27.4	21.8	27.6	23.7

3 Models

We represent users as sets of writings rather than sequences of writings. This is partly due to the intuition that the order of writings would not be significant in the context of forums, generally speaking. It is also due to the fact that treating writings sequentially would be cumbersome, especially if we consider training on all ten chunks. However, we do consider writings as sequences of words, as this is the main strength of RNNs. We therefore write a user u as the set of his m writings, $u = \{\mathbf{x}^{(1)}, \ldots, \mathbf{x}^{(m)}\}$. A given writing $\mathbf{x}^{(j)}$, is then a sequence of words, $\mathbf{x}^{(j)} = x_1^{(j)}, \ldots, x_\tau^{(j)}$, with τ being the index of the last word. Thus, $x_t^{(j)}$ is the t-th word of the j-th post for a given user.

3.1   Aggregating Predictions on Writings

Late Inter-Document Averaging. We set out to put together an approach that aggregates predictions made individually and sequentially on the writings of a user. That is, we read the different writings of a user in parallel and take the average prediction on them. This is our first model, Late Inter-Document Averaging (LIDA). Using the RNN architecture of our choice, we read each word of a post and update its *hidden state*,

$$h_t^{(j)} = f(x_t^{(j)}, h_{t-1}^{(j)}; \theta_{post}). \tag{1}$$

f is the *transition function* of the chosen RNN architecture, θ_{post} is the set of parameters of our particular RNN model and the initial state is set to zero,

$$h_0 = \mathbf{0}.$$

In practice, however, we take but a sample of users' writings and trim overlong writings (see Sect. 5). LIDA averages over the final state of the RNN, $h_\tau^{(j)}$, across writings,

$$a = \frac{1}{m} \sum_{j=1}^{m} h_\tau^{(j)} \tag{2}$$

This average is then projected into a binary prediction for the user,

$$p = \sigma(\mathbf{u}^\top \begin{bmatrix} a \\ 1 \end{bmatrix}), \tag{3}$$

using σ, the standard logistic sigmoid function, to normalize the output and a vector of parameters, $\mathbf{u}$. By averaging over all writings, rather than taking the sum, we ensure that the number of writings does not influence the decision. However, we suspect that regularizing on the hidden state alone will not suffice, as the problem remains essentially the same: gradient correction information will have to travel the entire length of the writings regardless of the corrections made as a results of other writings.

Continual Inter-Document Averaging. Our second model, Continual Inter-Document Averaging (CIDA), therefore aggregates the hidden state across writings at *every* time step, as opposed to only the final one. A first RNN, represented by its hidden state h_t, reads the writings as in Eq. 1. The resulting hidden states are averaged across writings and then fed as the input to a second RNN, represented by g_t,

$$a_t = \frac{1}{m} \sum_{j=1}^{m} h_t^{(j)}, \tag{4}$$

$$g_t = f(a_t, g_{t-1}; \theta_{user}). \tag{5}$$

g_τ is used to make a prediction similarly to Eq. 3.

3.2   Inter-document Attention

It stands to reason that averaging over the ongoing summary of each document would help in classifying a group of documents. Nonetheless, one would suspect that some documents would be more interesting than others to our task. Even if all documents were equally interesting, their interesting parts might not align well. Because we are reading them in parallel, we should try and prioritize the documents that are interesting at the current time step.

CIDA does not offer this possibility, as no weighting of terms is put in place in Eq. 4. Consequently, we turn to the *attention mechanism* [4] to provide this information. While several manners of both applying and computing the attention mechanism exist [8, 19, 26], we compute the variant known as *general* attention [19], which is both learned and content-dependent. In applying it, we introduce Inter-Document Attention (IDA), which will provide a weighted average to our previous model.

The computation of $h_t^{(j)}$, the post-level hidden state, remains the same, i.e. Eq. 1. However, these values are compared against the previous user-level hidden state to compute the relevant *energy* between them, $\hat{\alpha}_{jt}$

$$\tilde{\alpha}_t^{(j)} = g_{t-1} \mathbf{W}_{att} h_t^{(j)}, \tag{6}$$

where $\mathbf{W}_{att}$ is a matrix of parameters that learns the compatibility between the hidden states of the two RNNs. The resulting energy scalars, $\hat{\alpha}^{(j)}t$ are mapped to probabilities by way of softmax normalization,

$$\alpha_t^{(j)} = \frac{e^{\tilde{\alpha}_t^{(j)}}}{\sum_{k=1}^m e^{\tilde{\alpha}_t^{(k)}}}. \tag{7}$$

This probability is then used to weight the appropriate h_t,

$$a_t = \sum_{j=1}^m \alpha_t^{(j)} h_t^{(j)}. \tag{8}$$

g_t is given by Eq. 5. Through the use of this probability weighting, we can understand a_t as an expected document summary at position t when grouping documents together. As in the previous model, a prediction on the user is made from g_τ.

3.3   Intra-document Attention

We extend our use of the attention mechanism in the aggregation to the parsing of individual documents. Similarly to our weighting of documents in aggregation dependent on the current aggregation state, we compare the current input to past inputs to evince a *context* for it. This is known in the literature as *self-attention* [8]. We therefore modify the computation of h_t from Eq. 1 by adding a context vector, $c_t^{(j)}$, corresponding to the ongoing context in document j at time t:

$$h_t^{(j)} = f(x_t^{(j)}, c_t^{(j)}, h_{t-1}^{(j)}; \theta_{post}).$$ (9)

This context vector is computed by comparing past inputs to the present document-level hidden state,

$$\tilde{\alpha}_{t,t'}^{(j)} = h_t^{(j)} \mathbf{W}_{intra} x_{t'}^{(j)},$$ (10)

This weighting is normalized by softmax and used in adding the previous inputs together. We refer to this model as Inter- and Intra-Document Attention (InIDA).

This last attention mechanism arises from practical difficulties in learning long-range dependencies by RNNs. While RNNs are theoretically capable of summarizing sequences of arbitrary complexity in their hidden state, numerical considerations make learning this process through gradient descent difficult when the sequences are long or the state is too small [5]. This can be addressed in different manners, such as gating mechanisms [10,13] and the introduction of multiplicative interactions [24]. Self-Attention is one such mechanism where the context vector acts as a reminder of past inputs in the form of a learned expected context. It can be combined to other mechanisms with minimal parameter load.

4   Related Work

Choudhury et al. [11] used a more classical approach to classify Twitter users as being at risk of depression or not. They first manually crafted features that describe users' online behavior and characterize their speech. The measures were computed daily, so a user is represented as the time series of the features. Then, the training and predictions were done by a Support Vector Machine (SVM) with PCA for dimensionality reduction.

More similarly to our approach, Ive et al. [14] used Hierarchical Attention Networks [27] to represent user-generated documents. Sentence representations are learned using a RNN with an attention mechanism and are then used to learn the document's representation using the same network architecture. The computation of the attention weights they use is different from ours as it is non-parametric. Their equivalent of Eq. 6 would be

$$\tilde{\alpha}_t^{(j)} = h_t^{(j)\top} g_t^{(j)}$$ (11)

This means that the RNNs learn the attention weights along with the representation of the sequences themselves. This attention function has been introduced in [19] under that name of *dot*.

The *location-based function* [19] is a simpler version of the *general attention* that we used, that only takes into account the target hidden state. It is stated as such:

$$\tilde{\alpha}_t^{(j)} = W_{att} g_t^{(j)}$$ (12)

The *additive* function introduced in [4], has been improved in [19]. Luong et al. use a concatenation layer to combine the information of the hidden state and the context vector.

$$\tilde{\alpha}_t^{(j)} = \tanh(W_{att}g_t^{(j)} + W_{att}h_t^{(j)}) \tag{13}$$

Content-based addressing was developed as part of Neural Turing Machines [12], where the attention is focused on inputs that are similar to the values in memory.

$$\tilde{\alpha}_t^{(j)} = \text{cosine}[g_t^{(j)}, h_t^{(j)}] \tag{14}$$

5 Methodology

5.1 Preprocessing

As previously mentioned, documents are broken into words. The representation of these words is learned from the entirety of the training documents, all chunks included, using the skip-gram algorithm [20]. All words were turned to lowercase. Only the 40k most frequent words were kept. The embedded representation learned is of size 40, using a window of size five. The embeddings are shared by all models.

Documents are trimmed at the end at a length of 66 words, which is longer than 90% of the posts in the dataset. The number of documents varies greatly across user classes. We take small random samples without replacement of 30 documents per user at every iteration (epoch). We contend that sampling the user at every iteration allows us to train for longer as it is harder for the models to overfit when the components that make up each instance keep changing.

5.2 Model Configurations

We use the Multiplicative Long Short-Term Memory (mLSTM) [17] architecture as the post-level and user-level RNN, where applicable. The flexibility of the transition function in mLSTM has shown to be capable of arriving at highly abstract features on its own and can achieve competitive results in sentiment analysis [21]. Due to the limited number of examples, smaller models are required to avoid overfitting. We therefore set the embedded representation at 20 and the size of the hidden state of both RNNs to 80. Parameter counts are shown in Table 2.

5.3 Training

For our experiments, we reshuffle the original eRisk 2018 dataset, as the training and test sets do not have the same proportions among labels. To provide our models with more training examples, we divide the dataset 9:1, stratifying across labels. We use 10% of the training set as validation.

We train the models using the Adam optimizer [16], making use of 10% of the training data for validation. Having posited random intra-user sampling as a means of training longer, we set the training time to 30 epochs, taking the best model on validation over all epochs. As noted, the two classes are highly imbalanced. We use inverse class weighting to counteract this.

5.4   Evaluation

The nature of the task, which is to prioritize finding positive users, and the class imbalance in the dataset, we use the f1-score as a first metric in validation and in the final testing phase. The f1-score is useful to assess the quality of classification between unbalanced two unbalanced classes, one of which is designated as the positive class. It is defined as the harmonic mean between *precision* (out of all the positive examples, how many are correctly classified as positive) and *recall* (out of all examples classified as positive, how many were indeed in the positive class). Using True Positive (TP) as the number of positive examples correctly classified, False Positive (FP) the number of examples in the positive class incorrectly classified, and True Negative (TN) and False Positive (FP) for the negative class, we have the following equations.

$$precision = \frac{TP}{TP + FP} \tag{15}$$

$$recall = \frac{TP}{TP + FN} \tag{16}$$

$$f1\text{-}score = 2 \times \frac{precision \times recall}{precision + recall} \tag{17}$$

We evaluate our models on the best result on a validation set of 10% of the training data. These best results are selected over 30 epochs.

6   Results

Our preliminary results in validation are in accordance with our hypotheses. That is, continual aggregation surpasses late aggregation but falls short of the more sophisticated attention model. Moreover, the noticeable difference in performance has little to no cost in terms of parameter count.

Table 2. Parameter counts, precision, recall and f1-score (%) on the adapted test set for the eRisk 2018 corpus

Model	p.c	Precision	Recall	f1-score
LIDA	31k	39.7	51.2	45.6
CIDA	95k	41.7	69.8	52.2
IDA	101k	45.6	73.2	56.2
IIDA	175k	47.4	72.8	57.4

7 Conclusion

In this paper, we have put forward three RNN-based models that aggregate documents to make a prediction on their author. We applied this model to the eRisk 2018 dataset, which associates a user, as a sequence of online forum posts, to a binary label that identifies them as being at risk for depression or not.

With the goal of using RNNs to read the individual documents, we tested four methods of combining the resulting predictions, LIDA, CIDA, IDA and InIDA. We also introduced the inter-document attention mechanism. Our preliminary results show promise and confirm the parameter efficiency of the attention mechanism.

Future work could involve the use of dot-product alone, which, despite adding no parameters, has been found to be more effective for global attention [19]. An investigation into using late attention aggregation for all hidden states produced across all documents is also necessary.

References

1. CLEF eRisk Pilot Task. http://early.irlab.org/. Accessed 6 July 2018
2. CLPsych Shared Task. http://clpsych.org/shared-task-2017/. Accessed 6 July 2018
3. Ayers, J.W., Althouse, B.M., Allem, J.P., Rosenquist, J.N., Ford, D.E.: Seasonality in seeking mental health information on Google. Am. J. Prev. Med. (AJPM) 44(5), 520–525 (2013)
4. Bahdanau, D., Cho, K., Bengio, Y.: Neural machine translation by jointly learning to align and translate. arXiv preprint arXiv:1409.0473 (2014)
5. Bengio, Y., Simard, P., Frasconi, P., et al.: Learning long-term dependencies with gradient descent is difficult. IEEE Trans. Neural Netw. 5(2), 157–166 (1994)
6. Boak, A., Hamilton, H.A., Adlaf, E.M., Henderson, J.L., Mann, R.E.: The mental health and well-being of Ontario students, 1991–2017: Detailed Findings from the Ontario Student Drug Use and Health Survey, CamhOSDUHS (2016)
7. Canada, S.: Accessing Mental Health Care in Canada (2017). https://www150.statcan.gc.ca/n1/pub/11-627-m/11-627-m2017019-eng.htm
8. Cheng, J., Dong, L., Lapata, M.: Long short-term memory-networks for machine reading. In: Proceedings of the 2016 Conference on Empirical Methods in Natural Language Processing, pp. 551–561 (2016)
9. Chesney, E., Goodwin, G.M., Fazel, S.: Risks of all-cause and suicide mortality in mental disorders: a meta-review. World Psychiatry 13(2), 153–160 (2014)
10. Chung, J., Gulcehre, C., Cho, K., Bengio, Y.: Empirical evaluation of gated recurrent neural networks on sequence modeling. arXiv preprint arXiv:1412.3555 (2014)
11. De Choudhury, M., Gamon, M., Counts, S., Horvitz, E.: Predicting depression via social media. In: Seventh International AAAI Conference on Weblogs and Social Media (2013)
12. Graves, A., Wayne, G., Danihelka, I.: Neural Turing Machines. arXiv preprint arXiv:1410.5401 (2014)
13. Hochreiter, S., Schmidhuber, J.: Long short-term memory. Neural Comput. 9(8), 1735–1780 (1997)

14. Ive, J., Gkotsis, G., Dutta, R., Stewart, R., Velupillai, S.: Hierarchical neural model with attention mechanisms for the classification of social media text related to mental health. In: Proceedings of the Fifth Workshop on Computational Linguistics and Clinical Psychology: From Keyboard to Clinic, pp. 69–77 (2018)
15. Kessler, R., Berglund, P., Demler, O., et al.: The epidemiology of major depressive disorder: results from the national comorbidity survey replication (NCS-R). JAMA **289**(23), 3095–3105 (2003)
16. Kingma, D.P., Ba, J.: Adam: a method for stochastic optimization. arXiv preprint arXiv:1412.6980 (2014)
17. Krause, B., Lu, L., Murray, I., Renals, S.: Multiplicative LSTM for sequence modelling. arXiv preprint arXiv:1609.07959 (2016)
18. Losada, D.E., Crestani, F., Parapar, J.: Overview of eRisk: early risk prediction on the internet. In: Bellot, P., et al. (eds.) CLEF 2018. LNCS, vol. 11018, pp. 343–361. Springer, Cham (2018). https://doi.org/10.1007/978-3-319-98932-7_30
19. Luong, M.T., Pham, H., Manning, C.D.: Effective approaches to attention-based neural machine translation. arXiv preprint arXiv:1508.04025 (2015)
20. Mikolov, T., Sutskever, I., Chen, K., Corrado, G.S., Dean, J.: Distributed representations of words and phrases and their compositionality. In: Advances in Neural Information Processing Systems, pp. 3111–3119 (2013)
21. Radford, A., Jozefowicz, R., Sutskever, I.: Learning to generate reviews and discovering sentiment. arXiv preprint arXiv:1704.01444 (2017)
22. Reddit: Reddit. https://www.reddit.com/. Accessed 6 July 2018
23. Rodrigues, S., et al.: Impact of stigma on veteran treatment seeking for depression. Am. J. Psychiatr. Rehabil. **17**(2), 128–146 (2014)
24. Sutskever, I., Martens, J., Hinton, G.E.: Generating text with recurrent neural networks. In: Proceedings of the 28th International Conference on Machine Learning (ICML 2011), pp. 1017–1024 (2011)
25. Vermani, M., Marcus, M., Katzman, M.A.: Rates of detection of mood and anxiety disorders in primary care: a descriptive, cross-sectional study. Prim. Care Companion CNS Disord. **13**(2) (2011)
26. Xu, K., et al.: Show, attend and tell: neural image caption generation with visual attention. In: Bach, F., Blei, D. (eds.) Proceedings of the 32nd International Conference on Machine Learning. Proceedings of Machine Learning Research, vol. 37, pp. 2048–2057, PMLR, Lille, France, 07–09 July 2015. http://proceedings.mlr.press/v37/xuc15.html
27. Yang, Z., Yang, D., Dyer, C., He, X., Smola, A., Hovy, E.: Hierarchical attention networks for document classification. In: Proceedings of the 2016 Conference of the North American Chapter of the Association for Computational Linguistics: Human Language Technologies, pp. 1480–1489 (2016)

Short Papers

A Shallow Learning - Reduced Data Approach for Image Classification

Kaleb E. Smith[1]([✉]) and Phillip Williams[2]

[1] Florida Institute of Technology, Melbourne, FL 32901, USA
ksmith012007@my.fit.edu
[2] University of Ottawa, Ottawa, Canada

Abstract. Shepard Interpolation Neural Networks (SINN) lay a foundation addressing the flaws of deep algorithms, inspired by statistical interpolation techniques rather than biological brains it can be mathematically proven and the neuron interactions can be intuitively described. They also possess the ability to discriminate well with limited training data during the algorithm process. To enhance SINN from just regular vectorized images, we look to utilize hand designed and natural image features to help the SINN perform better on benchmark image classification data sets. We compare these input feature vectors using the SINN framework on three benchmark image classification test sets, showing comparable results to the state-of-the-art (SOTA) for a fraction of the computational and memory requirements due to SINN shallow learning ability.

1 Introduction

Deep neural networks have achieved ever increasing success and popularity as of late, achieving the state-of-the-art (SOTA) in domains such as image recognition and signal analysis. However, this success is gained at the cost of computational complexity [2]; current cutting edge models are often extremely large, requiring billions of parameters and extensive training time on massive clusters [2]. The aim of adapting Shepard Interpolation Neural Networks (SINN) to image recognition tasks is to exponentially reduce the size of the models needed for advanced classification tasks and possibly excel at this task when there is a lack of data.

In this paper, the issue of obtaining great performance result on image classification is addressed. We construct a SINN using hand designed HOG and SIFT features as its input and test them in the SINN framework on MNIST [3], CIFAR-10 and CIFAR-100 [7]. Then we look at naturally learned features by taking a significantly smaller CNN architecture to produce feature vectors utilized for the initialization of the SINN. These more advanced convolutional-SINN models are then tested on the same benchmark data sets. The proposed network demonstrates competitive accuracy on these data sets while being significantly smaller than the leading deep architectures on these data sets. We then look

© Springer Nature Switzerland AG 2019
M.-J. Meurs and F. Rudzicz (Eds.): Canadian AI 2019, LNAI 11489, pp. 345–351, 2019.
https://doi.org/10.1007/978-3-030-18305-9_28

to challenge both typical and SOTA classification algorithms when there is a lack of training data by downsizing the training data for these experiments by 50%, 75% and 90%. Our proposed methods outperforms all other classification methods when training data is limited.

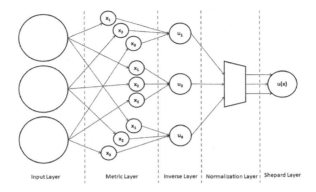

Fig. 1. Illustration of SINN architecture.

2   Related Works

The Shepard Interpolation Neural Network architecture has only been very recently developed, as such, there are only one published attempt at improving the approach outlined by Williams for image classification [1]. Smith et al. used convolutional neural network features to enhance image classification, which we look to advance in this paper [14]. However, Shepard Interpolation has been applied to tasks other than image classification. Smith et al. also used the SINN to classify time series data and showed SOTA performance in that field [12,13]. Ren et al. [4] outline the application of Shepard Interpolation to convolutional neural networks. The proposed method is the addition of a Shepard Interpolation layer as an augmentation on architectures used for low-level image processing tasks such as inpainting and super resolution.

Much of the foundation for SINN is outlined in [1]. The Shepard component of the architecture is comprised of several layers: the metric layer, inverse layer, normalization layer and the Shepard layer, seen in Fig. 1. By exploiting the Shepard data interpolation method, the architecture of the network can be designed rather than found through exploration, requiring very little hyper parameter tuning as well as providing increased efficiency. To further elaborate, several data points are selected from the each output class, and are used to deterministically initiate the weights of the layers. What is to be pointed out is the output of the Shepard Layer is then fed into a softmax function to give the network the ability to do multi-class classification problems. Also, the network will only increase in width, not depth. So the more metric nodes you have in the metric layer will widen the layer, which leads to more inverse nodes and so on, but the SINN's depth will not change.

3 A Feature Based Shepard Interpolation Neural Networks

Prior to the introduction of the feature based input layers to the Shepard interpolation architecture, the mapping from the image space to the feature space is simply flattening the image matrix to a vector; the key innovation proposed is to replace the image flattening step with more sophisticated feature extraction techniques (HOG, SIFT, CNN features), resulting in a much better overall accuracy for the model. These higher level nodes do not get "stacked" as seen in deep learning architectures, instead they work in only one layer, and the number of nodes in the layer can increase or decreases. This leads to the idea of a shallow learning process, which is quicker, more efficient, and competitive against SOTA deep methods.

For computer vision applications, the model can be imagined to be a transformation from an image space to a space of classification vectors. If the SINN is treated as a transform from the feature space to the output classification space, a second transform is needed from the image space to the feature space Eq. 1.

$$R^{i \times j} \rightarrow R^m \rightarrow R^n \tag{1}$$

where $R^{i \times j}, R^m, R^n$ are the image space, feature space and classification space respectively. Shepard Interpolation is a type of data interpolation which falls into the inverse distance weighting class utilizes a distance metric that describes the similarity of an original data point to an estimated functional data point [9]. The interpolation is dependent on the inverse weighting function

$$w_i(x) = \left(\frac{1}{d(x, x_i)} \right)^p \tag{2}$$

where $d(x, x_i)$ is some distance function. Shepard Interpolation is defined as

$$u(x) = \begin{cases} \frac{\sum_i^N w_i(x) u_i}{\sum_i^N w_i(x)} \\ u_i & d(x, x_i) = 0 \end{cases} \tag{3}$$

The exponential p typically is a free parameter to change according to the desired dimension of the polynomial fit. It is typical though to see $p = 2$ in Shepard's method, representing something similar to the euclidean distance.

4 Experiments

In order to validate that the our deep feature approaches present significant improvements over existing methods, a comparison of the different variants of the SINN versus the existing SOTA was done on the MNIST dataset [3] of 28×28 grayscale images of handwritten numbers, the CIFAR-10 and CIFAR-100 [7] data sets of 32×32 RGB images of various objects are explored.

Our experiment looks at the challenge of image classification when there is a lack of training data. For all three data sets, the training data is cut by 50%, 75%, and 90% (maintaining a even class distribution) to create a much smaller training set that is used for the algorithms. This means MNST training data goes from 60000 images to 30000, 15000, and 6000; while both CIFAR-10 and CIFAR-100's training set goes from 50000 to 25000, 12500, and 5000. We compare our results to a neural network (1 hidden layer of 100 nodes), a CNN (of size $32 \times (7 \times 7)$, $32 \times (5 \times 5)$, $32 \times (3 \times 3)$) and the SOTA deep methods. Each algorithm was run 100 times and the percent accuracy shown in Table 1 shows the average over all the test runs.

The SINN architecture contains only 100 Shepard nodes in the input layer. All of our implementations were coded in Python using Keras and sci-kit learn on a Linux based Dell Inspiron 15 7000 laptop with 8 GB RAM, 5th Gen Intel Core i7 Quad Core processor, and a NVIDIA GeForce 1050Ti GPU.

Table 1. Classification accuracy (in percentage correct) from our three data sets using an assortment of classification techniques and our proposed algorithm. The top contributor in each data set is in bold for each reduced training set.

Training data reduced by 50%									
	MNIST			CIFAR-10			CIFAR-100		
Feats:	HOG	SIFT	CNN	HOG	SIFT	CNN	HOG	SIFT	CNN
NN	93.45	94.62	94.25	**60.47**	71.58	71.75	31.49	34.91	41.87
CNN	92.24	91.85	93.14	50.78	80.34	81.58	35.89	60.47	68.58
SINN	**96.20**	**97.85**	**99.15**	52.85	**83.45**	**91.95**	**40.57**	**64.36**	**76.25**
Training data reduced by 75%									
	MNIST			CIFAR-10			CIFAR-100		
Feats:	HOG	SIFT	CNN	HOG	SIFT	CNN	HOG	SIFT	CNN
NN	77.85	78.31	78.46	33.75	54.85	59.14	24.96	28.85	27.07
CNN	81.67	81.74	83.25	34.85	67.13	66.78	34.20	39.34	37.85
SINN	**95.50**	**96.60**	**98.35**	**50.80**	**82.75**	**89.25**	**40.44**	**61.16**	**72.71**
Training data reduced by 90%									
	MNIST			CIFAR-10			CIFAR-100		
Feats:	HOG	SIFT	CNN	HOG	SIFT	CNN	HOG	SIFT	CNN
NN	41.85	43.14	49.17	24.98	26.74	28.69	20.41	20.89	20.12
CNN	43.50	47.60	51.25	32.85	54.62	51.85	21.85	24.96	27.20
SINN	**92.70**	**94.20**	**96.65**	**49.75**	**80.35**	**86.15**	**35.58**	**57.96**	**70.15**

In this experiment we wanted to look at how classification algorithms compare to our proposed method when training data is cut significantly. We tested each algorithm out on the feature sets used (HOG, SIFT, and CNN extracted

Table 2. Experimental results showing the SOTA algorithms when training data is downsized by 90%.

MNIST 10% of training data					
	HOG-SINN	SIFT-SINN	C-SINN	DMC [5]	CapsNet [10]
Accuracy %	92.70	94.20	**96.65**	44.30	41.25

CIFAR-10 10% of training data					
	HOG-SINN	SIFT-SINN	C-SINN	ART [6]	AutoAugment [11]
Accuracy %	42.75	80.35	**86.15**	40.35	41.85

CIFAR-100 10% of training data					
	HOG-SINN	SIFT-SINN	C-SINN	WRN [8]	AutoAugment
Accuracy %	35.58	57.96	**70.15**	25.84	30.24

features), then compared these results across the classifiers. Looking at Table 1 we can see an overwhelming shift in the other algorithms classification accuracy when the training data is limited while our proposed method maintains consistent accuracy. This is roughly 1500 images per class for MNIST, 1250 images for CIFAR-10 and only 125 images per class for CIFAR-100 and when the 90% case is considered training breaks down to 600 images per class for MNIST, 500 images per class for CIFAR-10, and 50 images per class for CIFAR-100. As can be seen from the large accuracy drops, these methods can not go through sufficient training to become a successful algorithm when there is little training data while the SINN excels. The success of the SINN is from its ability to infer new data in the feature space while going through training as it is described in Sect. 3 showing that even with a small amount of training data it can converge to a decision boundary that is optimal in image classification.

In Table 2 we can see the leading deep learning architectures versus the feature based SINN when 90% of the training data is taken away. This table strengthens the notion of these deep learning algorithms relying heavily on large datasets for training to be successful when used on typical testing data sets. The SOTA classifiers struggle when the training data is lost, losing accuracy of up to 45% in some cases. Even when AutoAugmentation is used these algorithms cannot generalize with little training data, showing that even though having the algorithm learn its optimal policies for maximal accuracy it is still highly reliant on the number of labeled training samples. This opens up more interesting cases as well, a AutoML SINN might break the one shot learning barrier being able to learn the right amount of metric nodes to have to optimize the hyperplane interpolation covering the most data samples.

5 Conclusion

In this paper, we outlined the shortcomings of the previous iteration of the SINN, as well as proposing viable improvements for adapting the algorithm to general purpose image classification tasks. We also show when the training data is cut by huge percentages, SINN still classify extremely well and outperform other classification methods (by upwards of 45%) including the SOTA. Possible future work could be to expand the architecture of the SINN to see if accuracy performance can increase (while of course decreasing computational cost). Finding the correlation between the number of samples used for training to accuracy (one-shot learning). There could be a better way to expand the activations functions in order to produce even greater results for data classification, this is one spot where AutoML might prevail. Also, looking to utilize SINN in other machine learning tasks, such as natural language processing, time series analysis, forecasting, or medical image analysis should be explored, since the approach is still new.

References

1. Williams, P.: SINN: shepard interpolation neural networks. In: Bebis, G., et al. (eds.) ISVC 2016. LNCS, vol. 10073, pp. 349–358. Springer, Cham (2016). https://doi.org/10.1007/978-3-319-50832-0_34
2. Le, Q.V.: Building high-level features using large scale unsupervised learning. In: 2013 IEEE International Conference on Acoustics, Speech and Signal Processing (ICASSP), pp. 8595–8598. IEEE, May 2013
3. LeCun, Y.: The MNIST database of handwritten digits (1998). http://yann.lecun.com/exdb/mnist/
4. Ren, J.S., Xu, L., Yan, Q., Sun, W.: Shepard convolutional neural networks. In: Advances in Neural Information Processing Systems, pp. 901–909 (2015)
5. Ciregan, D., Meier, U., Schmidhuber, J.: Multi-column deep neural networks for image classification. In: 2012 IEEE Conference on Computer Vision and Pattern Recognition (CVPR), pp. 3642–3649. IEEE, June 2012
6. Xie, S., et al.: Aggregated residual transformations for deep neural networks. In: 2017 IEEE Conference on Computer Vision and Pattern Recognition (CVPR) (2017)
7. Krizhevsky, A., Geoffrey, H.: Learning multiple layers of features from tiny images (2009)
8. Zagoruyko, S., Nikos, K.: Wide residual networks. arXiv preprint arXiv:1605.07146 (2016)
9. Shepard, D.: A two-dimensional interpolation function for irregularly-spaced data. In: Proceedings of the 1968 23rd ACM National Conference. ACM (1968)
10. Sabour, S., Frosst, N., Hinton, G.E.: Dynamic routing between capsules. In: Advances in Neural Information Processing Systems (2017)
11. Cubuk, E.D., et al.: AutoAugment: Learning Augmentation Policies from Data. arXiv preprint arXiv:1805.09501 (2018)
12. Smith, K.E., et al.: Shepard interpolation neural networks with k-means: a shallow learning method for time series classification. In: 2018 International Joint Conference on Neural Networks (IJCNN). IEEE (2018)

13. Smith, K.E., Williams, P.: Time series classification with shallow learning shepard interpolation neural networks. In: Mansouri, A., El Moataz, A., Nouboud, F., Mammass, D. (eds.) ICISP 2018. LNCS, vol. 10884, pp. 329–338. Springer, Cham (2018). https://doi.org/10.1007/978-3-319-94211-7_36

14. Smith, K.E., Williams, P., Chaiya, T., Ble, M.: Deep convolutional-shepard interpolation neural networks for image classification tasks. In: Campilho, A., Karray, F., ter Haar Romeny, B. (eds.) ICIAR 2018. LNCS, vol. 10882, pp. 185–192. Springer, Cham (2018). https://doi.org/10.1007/978-3-319-93000-8_21

Multi-class Ensemble Learning
of Imbalanced Bidding Fraud Data

Farzana Anowar[(⊠)] and Samira Sadaoui

University of Regina, Regina, SK, Canada
{fad469,sadaouis}@uregina.ca

Abstract. E-auctions are vulnerable to Shill Bidding (SB), the toughest fraud to detect due to its resemblance to usual bidding behavior. To avoid financial losses for genuine buyers, we develop a SB detection model based on multi-class ensemble learning. For our study, we utilize a real SB dataset but since the data are unlabeled, we combine a robust data clustering technique and a labeling approach to categorize the training data into three classes. To solve the issue of imbalanced SB data, we use an advanced multi-class over-sampling method. Lastly, we compare the predictive performance of ensemble classifiers trained with balanced and imbalanced SB data. Combining data sampling with ensemble learning improved the classifier accuracy, which is significant in fraud detection problems.

Keywords: Auction fraud · Fraud detection · Multi-class data ·
Hierarchical clustering · Data labeling · Data sampling ·
Ensemble classification

1 Introduction

E-auctions are gaining a huge popularity in numerous application domains. Nevertheless, the Internet Crime Complaint Center proclaimed that auction fraud is one of the top Internet crimes [1,2]. E-auctions represent a convenient way to conduct fraud due to several factors, such as the anonymity of users, flexibility of bidding, and affordable auctioning services. Auctions are vulnerable to three types of fraud: pre-auction (e.g. black-market goods), post-auction (e.g. non-shipment of goods), and in-auction (e.g. shill bidding and bid shielding). In this paper, we concentrate on shill bidding (SB) because, unlike the other two types of fraud, it does not leave any obvious evidence. SB is one of the most frequent auction fraud and also the most challenging to detect because it closely resembles usual bidding behavior [3]. To increase the seller's pay-off, a shill bidder (the seller himself and/or his collaborators) increases the price of the product (a good or a service) by submitting many bids via phony accounts and IP addresses. In this way, a genuine buyer can easily be cheated, and therefore loose money. Over the years, several sellers have been prosecuted for malicious bidding activities that led to large monetary losses for the winning bidders [4].

© Springer Nature Switzerland AG 2019
M.-J. Meurs and F. Rudzicz (Eds.): Canadian AI 2019, LNAI 11489, pp. 352–358, 2019.
https://doi.org/10.1007/978-3-030-18305-9_29

The large-scale SB detection problem can be effectively addressed with Machine Learning Algorithms (MLAs). Still, there are few SB classification studies owning to the following challenging problems:

- **Unavailability of SB data.** To produce SB samples, we need to carry out several laborious tasks as follows: (a) identify pertinent SB patterns, (b) determine robust algorithms to measure the SB patterns, (c) crawl a large number of auctions from a commercial website, (d) preprocess the raw auction dataset, and (e) evaluate the SB metrics against all the bidders of the auction dataset.
- **Unlabeled SB data.** Labeling SB samples with multi-dimensional features is a difficult operation that can be tackled with the help of MLAs, such as unsupervised learning (e.g. data clustering), semi-supervised learning, and active learning. Still, the human expertise is needed during the labelling process of the SB instances.
- **Imbalanced SB data.** A SB dataset is imbalanced, and the skewed class distribution has a negative impact on the classification algorithms. Additionally, the minority class (for example the fraud class), which usually possesses the highest misclassification cost, maybe misrepresented since it is viewed as noise. To handle the imbalanced classification problem, two main methods have been introduced: data sampling and cost-sensitive learning.

We represent our fraud detection problem as a three-class categorization model where a bidder is labeled as normal or suspicious or fraudulent in an auction. This categorization will help in increasing the ground truth of training data and in reducing the time of investigating the bidders. This present work emphasizes on the two problems of unlabeled and imbalanced three-class data. To conduct the empirical investigation of bidder classification, we utilize a high-quality SB dataset that has been developed by the authors in [4] from commercial auctions. To be able to label the SB data, we employ a robust hierarchical clustering technique to group users with similar bidding behavior, and the labeling approach proposed in [5] to determine the conduct of bidders. After that, we solve the uneven class distribution with an advanced multi-class over-sampling method. The imbalanced learning problem is even more complex when dealing with multi-class data because of the complicated relations between classes [6]. Based on two successful ensemble learners, AdaBoost (sequential learning) and Random Forest (parallel learning), we devise several fraud classifiers using imbalanced and balanced SB data. Ensemble methods, such as bagging, boosting and stacking, combine several learners to produce a much stronger meta-learner. Their objectives are to minimize the classification models' error rate and reduce the over-fitting of the training datasets. Random Forest is based on the Decision Tree learning, so we utilize the same baseline algorithm for AdaBoost. We search for the most fitting model for our SB dataset by comparing the performance of the generated SB classifiers for the suspicious and fraudulent classes.

2   Labeling Multi-class SB Data

In [4], the auctions of the hot product 'iPhone 7' has been scraped from eBay. iPhone 7 has been chosen because its auctions may have encouraged SB due to the high bid prices, long duration and large number of submitted bids. The raw dataset has gone through a rigorous preprocessing operation, and the generated dataset contains 807 auctions, 1054 bidders and 647 sellers [4]. Next, the authors implemented the metrics of eight unique SB predictors that have been found to occur frequently in fraudulent auctions. Then, they measured each SB metric against each bidder in each of the 807 auctions. The higher the metric value, the higher the level of doubt about the monitored bidder in an auction. The produced SB dataset possess a tally of 6321 samples. Table 2 exposes the eight SB patterns and their weights.

Table 1. Cluster distribution and labeling

Cluster #	Percentage	Label	Cluster #	Percentage	Label
Cluster 0	4.33%	Suspicious	Cluster 5	0.46%	Fraud
Cluster 1	30.28%	Normal	Cluster 6	1.80%	Fraud
Cluster 2	3.32%	Fraud	Cluster 7	23.98%	Normal
Cluster 3	13.95%	Normal	Cluster 8	8.35%	Normal
Cluster 4	5.72%	Suspicious	Cluster 9	7.78%	Normal

Most of the time, data labeling is done manually entirely, which is a tedious task. To label the SB data, we first utilize Hierarchical Clustering (HC) to group users with similar bidding behavior as well as two validation techniques to determine the optimal number of clusters [7]. HC is highly efficient for datasets of medium size and dimensionality, like our SB dataset. When using HC, "Complete Linkage" suitably partitioned the SB samples and provided much better results than Single, Average and Centroid distance functions. Based on Silhouette and Gap Statistics, we generate ten optimal clusters exposed in Table 1.

After data partitioning, we follow the labeling strategy introduced in [5] to determine whether the bidders in a given cluster behaved normally, suspiciously, or fraudulently. Firstly, we categorize the SB patterns in each cluster into two groups named "Low Category" and "High Category". We compute the mean of each fraud pattern for all the instances in a cluster. A pattern with the mean value from 0 to 0.50 is marked low, and from 0.51 to 1 high. To label a cluster, we manually examine the average and weight of the SB patterns. Table 2 presents three examples demonstrating the cluster labeling. In cluster 0, six out of eight patterns belong to the low category, but two patterns of medium and high weights fall in the high category. In this case, it is not possible to conclude about the conduct of the bidders, so we label them as suspicious. In cluster 1, all the SB patterns are in the low category, hence we label its bidders as normal. In cluster 3, five patterns are in high category, the high weighted pattern (Successive Outbidding) has a mean value of 0.969, and two medium weighted patterns

Table 2. SB data analysis and labeling

Cluster #	SB pattern	Weight	Category	Size	Label
Cluster 0			**Low value**	4.33%	Suspicious
	Bidder tendency	Medium	0.26608		
	Bidding ratio	Medium	0.429219		
	Auction bids	Low	0.082903		
	Last bidding	Medium	0.264267		
	Auction starting price	Low	0.048829		
	Early bidding	Low	0.17192		
			High value		
	Winning ratio	Medium	0.832261		
	Successive outbidding	High	0.817518		
Cluster 1			**Low value**	30.28%	Normal
	Bidding tendency	Medium	0.112438		
	Bidder ratio	Medium	0.133076		
	Successive outbidding	High	0.003396		
	Auction bids	Low	0.058353		
	Last bidding	Medium	0.124179		
	Auction starting price	Low	0.000269		
	Early bidding	Low	0.110384		
	Winning ratio	Medium	0.445501		
Cluster 2			**Low value**	3.32%	Fraud
	Bidder tendency	Medium	0.387139		
	Bidding ratio	Medium	0.344252		
	Auction bids	Low	0.375692		
			High value		
	Successive outbidding	High	0.969048		
	Last bidding	Medium	0.892746		
	Auction starting price	Low	0.70136		
	Early bidding	Low	0.862606		
	Winning ratio	Medium	0.899158		

(Last Bidding and Winning Ratio) have values greater than 0.89. Therefore, we label this cluster as fraud. We follow this strategy to classify the other seven clusters (Table 1). We observe that three clusters with approximately 5% of data indicate fraud, two clusters with approximately 10% of data are suspicious, and the rest of the clusters denoting around 84% of data is normal. To increase the ground truth of the training data, we examined thoroughly all the fraud instances (209 instances).

3   Sampling Multi-class SB Data

Any labeled fraud dataset is imbalanced, and in this case, classifiers will be influenced by the majority class, which means in our case the fraudulent bidders may be erroneously classified. In fraud detection problems, the situation is more unfortunate as it is the class of interest that is vital to detect since it carries a high cost of misclassification. Furthermore, a skewed class distribution often reduces the classifier performance [8]. In this study, we employ data over-sampling with SMOTE, which increase the representation of the minority classes [9]. The idea behind SMOTE is to overcome overtting and help classifiers to lessen the generalization error on untrained data [9]. Before the data sampling task, we first split the SB dataset into 70% training and 30% testing. We employ the stratification method to ensure that each of the tree classes is well represented in each subset. After applying SMOTE, we produce a balanced training dataset with a ratio of Normal to Suspicious to Fraudulent instances equals to 2:1:1. We chose this ratio (instead of 2:2:2) to not end up including many artificial samples. Also, [10] suggested that a ratio of 2:1 is preferable to achieve a balance between non-fraudulent and fraudulent data (Table 3).

Table 3. Over-sampling of 3-class SB training data

Imbalanced SB dataset			Balanced SB dataset		
Normal	Suspicious	Fraud	Normal	Suspicious	Fraud
3192	391	209	3192	1564	1672
84.18%	10.31%	5.51%	49.66%	24.33%	26.01%
Size: 3792			Size: 6428		
Class ratio (approx.) 8:2:1			Class ratio (approx.) 2:1:1		

4   Performance Evaluation

We first train AdaBoost (AB) and Random Forest (RF) with the balanced SB dataset, and then evaluate their accuracy on unseen SB data as shown in Table 4. RF did better that AB across all the metrics for the suspicious and fraudulent classes. For the fraud class, the gap between RF and AB is 5% for F-measure and 5% for FNR. Next, we apply RF and AB on the original SB dataset, and assess their performance on testing SB data as presented in Table 5. RF also did better that AB for the minority classes. We may observe that combining data sampling and ensemble learning is beneficial as it improved the predictive accuracy. For instance, for the class of interest, the performance gap between balanced and imbalanced data is 3% for the detection rate and 4% for the misclassification rate. This difference is significant in fraud detection problems.

Table 4. Classification performance on balanced SB data

Ensemble classifier	Label	Precision	Recall	F-measure	FPR	FNR
SMOTE+3-class RF	Normal	1	0.99	0.99	0	0.01
	Suspicious	0.95	0.99	0.97	0.05	0.01
	Fraud	0.98	0.99	0.98	0.02	0.01
SMOTE+3-class AB	Normal	0.99	0.99	0.99	0.01	0.01
	Suspicious	0.96	0.97	0.96	0.04	0.03
	Fraud	0.93	0.94	0.93	0.07	0.06

Table 5. Classification performance on imbalanced SB data

Ensemble classifier	Label	Precision	Recall	F-measure	FPR	FNR
3-class RF	Normal	0.99	0.99	0.99	0.01	0.01
	Suspicious	0.94	0.98	0.96	0.05	0.01
	Fraud	0.94	0.95	0.95	0.06	0.05
3-class AB	Normal	0.99	0.98	0.99	0.01	0.02
	Suspicious	0.95	0.95	0.95	0.05	0.05
	Fraud	0.92	0.91	0.92	0.08	0.09

5   Conclusion

To be able to minimize monetary losses for genuine buyers, it is necessary to make
e-auctions a fair environment by detecting fraudulent activities such as SB. The
research on SB classification is limited due to the complication of producing
SB training data. To label users as normal, suspicious or fraud, we combined a
hierarchical clustering method with a semi-automated labeling approach. Subsequently, we employed a multi-class over-sampling method because the labeled
SB dataset is imbalanced. The classification results demonstrated that ensemble classifiers, especially Random Forest, returned a high empirical performance
in terms of the detection and misclassification rates. Combining data sampling
with ensemble learning improved the accuracy. This improvement is significant
in fraud detection applications. When the SB classifier is deployed in real-world
scenarios, the auction admin. will launch an investigation for bidders that have
been detected as suspicious to confirm or reject the suspicion. For bidders classified as fraudulent, he will suspend their accounts since these users demonstrated
a strong likelihood of SB activities.

References

1. Arora, B.: Exploring and analyzing internet crimes and their behaviours. Perspect. Sci. **8**, 540–542 (2016)
2. Internet Crime Report I.: 2015 internet crime report, May 2015. https://pdf.ic3.gov/2015_IC3Report.pdf
3. Ford, B.J., Xu, H., Valova, I.: A real-time self-adaptive classifier for identifying suspicious bidders in online auctions. Comput. J. **56**(5), 646–663 (2012)
4. Alzahrani, A., Sadaoui, S.: Scraping and preprocessing commercial auction data for fraud classification. arXiv preprint arXiv:1806.00656 (2018)
5. Ganguly, S., Sadaoui, S.: Online detection of shill bidding fraud based on machine learning techniques. In: Mouhoub, M., Sadaoui, S., Ait Mohamed, O., Ali, M. (eds.) IEA/AIE 2018. LNCS (LNAI), vol. 10868, pp. 303–314. Springer, Cham (2018). https://doi.org/10.1007/978-3-319-92058-0_29
6. Wang, S., Yao, X.: Multiclass imbalance problems: analysis and potential solutions. IEEE Trans. Syst. Man Cybern. Part B (Cybern.) **42**(4), 1119–1130 (2012)
7. Anowar, F., Sadaoui, S., Mouhoub, M.: Auction fraud classification based on clustering and sampling techniques. In: 2018 17th IEEE International Conference on Machine Learning and Applications (ICMLA), pp. 366–371. IEEE (2018)
8. Ali, A., Shamsuddin, S.M., Ralescu, A.L.: Classification with class imbalance problem: a review. Int. J. Adv. Soft Comput. Appl. **7**(3), 176–204 (2015)
9. Fernández, A., Garcia, S., Herrera, F., Chawla, N.V.: Smote for learning from imbalanced data: progress and challenges, marking the 15-year anniversary. J. Artif. Intell. Res. **61**, 863–905 (2018)
10. Chang, W.H., Chang, J.S.: A novel two-stage phased modeling framework for early fraud detection in online auctions. Expert Syst. Appl. **38**(9), 11244–11260 (2011)

Towards Causal Analysis of Protocol Violations

Shakil M. Khan$^{(\boxtimes)}$ and Mikhail Soutchanski

Department of Computer Science, Ryerson University, Toronto, Canada
{shakilmkhan,mes}@scs.ryerson.ca

Abstract. When a protocol specified within a given system fails to ensure some desired properties, it is important to identify the actual causes of this failure. In this paper, we utilize a formal model of causal analysis in the situation calculus to show how one can specify the actual causes of such violations in non-deterministic protocols defined within dynamic systems. We show that our definition has some desirable properties.

1 Introduction

Reasoning about violations in protocols is essential for many applications where it is important to design protocols that adhere to certain desirable properties [4,9]. In case of a property violation, it is important to identify the actual causes of this failure. Such information can be used by the protocol designer to construct better protocols, e.g. by ensuring that certain execution paths are excluded. In this paper, we propose to utilize a formal model of causal analysis [2] in the situation calculus (**SC**) [12] to detect and reason about protocol violations.

We show how one can define the potential causes of protocol violations through the computation of causal chains within the SC. We make two assumptions: (1) there is a logical theory (with a complete initial state) that models how the system responds to actions, and (2) there is a non-deterministic protocol specified in the SC-based ConGolog programming language [5]. We are looking for events in all possible executions of the protocol to explain an observed effect.

Adopting a first-order language like the SC for causality analysis allows us to be more expressive. Namely, we can formulate quantified properties, model systems with infinite domains, and we can find violations in generic protocols specified over these systems. The underlying domain of objects in these systems can be infinite, e.g., it can include integer and real numbers with their standard interpretations. Furthermore, our formalization enables us to detect unwanted inter-component interactions in protocols, not just faulty component actions. We prove that our definition is sound and complete relative to a class of protocols.

© Springer Nature Switzerland AG 2019
M.-J. Meurs and F. Rudzicz (Eds.): Canadian AI 2019, LNAI 11489, pp. 359–365, 2019.
https://doi.org/10.1007/978-3-030-18305-9_30

2   Background

The Situation Calculus (SC). We use a version of the SC [12], where a dynamic domain is modeled using a basic action theory (**BAT**) $\mathcal{D}$ consisting of action precondition axioms (**APA**), successor-state axioms (**SSA**), initial state axioms, unique name axioms for actions (**UNA**), and domain-independent foundational axioms Σ. We also utilize the single-step regression operator ρ. Given a query "does ϕ hold in situation $do(\alpha, s)$?", ρ transforms it into an equivalent query "does ψ hold in s?", eliminating action α by compiling it into ψ. The expression $\rho[\phi, \alpha]$ denotes such an equivalent query obtained from the formula ϕ by replacing each fluent atom F in ϕ with the rhs of the SSA for F where the action variable a is instantiated with the ground action α, and then simplified using UNAs and constants. One can prove that given a BAT $\mathcal{D}$, a formula $\phi(s)$ uniform in s, and a ground action term α, we have that $\mathcal{D} \models \forall s.\ \phi(do(\alpha, s)) \leftrightarrow \rho[\phi(s), \alpha]$.

Example. We use the well-known dining philosophers problem [7] with three philosophers as our running example. The actions in this domain are $pickUp(p, f)$, $putDown(p, f)$, $eat(p)$, while the fluents are $hasFork(p, f, s)$, $thinking(p, s)$, and $eating(p, s)$. Most of these and the following axioms are self-explanatory; see [8] for details. We define a philosopher p is waiting in situation s as an abbreviation $waiting(p, s) \stackrel{\text{def}}{=} \neg(eating(p, s) \lor thinking(p, s))$. We use the relation $neighb$ to describe the seating arrangement and assume the fork F_{ij} is in between philosophers P_i, P_j. We sample a few axioms (all free variables are $\forall$-quantified at front):

$(a).\ Poss(pickUp(p, f), s) \leftrightarrow$
$\qquad \neg hasFork(p, f, s) \land \exists p'.(neighb(p, f, p') \lor neighb(p', f, p)) \land \neg hasFork(p', f, s),$

$(b).\ Poss(eat(p), s) \leftrightarrow \exists f, f'.\ f \neq f' \land hasFork(p, f, s)$
$\qquad\qquad\qquad\qquad \land\ hasFork(p, f', s) \land \neg eating(p, s),$

$(c).\ hasFork(p, f, do(a, s)) \leftrightarrow a = pickUp(p, f)$
$\qquad\qquad\qquad \lor\ (hasFork(p, f, s) \land \neg a = putDown(p, f)),$

$(d).\ eating(p, do(a, s)) \leftrightarrow a = eat(p) \lor (eating(p, s) \land \neg \exists f.\ a = putDown(p, f)),$

$(e).\ \forall p.\ thinking(p, S_0) \leftrightarrow p = P_1 \lor p = P_2 \lor p = P_3,$

$(f).\ \forall p, f.\ \neg eating(p, S_0) \land \neg hasFork(p, f, S_0).$

3   Actual Achievement Causes

Given a trace (a log), *actual achievement causes* are the key events responsible for achieving some effect. Here, we briefly review [2]. An effect is an SC formula $\phi(s)$ that is uniform in s. Given an effect $\phi(s)$, the actual causes of ϕ are defined relative to a *causal setting*, i.e., a BAT $\mathcal{D}$ representing the domain dynamics, and a narrative σ, representing the ground situation, where the effect was observed:

Definition 1. *A causal setting is a tuple $\langle \mathcal{D}, \sigma, \phi(s) \rangle$, where $\mathcal{D}$ is a BAT, σ is a situation term of the form $do([a_1, \cdots, a_n], S_0)$ with ground action functions $a_1, \cdots, a_n$ s.t. $\mathcal{D} \models executable(\sigma)$, and $\phi(s)$ is an SC formula uniform in s s.t. $\mathcal{D} \models \phi(\sigma)$.*

As the theory $\mathcal{D}$ does not change, we will often suppress $\mathcal{D}$. We require ϕ to hold by the end of the narrative σ. Following [2], we identify the potential causes of an effect ϕ with a set of pairs, each of which consists of a ground action term occurring in σ and the situation where this action was executed. The notion of the achievement condition suggests that if some action α mentioned in σ triggers the formula $\phi(s)$ to change its truth value from false to true relative to $\mathcal{D}$, and if there are no actions in σ after α that change the value of $\phi(s)$ back to false, then α is the actual cause of achieving $\phi(s)$ in σ. Batusov and Soutchanski [2] showed that when used together with the single-step regression operator ρ, this notion of achievement condition not only identifies the single action that brings about the effect of interest, but also captures recursively the actions that build up to it, i.e., the root causes. Additionally, one must include the preconditions under which these actions are executable. The following inductive definition formalizes this intuition. Let $\Pi_{apa}(\alpha, \sigma)$ be the right-hand side of the APA for action α with the situation term replaced by situation σ.

Definition 2. *A causal setting $\mathcal{C} = \langle \sigma, \phi(s) \rangle$ satisfies the achievement of ϕ via the situation term $do(\alpha^*, \sigma^*) \sqsubseteq \sigma$ iff there is an action α' and situation σ' s.t.:*

$$\mathcal{D} \models \neg\phi(\sigma') \wedge \forall s.\, do(\alpha', \sigma') \sqsubseteq s \sqsubseteq \sigma \rightarrow \phi(s),$$

and either $\alpha^* = \alpha'$ and $\sigma^* = \sigma'$, or $\sigma^* \sqsubseteq \sigma' \sqsubset \sigma$ and the causal setting $\langle \sigma', \rho[\phi(s), \alpha'] \wedge \Pi_{apa}(\alpha', \sigma') \rangle$ satisfies the achievement condition via the situation term $do(\alpha^*, \sigma^*)$. Whenever a causal setting $\mathcal{C}$ satisfies the achievement condition via situation $do(\alpha^*, \sigma^*)$, we say that the action α^* executed in situation σ^* is an achievement cause in the causal setting $\mathcal{C}$.

Since the process of discovering intermediary achievement causes using ρ cannot continue beyond S_0, it eventually terminates. Moreover, since the narrative σ is finite, the achievement causes of $\mathcal{C}$ also form a finite sequence of situation-action pairs, which we call the *achievement causal chain of $\mathcal{C}$*.

Example (cont'd). Let the philosophers P_1, P_2, P_3 sit around the table, with forks in between. Consider the trace $\sigma_1 = do([pickUp(P_1, F_{12}), pickUp(P_3, F_{23}), pickUp(P_1, F_{13}), eat(P_1)], S_0)$. We are interested in computing the actual causes of the effect $\phi_1 = eating(P_1, s)$. Then according to Definition 2, the causal setting $\langle \phi_1, \sigma_1 \rangle$ satisfies the achievement condition ϕ_1 via the situation $do(eat(P_1), S_3)$, where $S_3 = do([pickUp(P_1, F_{12}), pickUp(P_3, F_{23}), pickUp(P_1, F_{13})], S_0)$, so the action $eat(P_1)$ executed in S_3 is a (primary) achievement cause of ϕ_1.

Moreover, computing $\rho[eating(P_1, \sigma_1), eat(P_1)] \wedge Poss(eat(P_1), S_3)$ yields $\exists f,$ $f'.hasFork(P_1, f, s) \wedge hasFork(P_1, f', s) \wedge f \neq f' \wedge \neg eating(P_1, s)$ (let us call this formula ψ), and leads to a new causal setting $\langle S_3, \psi \rangle$. This satisfies the achievement condition via the action $pickUp(P_1, F_{13})$, so $pickUp(P_1, F_{13})$ executed in $S_2 = do([pickUp(P_1, F_{12}), pickUp(P_3, F_{23})], S_0)$ is a secondary achievement cause. Similarly, it can be shown that $pickUp(P_1, F_{12})$ executed in S_0 is also included in the causal chain. Notice that the action $pickUp(P_3, F_{23})$ is irrelevant.

We can also handle quantified queries, e.g. the actual causes of $\langle \sigma_2, \forall p. \ waiting(p, s) \rangle$, where $\sigma_2 = do([pickUp(P_1, F_{12}), pickUp(P_2, F_{23}), pickUp(P_3, F_{13})], S_0)$. Note that the integer-valued weight of pasta in the bowl can be easily modelled.

4   Causal Analysis of Protocol Violations

We model the behaviour of the system to be reasoned about as a BAT $\mathcal{D}$, while we encode an *observation* or *effect* using an SC formula $\phi(s)$ that is uniform in s, as above. For reasons explained below, we require the initial theory $\mathcal{D}_{S_0}$ to be complete both for relational and functional fluents. Let the protocol be specified using a ConGolog program δ [5], but we can work with any programming language defined on top of the SC. We are now ready to give our formal definition of the potential causes of a protocol violation in the SC:

Definition 3. *Given a system $\mathcal{DS} = \langle \mathcal{D}, \delta, \phi(s) \rangle$, the causes of violation of $\mathcal{DS}$ is the least set of causal chains $\mathcal{V}_{\mathcal{DS}}$ such that if there is a ground sequence of actions $\boldsymbol{a}$ for which $\mathcal{D} \models Do(\delta, S_0, do(\boldsymbol{a}, S_0)) \wedge \phi(do(\boldsymbol{a}, S_0))$, then $\mathcal{V}_{\mathcal{DS}}$ includes the causal chain relative to the causal setting $\langle \mathcal{D}, do(\boldsymbol{a}, S_0), \phi(s) \rangle$.*

Thus, the causes $\mathcal{V}_{\mathcal{DS}}$ of violating a non-deterministic protocol δ specified within a dynamic system $\mathcal{D}$ relative to a property $\phi(s)$ is the set of causal chains over all possible undesirable executions of δ, i.e. terminated executions over which $\phi(s)$ holds. Subsequently, we also call $\mathcal{V}_{\mathcal{DS}}$ a set of *conjectures*. Each conjecture specifies what actions in what situations should have been avoided by the executer, i.e. which paths in the execution tree of δ should have been prohibited by the protocol in an attempt to avoid failure $\phi(s)$. If the number of possible terminated executions of δ is finite, then $\mathcal{V}_{\mathcal{DS}}$ is also finite.

Note that Definition 3 may produce unintuitive results if the initial theory is incompletely specified. To see this, consider the non-deterministic program $(A|B)$, where the preconditions of A is $F(s)$ and that of B is $\neg F(s)$, and both A and B executed in S_0 have the effect that $\phi(s)$. Suppose that $\mathcal{D}$ does not specify the truth value of F in S_0. Although both A and B are the causes for $\phi(s)$, the theory $\mathcal{D}$ entails neither $executable(do(A, S_0))$, nor $executable(do(B, S_0))$, and therefore, according to our definition, the set of conjectures is empty, which is unintuitive. To avoid this issue, we require $\mathcal{D}$ to be initially complete.

Example (cont'd). Consider a simple protocol δ_1 specified in ConGolog:

$$(pickUp(P_1, F_{12}) \mid pickUp(P_2, F_{12})); (pickUp(P_1, F_{13}) \mid pickUp(P_2, F_{23}));$$
$$(eat(P_1) \mid eat(P_2) \mid \pi f. [Poss(pickUp(P_3, f), now)?; pickUp(P_3, f)]).$$

That is, first, either philosopher P_1 or P_2 non-deterministically picks up the fork F_{12} that is between them, then either of them picks up another available fork, and finally either P_1 eats, or P_2 eats, or P_3 picks up a fork. We would like to check if δ_1 violates the property that $\phi_3(s) = \neg\exists p.\ eating(p, s)$.

It is easy to see that there are only six possible executions of δ_1 and only in two of these cases, a philosopher is eating. For instance, no philosopher is eating in $do(\boldsymbol{a}_1, S_0)$, where $\boldsymbol{a}_1 = [pickUp(P_1, F_{12}), pickUp(P_2, F_{23}), pickUp(P_3, F_{13})]$. As such, $\mathcal{V}_{DS}$ for our example includes the causal chain relative to setting $\langle \mathcal{D}, do(\boldsymbol{a}_1, S_0), \phi_3(s)\rangle$. Note that the information provided by the causes of violation can be used by the protocol designer to reason about and improve on the protocol, in this case e.g. by ensuring that the second $pickUp$ action is only performed by the philosopher who is already holding another fork, etc.

Notice our approach can detect improperly synchronized inter-component interactions, as can be seen even in this simple example: while none of the philosophers failed to perform, in all four cases suggested by our causes of violation their actions are not synchronized relative to the fulfillment of $\neg\phi_3$.

We now show that our formalization has some intuitively desirable properties. First, a conjecture for a given complete execution of a protocol is unique:

Theorem 1. *If $\mathcal{K}_1$ and $\mathcal{K}_2$ are two conjectures of a particular execution $\boldsymbol{a}$ of protocol δ specified over a system $\mathcal{DS} = \langle \mathcal{D}, \delta, \phi(s)\rangle$, then $\mathcal{K}_1 = \mathcal{K}_2$.*

Moreover, the set of actions in a conjecture is sufficient for the effect to hold.

Theorem 2. *If $\mathcal{K}$ is a conjecture of an execution $\boldsymbol{a}$ of protocol δ specified over a system $\mathcal{DS} = \langle \mathcal{D}, \delta, \phi(s)\rangle$, and $\sigma_{\mathcal{K}}$ is the situation obtained by performing the actions in $\mathcal{K}$ in the order they appear in $\boldsymbol{a}$ starting from S_0, then $\mathcal{D} \models executable(\sigma_{\mathcal{K}}) \wedge \phi(\sigma_{\mathcal{K}})$.*

Theorems 1, 2, and Definition 3 together imply that our notion of causes of protocol violation is sound in the sense that each conjecture represents one or more undesirable executions of δ and correctly identifies the underlying reasons for the effect.

However, perhaps somewhat surprisingly, we can show that not every action in a conjecture is necessary for the effect to follow. Let $\sigma_{\boldsymbol{a}}^{\overline{a', s'}}$ denote the situation that can be obtained by executing the exact sequences of actions as in $\boldsymbol{a}$ starting in S_0, except for action a' in situation s'.

Theorem 3. *There is a system $\mathcal{DS} = \langle \mathcal{D}, \delta, \phi(s)\rangle$, an action a', and a situation s', s.t. if $\mathcal{K}$ is a conjecture in $\mathcal{V}_{DS}$ of a particular execution $\boldsymbol{a}$ of protocol δ and a' executed in s' is a cause in the conjecture $\mathcal{K}$, then: $\mathcal{D} \not\models \neg(executable(\sigma_{\boldsymbol{a}}^{\overline{a', s'}}) \wedge \phi(\sigma_{\boldsymbol{a}}^{\overline{a', s'}}))$.*

Thus, removing a cause a' in s' from the execution/trace $\boldsymbol{a}$ itself may not have any effect on $\phi(s)$ as it may be the case that another action on the trace restores the executability and/or brings about the effect, e.g. one that is currently being preempted by the cause. In fact this shows that our base framework does not choose an action as a cause when its effects are preempted by some earlier action.

Furthermore, we can show that the notion of modularity from [8] can be adapted to protocol violations, if one sub-divides the system into constituents.

Finally, we show that our notion of causes of protocol violation is complete with respect to a class of protocols, where each protocol δ_f has the following properties: each complete execution of δ_f is finite, and there is a finite number n of terminated executions of δ_f. The above assumptions can apply even if the underlying object domain is infinite, e.g., if in our example, there are fluents for the weight or the number of pasta in a bowl. If $\boldsymbol{a}$ is a sequence of actions, then let $\boldsymbol{a}_!$ denote any subsequence of this sequence that possibly omits some actions from $\boldsymbol{a}$ but does not alter the order of the actions in $\boldsymbol{a}$. Also, if $\mathcal{V}_{\mathcal{DS}}$ is the causes of violation of a system $\mathcal{DS}$, let $\mathcal{V}_{\mathcal{DS}}^{act}$ be the set that replaces each conjecture/causal chain in $\mathcal{V}_{\mathcal{DS}}$ with the sequential composition of the actions in the causal chain without changing the order of occurrence of these actions. We can prove the following:

Theorem 4 (Completeness). *If $\mathcal{V}_{\mathcal{DS}}$ is the causes of violation of a system $\mathcal{DS} = \langle \mathcal{D}, \delta_f, \phi(s) \rangle$, then there are no sequences of actions $\boldsymbol{a}$ and subsequence $\boldsymbol{a}_!$ such that $\mathcal{D} \models Do(\delta_f, S_0, do(\boldsymbol{a}, S_0)) \wedge \phi(do(\boldsymbol{a}, S_0))$ and $\boldsymbol{a}_! \notin \mathcal{V}_{\mathcal{DS}}^{act}$.*

5   Discussion

We emphasize that our formalization supports domains with infinitely many objects. This makes our work fundamentally different from approaches based on model checking [1]. Perhaps the closest work to ours that can be found in the literature is by Datta et al. [4], who proposed a framework for determining accountability of security violations for tasks and protocols such as authentication and key exchange. Like us, they also use actual causes to determine accountability. A key difference between our work and theirs is that while their analysis is tied to the underlying application (simple programs, threads, etc.), our work is based on a formal model of causality in the SC; thus, it is more general.

In addition, there has been work on automatic verification of partial correctness of (Con)Golog programs, e.g., [3,6,10]. These are mostly theoretical work. In contrast, our approach is more practical, since one can implement our causal analysis with the one-step regression operator using any off the shelf ConGolog interpreter that produces terminated executions.

Besides these, there has been practical work on checking partial correctness of Golog programs. For instance, [11] proposed mechanisms for automated verification of partial correctness of Golog programs using the notion of extended regression. The method has been implemented [11]. Examining how their approach would compare with our causal analysis-based approach is future work.

One limitation of our framework is that we assume that the initial state is completely specified. Also, currently we only deal with deterministic actions and discrete dynamic domains. Going beyond these limitations is future work.

Acknowledgement. This work was supported in part by the NSERC Canada.

References

1. Baier, C., Katoen, J.: Principles of Model Checking. MIT Press, Cambridge (2008)
2. Batusov, V., Soutchanski, M.: Situation calculus semantics for actual causality. In: Proceedings of AAAI Conference on Artificial Intelligence, pp. 1744–1752 (2018)
3. Claßen, J., Lakemeyer, G.: On the verification of very expressive temporal properties of non-terminating Golog programs. In: Proceedings of ECAI, pp. 887–892 (2010)
4. Datta, A., Garg, D., Kaynar, D.K., Sharma, D., Sinha, A.: Program actions as actual causes: a building block for accountability. In: Proceedings of IEEE Computer Security Foundations Symposium (CSF), pp. 261–275 (2015)
5. De Giacomo, G., Lespérance, Y., Levesque, H.J.: ConGolog, a concurrent programming language based on the situation calculus. Artif. Intell. **121**(1–2), 109–169 (2000)
6. De Giacomo, G., Lespérance, Y., Patrizi, F., Sardiña, S.: Verifying ConGolog programs on bounded situation calculus theories. In: Proceedings of AAAI Conference on Artificial Intelligence, pp. 950–956 (2016)
7. Hoare, C.A.R.: Communicating Sequential Processes. Prentice-Hall, Englewood Cliffs (1985)
8. Khan, S.M., Soutchanski, M.: Diagnosis as computing causal chains from event traces. In: Proceedings of AAAI Fall Symposium on Integrating Planning, Diagnosis, and Causal Reasoning (2018)
9. Leitner-Fischer, F., Leue, S.: Causality checking for complex system models. In: Giacobazzi, R., Berdine, J., Mastroeni, I. (eds.) VMCAI 2013. LNCS, vol. 7737, pp. 248–267. Springer, Heidelberg (2013). https://doi.org/10.1007/978-3-642-35873-9_16
10. Liu, Y.: A hoare-style proof system for robot programs. In: Proceedings of AAAI/IAAI, pp. 74–79 (2002)
11. Mo, P., Li, N., Liu, Y.: Automatic verification of Golog programs via predicate abstraction. In: Proceedings of ECAI, pp. 760–768 (2016)
12. Reiter, R.: Knowledge in Action. Logical Foundations for Specifying and Implementing Dynamical Systems. MIT Press, Cambridge (2001)

Lexicographic Preference Trees
with Hard Constraints

Sultan Ahmed$^{(\boxtimes)}$ and Malek Mouhoub

Department of Computer Science, University of Regina, Regina, Canada
{ahmed28s,mouhoubm}@uregina.ca

Abstract. The CP-net and the LP-tree are two fundamental graphical models for representing user's qualitative preferences. Constrained CP-nets have been studied in the past in which a very expensive operation, called dominance testing, between outcomes is required. In this paper, we propose a recursive backtrack search algorithm that we call Search-LP to find the most preferable feasible outcome for an LP-tree extended to a set of hard constraints. Search-LP instantiates the variables with respect to a hierarchical order defined by the LP-tree. Since the LP-tree represents a total order over the outcomes, Search-LP simply returns the first feasible outcome without performing dominance testing. We prove that this returned outcome is preferable to every other feasible outcome.

1 Introduction

Constraints and preferences coexist in many real world applications in which helping users by providing the most preferable feasible outcome is crucial. Quantitative representation of preferences is well-known in Multi-Attribute Utility Theory [6]. However, considering qualitative preferences is interesting as these preferences are natural, and easier to elicit from users. In this regard, Boutilier et al. [4] proposed the CP-net model that allows a graphical and compact representation of user's conditional *ceteris paribus* preference statements. Generally, the CP-net represents a partial order over the outcomes. Consequently, for Constrained CP-nets, various solving algorithms [1,3] have been proposed to find the set of feasible and preferentially non-dominated outcomes. In order to find this latter set, also called the Pareto set of outcomes, these algorithms require dominance testing between outcomes. Dominance testing is an expensive operation especially given that the Pareto set can be of exponential size.

On the other hand, Freuder et al. [5] studied a formalism, called Ordinal CSP, extending hard constraints to preferences that are represented as a lexicographic order over variables and domain values. To obtain the most preferable feasible outcome, the authors adopted the backtrack search in which they considered both lexical order (induced by the lexicographic preferences) and ordinary CSP heuristics to determine the variable order for instantiation, and provided appropriate trade-off between these two in terms of efficiency. One limitation of this formalism is that it does not consider conditional preferences. Consequently,

© Springer Nature Switzerland AG 2019
M.-J. Meurs and F. Rudzicz (Eds.): Canadian AI 2019, LNAI 11489, pp. 366–372, 2019.
https://doi.org/10.1007/978-3-030-18305-9_31

Wallace and Wilson [7] extended the above formalism, where the user provides: (1) a unique total order over the variables based on the variable importance, and (2) conditional preferences over the variable domain values. Given that the user provided variable order corresponds to the one induced by the parent-child relationship of the conditional preferences, the authors showed that the algorithms originally devised for the Ordinal CSP can be applied in this model as well. This model is similar to the CP-net provided that the lexicographic variable order guarantees a total order over the outcomes in the former one. Yet, this does not consider the fact that the variable order can also be conditioned given some value of other variables. Given this limitation, the Lexicographic Preference tree (LP-tree) [2] has been proposed as a more general formalism that considers that both lexicographic variable order and value order can be conditioned on the actual value of some more important variables. Instead of representing a unique total order over the variables, the LP-tree represents a set of hierarchical orders over the variables which are also total.

We extend the LP-tree graphical model to a set of hard constraints. To our best knowledge, this is the first time such model, that we call Constrained LP-tree, is proposed. In order to return the most preferable feasible outcome of the Constrained LP-tree, we have developed a recursive backtrack search algorithm that we call Search-LP. In some sense, solving the Constrained LP-tree is no harder than solving the underlying CSP, given that the expensive dominance testing is not needed as it is the case for the Constrained CP-net. Saying this, we cannot apply variable order heuristics [8] which often improve the performance of the search in practice. This is due to the fact that Search-LP instantiates the variables based on a hierarchical order of variables defined by the LP-tree.

2   Background

We assume a set of variables $V = \{X_1, X_2, \cdots, X_n\}$ with their corresponding domains $D(X_1), D(X_2), \cdots, D(X_n)$. We use $D(\cdot)$ to denote the domain of a set of variables as well. The decision maker wants to express preferences over the complete assignments on V. Each complete assignment can be seen as an outcome of the decision maker's action. The set of all outcomes is denoted by O. A preference order $\succ$ is a binary relation over O. For $o_1, o_2 \in O$, $o_1 \succ o_2$ indicates that o_1 is strictly preferred to o_2. The preference order is irreflexive, asymmetric and transitive. If for every $o_1, o_2 \in O$, either $o_1 \succ o_2$ or $o_2 \succ o_1$ holds, the preference order is a total order which we consider in this paper.

The size of O is exponential in the number of variables. Therefore, the direct assessment of the preferences is usually impractical. In this regard, the decision maker is often asked to provide preferences on the variables rather than the outcomes directly. In lexicographic precedence order, the decision maker specifies a total order on the variables first, and then a total order on the values of each variable. The variable order is based on the importance of the variables. A variable X is more important than another variable Y iff having a better value of X is preferred to having a better value of Y. Here, both variable and value order can depend on the actual value of more important variables. To represent such conditional lexicographic preferences, the LP-tree is a well-known model.

Definition 1. *[2] A Lexicographic Preference tree (LP-tree) L over V is a tree such that the following statements are true.*

1. *Every node is labelled with an attribute $X \in V$. $An(X)$ indicates the set of ancestor nodes of node X, while $De(X)$ indicates the set of descendant nodes of node X.*
2. *Every arc $\overrightarrow{(X,Y)}$ represents an importance relation, i.e., the parent node X is more important than the child node Y given the instantiation of $An(X)$.*
3. *Every arc $\overrightarrow{(X,Y)}$ is labelled with at least a value of X, indicating that the preferences represented by the subtree of the child node Y hold given these values of X and the instantiation of $An(X)$. The subtree of Y is unique to the subtrees of other children of X.*
4. *Every node X is labelled with a Conditional Preference Table (CPT) which represents a preference order on $D(X)$ for every instantiation of $An(X)$.*

The preference order between two outcomes of an LP-tree is defined below.

Definition 2. *[2] Let o_1 and o_2 be two outcomes of an LP-tree L. Let X be the node such that $An(X)$ gives the same value R to both o_1 and o_2, X gives two different values x_1 and x_2 to o_1 and o_2 correspondingly, and X exists in the subtree corresponding to $An(X) = R$. Given the instantiation of $An(X)$, if $CPT(X)$ represents $x_1 \succ x_2$, then we get $o_1 \succ o_2$. Similarly, if $CPT(X)$ represents $x_2 \succ x_1$, then we get $o_2 \succ o_1$.*

Lemma 1. *[2] Every LP-tree L represents a total order over the outcomes.*

The optimal outcome for an LP-tree is the most preferred outcome.

3   Constrained Optimization with LP-trees

We consider hard constraints with LP-trees, and propose a solving algorithm that returns the most preferable feasible outcome. In the algorithm, every time a set of variables are instantiated, the original LP-tree is updated to a reduced LP-tree by removing the instantiated variables. In this regard, first, we explain how we reduce the LP-tree given the instantiation. We call the reduced LP-tree a Compatible LP-tree because the preferences induced by the reduced LP-tree are also held by the original LP-tree given the instantiation.

3.1   Compatible LP-tree

Definition 3. *Let L be an LP-tree over V. Let $W \subset V$ be instantiated to w. An LP-tree L_c over $U = V - W$ is called compatible with respect to L given $W = w$*

if L_c is obtained from L in the following way. For every node $X \in W$ where X is instantiated to x, we apply the following:

1. *All subtrees for X except the subtree $X = x$ are deleted, and then the $CPT(Y)$ for every $Y \in De(X)$ is restricted to $X = x$.*
2. *If X has both a child and a parent, the arc $\overrightarrow{(Pa(X), X)}$ is replaced with $\overrightarrow{(Pa(X), Ch(X))}$, and then the arc $\overrightarrow{(X, Ch(X))}$ is deleted. If X has a parent but not a child, the arc $\overrightarrow{(Pa(X), X)}$ is deleted. If X has a child but not a parent, the arc $\overrightarrow{(X, Ch(X))}$ is deleted.*
3. *The node X and its CPT are deleted.*

Lemma 2. *Let L be an LP-tree over V and L_c be a Compatible LP-tree of L over $U \subset V$. The preference order induced by L_c is held by the preference order induced by L given $(V - U)$'s instantiation to p.*

Proof. Let o_{1c} and o_{2c} be two outcomes of L_c, and L_c induces $o_{1c} \succ o_{2c}$. Let the variable X gives two different values x_1 and x_2 to o_{1c} and o_{2c} correspondingly, and $An(X)$ in L_c, denoted as A_{L_c}, gives the same value z to o_{1c} and o_{2c}. By Definition 2, we can clearly see that $o_{1c} \succ o_{2c}$ implies $x_1 \succ x_2$ given $A_{L_c} = z$.

The definition of Compatible LP-tree indicates that $An(X)$ in L is a subset of, or equals to, $A_{L_c} \cup (V - U)$. Let $A_{L_c} \cup (V - U)$ is instantiated to q. Therefore, $An(X)$ in L gives the same value (let r) to the outcomes po_{1c} and po_{2c} of L. Given $An(X) = r$ in L, we have $x_1 \succ x_2$ which implies $po_{1c} \succ po_{2c}$. Thus, $o_{1c} \succ o_{2c}$ induced by L_c is also held by L for $V - U = p$. □

3.2 Constrained LP-tree

When some variables in an LP-tree are constrained, the optimal outcome might be infeasible. In this case, finding the most preferable feasible outcome is not trivial as the underlying CSP is itself NP-hard in general. For this purpose, we propose a recursive backtrack search algorithm that we call Search-LP. Search-LP has three parameters. The first parameter is a Compatible LP-tree L_c which is initially the original LP-tree L_{orig}. The second parameter is a set of hard constraints C which is initially the original set of constraints C_{orig}. The third parameter is a partial assignment K on $L_{orig} - L_c$ which is initially *null*. Search-LP returns the most preferable feasible outcome with respect to L_{orig} and C_{orig}.

Search-LP begins with the root node X and the preference order on $D(X)$. For every value x_i of X, the loop (lines 3–16) continues according to the preference order. In line 4, the current set of constraints C is strengthened by $X = x_i$ to get the new set of constraints C_i. If C_i is inconsistent, the branch $X = x_i$ is terminated (lines 5–7), and Search-LP continues with the next branch. Otherwise, the partial assignment K_i induced by $X = x_i$ and C_i is obtained. In line 9, the Compatible LP-tree L_{c_i} is obtained for the LP-tree L_c and instantiation K_i, using Definition 3. If L_{c_i} is empty, this indicates that all variables have been successfully instantiated. The outcome is stored in S, and Search-LP is stopped (line 10–12). If this termination criterion is not met, Search-LP is

---

Algorithm 1. Search-LP(L_c, C, K)

Input: Compatible LP-tree L_c which is initially the original LP-tree L_{orig}, a set of constraints C which is initially the original set of constraints C_{orig}, a partial assignment K on $L_{orig} - L_c$
Output: The most preferable feasible outcome with respect to L_{orig} and C_{orig}
  1: Let X be the root node of L_c
  2: Let $x_1 \succ x_2 \succ \cdots \succ x_l$ be the preference order on $D(X)$
  3: **for** $i = 1; i \le l; i = i + 1$ **do**
  4:     Strengthen C with $X = x_i$ to get C_i
  5:     **if** C_i is inconsistent **then**
  6:         **continue** with the next iteration
  7:     **end if**
  8:     Let K_i be the partial assignment induced by $X = x_i$ and C_i
  9:     Find the compatible LP-tree L_{c_i} given the instantiation K_i
 10:     **if** L_{c_i} is empty **then**
 11:         Set: $S = K \cup K_i$
 12:         **exit**
 13:     **else**
 14:         Search-LP$(L_{c_i}, C_i, K \cup K_i)$
 15:     **end if**
 16: **end for**
 17: Set: $S = null$
 18: **end**

---

called recursively until this is met (line 14). Every time, the partial assignment K_i is forwarded to the next call. Finally, if no feasible outcome exists, Search-LP saves *null* in S (line 17), and ends (line 18).

3.3   Formal Properties

We establish the correctness of Search-LP using the theorem below.

Theorem 1. *Let L be an LP-tree and C be a set of constraints over V. Search-LP returns the outcome o if and only if o is the most preferable feasible outcome with respect to L and C.*

Proof. Using Lemma 2, the preferences induced by the Compatible LP-tree in line 9 are also held by the original LP-tree given the instantiation. Therefore, in order to prove this theorem, we have to prove that: (1) for every $o_1 \succ o$, o_1 is infeasible with respect to C; and (2) o is preferred to every other feasible outcome o_2. We prove these below.

(1) This proof is by contradiction. Suppose o_1 is feasible with respect to C. Let X be the variable such that $An(X)$ gives the same value p to o_1 and o, and X gives x_1 and x to o_1 and o correspondingly. Therefore, we get that $o = pxo'$ and $o_1 = pxo_1'$. Since $o_1 \succ o$, we get $x_1 \succ x$. According to line 3 of Search-LP, the branch $X = x_1$ is created before the branch $X = x$ is created, in the search tree. It is easy to see that the outcome o_1 is associated with the branch $X = x_1$.

Since o_1 is feasible, Search-LP returns o_1 and is terminated. Search-LP does not return o which is a contradiction. Therefore, o_1 cannot be feasible.

(2) In this case, Search-LP returns o, and then stops searching other outcomes. o_2 is one of the feasible outcomes in the unsearched outcomes. Let o and o_2 differ on the values of variable X such that $An(X)$ gives the same value to both o and o_2. Let X gives x and x_2 to o and o_2 correspondingly. Since o is searched by Search-LP before even searching for o_2, we get that the branch $X = x$ is created before the branch $X = x_2$ is created, in the search tree. This gives us $x \succ x_2$ (using line 3), which also implies $o \succ o_2$. □

Although it is difficult to compare the complexity of an optimization problem with the complexity of a decision problem, in some sense, we can say that solving the Constrained LP-tree is no harder than the underlying CSP. However, a CSP can have an exponential number of outcomes. In contrast, the Constrained LP-tree has a single optimal outcome, and accordingly Search-LP is stopped as soon as this optimal outcome is obtained. A drawback of Search-LP is that the instantiation process occurs according to a defined variable and value order by the LP-tree, whereas the efficiency of the CSP can often be improved using various variable and value order heuristics [8].

4 Conclusion

We have proposed the Search-LP algorithm to obtain the most preferable feasible outcome for a Constrained LP-tree. Search-LP instantiates the variables and values according to the hierarchical orders of variables and values induced by the conditional relative importance relations and the conditional preferences which are defined in the LP-tree. In comparison with the Constrained CP-net, the Constrained LP-tree is superior in the fact that the latter does not require dominance testing which is a very expensive operation in CP-nets.

References

1. Ahmed, S., Mouhoub, M.: Constrained optimization with partial CP-nets. In: IEEE International Conference on Systems, Man, and Cybernetics, pp. 3361–3366 (2018)
2. Booth, R., Chevaleyre, Y., Lang, J., Mengin, J., Sombattheera, C.: Learning conditionally lexicographic preference relations. In: Proceedings of 19th European Conference on Artificial Intelligence, pp. 269–274 (2010)
3. Boutilier, C., Brafman, R., Domshlak, C., Hoos, H., Poole, D.: Preference-based constrained optimization with CP-nets. Comput. Intell. **20**, 137–157 (2004)
4. Boutilier, C., Brafman, R.I., Domshlak, C., Hoos, H.H., Poole, D.: CP-nets: a tool for representing and reasoning with conditional ceteris paribus preference statements. J. Artif. Intell. Res. (JAIR) **21**, 135–191 (2004)
5. Freuder, E.C., Wallace, R.J., Heffernan, R.: Ordinal constraint satisfaction. In: Fifth International Workshop on Soft Constraints (2003)
6. Keeney, R.L., Raiffa, H.: Decisions with Multiple Objectives: Preferences and Value Trade-Offs. Cambridge University Press, New York (1993)

7. Wallace, R.J., Wilson, N.: Conditional lexicographic orders in constraint satisfaction problems. In: Beck, J.C., Smith, B.M. (eds.) CPAIOR 2006. LNCS, vol. 3990, pp. 258–272. Springer, Heidelberg (2006). https://doi.org/10.1007/11757375_21
8. Yong, K.W., Mouhoub, M.: Using conflict and support counts for variable and value ordering in CSPs. Appl. Intell. **48**(8), 2487–2500 (2018)

Supervised Versus Unsupervised Deep Learning Based Methods for Skin Lesion Segmentation in Dermoscopy Images

Abder-Rahman Ali[1]($\boxtimes$), Jingpeng Li[1], and Thomas Trappenberg[2]

[1] Division of Computer Science and Mathematics,
University of Stirling, Stirling, UK
`abder@cs.stir.ac.uk, jingpeng.Li@stir.ac.uk`
[2] Faculty of Computer Science,
Dalhousie University, Halifax, Canada
`tt@dal.cs.ca`

Abstract. Image segmentation is considered a crucial step in automatic dermoscopic image analysis as it affects the accuracy of subsequent steps. The huge progress in deep learning has recently revolutionized the image recognition and computer vision domains. In this paper, we compare a supervised deep learning based approach with an unsupervised deep learning based approach for the task of skin lesion segmentation in dermoscopy images. Results show that, by using the default parameter settings and network configurations proposed in the original approaches, although the unsupervised approach could detect fine structures of skin lesions in some occasions, the supervised approach shows much higher accuracy in terms of Dice coefficient and Jaccard index compared to the unsupervised approach, resulting in 77.7% vs. 40% and 67.2% vs. 30.4%, respectively. With a proposed modification to the unsupervised approach, the Dice and Jaccard values improved to 54.3% and 44%, respectively.

Keywords: Deep learning · Dermoscopy · Melanoma

1 Introduction

Skin cancer is considered the most common cancer worldwide such that one in every three cancers diagnosed is a skin cancer. In Canada more than 80,000 skin cancer cases are diagnosed each year, of which more than 5,000 cases are melanoma, the deadliest form of skin cancer [1]. In the UK around 100,000 cases are diagnosed each year, accounting for 20% of all cancer cases. Essentially, the standard pipeline in automatic dermoscopic image analysis consists of three stages: image segmentation, feature extraction and selection, and lesion classification. Image segmentation is considered a crucial stage since it affects the accuracy of subsequent stages. However, segmentation is difficult due to the great variety of lesion shapes, sizes, textures and colors. Other difficulties are

© Springer Nature Switzerland AG 2019
M.-J. Meurs and F. Rudzicz (Eds.): Canadian AI 2019, LNAI 11489, pp. 373–379, 2019.
https://doi.org/10.1007/978-3-030-18305-9_32

related to the presence of dark hair covering the lesions and the existence of specular reflections.

Unlike the studies proposed in literature, we utilize a commonly used architecture in the medical imaging domain for the supervised part of the paper (i.e. U-Net). Taking that further, we also employ an unsupervised deep learning based approach, which could be considered a good approach when a shortage of data and its ground truth is faced, which is a common challenge in the medical imaging domain. No pre-processing or post-processing operations are carried out. This study is specifically about comparing the relative performance of supervised versus unsupervised learning. We are interested in the unsupervised method as labeled data are often scares. We therefore directly compare two methods, a baseline U-Net architecture that is prominent for medical image data segmentation, and a recently proposed unsupervised method. In this paper we also propose some modifications to the unsupervised method to improve its performance.

U-Net is described in Sect. 2 as a supervised deep learning based method, the unsupervised deep learning based method is described in Sect. 3, results are depicted and discussed in Sect. 4, and the paper is concluded in Sect. 5.

2 Supervised Deep Learning Based Segmentation

U-Net is an end-to-end encoder-decoder network for semantic segmentation which was firstly used in medical image segmentation. The architecture is composed of left (down) and right (up) sides. The *down* part, which follows the typical convolutional network architecture, is the encoder part where convolution blocks are applied followed by max pooling in order to encode the input image into feature representations at multiple levels. The number of feature channels are doubled at each downsampling step. The *up* part consists of upsampling the feature map followed by a convolution operation that brings the number of feature channels to half. A concatenation with the corresponding cropped feature map from the *down* part occurs, followed by two 3 × 3 convolutions which are also followed by two ReLU operations and one 2 × 2 max-pooling operation with stride 2 used for downsampling. The cropping process is essential as border pixels are lost at each convolution. The resulting architecture is that the expansive path is symmetric to the concatenating path, yielding a u-shaped architecture. The final layer of U-Net uses a 1 × 1 convolution to map each 64 feature vector to the desired number of classes. The network is composed of 23 convolutional layers in total, provided that it does not have any fully connected layers and uses only the valid part of each convolution. For the border region of the image, the pixels are predicted by an *overlap-tile strategy* such that the missing context is extrapolated by mirroring the input image, thus allowing the U-Net network to be applied on large images. The outcome of this stage is the segmented image, where the foreground (in white) represents the skin lesion and the background (in black) represents the skin.

3   Unsupervised Deep Learning Based Segmentation

In this study, we use a CNN for image segmentation in a fully unsupervised manner as proposed by Kanezaki [4]. No training data and labels thus need to be prepared beforehand. This unsupervised setting could be crucial when it comes to skin lesion images that suffer from shortage in sufficient amount of data with defined ground truth that can be used for training a CNN.

Three constraints need to be met for the prediction of the label $\{c_n\}$: (i) feature similarity, (ii) spatial continuity, and (iii) number of unique cluster labels. In the *feature similarity* constraint, pixels with similar features are assigned the same label, where the feature vectors are clustered into q clusters and each pixel is assigned to the closest point amongst the q representative points (grouping similar pixels into clusters). Based on the second constraint, the clusters of image pixels have to be *spatially continuous* (cluster labels of neighboring pixels are identical). Here, K superpixels (a group of connected pixels with similar colors) $\{S_k \in \mathbb{R}^n\}_{k=1}^K$ are extracted from the input image using SLIC [5], where S_k refers to the set of pixel indices that belong to the kth superpixel, and each pixel belonging to some superpixel will have the same cluster label. The most frequent cluster label c_{max} for all $c_n \in \{1, ..., q\}$ is then selected. While the first two criteria aid in the grouping of pixels, the *third* criterion poses a constraint on the number of unique cluster labels, avoiding by that any undersegmentation (i.e. only one cluster) as this criterion gives preference to a large number of clusters. Figure 1 illustrates the unsupervised deep learning approach, and Fig. 2 shows a sample of image clusters before (upper row) and after (lower row) superpixel refinement, respectively, where clusters become more compact after refinement.

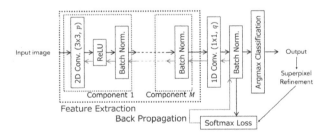

Fig. 1. Unsupervised deep learning based approach [4]: the network starts by predicting the cluster labels with fixed parameters in the forward process. A linear classifier (argmax classification) is then applied to obtain a response map, eventually resulting in a cluster label for each pixel. The superpixel refinement step makes sure that the cluster labels of neighboring pixels are identical. The network parameters are then trained with the fixed predicted cluster labels, and the softmax loss is calculated between the response and refined cluster labels (resembles supervised learning). The process is repeated T times to obtain the final cluster labels predictions

4   Results and Discussion

The U-Net architecture is trained on 2344 dermoscopy images along with their corresponding ground truth response masks from the "ISIC 2018: Skin Lesion Analysis Towards Melanoma Detection"[1] grand challenge datasets. To make the most out of the training data, augmentation using some transformations has been applied (such as rotation and shifting horizontally and vertically). Images used have been resized to 512 × 512. The U-Net model was trained for 20 epochs on a Tesla P100 GPU. Different parameters have been set to the unsupervised deep learning approach (number of channels 100, iterations 1000, clusters 2, learning rate 0.1, convolutional layers 2, super-pixels 10000, and compactness of superpixels 100), and the approach has also been run on a Tesla P100 GPU.

The two approaches were tested on 250 images for segmentation. Results in Fig. 3(a) depict that the unsupervised approach shows less quality results than those of the U-Net. This might be due to the close intensity values between the mole and the background which is apparent in the second image (counting from the left); this applies to the other images with different degrees of closeness. The presence of artifacts (hair) as shown in the fourth image could also be another reason of such poor performance. Low resolution images also affect the outcomes of the approach. The results could be improved by adding a pre-processing stage (i.e. image enhancement, hair removal) before running the algorithm. However, this is not the case with some other test images where the unsupervised approach could be better in detecting the fine structure of the skin lesion. Figure 3(b) highlights some examples on such cases. In particular, the unsupervised approach seems to work better in situations where there is a clear distinction between the mole (darker foreground) and the skin (brighter background), and where there are not much color intensity variations.

We use two metrics (Dice coefficient and Jaccard Index) to measure the similarity between two images. Tables 1, 2, 3 and 4 depict the results.

The average Dice coefficient and Jaccard index values for the two approaches when applied on 250 test images are summarized in Table 5, which shows that U-Net performs better than the unsupervised approach.

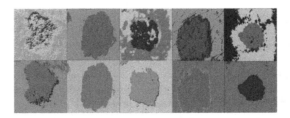

Fig. 2. The upper row shows the image clusters of the unsupervised approach before superpixel refinement, and the bottom row shows the clusters after refinement

---

[1] https://challenge2018.isic-archive.com/

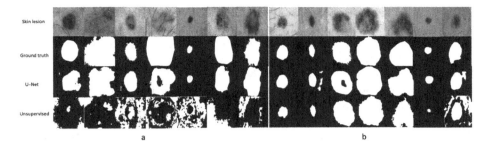

Fig. 3. (a) The unsupervised deep learning based approach results (b) U-Net results

Table 1. Dice coefficient values for the U-Net and unsupervised results shown in Fig. 3(a). The numbers represent the ordering of the images as read from left to right

Method	Image-1	Image-2	Image-3	Image-4	Image-5	Image-6	Image-7
U-Net	83.2%	86.2%	86.7%	77.5%	90.7%	78.5%	88.4%
Unsupervised	7.3%	13.1%	19.6%	33%	0%	18.3%	0.3%

Table 2. Jaccard index values for the U-Net and unsupervised results shown in Fig. 3(a)

Method	Image-1	Image-2	Image-3	Image-4	Image-5	Image-6	Image-7
U-Net	71.2%	75.7%	76.5%	63.3%	83%	64.6%	79.2%
Unsupervised	3.8%	7%	10.8%	19.7%	0%	10%	0.2%

Table 3. Dice coefficient values for the U-Net and unsupervised results shown in Fig. 3(b)

Method	Image-1	Image-2	Image-3	Image-4	Image-5	Image-6	Image-7
U-Net	94.7%	68.6%	89.8%	96.9%	91.4%	92.4%	85%
Unsupervised	75.5%	85.4%	94.4%	97%	89.8%	37.4%	52.4%

Table 4. Jaccard index values for the U-Net and unsupervised results shown in Fig. 3(b)

Method	Image-1	Image-2	Image-3	Image-4	Image-5	Image-6	Image-7
U-Net	89.9%	52.2%	81.4%	94%	84.2%	85.8%	74%
Unsupervised	60.6%	74.5%	89.4%	94.3%	81.5%	23%	35.5%

Table 5. Average Dice and Jaccard values for U-Net and the unsupervised deep learning based approach

Method	Dice	Jaccard
U-Net	77.7%	67.2%
Unsupervised	40%	30.4%

We have tried different variations of the two approaches and tested those on the 250 test images, for which the evaluations are depicted in Table 6. For the *unsupervised based approach*, two variations have been made. The first variation consists of: (i) sigmoid used instead of ReLU as an activation function in the convolutional components; (ii) quick shift clustering [6] used instead of SLIC; (iii) adam used as an optimization function instead of the stochastic gradient descent. The second variation consists of: (i) SELU (Scaled Exponential Linear Unit) [8] used instead of ReLU; (ii) using Felzenszwalb's efficient graph based image segmentation method [7] instead of SLIC. For *U-Net*, two variations have been made. The first variation is using tanh instead of ReLU as the activation function in the convolutional layers on both sides of the architecture (i.e. down and up). The second variation is using SELU as an activation function.

Table 6. Average Dice and Jaccard values for different variations of U-Net and the unsupervised deep learning based approach

Method	Dice	Jaccard
Unsupervised (sigmoid-quickshift-adam)	52.3%	41.8%
Unsupervised (selu-felzenszwalb-sgd)	54.3%	44%
U-Net (tanh)	68%	57.4%
U-Net (selu)	67.2%	55%

5   Conclusion

The unsupervised approach is able to detect fine structures in skin lesions better than U-Net in some test samples. However, U-Net shows to provide much better accuracy in terms of Dice coefficient and Jaccard index. A proposed modification to the unsupervised approach caused the Dice and Jaccard values to improve. As a future work, we plan to combine both supervised and unsupervised deep learning based approaches (ensemble) for skin lesion segmentation, such that the output of one approach could be used as an input to the other approach. We also plan to improve the unsupervised approach to avoid poor results as demonstrated in Fig. 3(a).

Acknowledgment. The work was supported by Natural Science Foundation of China (Grant No. 71571076).

References

1. About Skin Cancer. http://www.canadianskincancerfoundation.com/about-skin-cancer.html. Accessed 17 Dec 2018
2. Ronneberger, O., Fischer, P., Brox, T.: U-Net: convolutional networks for biomedical image segmentation. In: Navab, N., Hornegger, J., Wells, W.M., Frangi, A.F. (eds.) MICCAI 2015. LNCS, vol. 9351, pp. 234–241. Springer, Cham (2015). https://doi.org/10.1007/978-3-319-24574-4_28

3. Badrinarayanan, V., Kendall, A., Cipolla, R.: Segnet: a deep convolutional encoder-decoder architecture for image segmentation. arXiv:1511.00561v2 [cs.CV] (2015)
4. Kanezaki, A.: Unsupervised image segmentation by backpropagation. IEEE SigPort (2018). http://sigport.org/2710. Accessed 18 Dec 2018
5. Achanta, R., Shaji, A., Smith, K., Lucchi, A., Fua, P., Süsstrunk, S.: Slic superpixels compared to state-of-the-art superpixel methods. IEEE Trans. Pattern Anal. Mach. Intell. **34**(11), 2274–2282 (2012)
6. Vedaldi, A., Soatto, S.: Quick shift and Kernel methods for mode seeking. In: Forsyth, D., Torr, P., Zisserman, A. (eds.) ECCV 2008. LNCS, vol. 5305, pp. 705–718. Springer, Heidelberg (2008). https://doi.org/10.1007/978-3-540-88693-8_52
7. Felzenszwalb, P., Huttenlocher, D.: Efficient graph-based image segmentation. Int. J. Comput. Vis. (IJCV) **59**(2), 167–181 (2004)
8. Klambauer, G., Unterthiner, T., Mayr, A, Hochreiter, S.: Self-normalizing neural networks. arXiv preprint arXiv:1706.02515 (2017)

Lifted Temporal Maximum
Expected Utility

Marcel Gehrke[(✉)] ⓘ, Tanya Braun ⓘ, and Ralf Möller

Institute of Information Systems, University of Lübeck, Lübeck, Germany
{gehrke,braun,moeller}@ifis.uni-luebeck.de

Abstract. The dynamic junction tree algorithm (LDJT) efficiently answers exact filtering and prediction queries for temporal probabilistic relational models by building and then reusing a first-order cluster representation of a knowledge base for multiple queries and time steps. To also support sequential online decision making, we extend the underling model of LDJT with action and utility nodes, resulting in parameterised probabilistic dynamic decision models, and introduce meuLDJT to efficiently solve the exact lifted temporal maximum expected utility problem, while also answering marginal queries efficiently.

1 Introduction

Areas such as healthcare and logistics deal with probabilistic data including relational and temporal aspects and need efficient exact inference algorithms. These areas involve many objects in relation to each other with changes over time and uncertainties about object existence, attribute value assignments, or relations between objects. More specifically, healthcare systems involve electronic health records (relational) for many patients (objects), streams of measurements over time (temporal), and uncertainties due to, e.g., missing information. In this paper, we study the problem of supporting exact decision making and query answering in temporal probabilistic relational models.

We [3] present parameterised probabilistic dynamic models (PDMs) to represent temporal probabilistic relational behaviour and propose the lifted dynamic junction tree algorithm (LDJT) to efficiently answer multiple *filtering* and *prediction* queries. LDJT combines the advantages of the interface algorithm [5] and the lifted junction tree algorithm [2]. Specifically, this paper contributes (i) parameterised probabilistic dynamic decision models (PDDecMs) by adding action and utility nodes to PDMs and (ii) meuLDJT to efficiently solve the temporal maximum expected utility (MEU) problem using PDDecMs, while also answering marginal queries efficiently.

Nath and Domingos [6] perform first steps to formally define action and utility nodes for static probabilistic relational models. Further, Apsel and

This research originated from the Big Data project being part of Joint Lab 1, funded by Cisco Systems Germany, at the centre COPICOH, University of Lübeck.

© Springer Nature Switzerland AG 2019
M.-J. Meurs and F. Rudzicz (Eds.): Canadian AI 2019, LNAI 11489, pp. 380–386, 2019.
https://doi.org/10.1007/978-3-030-18305-9_33

Brafman [1] propose an exact lifted static solution to the MEU problem based on [6]. We [4] propose a solution based on LJT to combine decision support and efficient marginal query answering. However, in this paper, we propose to include sequential decision making. Research on sequential decision making relates to first-order (partially observable) Markov decision processes (FO (PO)MDPs) [7]. In contrast to FO POMDPs, which support *offline* decision making, we propose to support probabilistic *online* decision making, which allows for reacting to observations as well as for query answering.

In the following, we present PDDecMs including the MEU problem. Afterwards, we introduce meuLDJT to solve the lifted temporal MEU problem efficiently.

2   Lifted Temporal Maximum Expected Utility

We recapitulate PDMs [2,3] and then define PDDecMs with action and utility nodes to support decision making. Finally, we define the temporal MEU problem for PDDecMs.

2.1   Parameterised Probabilistic Models

A parameterised probabilistic model (PM) combines first-order logic with probabilistic models, representing first-order constructs using logical variables (logvars) as parameters. We would like to remotely infer the condition of patients with regard to retaining water. To determine the condition of patients, we use the change of their weights and additionally use the change of weights of people living with a patient to reduce the uncertainty for inferring conditions. The cause of an increase in weight could either be overeating or retaining water. In case both persons gain weight, overeating is more likely. Otherwise, only one person gains weight and retaining water is more likely.

Definition 1. *Let* $\mathbf{L}$ *be a set of logvar names,* Φ *a set of factor names, and* $\mathbf{R}$ *a set of (randvar) names. A parameterised randvar (PRV)* $A = P(X^1, ..., X^n)$ *represents a set of randvars behaving identically by combining a randvar* $P \in \mathbf{R}$ *with logvars* $X^1, ..., X^n \in \mathbf{L}$. *If* $n = 0$, *the PRV is parameterless. The domain of a logvar* L *is denoted by* $\mathcal{D}(L)$. *The term* $range(A)$ *provides possible values of a PRV* A. *Constraint* $(\mathbf{X}, C_{\mathbf{X}})$ *allows for restricting logvars to certain domain values and is a tuple with a sequence of logvars* $\mathbf{X} = (X^1, ..., X^n)$ *and a set* $C_{\mathbf{X}} \subseteq \times_{i=1}^{n} \mathcal{D}(X^i)$. *The symbol* $\top$ *denotes that no restrictions apply and may be omitted. The term* $lv(Y)$ *refers to the logvars,* $rv(Y)$ *to the randvars,* $gr(Y|C)$ *denotes the set of instances of* Y *with all logvars in* Y *grounded w.r.t. constraint* C.

To model the example, we use the randvar names C, LT, S, and W for Condition, LivingTogether, ScaleWorks, and Weight, respectively, and the logvar names X and X'. From the names, we build PRVs $C(X)$, $LT(X, X')$, $S(X)$, and $W(X)$. The domain of X and X' is $\{alice, bob, eve\}$. The range of $C(X)$ is $\{ok, bad\}$, $LT(X, X')$ and $S(X)$ have range $\{true, false\}$, and $W(X)$ has range $\{ok, high\}$.

382 M. Gehrke et al.

Definition 2. *We denote a parametric factor (parfactor) g with $\forall \mathbf{X} : \phi(\mathcal{A}) \mid C$, $\mathbf{X} \subseteq \mathbf{L}$ being a set of logvars over which the factor generalises, and $\mathcal{A} = (A^1, ..., A^n)$ a sequence of PRVs. We omit ($\forall \mathbf{X} :$) if $\mathbf{X} = lv(\mathcal{A})$. Function $\phi : \times_{i=1}^n range(A^i) \mapsto \mathbb{R}^+$ with name $\phi \in \Phi$ is identical for all grounded instances of $\mathcal{A}$. A list of all input-output values is the complete specification for ϕ. C is a constraint on $\mathbf{X}$. A PM $G := \{g^i\}_{i=0}^{n-1}$ is a set of parfactors and semantically represents the full joint probability distribution $P_G = \frac{1}{Z} \prod_{f \in gr(G)} f$ with Z as a normalisation constant.*

Now, we build the PM G^{ex} with: $g^0 = \phi^0(C(X), S(X), W(X)) \mid \top$ and $g^1 = \phi^1(C(X), C(X'), LT(X, X')) \mid \kappa^1$. The constraint κ^1 of g^1 ensures that $X \neq X'$ holds.

We define PDMs based on the first-order Markov assumption and a stationary process. A PDM is composed of a PM G_0 for the initial time step and $G_\rightarrow$, which connects two PMs with inter-slice PMs to model the temporal behaviour. Figure 1 shows a PDM with PRVs connected to parfactors and inter-slice parfactors g^{LT}, g^C, and g^S.

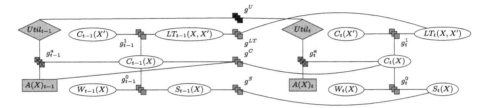

Fig. 1. Retaining water example with action and utility nodes

2.2 Parameterised Probabilistic Decision Models

Let us extend PDMs with action and utility nodes, resulting in PDDecMs.

Definition 3. *We represent actions and utilities by PRVs. Let Φ^u be a set of utility factor names. The range of action PRVs is disjoint actions and the range of utility PRVs is $\mathbb{R}$. A parfactor with a utility PRV U is a utility parfactor. We denote a utility parfactor u with $\forall \mathbf{X} : \mu(\mathcal{A}) \mid C$, where $U \in \mathcal{A}$ and C a constraint on $\mathbf{X}$. Function $\mu : \times_{i=1}^n range(A^i) \mapsto \mathbb{R}$, $A^i \in \mathcal{A}$, with name $\mu \in \Phi^u$ is defined identically for all grounded instances of $\mathcal{A}$ and its output is the additive change of U's value. Therefore, after the evaluation of a μ function, the initial value of U, i, is changed by the output value, j, resulting in the new value of U, which is $i + j$. The default initial utility value is 0. A parameterised probabilistic decision model (PDecM) G extends a PM with an additional set G^u of utility parfactors. Let $rv(G^u)$ refer to all probability randvars in G^u. Semantically G^u represents the combination of all utilities $U_G = \sum_{f \in gr(G^u)} f$.*

The μ functions output a utility, i.e., a scalar, which makes comparing utility values easy. Hence, we can easily test how discriminable actions are.

For a PDDecMs, which extends a PDDecM to the temporal case similar to a PDM with a PM, to connect two utility PRVs, we define a utility transfer function.

Definition 4. *A utility transfer function λ has utility PRVs $\mathbf{U}$ as input and one utility PRV U_o as output. Additionally, λ can have non utility PRVs as input. λ specifies how the value of U_o is additively changed, possibly depending on non utility PRVs. In this fashion, PDDecMs transfer utility values to the next time step and allow for discounting.*

Figure 1 shows one action (square) and one utility (diamond) PRV in grey, utility parfactors (crosses), and a utility transfer parfactor g^U in black. Assume actions are: A^1 is *visit patient* and A^2 is *do nothing*. For example, patients with a chronic heart failure might tend to retain water. In case water retention is detected early on, treatment is easier. However, if this water retention remains undetected, water can also retain in the lung, which can lead to a pulmonary edema, making a treatment more costly. More importantly, pulmonary edema is an acute life-threatening condition. A^1 also influences the utility as for a doctor, with limited time, visiting a patient is expensive. Thus, one always needs to consider that alerting a doctor too early generates unnecessary appointments and alerting a doctor too late can have serious consequences for patients.

2.3 Maximum Expected Utility in PDDecMs

PDDecMs encode trade-offs in utility parfactors. Connecting $Util_{t-1}$ and $Util_t$ with a utility transfer parfactor g^U makes utility PRVs time-dependent and allows for discounting. For example, g^U specifies that the value of $Util_{t-1}$ is reduced by 5 and then added to $Util_t$. To select an action, we begin by defining probability and utility queries.

Definition 5. *Given a PDecM G, a ground PRV Q and grounded PRVs with fixed range values $\mathbf{E}$, the expression $P(Q|\mathbf{E})$ denotes a probability query w.r.t. P_G and the expression $U(Q, \mathbf{E})$ refers to a utility w.r.t. U_G.*

Semantically, the expected utility of a PDecM or a PDDecM G, with utility part G^u, under assignment to action PRVs $\mathbf{a}$ is given by:

$$eu(G|\mathbf{a}) = \sum_{r \in range(rv(G^u))} P(r|\mathbf{a}) \cdot U(r, \mathbf{a}) \qquad (1)$$

We calculate a belief state and combine the belief state with corresponding utilities. By eliminating all randvars, one obtains the expected utility. We define the MEU problem as:

$$meu[G] = (\arg \max_{\mathbf{a}} eu(G|\mathbf{a}), \max_{\mathbf{a}} eu(G|\mathbf{a})) \qquad (2)$$

Equation (2) defines how to calculate the MEU for a PDecM and a PDDecM. Using a utility transfer function, we see the problem as an iterative filtering problem. The expected utility is calculated for one time step and then the utility value is transfered to the next time step. Therefore, the utility value of the latest time step, is the overall utility value. Due to the inherent uncertainty of PDDecMs, calculating the best actions is only feasible for a finite horizon, as one needs to iterate over all possible action assignments.

3 Solving the MEU Problem with MeuLDJT

We illustrate how meuLDJT incorporates utilities and solves the MEU problem.

Including Utilities. Algorithm 1 outlines how meuLDJT includes utilities. Similar to LDJT with PDMs (c.f. [3]), meuLDJT first builds an (FO jtree) from a PDDecM. Allowing utility parfactors in parclusters is straight forward. While constructing the FO jtree, meuLDJT treats the utility parfactor in the same way as probability parfactors. With the parclusters, meuLDJT distributes local information by message passing. To calculate the probability messages, meuLDJT excludes utility parfactors as they do not influence the probability distributions. Using the probability messages, meuLDJT can calculate the current utility value and distribute the value as long as the utility PRV is in the separator. To calculate utilities, utility parclusters need to know the probability distributions, which is distributed during message passing to each parcluster. Using the probability distributions, meuLDJT calculates for each group a utility value and multiplies the value by the number of groundings. The new utility value is then added to the old utility value. Further, the utility transfer function ensure the transfer of utility values, while LDJT's behaviour ensures preserving the current state.

Algorithm 1. meuLDJT for a PDDecM G, Queries $\{\mathbf{Q}\}_{t=0}^{T}$, and Evidence $\{\mathbf{E}\}_{t=0}^{T}$

---

    procedure MEULDJT($G_0, G_{\rightarrow}, \{\mathbf{Q}\}_{t=0}^{T}, \{\mathbf{E}\}_{t=0}^{T}$)
        ($J_0, J_t, \mathbf{I}_t$) := DFO-JTREE($G_0, G_{\rightarrow}$)
        while $t \neq T + 1$ **do**
            J_t := LJT.EnterEvidence($J_t, \mathbf{E}_t$)
            J_t := LJT.PassProbMessages(J_t)
            J_t := LJT.PassUtilMessages(J_t)
            AnswerQueries($J_t, \mathbf{Q}_t$)
            ($J_t, t, \alpha[t-1]$) := ForwardPassJ_t, t)

Answering MEU Queries. meuLDJT can answer MEU queries for a finite horizon. The horizon defines how far meuLDJT predicts into the future. For a given horizon, meuLDJT tests all action sequences to find the best action sequence. Hence, meuLDJT constructs all action sequences for a horizon. In general, meuLDJT constructs r^{h+1} action sequences, where r is the number of actions and h the horizon. For each action sequence, meuLDJT enters the sequence as evidence and answers the expected utility query for that sequence. Finally, after having tested all action sequences, meuLDJT returns the best action sequence and the expected utility value.

Assume that $t = 3$ and we only have a horizon of 1 to answer an MEU query, then meuLDJT constructs the action sequences. In this case, there are four action sequences. For example, the first action sequence is $A_3(X) = A^1$ and $A_4(X) = A^1$. Now, meuLDJT enters $A_3(X) = A^1$ in the FO jtree for time step 3. Message passing on the FO jtrees is now performed in two steps. The first step is to calculate probability messages. The second step is to calculate utility messages. meuLDJT uses the probability messages and the evidence, which includes the current action, and distributes the utility through the FO jtree. After message passing, each parcluster can answer queries about its PRVs. Thus, meuLDJT can answer multiple probability and utility queries efficiently, as it can reason over representatives. To proceed in time, meuLDJT uses the *out-cluster* to calculate an α message over the interface PRVs, which now includes $Util_t$. Hence, the current belief state and utility value is stored in the α message and then added to the *in-cluster* of the next time step, in this case $\mathbf{C}_4^3$. Using the utility transfer function, the current FO jtree contains the overall utility value. Now, meuLDJT enters $A_4(X) = A^1$ in the FO jtree for time step 4 and performs the two message passes. Finally, meuLDJT calculates the expected utility value and proceeds with the next action sequence, until meuLDJT has the expected utility values for all four sequences.

Theorem 1. *meuLDJT is sound, i.e., it produces the same result as a ground algorithm.*

Proof sketch. LDJT is sound [3] and meuLDJT uses LDJT for the probability calculations. Thus, marginal *filtering* and *prediction* queries are still sound. To calculate the new utility value, meuLDJT efficiently calculates the utility value and adds it to the old utility value. While calculating the utility value, meuLDJT accounts for the groundings, namely, it calculates the utility value for one representative and multiplies it by the number of groundings. As all instances behave the same, they instances each would contribute the same utility value in a ground model. Hence, by calculating a utility for one representative and multiplying the utility by the number of groundings, meuLDJT obtains the same result a ground algorithm would obtain. Additionally, the message passing inside of a FO jtree ensures that the current utility value is known at all relevant parclusters and the transfer function preserves the value over time.

4   Conclusion

We present PDDecMs, an extension to PDMs, and meuLDJT for sequential probabilistic online decision support by calculating a solution to the lifted temporal MEU problem. Areas such as healthcare could benefit from the lifting idea for many patients in combination with the efficient handling of temporal aspects of meuLDJT and the support of different kinds of queries. By extending the underlying model with action and utility nodes, complete healthcare processes including treatments can be modelled. Additionally, by maximising the expected utility, meuLDJT can calculate a best action. Further, meuLDJT can efficiently answer a combination of *expected utility*, *filtering*, and *prediction* queries. Thus, meuLDJT can support decision support as well as help to understand the suggested decision by also efficiently answering multiple marginal queries.

We currently check whether meuLDJT can reuse computations from previous expected utility calculations by, e.g., identifying dominant actions for belief state regions.

References

1. Apsel, U., Brafman, R.I.: Extended lifted inference with joint formulas. In: Proceedings of the 27th Conference on Uncertainty in Artificial Intelligence, pp. 11–18. AUAI Press (2011)
2. Braun, T., Möller, R.: Lifted junction tree algorithm. In: Friedrich, G., Helmert, M., Wotawa, F. (eds.) KI 2016. LNCS (LNAI), vol. 9904, pp. 30–42. Springer, Cham (2016). https://doi.org/10.1007/978-3-319-46073-4_3
3. Gehrke, M., Braun, T., Möller, R.: Lifted dynamic junction tree algorithm. In: Chapman, P., Endres, D., Pernelle, N. (eds.) ICCS 2018. LNCS (LNAI), vol. 10872, pp. 55–69. Springer, Cham (2018). https://doi.org/10.1007/978-3-319-91379-7_5
4. Gehrke, M., Braun, T., Möller, R., Waschkau, A., Strumann, C., Steinhäuser, J.: Lifted maximum expected utility. In: Koch, F., et al. (eds.) AIH 2018. LNCS (LNAI), vol. 11326, pp. 131–141. Springer, Cham (2019). https://doi.org/10.1007/978-3-030-12738-1_10
5. Murphy, K.P.: Dynamic bayesian networks: representation, inference and learning. Ph.D. thesis, University of California, Berkeley (2002)
6. Nath, A., Domingos, P.: A language for relational decision theory. In: Proceedings of the International Workshop on Statistical Relational Learning (2009)
7. Sanner, S., Kersting, K.: Symbolic dynamic programming for first-order POMDPs. In: Proceedings of the Twenty-Fourth AAAI, pp. 1140–1146. AAAI Press (2010)

Automatic Generation of Video Game Character Images Using Augmented Structure-and-Style Networks

Matthew T. Mann[(✉)] and Howard J. Hamilton

University of Regina, Regina, Canada
mattmann624@gmail.com

Abstract. We propose a fast, flexible approach to automatically generating two-dimensional character images for video games. We treat the generation of character images as a two-part machine learning problem. The first task is to generate structured images that represent common attributes of characters as images. The second task is to add details to the structured images that appear to fit with some overall theme for each image. For both tasks, we employ generative adversarial network architectures, with modifications to improve their performance for their respective tasks. The resulting As^2-GAN approach generates character images that are as realistic as those generated with the DCGAN approach, and are more consistent in quality and structural resemblance to images from the dataset. The As^2-GAN approach also provides image creators with more control over images than typical one-step methods, while being able to generate high quality images using small datasets.

Keywords: Machine learning · Generative adversarial network · Image generation · Structure and style

1 Introduction

Software capable of generating 2D images subject to constraints has many practical applications. One particular application is in the large-scale generation of characters for video games. We propose applying machine learning to the task of real-time generation of a set of distinct video game character images that are stylistically consistent with an initial set of character images called the *base set*. After off-line learning, a learned model, also called an *agent*, can be applied in a video game to quickly generate unique character images that fit the stylistic constraints of that game. As an instance of this task, we consider the problem of generating convincing images of novel characters suited to the popular television and video game series, Pokémon. The remainder of this paper is organized as follows. Section 2 describes related work on image generation. Section 3 describes our proposed As^2-GAN approach. Section 4 describes the results of an evaluation of our approach and Sect. 5 provides conclusions.

© Springer Nature Switzerland AG 2019
M.-J. Meurs and F. Rudzicz (Eds.): Canadian AI 2019, LNAI 11489, pp. 387–393, 2019.
https://doi.org/10.1007/978-3-030-18305-9_34

2 Related Work

To approach the problem of image generation, many have turned to machine learning, and, specifically to Generative Adversarial Networks (*GAN*) (Fig. 1).

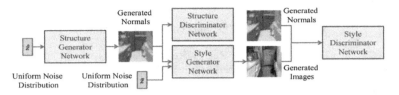

Fig. 1. s^2-GAN generative pipeline, as portrayed in [11].

Generative Adversarial Networks. A GAN [1] is a machine learning architecture that contains a generator (G) and a discriminator (D) which train adversarially in a minimax game. D attempts to distinguish between real and fake samples from a dataset, outputting a single value for each sample. G attempts to fool D into thinking the samples it generates are real. G generates samples by transforming a *latent vector* of random numbers into data. D's objective is to minimize its error when predicting the validity of a sample. G's objective is cause D to misidentify samples made by the generator as valid. D and G train together in an attempt to find the Nash Equilibrium, where G generates samples with such quality that D is unable to distinguish valid and invalid samples, thus producing high quality data. We define a training step as using each training configuration of an architecture exactly once to train an agent.

Various GAN architectures optimize for quality, diversity, efficiency, versatility, etc. Some of these architectures are DCGAN [8], LSGAN [6], s^2-GAN [11], and α-GAN [10]. Domain transfer GAN architectures, e.g., pix2pix [3] and BicycleGAN [12], allow a sample to be mapped from one style or domain to another. The most relevant architectures to this research are DCGAN, s^2-GAN, α-GAN, and BicycleGAN. We describe GANs in the context of image generation, where every sample is a 2D image.

Structure and Style Adversarial Networks. An s^2-GAN [11] is an approach to image generation that attempts to decompose the learning task into two subtasks performed by separate GANs (Fig. 1). The first GAN is the Structure GAN, which attempts to generate images that represent the structure of an image, rather than a realistic image in itself. The second GAN is the Style GAN, which then attempts to translate the structured images produced by the Structure GAN into realistic, detailed images. Each generator is trained adversarially with a discriminator, similar to in a typical GAN.

3 The As²-GAN Approach

The Augmented Structure-and-Style GAN (As²-GAN) approach to generating game character images is to decompose the task into two steps, generating structured images containing high-level features and then transforming those structured images to styled images containing low-level features. This two-step approach required solving two learning problems: learning to generate structured character images and learning to fill in details for such images. The two-step approach simplifies the overall learning approach, since it gives a concrete suggestion for decomposing the learning problem. Our contribution is an improvement to versatility, diversity, and flexibility during the generation of both structured and styled images, as opposed to the original s^2-GAN.

Generating High-Level Details. The first learning problem is: given a set of structured character images, train a model that generates structured images that are consistent with the underlying constraints on the structured images. The learning method consisted of directly applying a modified version of α-GAN [10] to the structured images. This consists of the four training configurations illustrated in Fig. 2. The Discriminator (a) and Adversarial (c) configurations are typical GAN configurations. To increase diversity in the image generation process, we include an Encoder (E), with two training configurations. The Latent Regressor (LR) (b) generates an image from a random latent vector using G. Then, E learns to predict what the numbers in that random latent vector were. The Variational Autoencoder (VAE) [4] (d), uses E to reduce an image to a latent vector, then learns to reconstruct the image from that latent vector using G. These configurations both ensure that G is able to use its vector input to reproduce images similar to any sample from the base set, thus ensuring diversity. The influence of a training configuration relative to the Discriminator and Adversarial configurations is represented using λ. Over training time, we lower the influences of the VAE and LR configurations (λVAE and λLR) relative to the influences of the other configurations. Thus, G begins by learning to create diverse images, but over time, due to adversarial training, produces more realistic images.

Training Details. We chose an initial learning rate of 0.0002, which, based on the work of Suki Lau [5], decreases by 10% after every 10 000 steps during training. The influences of the VAE and LR configurations (λVAE and λLR) are 50% at the beginning of training, and decrease by 20% every 10 000 steps, rather than by 10%. We chose to use He initialization [2] on all convolution layers because it provided extra training stability and sample quality during preliminary experiments. D's output has a linear activation function, and uses a mean squared error cost function, as an LSGAN [6].

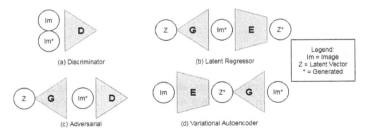

Fig. 2. Training configurations for the Structure GAN of As^2-GAN used to generate high-level structured images. Green models are trained in the configurations; blue models remain static. (Color figure online)

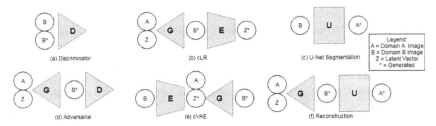

Fig. 3. Training configurations for the Style GAN in As²-GAN used to generate low-level styled images. Green models are trained in the configurations; blue models remain static. (Color figure online)

Generating Low-Level Details. The second learning problem is: given a set of structured images, transform them into styled images that retain the structure of the structured images. For this problem, we designed a new GAN architecture inspired by BicycleGAN [12]. We obtain the discriminator, adversarial, cVAE and cLR configurations from Zhu et al. [12]. The cVAE and cLR configurations work similarly to the VAE and LR configurations in Sect. 3, but additionally use a structured image as input to the Generator. Then, we added a U-Net segmentation model [9] (c, f), which attempts to reconstruct the original structured images from the generated styled images. Instead of simply reconstructing the image, it attempts to categorize each pixel as it was in the original structured images. By updating the generator with a categorical loss function, we intended to cause the structure of generated samples to be visually clear. All six training configurations can be seen in Fig. 3. As with BicycleGAN and pix2pix [3], the discriminator (*D*) was used to determine the realism of overlapping patches of an image, rather than an entire image. We used patches of 32×32 pixels.

Training Details. We chose to use a learning rate of 0.0002, which decreases by 10% every 10 000 steps during training. The influence of the cVAE and cLR configurations (λcVAE and λcLR) are 50% at the beginning of training, and decrease by 20% every 10 000 steps. The influence of the Reconstruction configuration (λReconstruction) is set to 10% until the final 20 000 training steps, where it is increased to 100%. We chose to use He initialization [2] on all convolution layers because it provides additional training stability and sample quality. D's output has a linear activation function, and uses a mean squared error cost function, as an LSGAN [6].

4   Experimental Results

We conducted three sets of experiments aimed at training an agent that generates convincing novel character images suited to the popular television and video game series, Pokémon [7]. For training, we used a dataset of 721 Pokémon images and where applicable, 249 structured Pokémon images. We prepared the structured images by classifying each pixel into a one-hot vector, in which each pixel was labelled as eye (red), body (green), or background (blue). All Pokémon images are subject to copyright

(a) DCGAN

(b) s2-GAN

(c) As2-GAN

(d) Real Images

Fig. 4. Sample generated images and real character images.

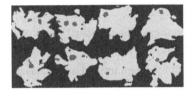

Fig. 5. Generated structured images based on structured Pokémon images. (Color figure online)

Fig. 6. Samples generated on a small dataset of 80 pairs of images.

by the Pokémon Company, and are included under Fair Use. We conducted experiments to evaluate the quality of images produced by DCGAN [8], s^2-GAN [11] and As^2-GAN, respectively. Figure 4 summarizes our results by showing eight selected character images produced by each method, as well as eight real Pokémon images. Table 1 shows the number of training steps taken and learning rate used during the experiments. Figure 5 shows eight typical high-level structured images generated by the structure generation step of As^2-GAN.

s^2-GAN overfit to the styles of the images in the dataset and failed to remain consistent with the structure provided in the generated structured images. This caused samples with the same structured image to all be visually similar (not shown), because the Style GAN completely ignored its latent vector input. As^2-GAN generated images that had similar levels of realism to DCGAN, but were superior because they consistently contained structural components, especially eyes and body shapes. As with s^2-GAN, the two-step approach used in As^2-GAN makes the generative process flexible. It splits the generation process into two steps, allowing for manual generation of structured images or manual stylization. To further demonstrate the versatility of the

Table 1. Training parameters during experimentation.

Approach	Training steps	Learning rate
DCGAN	250 000	0.0002
s^2-GAN	150 000	0.00015
As^2-GAN	200 000	0.0002

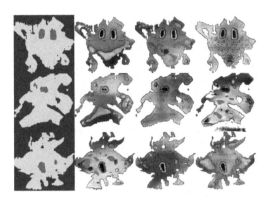

Fig. 7. Generated structured images with multiple corresponding generated styled images.

As^2-GAN, Fig. 6 shows images produced by As^2-GAN when trained on a dataset of 80 pairs of structured and styled images for 100 000 steps. Additionally, one structured image can be used to create multiple different styled images, as demonstrated in Fig. 7.

5   Conclusion

The As^2-GAN approach has the potential to contribute to the cost-effective generation of 2D video game character images. By using our approach, an artist needs to make only a small number of character images, and then let an agent generate many character images after offline training.

References

1. Goodfellow, I.J., et al.: Generative adversarial nets. In: NIPS, vol. 27 (2014)
2. He, K., Zhang, X., Ren, S., Sun, J.: Delving deep into rectifiers: surpassing human-level performance on ImageNet classification. In: IEEE International Conference on Computer Vision, pp. 1026–1034 (2015)
3. Isola, P., Zhu, J.-Y., Zhou, T., Efros, A.A.: Image-to-image translation with conditional adversarial networks. In: IEEE Conference on Computer Vision and Pattern Recognition, pp. 5967–5976 (2016)
4. Kingma, D.P., Welling, M.: Auto-encoding variational Bayes. CoRR (2013)

5. Lau, S.: Learning rate schedules and adaptive learning rate methods for deep learning (2017). https://towardsdatascience.com/learning-rate-schedules-and-adaptive-learning-rate-methods-for-deep-learning-2c8f433990d1

6. Mao, X., Li, Q., Xie, H., Lau, R.Y., Wang, Z., Smolley, S.P.: Least squares generative adversarial networks. In: IEEE International Conference on Computer Vision (2016)

7. The Pokémon Company. https://www.pokemon.com/

8. Radford, A., Metz, L., Chintala, S.: Unsupervised representation learning with deep convolutional generative adversarial networks. CoRR (2015)

9. Ronneberger, O., Fischer, P., Brox, T.: U-Net: convolutional networks for biomedical image segmentation. In: Navab, N., Hornegger, J., Wells, W.M., Frangi, A.F. (eds.) MICCAI 2015. LNCS, vol. 9351, pp. 234–241. Springer, Cham (2015). https://doi.org/10.1007/978-3-319-24574-4_28

10. Rosca, M., Lakshminarayanan, B., Warde-Farley, D., Mohamed, S.: Variational approaches for auto-encoding generative adversarial networks. CoRR (2017)

11. Wang, X., Gupta, A.: Generative image modeling using style and structure adversarial networks. In: Leibe, B., Matas, J., Sebe, N., Welling, M. (eds.) ECCV 2016. LNCS, vol. 9908, pp. 318–335. Springer, Cham (2016). https://doi.org/10.1007/978-3-319-46493-0_20

12. Zhu, J.-Y., et al.: Toward multimodal image-to-image translation. In: NIPS (2017)

Weaving Information Propagation: Modeling the Way Information Spreads in Document Collections

Charles Huyghues-Despointes[1,2]([✉]), Leila Khouas[2], Julien Velcin[1], and Sabine Loudcher[1]

[1] Université de Lyon, Lyon 2, ERIC EA 3083, Lyon, France
{charles.huyghues-despointes,julien.velcin,sabine.loudcher}@univ-lyon2.fr
[2] Bertin IT, Montpellier, France
leila.khouas@bertin.fr

Abstract. Information usually spreads between people by the mean of textual documents. During such propagation, a piece of information can either remain the same or mutate. We propose to formulate information spread with a set of time-ordered document chains along which some information has likely been transmitted. This formulation is different from the usual graph view of a transmission process as it integrates a notion of lineage of the information. We also propose a way to construct a candidate set of document chains for the information propagation in a corpus of documents. We show that most of the chains have been judged as plausible by human experts.

Keywords: Information lineage · Information propagation · Document chains · Textual data stream

1 Introduction

Internet came with a burst in document publication and accessibility. Nowadays, one can get far more news articles, videos or audio podcasts than one can digest on a daily basis. This has fostered new tool-assisted methods that take advantage of computer fast processing capabilities. Most of these documents convey information expressed in natural language. To analyze this information, the user mostly use search tools with a preconceived idea of what he is searching through a question or keywords. The user may need tools to understand how information propagates without using these preconceptions.

There are research efforts to study why and how information propagates in a corpus. Some works are interested on modeling the diffusion process between authors of documents in order to recover the propagation network as [1,2]. Whereas these methods give us insights on the diffusion process between authors, they do not express the diffusion between documents. The difficulty of uncovering the diffusion over documents is pointed in [3,4] where authors try to trace a piece

© Springer Nature Switzerland AG 2019
M.-J. Meurs and F. Rudzicz (Eds.): Canadian AI 2019, LNAI 11489, pp. 394–399, 2019.
https://doi.org/10.1007/978-3-030-18305-9_35

of information back to its primary source. Most of these models consider the diffusion as an information exchange [5] with no modification. However, during the information diffusion over documents, the specific context a piece of information appears in, may slightly make its meaning mutate, even when the information syntax does not evolve [6]. Moreover, those models expect pieces of information to be explicitly defined beforehand, which makes it difficult to express mutations. Furthermore, if two pieces of information share a similar context, as in Fig. 1a, knowing their diffusion history could give us insights on how and why they differ. We propose to study how information propagates through textual document streams. We consider that a succession of documents, that we call a *chain*, is the basic structure for describing how information propagates. A set of chains is represented in Fig. 1b.

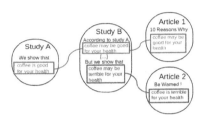 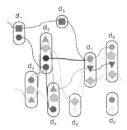

(a) There is a lineage from Study A and Study B to Article 1, and a lineage from Study B to Article 2 only.

(b)   The colored symbols stand for (up to now unknown) pieces of information.

Fig. 1. Propagation chains, a concrete (left) and an abstract (right) representations

We never know how the propagation phenomenon unfolds precisely. However, we consider that if information flows from a document to another, there should be some semantic similarity between these two documents. Given this assumption, we propose an algorithm to construct chains of similar documents as first candidates for propagation chains. Then, we present the results and our methodology to evaluate these chains. We show that human experts reach a consensus on what is a plausible chain of propagation and what is not, and that our calculated chains match with this evaluation. We conclude in giving several applications that can be built on top of this general chain propagation model.

2   Modeling Information Propagation with Chains

We do not formally define what one piece of information is. It can be expressible facts, ideas, opinions, reasoning or even sensations, but it also may be some complex discrete representation of information. We denote K the set of all pieces of information, and D a corpus of documents. An interesting property of K elements is that they may be more or less similar. Statements like "It may rain today" and "It will rain today" are not exactly the same, still they share some semantic. We set $sem_K : K \times K \to [0,1]$ a semantic similarity metrics over K.

During propagation between the two documents d_i and d_j, a piece of information k_0 of d_i may mutate, resulting in a possibly different k_1 in d_j. We note such mutation as a pair (k_0, k_1). We can express the propagation event between d_i and d_j by enumerating all the mutations that occur between them, noted as $M_{i,j}$. We define a propagation event e as a triplet $e = (d_i, d_j, M_{i,j})$ for $d_i, d_j \in D$. There exists a non-trivial threshold ϵ such that: if $(d_i, d_j, M_{i,j})$ is a propagation event then $(k_i, k_j) \in M_{i,j} \implies sem_k(k_i, k_j) \geq 1 - \epsilon$.

This propagation model can express historical modifications of information. Given two propagation events $(d_1, d_2, \{(k_1, k_{1'})\})$ and $(d_2, d_4, \{(k_{1'}, k_{1''})\})$, we can flow back the origin of $k_{1''}$ in d_2, as k_1 in d_1. Thus, information has propagated along the path $d_1 d_2 d_4$. Such path of documents is what we call a *propagation chain*. Given P a set of propagation events over documents, a chronologically ordered sequence of documents $c = d_0 d_1 \ldots d_n$ is a propagation chain if:

$$\forall i \in \{1, 2, \ldots, n\}, \exists\, e_i, e_{i+1} \in P / \begin{cases} e_i = (d_{i-1}, d_i, M_{i-1,i}), \\ e_{i+1} = (d_i, d_{i+1}, M_{i,i+1}) \\ \exists (k, k') \in M_{i-1,i} \wedge \exists (k', k'') \in M_{i,i+1} \end{cases}$$

We denote the set of propagation chains by T, which stands for trajectory. Conceptually, each transition of a propagation chain must keep a common endpoint (document d_i) and a common semantic endpoint (piece of information k'). Note that T has the following property: if $c = d_0 d_1 \ldots d_n$ is a propagation chain of T, then $\forall 1 \leq i < j \leq n, c[i, j] = d_i \ldots d_j$ is also a propagation chain of T. Furthermore, we say that $c[i, j]$ is a sub-chain of c.

3   An Approach to Propagation Chain Approximation

In this section, we assume known a corpus D. We denoted by T_D the chains only composed of documents from D. We do not explicitly know the information pieces contained in documents. Instead, we have a similarity function sim between documents. In order to compute a good approximation of T_D, we construct coherence metrics for propagation chains based on that similarity. Then, we compute the chains that satisfy this metric up to a given threshold.

It seems reasonable that most of the propagation chains should be coherent chains. We model the coherence using a metric, denoted by coh that assigns a number between 0 and 1 to a chain. We say that a chain c is *coherent* if $coh(c) > 1 - \epsilon$, with ϵ a given coherence threshold. In order to construct every document chain satisfying our coherence criterion, we make use of the property stipulating that if c is a propagation chain, then every sub-chain c is also a propagation chain. Extending this proposition to coherent chains implies that every sub-chain of c must satisfy our coherence criterion. This allows us to use a dynamic programming approach. We define $FinishIn(d)$ as the set of coherent chains finishing by d. In order to finish in d, a document chain must be the concatenation (operator .) of a chain $d'd$ and at most one chain from $FinishIn(d')$. We can then construct the Candidates for d and the coherent chains T_{coh}:

$$Candidates(d) = \bigcup_{d'/coh(d'd)>1-\epsilon} \{c.d/c \in FinishIn(d') \cup \{d'\}\}$$

$$FinishIn(d) = \{c \in Candidates(d)/coh(c) > 1 - \epsilon\}$$

$$T_{coh} = \bigcup_{d \in D} FinishIn(d)$$

We can solve this problem using a bottom-up strategy, starting from oldest to newer documents, since a chain always respects the publication chronology. This approach has some complexity issues due to the potentially exponential number of coherent chains. For that purpose, we introduce a constant safeguard threshold on the maximal number of coherent chains finishing in every document. This ensures a linear memory complexity.

4 Experiments

We sampled two datasets of 150 documents. The first one comes from the Citation Network Dataset V1 AMINER[1] constructed by [7]. This is mostly abstracts from scientific papers extracted from ACM and DBLP. Our second dataset is sampled from the full set of news articles posted on the US version of the Huffington Post from 1st July to 30th November 2016. For document similarity, we used a TFIDF cosinus similarity where we enriched the bag of words with n-grams of size 2 to 4. For chain coherence, we used the minimal similarity and the arithmetic mean of similarity between document pairs inside the chain. Three different coherence thresholds are considered: 10%, 20% and 50%. It yields six coherent chain sets for each dataset (81 and 149 chains to evaluate, respectively).

For the sake of annotation, we gathered experts who have a professional level in English and a good knowledge of reading scientific papers. We spread chains between experts such that every chain is at least evaluated twice and two experts are not exposed to the same succession of chains. For each document chain, an expert evaluates links between two successive documents (determining if there is a strong, weak or no semantic link). Then, he evaluates sub-chains from the first document to every other determining if it is strongly, weakly or not plausible that some information propagated.

Agreement ratio is computed as the proportion of paired evaluations with the same conclusion over the total number of paired evaluations. We speak of presence if the chain or the link is annotated as strongly or weakly plausible. Experts agreed on more than 70% cases on the semantic link presence for both datasets and it exceeds 80% for the presence of a plausible chain. This result reinforces the intuition that it is easier to reach a consensus when we have more context. This shows that experts can have an intuition of chain plausibility with consistency, which means that detecting coherence in a chain is a feasible task.

---

[1] The full dataset can be obtained at: https://aminer.org/citation.

Now, we label the annotation for each chain with five categories for both the semantic link task and the chain plausibility task. Each category represents the majority of annotations. **Category 1** stands for **strong intensity**. **Category 2** is for **weak intensity**. **Category 3** stands for **presence** but with no intensity agreement. **Category 4** is for **absence**. **Category 5** is finally for the case where there is no majority agreement. Results on AMINER are good with nearly 70% of plausible chains (Cat. 1, 2 and 3) and 75% of linked chains. On the other hand, Huffpost results are much weaker. 64% of evaluated links are judged non-existent and 75% of chains are judged non plausible. We will show that these non relevant chains come from a low coherence threshold.

Annotated chains serve as a ground truth allowing us to compare different coherence criterion. Studied similarity metrics are: TFIDF with a cosine similarity; Doc2Vec with a cosine similarity, with vectors of size 20 learnt on the dataset; RWR: a random-walk-based approach [8], parametrized with a 99% restart probability. For the coherence, we use the arithmetic mean coherence which is computed as the mean similarity between document pairs from the same chain. We consider the coherence as a good discrimination function if there is a small or even null intersection over coherence range (defined as the mean plus or minus the standard deviation) for strongly plausible and non-plausible chain categories. These intervals are presented in Fig. 2. On both datasets, the Doc2Vec-based coherence is the most discriminating metric. Generally, strongly and weakly plausible chains have higher coherence than non-plausible ones. For Huffpost, the results explain the proportion of bad chains observed. A typical HuffPost bad chain has a TFIDF based coherence under 0.2. It means that those chains mostly come from the trajectory computed with a 0.1 coherence threshold. These coherence metrics prove that capturing human judgment over document chains is partly possible by using well-known similarities.

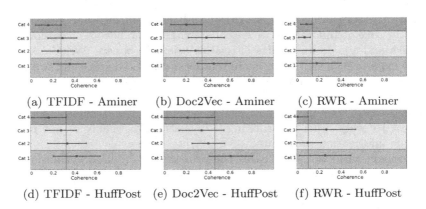

(a) TFIDF - Aminer   (b) Doc2Vec - Aminer   (c) RWR - Aminer

(d) TFIDF - HuffPost   (e) Doc2Vec - HuffPost   (f) RWR - HuffPost

Fig. 2. Mean and standard deviation for different document similarities by annotation category. A vertical line marks the absence category (Cat 4) overlap.

5   Conclusion

Considering the information propagation through trajectories over a textual document network is a novel idea. An important advantage over other methods based on the propagation graph lies in a better understanding of the history of information propagation. We proposed an approximation of trajectory by coherent chains that can be solved through dynamic programming. To qualify the computed chains, we realised a human evaluation campaign. This campaign had two benefits. First, we saw that human evaluations were consistent between themselves, which suggests that recognizing a plausible propagation chain is feasible. Second, we used evaluations as a ground truth for testing different coherence criterion. We saw that criterion based on well-known metrics succeed in capturing human judgment. We consider this result as a first proof that this task may be solved using an automatic process.

For future work, we plan to overcome the necessity of guessing a correct coherence threshold. We also plan to automatically identify the pieces of information that propagate along the chains. We foresight multiple use cases for a good trajectory approximation. One use case is to easily navigate in the document space for an analyst user by following an interesting subject flow, or to give him a good understanding of the information flow by summing up chains into a metromap of information. It may also be an interesting framework to study how information pieces interact with each other along the chains.

References

1. Gomez-Rodriguez, M., Balduzzi, D., Schölkopf, B.: Uncovering the temporal dynamics of diffusion networks. In: Proceedings of the 28th International Conference on Machine Learning, ICML 2011, Washington, USA, pp. 561–568 (2011)
2. Zarezade, A., Khodadadi, A., Farajtabar, M., Rabiee, H.R., Zha, H.: Correlated cascades: compete or cooperate. In: Proceedings of the Thirty-First AAAI Conference on Artificial Intelligence, San Francisco, California, USA, pp. 238–244 (2017)
3. Pinto, P.C., Thiran, P., Vetterli, M.: Locating the source of diffusion in large-scale networks. Phys. Rev. Lett. **109**, 068702 (2012)
4. Farajtabar, M., Gomez-Rodriguez, M., Zamani, M., Du, N., Zha, H., Song, L.: Back to the past: source identification in diffusion networks from partially observed cascades. In: Proceedings of the Eighteenth International Conference on Artificial Intelligence and Statistics, AISTATS 2015, San Diego, California, USA (2015)
5. Zafarani, R., Ali Abbasi, M., Liu, H.: Social Media Mining, An Introduction. Cambridge University Press, New York (2014)
6. Leskovec, J., Backstrom, L., Kleinberg, J.: Meme-tracking and the dynamics of the news cycle. In: Proceedings of the 15th ACM SIGKDD International Conference on Knowledge Discovery and Data Mining. KDD 2009, pp. 497–506. ACM, New York (2009)
7. Tang, J., Zhang, J., Yao, L., Li, J., Zhang, L., Su, Z.: Arnetminer: extraction and mining of academic social networks. In: KDD 2008, pp. 990–998 (2008)
8. Shahaf, D., Guestrin, C.: Connecting the dots between news articles. In: Proceedings of the 16th ACM SIGKDD International Conference on Knowledge Discovery and Data Mining. KDD 2010, pp. 623–632. ACM, New York (2010)

Mitigating Overfitting Using Regularization to Defend Networks Against Adversarial Examples

Yoshimasa Kubo$^{(\boxtimes)}$ and Thomas Trappenberg

Faculty of Computer Science, Dalhousie University, Halifax, Canada
yoshi.with.ai@gmail.com, trappenberg@gmail.com

Abstract. Recent work has shown that neural networks are vulnerable to adversarial examples. There is an discussion if this problem is related to overfitting. While many researcher stress that overfitting is not related to adversarial sensitivity, Galloway et al. [4] showed that mitigating overfitting improves the accuracy on adversarial examples. In this study we add to this view that overfitting is a factor in adversarial sensitivity. To make this argument, we include two directions in our study, the first is to evaluate several standard regularization techniques with adversarial attacks and to the second is to evaluate binarized stochastic neural networks on adversarial examples. We report that strong regularizations including binarized stochastic neural networks do not only improve overfitting but also help the networks in fighting against adversarial attacks. Supplemental materials are available at https://github.com/ykubo82/ovf.

1 Introduction

Adding small designed perturbations to images can drastically change the classification accuracy of machine learning models. These perturbed examples are called *adversarial examples* [12]. The study of adversarial sensitivity of networks is important not only because these examples could pose a potential security threat for practical machine learning applications, but also to shed light on the representational capacity of deep networks [8].

The cause of adversarial example is still debated. Goodfellow et al. [6] argued that an example generated for one model is often misclassified by other models which consist of different architectures that were trained on disjoint training sets. Because of these behaviors, the authors defined that overfitting is *not* related to the cause of adversarial examples.

On the other hand, Galloway et al. [4] discuss that researchers falsely concluded that overfitting is not related to the cause of adversarial examples. The authors show that pruning overfitting could be the reason to improve the network accuracy on adversarial examples using strong L2 weight decay. In their paper, the authors suggest that starting with smaller models that can handle

© Springer Nature Switzerland AG 2019
M.-J. Meurs and F. Rudzicz (Eds.): Canadian AI 2019, LNAI 11489, pp. 400–405, 2019.
https://doi.org/10.1007/978-3-030-18305-9_36

strong weight decay is worth exploring as a natural defense against the adversarial examples. In this paper we add to these arguments that overfitting is at least contributing to adversarial sensitivity. If overfitting is not problematic, decision boundaries for well generalized networks and overfitted networks would not change much their accuracy performance on adversarial examples. We hence study the performance on networks with various forms of regularization on adversarial attacks.

More specifically, in this paper we apply dropout [11] to shallow convolutional neural networks (CNN) with low to high dropout probabilities. In [6], the authors applied dropout to the maxout networks and sowed that they reduce the error. However, they did not check whether or not changing the strength of regularization affects the network performance on adversarial examples. We check the relationship between the overfitting rate and the accuracies with changing the probability on adversarial examples. Furthermore, we also study dropout as regularizer with adversarial examples similar to Wang et al. [13]. However, different to Wang et al., we do not use dropout during testing and specifically check if overfitting is a problem. Furthermore, we applied L1 and L2 weight decay to the shallow CNN. Galloway et al. [4] have already shown that L2 weight decay improves the network performance on adversarial examples, but we add here to check a larger range of λ from weak to strong in response to Papernot et al. [10], and we show the accuracy is improved when λ is well tuned.

Finally, we apply binarized stochastic neural networks to adversarial examples. Galloway et al. [3] have already shown that the binary neural networks with straight-through estimation (ST) [1] improved robustness against some adversarial examples. In our study, we investigate the networks with the special case of REINFORCE estimation [1] compared to ST and deterministic binarized neural networks (only activation is binarized with 0 or 1).

2   Experiments

2.1   Parameter Setting

We start by applying a shallow convolutional neural network (CNN) to the MNIST data with changing dropout probabilities. This network consists of one convolutional layer with varying dropout probabilities that we set in the experiments to 10, 30, 50, 70, or 90%, and two deterministic dense layers. We compare this to a network without any regularizations (dropout is 0%). The second experiment is the same as the first, but we use the L1 and L2 weight decay instead of dropout. The network condition is the same as the first one. The hyperparameter λ are 1e−5, 1e−4, 1e−3, 1e−3, and 1e−1. Finally, we apply three versions of a binarized MLP, one with a straight through estimator and the slope-annealing trick [2] (ST-MLP), one based on the REINFORCE algorithm (REINFORCE-MLP), and one binarized deterministic MLP (BDN-MLP). These MLPs are combined with one (1CNN) or two (2CNN) convolutional layers. On testing, we use white-box adversarial examples and do not use adversarial training [6]. All convolutional layersof the MLP have rectified linear activations, while the first dense layer of BSN and BDN use the logistic activation function. The networks are trained on the softmax cross entropy loss function, and we use the Adam optimizer [7] and Xaiver initialization [5] for all of the networks. The results are averaged over the 10 runs. Experiments are conducted on the MNIST dataset [9] with the regular 60000 28×28 grayscale images as training set and 10000 for testing set. The size of the networks and some additional parameters are specified in the supplemental materials.

2.2   Fast Gradient Sign Method (FGSM)

The Fast gradient sign method is one of the methods to craft adversarial examples [6]. After training a model on the dataset, the parameters of the models are frozen. The derivative of the cost function with respect to the inputs $\text{Loss}(w; x, y)$ is used for crafting adversarial examples $\tilde{x}$ with sign function as $\tilde{x} = x + \epsilon \cdot \text{sign}(\nabla_x \text{Loss}(w; x, y))$, where x is the input, y is the target, and ϵ is the hyperparameter to change the perturbation level or magnitude. In our experiments, we checked the results with ϵ from 0.01 to 0.25. For each model, the accuracies are calculated using all perturbed testing data.

2.3   Results for Changing Dropout Probability

The network without dropout is a weak model against the adversarial examples (Fig. 1). The network is well generalized on the clean dataset (the accuracies are more than 98% on the testing set), but the network is more vulnerable to the examples even small perturbations compared to the network with dropout. The high dropout probabilities help the network prevent overfitting. It is common to use a dropout probability 50%, but in this study, the 90% dropout probability is the optimal to mediate adversarial examples. Compared to the network without dropout, the network with 90% dropout probability is more robust, especially at the mild perturbation level (at $\epsilon = 0.1$, the accuracy difference between the network is around 40%). Furthermore, we observe the errors for each model between training and testing set after the training, $Error = (100\text{-}accuracy)$ to check if overfitting occurs for each model.

2.4   Results for Changing L1 and L2 λ parameters

Figure 2a shows strong L1 weight decay helps the network for defending against the adversarial examples, but a too strong decay might not be helpful in this case. The model with the strongest λ is underfitting on the clean dataset. Interestingly, this underfitting model is better than overfitting. The overfitted models are more vulnerable on even small perturbation levels. Figure 2b shows that the results of the network with L2 weight decay are similar results to the L1 results. The accuracies on $\lambda < 1e-2$ are not affected by the weight decay while on $\lambda \geq 1e-2$ are increased.

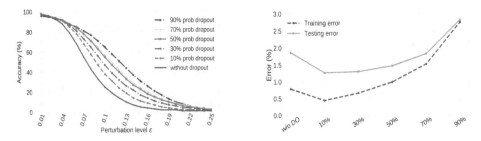

Fig. 1. Results of the dropout experiments Left: Accuracies for the network with dropout on perturbed testing set. Right: errors for the networks with dropout and without dropout (w/o DO) on training and testing set after the training for checking the overfitting levels on each model.

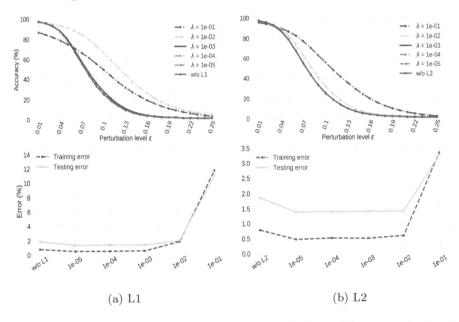

(a) L1   (b) L2

Fig. 2. Results of the L1 weight decay experiments: Left-top: The accuracies for the network with L1 weight decay on perturbed testing set. L1 weight decay - the strong λ is helpful for defending the adversarial example, but the too strong λ could not be helpful.

2.5   Results for Binarized Stochastic Neural Network

The previous experiments already demonstrated that strong regularization helps the network to defend against the adversarial examples. We conduct similar experiments with binarized stochastic networks (ST and REINFORCE) and deterministic (BDN). Figures 3a to c show that the accuracies for MLPs, 1CNN, and 2CNN with both binarized stochastic and deterministic neurons on adver-

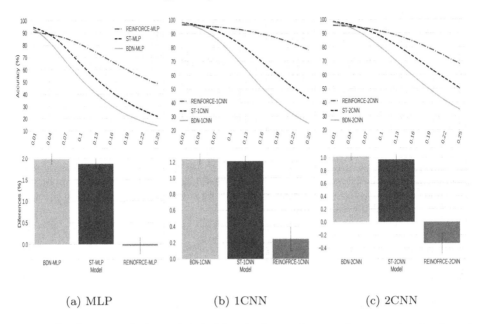

(a) MLP   (b) 1CNN   (c) 2CNN

Fig. 3. Stochastic and deterministic binarized neural network accuracies (top) and accuracy differences (bottom) between training and testing set (clean dataset). a, b and c REINFORCE is the best performance in each condition in proportion to the accuracy differences. Every REINFORCE-based algorithm does generalized well.

sarial examples with perturbation levels. The figures show that both stochastic network performances are always better than BDN. The bottom of the figures shows the differences between training and testing sets to check the ovefitting levels for each network.

3   Discussion and Conclusion

Regularizations drastically improves accuracies on adversarial examples. We have shown this here for several regularization techniques, including dropout, weight decay, and binarized neurons. This clearly demonstrates that overfitting contributes to the cause of adversarial sensitivity. Such regularization techniques usually help to smooth decision, thereby moving them further from training examples.

Strong regularization could be the cause of underfitting such as L1 weight decay. However, we have seen that in our L1 weight decay experiment, the accuracy with overfitting is worse than underfitting. This is another indications that overfitting is related to the cause of adversarial examples. While weight decay and dropout help with protecting against adversarial attacks, we suggest to use binarized stochastic neural networks for such protection because of their high performance and confidence levels (available in the supplemental material) for the correct classes even at high perturbation levels. This is especially the case with the REINFORCE training of binarized neurons.

Although the networks with these regularizers cannot protect themselves on high perturbation levels, this study reports the improvement of accuracies in mild perturbation levels on adversarial examples could be an evidence of the relation to the cause of adversarial examples. Also, binarized stochastic neural networks defend themselves against the adversarial examples even with strong perturbation levels.

References

1. Bengio, Y., Léonard, N., Courville, A.C.: Estimating or propagating gradients through stochastic neurons for conditional computation. arXiv preprint arXiv:1308.3432 (2013)
2. Chung, J., Ahn, S., Bengio, Y.: Hierarchical multiscale recurrent neural networks. arXiv preprint arXiv:1609.01704 (2016)
3. Galloway, A., Taylor, G.W., Moussa, M.: Attacking binarized neural networks. In: International Conference on Learning Representation (2018)
4. Galloway, A., Taylor, G.W., Moussa, M.: Predicting adversarial examples with high confidence. arXiv preprint arXiv:1802.04457 (2018)
5. Glorot, X., Bengio, Y.: Understanding the difficulty of training deep feedforward neural networks. In: Proceedings of the International Conference on Artificial Intelligence and Statistics (AISTATS 2010). Society for Artificial Intelligence and Statistics (2010)
6. Goodfellow, I.J., Shlens, J., Szegedy, C.: Explaining and harnessing adversarial examples. arXiv preprint arXiv:1412.6572 (2014)
7. Kingma, D.P., Ba, J.: Adam: a method for stochastic optimization. arXiv preprint arXiv:1412.6980 (2014)
8. Kurakin, A., Goodfellow, I.J., Bengio, S.: Adversarial examples in the physical world. arXiv preprint arXiv:1607.02533 (2016)
9. LeCun, Y., Cortes, C.: MNIST handwritten digit database (1998). http://yann.lecun.com/exdb/mnist/
10. Papernot, N., McDaniel, P.D., Wu, X., Jha, S., Swami, A.: Distillation as a defense to adversarial perturbations against deep neural networks. In: IEEE Symposium on Security and Privacy, pp. 582–597. IEEE Computer Society (2016)
11. Srivastava, N., Hinton, G.E., Krizhevsky, A., Sutskever, I., Salakhutdinov, R.: Dropout: a simple way to prevent neural networks from overfitting. J. Mach. Learn. Res. **15**(1), 1929–1958 (2014). http://www.cs.toronto.edu/~rsalakhu/papers/srivastava14a.pdf
12. Szegedy, C., et al.: Intriguing properties of neural networks. In: International Conference on Learning Representation (2014)
13. Wang, S., et al.: Defensive dropout for hardening deep neural networks under adversarial attacks. arXiv preprint arXiv:1809.05165 (2018)

Self-training for Cell Segmentation and Counting

J. Luo[1]([✉]), S. Oore[1,3], P. Hollensen[2], A. Fine[2], and T. Trappenberg[1]

[1] Faculty of Computer Science, Dalhousie University, Halifax, Canada
junliangluo15@gmail.com, osageev@gmail.com, trappenberg@gmail.com
[2] Alentic Microscience Inc, Halifax, Canada
{phollensen,afine}@alentic.com
[3] Vector Institute for Artificial Intelligence, Toronto, Canada

Abstract. Learning semantic segmentation and object counting often need a large amount of training data while manual labeling is expensive. The goal of this paper is to train such networks on a small set of annotations. We propose an Expectation Maximization(EM)-like self-training method that first trains a model on a small amount of labeled data and adds additional unlabeled data with the model's own predictions as labels. We find that the methods of thresholding used to generate pseudo-labels are critical and that only one of the methods proposed here can slightly improve the model's performance on semantic segmentation. However, we also show that the induced value changes in the prediction map helped to isolate cells that we use in a new counting algorithm.

1 Introduction

Deep neural networks typically need to be trained on large amounts of labeled data such as pixel-wise annotations for semantic segmentation models and position annotations for counting objects. Human annotations are time consuming and sometimes inaccurate. Here we study a semi-supervised learning method that uses only a small set of labeled data together with unlabeled data. The basic idea is to use an EM-like approach in which an initial model from a small labeled data set will be used to produce labels for further unlabeled data (E-step), which in turn will be used to optimize the model (M-step). Recent works have shown good performance with few and/or coarsely labeled data for classification and semantic segmentation [4,7]. Inspired by these previous works, we investigate an EM-like self-training method for segmenting microscopic images with the purpose of cell counting. We apply our methods to a dataset of microscopic human blood images from Alentic Microscience Inc.

Our semantic segmentation network is based on the U-Net architecture [8]. We initially train the model on a small amount of annotated images and treat the rest of the images in training set as unlabeled data. During the self-training iteration, the model estimates the labels for the next batch of unlabeled images. We then employ a thresholding mechanism which attempts to incorporate the

© Springer Nature Switzerland AG 2019
M.-J. Meurs and F. Rudzicz (Eds.): Canadian AI 2019, LNAI 11489, pp. 406–412, 2019.
https://doi.org/10.1007/978-3-030-18305-9_37

Table 1. Confidence threshold methods and mathematical equations

Confidence threshold	Equation
1. Mean probability	$L(p) = \overline{pb_1, ..., pb_T}$
2. Mean binary threshold	$L(p) = \begin{cases} 1, & \text{if } \overline{pb_1, ..., pb_T} > \tau \\ 0, & \text{otherwise} \end{cases}$
3. Mean probability threshold	$L(p) = \begin{cases} 1, & \text{if } \overline{pb_1, ..., pb_T} > \tau \\ \overline{pb_1, ..., pb_T}, & \text{otherwise} \end{cases}$
4. Nearby pixels threshold	$L(p) = \begin{cases} 1, & \text{if any } t \in T \;\; \sum pb_t(pt_{3\times3}) > \tau \\ \overline{pb_1, ..., pb_T}, & \text{otherwise} \end{cases}$

model's estimate and confidence in order to obtain a self-determined label for each image, which we refer to as a pseudo-label to distinguish it from the ground truth label. This batch of unlabeled images is added into the training set along with their pseudo-labels. Images that have already been added into the training set with pseudo-labels are reassigned new labels based on new estimates. This stage is repeated for multiple iterations.

We tried four different thresholding mechanisms to map estimates to pseudo-labels, each based on a different methods of estimating confidence in classification (Sect. 2). Our empirical results suggest that only one of those mechanisms (sampled *mean probability threshold*) leads to a small improvement when self-training the model on the pseudo-labels. While the other methods reduce performance in terms of pixel-wise overlap measure (IoU), we noticed an secondary effect: those methods that lowered the IoU accuracy lead to a rapid intensity change in the per-pixel prediction map at locations associated with edges, and that sharper visual gradient in fact helps *count* the cells. This is noteworthy, because there are many cases in which counting cells is important.

2   Semantic Segmentation

We investigate how to convert the probability maps provided by the last sigmoid layer of a neural network into labels. We would like the network to self-train on those pixels on which the network is sufficiently confident in its predictions. For this reason, we need a confidence measure for thresholding. We use Monte Carlo (MC) Dropout [5] to calculate pixel-wise image uncertainty estimation by computing the mean and variance of T runs. Based on this map we define several methods to generate pseudo labels (see Table 1).

We calculate the mean of T Monte Carlo samples of the sigmoid layer output probabilities $(pb_1, ..., pb_T)$. In method 1 (*Mean probability*) this mean is directly used as the label (L) of the pixel (p). In methods 2 and 3, if the mean is higher than a constant threshold τ, then the pseudo-label $L(p)$ is set to 1, otherwise it is set to 0 (for *mean binary threshold*); or it is left as the mean probability value (for *mean probability threshold*). For *nearby pixels threshold*, the label $L(p)$ depends on the sum of the sigmoid probabilities of the 3×3 pixel patch $(pt_{3\times3})$, that contains the pixel under consideration. For each of the T Monte Carlo samples, a threshold τ is compared against the sum. If the sum is greater than τ, $L(p)$ is set to 1. If the sum is smaller for all T samples, it is set to the mean probability value.

The loss function used is a weighted pixel-wise cross entropy (CE), which utilizes the variance of T predictions to help the classifier assess the trustworthiness of the label of a pixel. Let i, j denote the coordinates of an image pixel. Further, $\hat{y}_{p_{ij}}$ is the predicted class of the pixel and $y_{p_{ij}}$ is the true label of the pixel. For each pixel, $\widetilde{\sigma^2_{ij}}$ denotes the variance of its T predicted probabilities min-max scaled to interval $[1,100]$ over all variances of pixels in this image. The weight is the reciprocal of the $\widetilde{\sigma^2_{ij}}$. Therefore, pixels with high variance predictions are weighted less during the training, reducing the effect of learning from pixels with uncertain predictions.

$$Loss = \sum_{i,j} \frac{1}{\widetilde{\sigma^2_{ij}}} * CE(y_{p_{ij}}, \hat{y}_{p_{ij}})$$

The self-training process can be interpreted as a Classification Expectation-Maximization (CEM) process defined in [2] as the log likelihood shows below.

$$logL = \sum_{c=1}^{C} \Big\{ \sum_{n \in I_n} \log\{\pi_c \cdot f_c(x_n, \theta_c)\} + \sum_{m=N+1}^{N+M} t_{c,m} \log\{\pi_c \cdot f_c(x_m, \theta_c)\} \Big\}$$

Suppose we have N labeled data (with indices I_n) and M unlabeled data, belonging to C classes. Here f_c are the parametric densities with unknown parameters $\theta_0, ..., \theta_C$. The unknown data are drawn from a mixture of densities with C components in unknown proportions $\pi_1, ..., \pi_C$. The $t_{c,m} = (t_{1,m}...t_{c,m})$ is the class indicator (t_c) vector for a unlabeled data m. The t_c is fixed for labeled data. In our model, the E step is an initial supervised stage that guides the model to create an expectation $E(t_c|x)$. The prediction and re-prediction for generating and updating pseudo-labels in our model corresponds to the classification (C) and M steps: assign new t_c and estimate new π_c and θ_c.

Our experiments are done with a human blood sample dataset, taken with a lensless microscope [1]. The dataset contains 504 training (assumed most unlabeled) and 112 testing (labeled) images, where each (128×128 RGB) image contains 150+ cells. The self-training approach relied on a sufficient preliminary accuracy not to amplify errors. We thus investigated finding sufficient data for initial supervised training. Figure 1 summarizes the result. We found that training on *one* 128×128 labeled image gives sufficient accuracy for the supervised stage. For unsupervised stage, we use $T = 100$ forward passes and a dropout rate of 50% for MC dropout inference on unlabeled data. For each iteration, the model is trained on the current training set for 100 epochs and predicts the label map of the next *one* unlabeled image. The results are presented in Table 2.

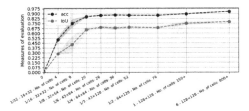

Fig. 1. Average testing accuracies and IoUs over 5 runs (300 iterations) for training the network with different initial training sizes in configurations given by: [percent of a full image, pixels, number of objects]

Table 2. Summary of mean and variance of segmentation accuracies (acc), Intersection over Union (IoU), positive predictive value (PPV), true positive rate (TP), true negative rate (TN), prediction rate (P) over 5 runs. Left: Performance measures in evaluation after training on labeled data (before self-training). Right: After self-training

Threshold	iters	Measures					
		acc	IoU	PPV	TP	TN	P
Mean probability	15	0.889→0.720 (±0.010)(±0.017)	0.734→0.349 (±0.020)(±0.029)	0.886→0.971 (±0.008)(±0.006)	0.814→0.269 (±0.021)(±0.045)	0.936→0.995 (±0.004)(±0.001)	0.349→0.105 (±0.007)(±0.017)
Mean binary threshold (τ : 0.5)	28	0.889→0.851 (±0.013)(±0.020)	0.735→0.621 (±0.029)(±0.052)	0.880→0.970 (±0.011)(±0.005)	0.818→0.628 (±0.028)(±0.056)	0.932→0.988 (±0.006)(±0.003)	0.353→0.246 (±0.009)(±0.023)
Mean probability threshold (τ : 0.5)	28	0.885→0.910 (±0.008)(±0.014)	0.723→0.759 (±0.021)(±0.091)	0.889→0.885 (±0.007)(±0.024)	0.796→0.878 (±0.022)(±0.056)	0.939→0.929 (±0.005)(±0.019)	0.334→0.377 (±0.010)(±0.032)
Mean probability threshold (τ : 0.55)	28	0.887→0.902 (±0.015)(±0.014)	0.728→0.764 (±0.034)(±0.030)	0.885→0.880 (±0.014)(±0.021)	0.807→0.858 (±0.035)(±0.020)	0.936→0.928 (±0.008)(±0.013)	0.346→0.370 (±0.013)(±0.008)
Nearby pixels threshold (τ : 8)	15	0.889→0.745 (±0.007)(±0.023)	0.732→0.389 (±0.015)(±0.045)	0.889→0.974 (±0.023)(±0.004)	0.809→0.338 (±0.026)(±0.063)	0.938→0.994 (±0.016)(±0.002)	0.346→0.132 (±0.018)(±0.025)

In our tests with *mean probabilities* and *mean binary threshold* as pseudo-labels, the accuracies decrease and the IoU and TP become lower. Reaching a slight improvement of accuracy and IoU can be achieved by using *mean probability threshold* pseudo-labels. Pixel accuracy increases 1–2% and IoU increases 3–4%, reaching a 5–7% TP improvement. This non-binary threshold scheme guides the network to emphasize confident high-value pixels, leading the network to make more TP predictions. The network re-adjusts itself and results in a 3–4% more positive predictions with only slightly less TN (<1%). For *nearby pixels threshold* pseudo-label, the accuracy and IoU decrease and result in a very low TP of 33.8%. Notably, the speed of decrease in the values of predictions varies. In particular, the predicted value of pixels near cell walls change rapidly as Fig. 2 shows, a fact that we use for our counting algorithm.

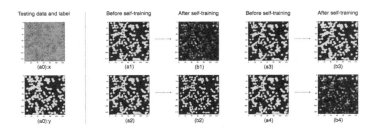

Fig. 2. Inference results: (a0)x, (a0)y: Input data and label. (a1)–(a4): After training on labeled data (before self-training). (b1)–(b4): After self-training for iterations in Table 2 using mean probability; mean binary threshold (0.5); mean probability threshold (0.5); nearby pixels threshold (8) pseudo-labels

3   Counting

From the self-trained results, we noticed that the cell boundaries become somewhat clearer after applying self-training with *nearby pixels threshold*. Therefore, we propose a counting method based on these pixel value changes to test this hypothesis. We first apply a Laplacian filter: a 2-D isotropic measure of the 2nd spatial derivative of an image [6], on the predicted map (M) of the self-trained model to get a map (∇M) indicating rapid changes. The values of the ∇M are normalized to remove the mean and scaled to unit variance. We choose two thresholds $c1$ and $c2$. Let $map1$ be a binary map with 1's only at those pixels where ∇M had a value less than $c1$. $map1$ corresponds to the regions where the change in M was small. Similarly, let $map2$ be a binary map with 1's only at those pixels where ∇M has a value larger than $c2$. We subtract $map2$ from a $map1$ to get our pixels of interest ($diff$): isolated pixel components corresponding to individual cells. Since $map1$ and $map2$ do not overlap, $map2$ is dilated by a 2×2 kernel before subtraction. The resulting image ($diff$) is then cleaned by being scanned by a $n \times n$ window, replacing small non-connected pixel components with a single dot denoting a cell.

We conduct experiments on our cells dataset and a synthetic 2D-Gaussian dataset (More challenging since overlap. Description and result details can be found at: https://github.com/JunLLuo/Synthetic-2d-GD). In our test with cell images, We validate on 50 testing images (128×128) with manual dot annotations where each dot corresponds to one cell. Since the predicted dots are not required to be at the centre of the cells, we evaluate the dot prediction by doing an injective function. Each dot in the prediction map is mapped to at most one dot in ground truth with a restriction of a limited distance (≤ 7 Euclidean distance of pixels). The dots that are not mapped are labeled as FP, while in ground truth are FN. The results are presented in Table 3. A higher $c2$ increases the number of P, leading to a higher TPR but lower PPV. A lower $c1$ does the opposite. We also use Circle Hough Transform (CHT) algorithm [3], to detect

Table 3. Evaluation of counting averaged over 5 runs on 50 testing images of cells and (same injective evaluation) on 10 testing images of synthetic 2D-Gaussian. The numbers in columns: P (positive), GT (ground truth), TP (true positive), FP (false positive), FN (false negative) correspond to the number of cells/Gaussians. In columns: TPR (true positive rate), PPV, rates are calculated from the whole 50/10 image samples

Dataset	Method	c1	c2	Evaluation measure						
				P	GT	TP	FP	FN	TPR	PPV
Cells	Our method (n = 5)	−1.00	0.45	140.9	144.7	131.8	9.0	12.8	0.911	0.936
				(±9.6)	(±10.2)	(±9.1)	(±3.3)	(±4.5)		
		−1.00	0.85	147.3	144.7	134.8	12.5	9.9	0.932	0.915
				(±9.9)	(±10.2)	(±9.6)	(±3.6)	(±3.5)		
		−1.45	0.45	130.7	144.7	125.2	5.4	19.4	0.866	0.958
				(±8.5)	(±10.2)	(±8.3)	(±2.4)	(±5.7)		
	Circle Hough Transform: radius: (4, 5)			132.0	144.7	122.8	9.3	21.8	0.849	0.930
				(±9.6)	(±10.2)	(±8.5)	(±3.4)	(±5.7)		
Synthetic Gaussians	Our method (iter = 8 n = 6)	−2.00	1.00	196.6	200.0	180.3	16.3	19.7	0.902	0.917
				(±9.3)	(±0.0)	(±5.0)	(±5.7)	(±5.0)		
		−2.00	2.00	209.6	200.0	183.0	26.6	17.0	0.915	0.873
				(±11.2)	(±0.0)	(±5.0)	(±8.1)	(±5.0)		

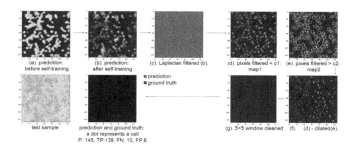

Fig. 3. A test sample for the counting process on cells images: (a) Prediction after training on 1 labeled image. (b) Prediction after 7 self-training iterations using nearby pixels pseudo labels. (c) Laplacian filtered map (∇M) (d) *map1*. (e) *map2*. (f) *diff*. (g) Dots prediction

circular objects (cells) in the segmentation map for comparison with our algorithm. Our algorithm outperforms CHT by a 7% greater TPR with c1 of −1 and c2 of 0.45. Figure 3 shows an example of the counting process on a cells image.

4   Conclusion and Future Work

While we originally aimed to improve segmentation accuracies for datasets with small numbers of examples using self-training, we found that results critically depend on the thresholding method, and we only found a small improvement in that regard for one of the methods proposed here. However, we found that visually modifying the data based on the pseudo-labels helped with the identification of cell boundaries. This in turn helped with the identification of cells and, ultimately, with the corresponding counting. This is a promising new direction for using model predictions in combinations with uncertainty measure.

References

1. Alentic Microscience Inc. (2019). http://www.alenticmicroscience.com/
2. Amini, M.R., Gallinari, P.: Semi-supervised logistic regression. In: ECAI, pp. 390–394 (2002)
3. Atherton, T.J., Kerbyson, D.J.: Size invariant circle detection. Image Vis. Comput. **17**(11), 795–803 (1999)
4. Bank, D., Greenfeld, D., Hyams, G.: Improved training for self training by confidence assessments. In: Arai, K., Kapoor, S., Bhatia, R. (eds.) SAI 2018. AISC, vol. 858, pp. 163–173. Springer, Cham (2019). https://doi.org/10.1007/978-3-030-01174-1_13
5. Gal, Y., Ghahramani, Z.: Dropout as a bayesian approximation: representing model uncertainty in deep learning. In: International Conference on Machine Learning, pp. 1050–1059 (2016)
6. Juneja, M., Sandhu, P.S.: Performance evaluation of edge detection techniques for images in spatial domain. Int. J. Comput. Theory Eng. **1**(5), 614 (2009)
7. Papandreou, G., Chen, L.C., Murphy, K.P., Yuille, A.L.: Weakly-and semi-supervised learning of a deep convolutional network for semantic image segmentation. In: Proceedings of the IEEE International Conference on Computer Vision, pp. 1742–1750 (2015)
8. Ronneberger, O., Fischer, P., Brox, T.: U-net: convolutional networks for biomedical image segmentation. In: Navab, N., Hornegger, J., Wells, W.M., Frangi, A.F. (eds.) MICCAI 2015. LNCS, vol. 9351, pp. 234–241. Springer, Cham (2015). https://doi.org/10.1007/978-3-319-24574-4_28

Enhancing Unsupervised Pretraining with External Knowledge for Natural Language Inference

Xiaoyu Yang[1]([✉]), Xiaodan Zhu[1], Huasha Zhao[2], Qiong Zhang[2], and Yufei Feng[1]

[1] Queen's University, Kingston, Canada
{17xy44,xiaodan.zhu,feng.yufei}@queensu.ca
[2] Alibaba Group, San Mateo, CA, USA
{huasha.zhao,qz.zhang}@alibaba-inc.com

Abstract. Unsupervised pretraining such as BERT (Bidirectional Encoder Representations from Transformers) [2] represents the most recent advance on learning representation for natural language, which has helped achieve leading performance on many natural language processing problems. Although BERT can leverage large corpora, we assume it cannot learn all needed semantics and knowledge for natural language inference (NLI). In this paper, we leverage human-authorized external knowledge to further improve BERT, and our results show that BERT, the current state-of-the-art pretraining framework, can benefit from external knowledge.

Keywords: BERT · Natural Language Inference · External knowledge

1 Introduction

Natural language inference (NLI) is a research problem that has attracted extensive attention recently. Specifically, NLI is concerned with determining whether a natural-language hypothesis h can be inferred from a premise p. Neural networks are very effective in modelling NLI and they often focus on end-to-end training by assuming that all inference knowledge is learnable from the provided training data. We believe complicated NLP problems such as NLI will benefit from leveraging knowledge available outside the training data.

This paper explores and merges two typical types of approaches: those explicitly incorporating human-authorized knowledge, and those based on unsupervised pretraining [2,5,6]. To the best of our knowledge, this is the first attempt to integrate these two approaches in NLI. We perform a study with BERT [2], which represents the most recent advance on learning representation for natural language. We incorporate external knowledge into both the overall inference of the final layer of network and also the self-attention components of BERT. Our models are evaluated on two NLI datasets, i.e., Multi-Genre Natural Language Inference dataset [9] (MultiNLI) and the Glockner dataset [3], and the results show that the proposed models achieve better results.

© Springer Nature Switzerland AG 2019
M.-J. Meurs and F. Rudzicz (Eds.): Canadian AI 2019, LNAI 11489, pp. 413–419, 2019.
https://doi.org/10.1007/978-3-030-18305-9_38

2   Related Work

Pretrained Models in NLI and Other Tasks. In this paper we view unsupervised pretraining as a way to learn and incorporate external semantics and knowledge. Unsupervised pretraining has recently shown to be very effective in a wide range of NLP tasks. Generative Pretrained Transformer (GPT) [6] and BERT [2] are two typical finetune-based pretrained models which use pretraining-then-finetuning architectures on downstream tasks. In this paper we further show that human-created knowledge can further complement the pretraining models in achieving better NLI prediction.

Evaluation of Natural Language Inference Models. Previous research has paid attention to judging if existing NLI systems have learned NLI-related knowledge. For example, [8] uses a *swapping* evaluation method, by switching a premise and its hypothesis to test the robustness of a model. Efforts have also been made to propose new test dataset, such as [3], in which premises are taken from the SNLI training set and then for each premise, hypotheses of different inference categories are generated by replacing a single word in premise sentence.

3   Models

This section describes the models we propose to incorporate external knowledge. The overall framework of our approaches is depicted in Fig. 1.

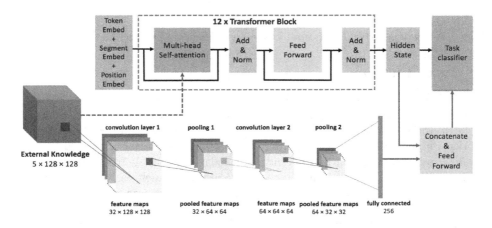

Fig. 1. Overview of BERT enhanced with external knowledge models.

As shown in Fig. 1, the upper part is the main component of a pretraining architecture (here BERT). We incorporate external knowledge into BERT with two basic approaches. In this study we use the official pretrained models and code[1].

---

[1] https://github.com/google-research/bert.

3.1   External Knowledge

To carry out our study on enriching the state-of-the-art BERT pretrained models, we follow [1] to use the existing lexical knowledge, which is accumulated from WordNet [4]. Given a pair of words in WordNet, we use five basic semantic relations: **hypernymy, hyponymy, co-hyponyms, antonymy** and **synonymy**. The relation of hypernymy and hyponymy might imply entailment, while antonymy and co-hyponyms may be related to contradiction. Given a pair of sentences with maximum length of n, we can extract from WordNet a feature matrix of $n \times n \times 5$, and we process it with the method proposed in [1].

3.2   Incorporating External Knowledge in Self-Attention

External knowledge may help align inference-related concepts between a premise and hypothesis, so we combine external knowledge with the self-attention weight or value for each head of multi-head attention in BERT. BERT is composed of stacked Transformer blocks of identical shape. The self-attention layer in Transformer block contains multiple independent parallel heads.

For input matrix $H = [\boldsymbol{h_1}, \boldsymbol{h_2}, \ldots, \boldsymbol{h_t}, \ldots, \boldsymbol{h_n}]$, $\boldsymbol{h_t}$ denotes the input vector at position t, and n represents the maximum length of input sequences. First, the input is transformed into Key(K), Query(Q) and Value(V), respectively:

$$(Q^i)^T = [\boldsymbol{q_1}, \boldsymbol{q_2}, ..., \boldsymbol{q_t}, ..., \boldsymbol{q_n}] = W_q^i[\boldsymbol{h_1}, \boldsymbol{h_2}, ..., \boldsymbol{h_t}, ..., \boldsymbol{h_n}] \qquad (1)$$

$$(K^i)^T = [\boldsymbol{k_1}, \boldsymbol{k_2}, ..., \boldsymbol{k_t}, ..., \boldsymbol{k_n}] = W_k^i[\boldsymbol{h_1}, \boldsymbol{h_2}, ..., \boldsymbol{h_t}, ..., \boldsymbol{h_n}] \qquad (2)$$

$$(V^i)^T = [\boldsymbol{v_1}, \boldsymbol{v_2}, ..., \boldsymbol{v_t}, ..., \boldsymbol{v_n}] = W_v^i[\boldsymbol{h_1}, \boldsymbol{h_2}, ..., \boldsymbol{h_t}, ..., \boldsymbol{h_n}] \qquad (3)$$

Then we compute H'^i, the output of Transformer block for head i as follows. $\boldsymbol{h_t'}$ represents the output for the tth position in output sequence. The output H'^i is calculated as a weighted sum of V^i according to the attention weight matrix A^i:

$$A^i = \{\alpha_{jm}\} = softmax(\frac{Q^i K^{iT}}{\sqrt{d_k}}) \qquad (4)$$

$$H'^i = [\boldsymbol{h_1'}, \boldsymbol{h_2'}, ..., \boldsymbol{h_t'}, ..., \boldsymbol{h_n'}]^T = A^i V^i \qquad (5)$$

We use $\sqrt{d_k}$ as a scaling factor when calculating the attention weight matrix A^i same as [7], and α_{jm} is related to $\boldsymbol{q_j}$ and $\boldsymbol{k_m}$. The final block output H' is the concatenation of all the heads along feature dimension as follows:

$$H' = [H'^1 \oplus H'^2 \oplus ... \oplus H'^n] \qquad (6)$$

Knowledge-Enhanced Self-Attention Weight. Given a token sequence of $w_1, w_2, ...w_n$, to clarify our mathematical notations, we use H and H' to represent the input and output token sequence, respectively. In BERT, the premise and hypothesis are concatenated into one sequence (possibly zero-padded) of maximum length n. For an arbitrary word pair (w_i, w_j), E is an $n \times n \times 5$ tensor, among which the vector e_{ij} at position (i, j) represents the knowledge feature of (w_i, w_j). To merge external knowledge into self-attention, for a sequence of n tokens, we calculate the $n \times n$ self-attention weight matrix A'^i and the output of Transformer block H'^i in i^{th} head as follows:

$$A'^i = f([A^i \oplus E]) \tag{7}$$
$$H'^i = [h'_1, h'_2, ..., h'_t, ..., h'_n]^T = A'^i V^i \tag{8}$$

Here A^i is the original dot product weight matrix in the transformer, $\oplus$ denotes concatenation over feature depth, and the trainable function f maps each feature vector at position (j, m) to a scalar attention weight, where $j, m = 1, \ldots, n$.

Knowledge-Enhanced Value Vectors. We also add the external knowledge to value vector V as follows. The attention weight matrix $A^i = \{\alpha_{jm}\}$ for i^{th} head remains the same as original model in Eq. (4), and for the ith head, we calculate the output matrix H'^i as follows:

$$h'_j = \sum_{m=0}^{n} \alpha_{jm}[v_m \oplus e_{jm}] \tag{9}$$
$$H'^i = [h'_1, h'_2, \ldots, h'_j, \ldots, h'_n]^T \tag{10}$$

Here j denotes the position of output token vector, and v_m is the value vector at position m. The external knowledge feature e_{jm} at position (j, m) in E is concatenated with the value vector v_m before the weighted summation.

3.3   Encoding External Knowledge into the Global Inference Decision

In the second approach, we incorporate external knowledge to help the global inference aggregation. Specifically, we encode external knowledge with a convolutional neural network (CNN) and then combine the resulting representation into the global inference stage. The CNN is composed of two convolutional layers, and each layer is followed by a pooling layer as shown in Fig. 1.

In contrast to the pre-trained transformer network, the convolutional parameters are randomly initialized and trained during downstream task. The last dimension of external knowledge matrix E is treated as input channels of CNN, and CNN(E) is a fixed-length vector extracted from the original external knowledge matrix with CNN. Take h' as the output hidden state vector of BERT and h'' is the new vector used for task classifier.

$$h'' = f([h' \oplus CNN(E)]) \tag{11}$$

Here f represents the feed forward layer in Fig. 1 to merge CNN(E) into h'.

4 Experiments

Datasets. We run our experiments on two NLI datasets: the Multi-Genre Natural Language Inference (MultiNLI) dataset [9] (the matched test) and the Glockner dataset [3]. For the MultiNLI dataset, the models are trained and tested on the official split of training, development, and test sets. Since the Glockner dataset only provides a test set which emphasizing on lexical relations, same as the previous work [3], our models are trained on SNLI and evaluated on the Glockner test set.

Processing External Knowledge. As discussed above, we extract five types of relational features between word pairs from WordNet. Table 1 shows the details of the external knowledge used in the experiments. For each type of relation (a row in the table), we list how many pairs we found in WordNet (second column). For example, 753,086 is the total number of hypernym pairs we obtain from WordNet. Since not all of these pairs will be used in our experiments, we further list how many pairs are used in the MultiNLI and SNLI/Glockner experiments. For example, among the above 753,086 hypernym pairs, 168,006 pairs have both words appearing in the MultiNLI vocabulary.

Table 1. Statistics for different types of relations.

Feature	WordNet #Pairs	MultiNLI #Pairs	SNLI & Glockner #Pairs
Hypernymy	753,086	168,006	124,511
Hyponymy	753,086	168,006	124,511
Co-hyponyms	3,674,700	378,204	310,384
Antonymy	6,617	1,320	1,028
Synonymy	237,937	39,219	31,386

Results. During the fine-tuning procedure, we use the officially released pretrained model and parameters of $BERT_{BASE}$ and initialize the parameters of the classification layer; then all parameters are fine-tuned with the datasets of our downstream tasks. As for the CNN, we used two convolutional layers followed by pooling layers. The filters for the first convolutional layer have kernel sizes of 25×25 with 32 channels, and the stride is 1. This is followed by a stride 2 max pooling layer. The filter for the second convolutional layer has size 5×5, with 64 channels and a stride of 1. The last pooling layer is applied with a stride of 2.

Table 2 shows the results of our experiments. We report the performance of our models on the MultiNLI (the matched set) and Glockner datasets. The table shows that our model that adds external knowledge into BERT at the global inference layer outperforms BERT on both the MultiNLI and Glockner datasets with the accuracy of 84.72% and 95.20%, respectively. Particularly, on the Glockner dataset, we observed the accuracy increases from 94.74% to 95.20%.

Table 2. Accuracies of models on MultiNLI (the matched set) and Glockner test set. Same as in the previous work, the model tested on Glockner is trained on the SNLI dataset.

Model	MultiNLI accuracy (%)	Glockner accuracy (%)
BERT	84.6	94.74
BERT + External (Global Infer.)	**84.72**	**95.20**
BERT + External (Attn. Weights)	82.70	94.40
BERT + External (Value Vectors)	81.82	93.32

To further analyze the impact of external knowledge on BERT, instead of using the external knowledge feature vectors, we also concatenate the hidden state with vectors initialized from Gaussian distribution. Random vectors are the same length as our external knowledge feature vectors, and their performances are inferior to our proposed model.

We also found that adding external knowledge to self-attention components in Transformer blocks did not bring further improvement, as shown in Table 2. We further combined these models with the above model that uses external knowledge in global inference but did not observe additional improvement. These direct modifications made inside Transformer blocks may potentially perturb the BERT networks and what BERT has learned in pre-training. Further studies on exploring external knowledge in pretrained models, however, is still desirable.

5   Conclusions

BERT has recently achieved considerable success on learning representation for natural language, improving the state-of-the-art performance on a wide range of NLP tasks. Modelling natural language inference is notoriously challenging but a basic problem towards true natural language understanding. In this paper we propose models that explore existing external knowledge, which complements what BERT learned from large pretraining corpora, and achieve better results on two NLI datasets. The experiments show that the improvement is achieved by encoding external knowledge and incorporating it into the global inference layer. Adding external knowledge into the self-attention components of BERT does not bring further improvement, while more studies on exploring external knowledge in pretrained models are further desirable as future work.

References

1. Chen, Q., Zhu, X., Ling, Z.H., Inkpen, D., Wei, S.: Neural natural language inference models enhanced with external knowledge. In: ACL (2018)
2. Devlin, J., Chang, M.W., Lee, K., Toutanova, K.: BERT: pre-training of deep bidirectional transformers for language understanding. arXiv:1810.04805 (2018)

3. Glockner, M., Shwartz, V., Goldberg, Y.: Breaking NLI systems with sentences that require simple lexical inferences. arXiv preprint arXiv:1805.02266 (2018)
4. Miller, G.A.: WordNet: a lexical database for english. Commun. ACM **38**(11), 39–41 (1995)
5. Peters, M.E., et al.: Deep contextualized word representations. arXiv:1802.05365 (2018)
6. Radford, A., Narasimhan, K., Salimans, T., Sutskever, I.: Improving language understanding by generative pre-training. https://s3-us-west-2.amazonaws.com/openai-assets/research-covers/language-unsupervised/language_understanding_paper.pdf (2018)
7. Vaswani, A., et al.: Attention is all you need. In: NIPS (2017)
8. Wang, H., Sun, D., Xing, E.P.: What if we simply swap the two text fragments? A straightforward yet effective way to test the robustness of methods to confounding signals in nature language inference tasks. arXiv:1809.02719 (2018)
9. Williams, A., Nangia, N., Bowman, S.: A broad-coverage challenge corpus for sentence understanding through inference. In: ACL (2018)

An Experiment for Background Subtraction in a Dynamic Scene

Ting-Yuan Lin[1], Jeng-Sheng Yeh[2], Fu-Che Wu[3(✉)], Yung-Yu Chuang[1],
and Andrew Dellinger[4]

[1] National Taiwan University, Taipei, Taiwan
ro6842@gmail.com, cyy@csie.ntu.edu.tw
[2] Ming Chuan University, Taoyuan, Taiwan
jsyeh@mail.mcu.edu.tw
[3] Providence University, Taichung, Taiwan
fcwu@pu.edu.tw
[4] Elon University, Elon, USA
adellinger@elon.edu

Abstract. This paper aims to analyze a background subtraction algorithm. Different from tradition methods, we feed the trained network with the target and background images. The paper focuses on how to get background images without using the temporal median filter. We use Gaussian mixture models to produce background images. In this way, the accuracy of background images increases. We also study the difference between grayscale and RGB images, and adding the foreground masks from the convolutional Neural Networks to the Gaussian mixture models. Experiments lead on the 2014 ChangeDetection.net dataset show that our proposed method outperforms several state-of-the-art methods, including IUTIS-5, PAWCS, SuBSENSE and so on.

Keywords: Background subtraction · Convolutional neural networks ·
Gaussian mixture models · Temporal median filter

1 Introduction

Detection of foreground objects has been developed in many ways for decades. Background elimination is one of the well-known methods, which requires the camera to be stationary, with foreground objects detected by the difference between the target and background frames. Therefore, background frames, also known as background modules, are of particular importance. The easiest way to accomplish background elimination is to filter through the median of the timings, but the background that this method produces is not reliable in some cases, such as dynamic backgrounds, and it limits the ratio of background to more than half the time axis. Therefore, we propose to use a Gaussian mixture module to generate a background module, so that the background of the reference will be more reliable.

© Springer Nature Switzerland AG 2019
M.-J. Meurs and F. Rudzicz (Eds.): Canadian AI 2019, LNAI 11489, pp. 420–425, 2019.
https://doi.org/10.1007/978-3-030-18305-9_39

Inspired by Braham and Droogenbroeck [1], we throw the computation of background elimination to the convolutional neural network (Fig. 1). In contrast to traditional background cancellation algorithms, deep learning makes foreground object detection more accurate, as we showed in the comparison to other methods in the dataset of the 2014 ChangeDetection.net repository [2].

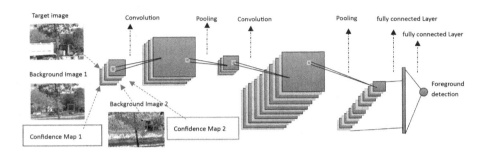

Fig. 1. Convolution neural network architecture and training process: The inputs of the network are the target image, two background images, and two confidence maps. The architecture is similar to LeNet-5 [3], which consists of two rounds of convolution and pooling layers and the last two fully connected layers to predict the foreground object.

In the paper by Marc Braham and Marc Van Droogenbroeck, only two channels, target and background images, are used for training. However, a neural network is very flexible and can use multiple channels as its input. Therefore, we want to provide more information for the training. Thus, instead of only one background image, two background images with confidence maps are input and are expected to produce more accurate training results.

2   Previous Work

Although convolutional neural networks have received much attention in recent years, neural networks have been developed for a long time. Gil-Jimenez, Maldonado-Bascon, et al. [4] proposed method tries to classify each zone of the image into one of the following groups: Static background, Noisy background, or Impulsive background. Classification is according to the zone's behavior along the last processed frames.

Later papers are mostly about different forms of treatment on the input to train the neural network to determine the attributes of pixels. Wang, Bao, Zhang [5] convert each pixel in HSV color space as the input layer, and the output layer can determine whether this pixel is foreground or background. The reason for using HSV is that HSV is closer to human-perceived color than RGB. In this case, their work is a type of neural network for perception. However, unlike other neural networks, it combines a probabilistic neural network (PNN), and a winner take all (WTA) network. So their proposed work has a more powerful discernment ability that can efficiently detect sudden changes in the background.

With the improvement of computing power, the architecture of a neural network has also evolved. Do and Huang [6] propose a novel motion detection approach based on a RBF through artificial neural networks. Their approach involves two important modules: a multi-background generation module and a moving object detection module. The former module effectively generates a flexible multi-background model which can effectively express the dynamic range of each pixel within the background. After a high-quality multi-background model is generated, the latter module detects the pixels of moving objects to form the moving objects mask.

Also at that time, artificial intelligence was very hot so Shobha and Satish Kumar [7] presented an Artificial Neural Network (ANN) approach of using the texture properties to model the background and detect the moving objects in video sequence. Ramirez-Quintana and Chacon-Murguia [8] proposed the Retinotopic SOM (RESOM)- a neural network based on Self-Organizing Retinotopic Maps, which was applied in dynamic segmentation using background modeling. Athilingam, Kumar, and Kavitha [9] propose an approach based on neuronal mapping for segmentation of targets with hybrid background subtraction and adaptive mean shift filtering. With this method, scenes containing moving backgrounds and robust illumination changes can be considered effectively. De Gregorio and Giordano [10] proposed a pixel?based Weightless Neural Network (WNN) method to face the problem of change detection in the field of view of a camera. In order to feed the discriminators with the right input, they create one discriminator for each pixel of the video frame. Guo, Qi [11] and Xu et al. [12] both adopt a Restricted Boltzmann Machine (RBM) to adaptively generate the background image.

Also, we are inspired by Stauffer and Grimson's [13] work by using the Gaussian Mixture module to generate the background. The experimental results show that this indeed enhances the results. The details will be mentioned later in the paper.

3   Experiments

The experiments run on a workstation with Intel Xeon E5-2650 v3 CPU (2.30 GHz), Nvidia Tesla K80 GPU, and 128 GB memory. The primary purpose of this study is to improve the algorithm proposed by Braham and Droogenbroeck and to modify the background module to achieve better results. The experimental data also is the 2014 ChangeDetection.net database, which includes many special cases such as night landscapes, thermal images, dynamic backgrounds, camera vibrations, low-resolution images and much more.

The results of the study by Braham and Droogenbroeck are not so good in the case of a dynamic background, even worse than St-Charles et al.'s [14] proposed PAWCS method. Therefore, we try to improve the result in this case. The experimental results are shown in Table 1. It can be seen from the table

Table 1. The F-measures of different methods for dynamic backgrounds

Method	F-measure	Method	F-measure
Our method	0.9199	PAWCS [14]	0.8965
ConvNet-GT [1]	0.8943	GMM [13]	0.7085

that our F-measure has been raised to 0.9199, outperforming all the algorithms proposed previously. However, if we discuss the precision and recall separately, we can see that compared with ConvNet-GT [1], the F-measure of the Fall dataset improved 0.0298, which is 0.0214 higher than the Boats dataset. It can be inferred that the improvement of the background module influences the Fall dataset more. Also, the improvement of the background module has a good influence on the precision no matter whether it is the Fall or the Boats dataset. But the recall dropped slightly.

Figures 2 show the results by using different approaches to process selected images from the Fall sequence in the 2014 ChangeDetection.net repository. The probability of falsely classifying the tree as the foreground becomes less, which also is the main reason why our precision rate can increase. Figures 3 show the results of the ship sequence. Although the accuracy and recovery results are somewhat improved compared to the ConvNet-GT method, they are not as obvious as the Fall sequence because the false classification in their methods are less. Thus, the overall outcome in these images did not improve so much.

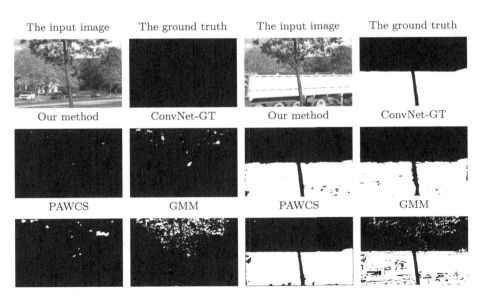

Fig. 2. Comparing the results by different approaches from the Fall sequences

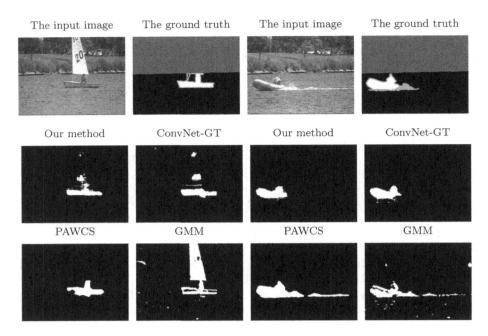

Fig. 3. Comparing the results by different approaches from the Boats sequences

Our experiment mainly focuses on the dynamic background. Experiments show that the improvement of the background module will make the result better. However, to verify whether this approach still works on other categories of data, we also compare the result with the general data. The datasets included in the general background are Highway and Pedestrians. The experimental results show that the F-measure for our method is 0.9630 and for ConvNet-GT is 0.9664, which is not far away from each other. If we discuss the precision and recall separately, we can see that compared with ConvNet-GT, both precision and recall are not much different. From this, we can see that our method will significantly improve the situation with the dynamic background, but the improvement in the general background is not so noticeable.

4   Conclusion

In this study, we eliminate the background with a convolution neural network and use the GMM to generate a reliable background module for training. The experimental results in the 2014 ChangeDetection.net repository show good results. And the improvement by our background module for different test data sets will have different performance. Usually, the degree of improvement and background changes showed a positive correlation.

Acknowledgments. This work was supported in part by the Ministry of Science and Technology, Taiwan, R.O.C., under grant no. MOST 107-2221-E-126-005.

References

1. Braham, M., Droogenbroeck, M.V.: Deep background subtraction with scene-specific convolutional neural networks. In: IEEE International Conference on Systems, Signals and Image Processing (IWSSIP), Bratislava, Slovakia, May 2016, pp. 1–4 (2016). https://doi.org/10.1109/IWSSIP.2016.7502717
2. Goyette, N., Jodoin, P.-M., Porikli, F., Konrad, J., Ishwar, P.: A novel video dataset for change detection benchmarking. IEEE Trans. Image Process. **23**, 4663–4679 (2014)
3. LeCun, Y., Bottou, L., Bengio, Y., Haffner, P.: Gradient-based learning applied to document recognition. Proc. IEEE **86**, 2278–2324 (1998)
4. Jiménez, P.G., Maldonado-Bascón, S., Gil-Pita, R., Gómez-Moreno, H.: Background pixel classification for motion detection in video image sequences. In: Mira, J., Álvarez, J.R. (eds.) IWANN 2003. LNCS, vol. 2686, pp. 718–725. Springer, Heidelberg (2003). https://doi.org/10.1007/3-540-44868-3_91
5. Wang, Z., Bao, H., Zhang, L.: PNN based motion detection with adaptive learning rate. In: International Conference on Computational Intelligence and Security, CIS 2009, Beijing, December 2009
6. Do, B., Huang, S.: Dynamic background modeling based on radial basis function neural networks for moving object detection. In: International Conference on Multimedia and Expo, ICME 2011, Barcelona, Spain, July 2011
7. Shobha, G., Satish Kumar, N.: Adaptive background modeling and foreground detection in video sequence using artificial neural network. In: International Conference on Intelligent Computational Systems, ICICS 2012, Dubai, January 2012
8. Ramirez-Quintana, J., Chacon-Murguia, M.: Self-organizing retinotopic maps applied to background modeling for dynamic object segmentation in video sequences. In: International Joint Conference on Neural Networks, IJCNN 2013, August 2013
9. Athilingam, R., Kumar, K., Kavitha, G.: Neuronal mapped hybrid background segmentation for video object tracking. In: International Conference on Computing, Electronics and Electrical Technologies, ICCEET 2012, pp. 1061–1066 (2012)
10. De Gregorio, M., Giordano, M.: Change detection with weightless neural networks. In: IEEE Change Detection Workshop, CDW 2014, June 2014
11. Guo, R., Qi, H.: Partially-sparse restricted Boltzmann machine for background modeling and subtraction. In: International Conference on Machine Learning and Applications, ICMLA 2013, pp. 209–214 (2013)
12. Xu, L., Li, Y., Wang, Y., Chen, E.: Temporally adaptive restricted Boltzmann machine for background modeling. In: AAAI 2015, Austin, Texas USA, January 2015
13. Stauffer, C., Grimson, E.: Adaptive background mixture models for real-time tracking. In: IEEE International Conference on Computer Vision and Pattern Recognition (CVPR), vol. 2, Fort Collins, Colorado, USA, pp. 246–252, June 1999
14. St-Charles, P.-L., Bilodeau, G.-A., Bergevin, R.: A self-adjusting approach to change detection based on background word consensus. In: IEEE Winter Conference on Applications of Computer Vision (WACV), Waikoloa Beach, Hawaii, USA, pp. 990–997, January 2015

Machine Translation on a Parallel
Code-Switched Corpus

M. A. Menacer$^{(\boxtimes)}$, D. Langlois, D. Jouvet, D. Fohr, O. Mella, and K. Smaïli

LORIA, Campus Scientifique, BP 239, 54506 Vandoeuvre-lès-Nancy, France
`mohamed-amine.menacer@loria.fr`

Abstract. Code-switching (CS) is the phenomenon that occurs when a speaker alternates between two or more languages within an utterance or discourse. In this work, we investigate the existence of code-switching in formal text, namely proceedings of multilingual institutions. Our study is carried out on the Arabic-English code-mixing in a parallel corpus extracted from official documents of United Nations. We build a parallel code-switched corpus with two reference translations one in pure Arabic and the other in pure English. We also carry out a human evaluation of this resource in the aim to use it to evaluate the translation of code-switched documents. To the best of our knowledge, this kind of corpora does not exist. The one we propose is unique. This paper examines several methods to translate code-switched corpus: conventional statistical machine translation, the end-to-end neural machine translation and multitask-learning.

Keywords: Code-switching · Machine translation ·
Statistical machine translation · Neural machine translation

1 Introduction

Code-switching (CS) can be defined as the use of more than one language by a speaker within an utterance. This phenomenon occurs generally in multilingual communities where speakers are known for their ability to code switch their languages during the communication.

CS is used both in informal (tweets and online content) and formal (proceedings of multilingual institution) texts. This makes the code-switched data a challenging issue for Natural Language Processing (NLP). Even if there were several linguistic studies on mixed languages [4,11], the computational processing of this kind of data remains relatively weak [2,10].

One of the challenges of the NLP concerning code-switched texts is the lack of data and tools. That is why, our first objective is to provide a code-switched dataset. Code-switched corpora are either generated in an artificial way [12,13] or collected from social media and/or online texts [1]. Once the data are collected, the processing of code-switching is carried out by adapting the existing models and tools or by proposing new ones.

© Springer Nature Switzerland AG 2019
M.-J. Meurs and F. Rudzicz (Eds.): Canadian AI 2019, LNAI 11489, pp. 426–432, 2019.
https://doi.org/10.1007/978-3-030-18305-9_40

Recent efforts related to CS are more about data collection and analysis [7,14]. Few works have explored downstream tasks as Language Modeling (LM) [8], Automatic Speech Recognition (ASR) [12] and Machine Translation (MT) [5].

In this article, we focus on the machine translation of Arabic-English code-switched documents. This is a challenging task for two reasons. In fact, the development of a MT system needs parallel corpora. This kind of data is usually available for pure languages, however, for mixed languages, there is no available resource. Thus, our first objective is to build a code-switched corpus with two reference translations, one for each language. This resource will mainly be used for tuning and testing the performance of the translation system. The second reason that makes difficult to translate code-switched documents is that models do not exist for this kind of data. Our approach in this work consists of using and adapting existing models in the aim to study the impact of the code-switched data on the machine translation.

2   How to Build a Code-Switched Parallel Corpus?

Our main objective is to translate mixed Arabic-English documents to pure Arabic and/or to pure English forms. To reach this objective, we decided to investigate the United Nations (UN) documents, because they mix English with Arabic in the Arabic official documents.

The parallel Arabic-English corpus extracted from UN documents corresponds to the period between January 2000 and September 2009 (MultiUN [6]). Table 1 sets out some statistics about this corpus after tokenization, truecasing and cleaning the Arabic and English corpora.

Table 1. Statistics about the parallel corpus.

Language	#sentences	#words	#unique words
Arabic	9.7 M	232.7 M	690 k
English		275.3 M	388 k

The particularity of this corpus is that the Arabic sentences can include English segments but also segments from French, Spanish or other languages. All these segments are written in Latin script.

All code-switched sentences are extracted from the entire corpus (i.e. sentences containing words in Arabic characters and words in Latin characters). This leads to a parallel code-switched corpus representing only 3% from the original corpus. Among these mixed sentences, we kept only those without URL links, email addresses and those that do not contain acronyms, since acronyms are generally not translated and are kept as they are.

Due to the parallel nature of the corpus, the English translation of each code-switched sentence can be easily recovered. For these remaining code-switched sentences, obviously Arabic is the dominant language. However, words in Latin script represent 12% of the total number of this corpus.

Code-switched sentences can be translated into pure English (for an English-speaking people) or pure Arabic (for an Arabic-speaking people). The English translation is already available but we do not have a pure translation into Arabic. To do so, we propose to produce it automatically. All the Arabic segments in a code-switched sentence are kept as they are and the English segments are translated into Arabic by using Google translate API[1]. The addition of the Arabic segments and the translated English segments into Arabic constitutes a Pure Arabic Sentence, named *PAS*.

Table 2 describes some statistics about the resulting parallel code-switched corpus.

Table 2. Statistics about the parallel code-switched corpus.

Language	#sentences	#words	#unique words
CS	37 k	1.09 M	99 k
English ref		1.14 M	64 k
Arabic ref		1.06 M	88 k

An example of a code-switched sentence and its reference translations (pure Arabic and pure English) is set forth in Table 3.

Table 3. Examples of a code-switched sentence with their Arabic/English translation reference (Arabic segments are written in buckwalter transliteration).

Code-switched sentence	*AmA bAl<nklyzyp fhy mrkbp mn Al>Hrf Al>wlY mn AlklmAt AltAlyp* boosting and inspiring dynamic youth achievement
Arabic reference translation	*AmA bAl<nklyzyp fhy mrkbp mn Al>Hrf Al>wlY mn AlklmAt AltAlyp* tEzyz w<lhAm Al<njAz AldynAmyky ll\$bAb
English reference translation	It is also an acronym for Boosting and Inspiring Dynamic Youth Achievement

The Arabic reference translation is a mix of human produced data (original Arabic sentences) and Google machine translation produced data (the CS parts

---

[1] https://github.com/ssut/py-googletrans.

are translated back into Arabic); hence one needs to ensure the quality of the generated translation. That is why, we decided to evaluate manually the Arabic reference translation.

To investigate the effect of different sources of information on the evaluation procedure, we asked 6 people to evaluate 1200 sentences by using the five-point scale shown in Table 4 according to two evaluation scenarios:

Scenario 1: target only the user evaluate only *PAS* without any information about the source code-switched sentence.

Scenario 2: source+target participant has both *PAS* and the source code-switched sentence.

Table 4. Evaluation scale used for each evaluation scenario.

	Scenario 1	Scenario 2
1	incomprehensible	no relation between source and target
2	some segments are understandable	some segments are correctly translated
3	understandable but non-native Arabic	understandable translation but non-native Arabic
4	very understandable	understandable translation
5	excellent	good translation

The final score is calculated by computing the sample mean of all sentence judgments. The results are presented in Table 5.

Table 5. Average score of human evaluation of the Arabic reference translations.

Participants	p1	p2	p3	p4	p5	p6	Mean
Scenario 1	2.86	3.14	3.21	3.32	3.49	3.51	3.26
Scenario 2	4.09	4.10	4.53	4.04	3.53	4.00	4.05

Concerning the first scenario, all participants except *p1* consider globally that the produced *PASs* are understandable but suffer from a bad style in the form of the Arabic structure. For the second scenario, the participants consider that the produced *PASs* are understandable. This is probably due the influence of the users by the source sentence, which is was available in the second evaluation scenario.

For further information about how the participants have evaluated the translations, we decided to measure the inter rater reliability. To do so, participants should evaluate the same set of sentences, and thus another set of 100 sentences are selected randomly from the entire corpus and they are evaluated by each participant. The inter rater reliability is measured by using Fleiss' kappa coefficient [9]. The obtained value was 0.41 with a standard error of 0.02 which can be interpreted as there is a moderate agreement between participants. This is an expected result since the two categories *very understandable/excellent* in scenario 1 and *understandable translation/good translation* in scenario 2 are so close, hence for some sentences, this could easily be confused.

3   The Impact of Code-Switched Data on Machine Translation

In this section, we present several scenarios of machine translation. For each of them, we investigate both the conventional Statistical Machine Translation (SMT) and the Neural Machine Translation (NMT) approaches. The four scenarios are the following:

Baseline System. With 1 M of parallel sentences (pure Arabic and pure English), we trained machine translation systems and we tested them on a code-switched test corpus.

With Output Copy (WOC) System. The baseline system is used except that the English segments in the code-switched test corpus are directly copied into the output.

Translation Based on Multilingual Training Corpus. In this case, we trained a translation system with a code-switched corpus built automatically by replacing, in the Arabic corpus, some segments by their English translation. The translations are extracted from the phrase table of the baseline system.

Multitask Learning. Since the source corpus to translate is a mixture of Arabic and English segments, our idea is to train one model that performs translation in the two directions, from Arabic to English and vice versa.

Experiments were performed on 5 k code-switched sentences extracted from our resource produced in Sect. 2. While the reference translation is available for both languages, the translation is just carried out from code-switched sentences to pure English. Table 6 sets forth the evaluation of translation in terms of BLEU metric and the Out-of-vocabulary rate (OOV).

In the case where the test corpus is translated into pure English by using the baseline system, all segments with Latin script are considered as out-of-vocabulary; this explains the OOV rate of 14.09%. Besides, the NMT approach performs poorly on code-switched data compared to the SMT approach. This is due to the OOV rate. In fact, NMT performs on constraint vocabulary, all words that do not exist in the vocabulary are replaced by a special token. Even by substituting unknown words with source words that have the highest attention

Table 6. The evaluation of the machine translation systems

Attempts	SMT	NMT	OOV (%)
Baseline	29.89	24.09	14.09
WOC	31.06	**33.11**	14.09
Multilingual	**32.06**	31.07	5.67
Multitask	/	28.36	14.09

weight, the translation is not improved. One explanation for this would be that the attention model does not functionally play the role of a word alignment between the source and the target sentences. It provides only a soft-alignment to help the decoder to decide parts of the source sentence to pay attention to [3].

For this reason, preventing the translation of foreign segments, and copying them directly into the output (WOC in Table 6) improves the translation comparing with the baseline system and specially in the NMT approach.

Training the translation systems on an artificial code-switched corpus (multilingual in Table 6) improves the translation compared with the baseline system where we used a monolingual parallel corpus. This approach outperforms also the second system (WOC), where segments in foreign language are not translated but copied directly into the output. These improvements are justified by the decrease of the OOV rate (5.67% against 14.09%) and by the fact that the training corpus is approaching the test corpus. The same observation holds true for the NMT approach, except that training the neural network on an artificial multilingual corpus does not outperform the second system.

Ultimately, training one model that performs the translation in the two directions (multitask learning) yields a gain of 4% BLEU points over the baseline system where a single task is learned.

4   Conclusion

The study carried out in this work focused on the impact of code-switching on the translation system. We firstly investigated whether the code-switching occurs in formal text. We accomplished this through the Arabic-English parallel corpus extracted from the official documents of United Nations. From this corpus, we built and evaluated a parallel code-switched resource, which is available for free access[2]. It also provides a valuable resource for studying multilingual practices in other works.

Several training/translation strategies were investigated for both the SMT and the NMT approaches. Results showed that the conventional SMT reaches best performance if a multilingual model is trained on mixed languages. Besides, we found that avoiding translation of segments in foreign language is the best

---

[2] https://smart.loria.fr/Fichiers/MTCS.rar.

strategy for the end-to-end model. We also found that training the neural network on two translation tasks instead of one improved the translation of code-switched sentences.

Furthermore, in our work, the whole code-switched sentence is translated directly with the different systems; it would be interesting to identify implicitly the language of the foreign segments and carry out the translation with an appropriate translation system as it was carried out in [13] for the recognition of mixed speech.

References

1. Abidi, K., Menacer, M.A., Smaïli, K.: Calyou: a comparable spoken Algerian corpus harvested from Youtube. In: 18th Annual Conference of the International Communication Association (Interspeech) (2017)
2. Abidi, K., Smaïli, K.: An empirical study of the Algerian dialect of social network. In: ICNLSSP 2017 - International Conference on Natural Language, Signal and Speech Processing. Casablanca, Morocco, December 2017. https://hal.inria.fr/hal-01659997
3. Bahdanau, D., Cho, K., Bengio, Y.: Neural machine translation by jointly learning to align and translate. arXiv preprint arXiv:1409.0473 (2014)
4. Bullock, B.E., Hinrichs, L., Toribio, A.J.: World englishes, code-switching, and convergence. Oxford Handbook of World Englishes (2014)
5. Carpuat, M.: Mixed language and code-switching in the canadian hansard. In: Proceedings of the First Workshop on Computational Approaches to Code Switching, pp. 107–115 (2014)
6. Eisele, A., Chen, Y.: MultiUN: a multilingual corpus from united nation documents. In: Tapias, D., et al., (eds.) Proceedings of the Seventh Conference on International Language Resources and Evaluation, pp. 2868–2872. European Language Resources Association (ELRA), May 2010
7. Gambäck, B., Das, A.: Comparing the level of code-switching in corpora. In: Chair, N.C.C., et al. (eds.) Proceedings of the Tenth International Conference on Language Resources and Evaluation (LREC 2016). European Language Resources Association (ELRA), Paris, France, May 2016
8. Garg, S., Parekh, T., Jyothi, P.: Dual language models for code mixed speech recognition. CoRR abs/1711.01048 (2017), http://arxiv.org/abs/1711.01048
9. Landis, J.R., Koch, G.G.: The measurement of observer agreement for categorical data. Biometrics pp. 159–174 (1977)
10. Molina, G., et al.: Overview for the second shared task on language identification in code-switched data. In: Proceedings of the Second Workshop on Computational Approaches to Code Switching, pp. 40–49 (2016)
11. Poplack, S.: Sometimes i'll start a sentence in spanish y termino en espanol: toward a typology of code-switching1. Linguistics **18**(7–8), 581–618 (1980)
12. Toshniwal, S., et al.: Multilingual speech recognition with a single end-to-end model. arXiv preprint arXiv:1711.01694 (2017)
13. Watanabe, S., Hori, T., Hershey, J.: Language Independent End-to-End Architecture for Joint Language Identification and Speech Recognition, pp. 265–271, December 2017
14. Yoder, M., Rijhwani, S., Rosé, C., Levin, L.: Code-switching as a social act: the case of Arabic Wikipedia talk pages. In: Proceedings of the Second Workshop on NLP and Computational Social Science, pp. 73–82 (2017)

Weighting Words Using Bi-Normal Separation for Text Classification Tasks with Multiple Classes

Jean-Thomas Baillargeon$^{(\boxtimes)}$, Luc Lamontagne, and Étienne Marceau

Université Laval, Québec, Canada
{jean-thomas.baillargeon,luc.lamontagne}@ift.ulaval.ca,
etienne.marceau@act.ulaval.ca

Abstract. An important usage of natural language processing is creating vector representations of documents as features in a classification task. The traditional bag-of-word approach uses one-hot vector representations of words that aggregate into sparse vector document representation. This representation can be enhanced by weighting words that contribute the most to a classification task. In this paper, we propose a generalization of the Bi-Normal Separation metric that enhances vector representations of documents and outperforms TF-IDF scaling algorithms for one-of-m classification tasks.

Keywords: Machine learning · Natural language processing ·
Text classification · Bi-Normal Separation · Risk classification

1 Introduction

Classification is a supervised learning task that allows decision automation in the form of selecting the right answer from a finite and discrete class set. Classification applications using natural language processing (NLP) are solving problems such as, but not limited to, sentiment analysis, language identification and categorization of documents.

Our current work pertains to the classification of documents for the commercial insurance sector. The classification task is to determine the risk class to which a company seeking insurance belongs. In a traditional insurance company, this task is accomplished by underwriters asking various questions to the company representative. However, to automate the process, one would need to get data from external sources. Sources of corporate information, such as Google Places or Yellow Pages can be exploited for this purpose. They provide short descriptions and the main keywords of company activities.

Documents containing company descriptions are aggregated into input vectors and are fit into a feature-based classifier such as Naive Bayes, Logistic Classifier and Support Vector Machines. One challenge in classification tasks is to use features adequately to accurately allocate documents within different

© Springer Nature Switzerland AG 2019
M.-J. Meurs and F. Rudzicz (Eds.): Canadian AI 2019, LNAI 11489, pp. 433–439, 2019.
https://doi.org/10.1007/978-3-030-18305-9_41

classes. The bag-of-word approach produces vectors that lacks the discriminating power to obtain a satisfying classification performance. Highlighting important features can be achieved by using scaling algorithms such as the popular Term Frequency-Inverse Document Frequency (TF-IDF) from [7] or the overperforming, but limited to binary classification, Bi-Normal Separation (BNS) as proposed by [2].

BNS was presented as an overperforming feature selection metric. It was used afterward as a feature scaling metric in [3] to alter vectors in a bag-of-word approach in binary classification tasks. In [4], the authors used BNS in a multi-class context, without providing details on the computation of the optimal multiclass score, making their approach not reproducible. Although BNS metric as feature scaling is an improvement over the TF-IDF metric, it is seldom encountered in the literature, mostly due to its limitation to binary classification.

As we shall see, the problem of extending BNS to a multiclass setting has never been published in the literature and its contribution to the representation of sparse documents vectors has never been evaluated in the literature.

In this paper, we propose an approach to extend the BNS feature scaling to multiclass problems. We also provide a python implementation[1], using the `TransformerMixin` interface from `scikit-learn`. The structure of the paper is as follows. We begin by presenting two scaling algorithms in Sect. 2, we present the extention to the bi-normal separation scaling algorithms in Sect. 3 and we evaluate our extention in Sect. 4.

2   Feature Scaling

When using a bag-of-word approach, text representation can be enhanced with feature scaling algorithms, such as TF-IDF and BNS. These scaling schemes alter the vector representation of a document by increasing the relative importance of useful words.

The most popular TF-IDF metric improves the representation of documents by penalizing words that are found in many documents within a corpus and increases the score of words that are concentrated in a few.

As an alternative, the BNS metric increases the score of features that provides more true positives than false positives in a binary classification task. In other words, a word that is found in only one class will get a higher score than a word that is found in all classes, as the former will only yield true positives and no false positives.

Given the corpus D, with $C \subset D$ the set of documents that belong to the positive class and $C' = D \setminus C$ the set of documents that belong to the negative class, BNS weighting for each word $w_i, i \in 1, ..., |V|$ in the vocabulary V is defined by

$$BNS(w_i, D) = abs(\Phi^{-1}(TPR_C(w_i, D)) - \Phi^{-1}(FPR_C(w_i, D))), \qquad (1)$$

---

[1] Code repository: https://github.com/jtbai/extended-bns.

where abs corresponds to the absolute value function and Φ^{-1} is the quantile function of a normal random variable. Furthermore, TPR_C and FPR_C are respectively the true positive rate and false positive rate of classification for the positive class C, and

$$TPR_C = \frac{|d_j : w_i \in d_j, d_j \in C|}{|d_j \in C|}; \quad FPR_C = \frac{|d_j : w_i \in d_j, d_j \notin C|}{|d_j \notin C|}.$$

Feature scaling scheme BNS is only usable in binary classification tasks. Since the problem of classifying company in a risk category requires a finer granularity, this methodology need to be extended to cope with multi-class problems. Our approach to generalize BNS metric to one-of-m classification problems is presented in Sect. 3.

3   Extending Bi-Normal Separation Scaling

As previously stated, BNS feature scaling is only applicable to binary classification. From (1), one can see that scoring is computed using the true and false positive rate of class detection using a word. This implies that only two classes can be used in (1).

In this section, we present the methodology and the experiment that lead us to conclude that the best formulation for the Extended Bi-Normal Separation (EBNS) feature scaling for a one-of-m classification task is given by

$$EBNS(w_i, D) = max_j\big(\Phi^{-1}(TPR_{C_j}(w_i, D)) - \Phi^{-1}(FPR_{C_j}(w_i, D)))\big), j = 1, ..., m. \tag{2}$$

3.1   Symmetry Hypothesis Underlying BNS

The first modification to make is to reconsider the usage of an absolute value operator in the formulation of BNS. As presented in (1), the absolute value is used as a computational shortcut because the inner BNS score function is symmetric. The usage of this shortcut is possible because of the symmetry of two binary classes. We illustrate this symmetry with the example provided in Fig. 1.

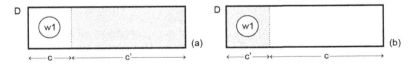

Fig. 1. Binary classification corpus with word w_1 being present in the minority class

Figure 1 presents two scenarios. Both scenarios contains the corpus D divided in two sets. A minority set containing d documents out of total $|D|$ documents from the corpus and a majority set containing the other $|D| - d$ documents. The word w_1 is present in n documents, all belonging to the minority class. Figure 1(a) represents the scenario where the minority class is chosen to be the positive (C) class. Figure 1(b) represents the opposite, where the majority class is the positive class.

If we were to compute the BNS score of a word w_1 for the scenarios s_a and s_b, (1) becomes

$$BNS_{s_a}(w_1, D) = abs\big(\Phi^{-1}\big(\frac{n}{d}\big) - \Phi^{-1}\big(\frac{0}{|D| - d}\big)\big); \tag{3}$$

$$BNS_{s_b}(w_1, D) = abs\big(\Phi^{-1}\big(\frac{0}{|D| - d}\big) - \Phi^{-1}\big(\frac{n}{d}\big)\big). \tag{4}$$

From (3) and (4), it follows that

$$\begin{aligned} BNS_{s_a}(w_1, D) &= abs(\Phi^{-1}\big(\frac{n}{d}\big) - \Phi^{-1}(0)) = abs(\Phi^{-1}(0) - \Phi^{-1}\big(\frac{n}{d}\big)) \\ &= BNS_{s_b}(w_1, D). \end{aligned}$$

One can easily show that the symmetry hypothesis holds to scenarios where word w_1 is included in both classes because of the symmetric property of the normal distribution.

3.2   Extending BNS

The example presented in Sect. 3.1 cannot be extended to classification problems with $m > 2$ classes. Being correlated with class C_1 does not imply a symmetric inverse correlation with each other class $C_2, C_3, .., C_m$. Since the use of the absolute value is only possible when the correlation symmetry hypothesis is verified, (1) has to be modified to alleviate this hypothesis to be used in one-of-m classification tasks.

By removing the symmetry hypothesis, a BNS subscore is required for each class. As opposed to the binary case where one calculation was required to get both the subscores s and $-s$, the m BNS subscores of a one-of-m classification task have to be individually calculated using an evaluation scheme. We selected the one-vs-all approach. In turn, each class becomes the positive class and the grouping of the other ones become the negative class. This approach is illustrated in Fig. 2.

A BNS subscore vector $\boldsymbol{BNS(w_i, D)}^m$ is calculated for each word w_i of the corpus D using (1) for each class $C_i, i \in 1, ..., m$. The vector containing all BNS subscores needs to be aggregated to obtain a single value for each word to be used as a scaling metric.

Multiple aggregation operators, such as the mean, the minimum and the maximum could be used. To select the optimal operator, we determine empirically which would allow better classification performance. Our intuition is that the maximum function would be the optimal aggregation function. As indicated

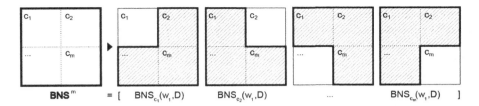

Fig. 2. Evaluating BNS subscore vector using a one-vs-all approach

previously, one important feature of the BNS score is to highlight correlation between a word and a document class. The maximum function keeps the greatest correlation between a word and a class and thus supports the aforementioned feature. Furthermore, using the maximum function on a binary classification gives the same score as (1).

While being potential candidates, we also believe that the mean and minimum function would not perform as well as the maximum. The mean function would level any outstanding correlation and would flat out the discriminating power of a word. We believe the minimum function would underperform in a multi-class setting since, although a very low score points to a strong inverse correlation, it does not provide any indication of how the word correlates individually with classes from the negative group.

4   Evaluation of EBNS

In this section, we present the data and the experiment used to validate our approach to EBNS and the results we obtained.

4.1   Datasets

To test our EBNS scheme, we used three labelled corpora. The first dataset, referred to as **Wikipedia**, contains the first paragraph of company articles from Wikipedia. These short descriptions were hand-labelled to reflect the type of activity of a company. The second dataset, referred to as **Yellow Pages**, has 1900 documents divided into 180 classes. The documents are Yellow Pages keywords related to a company activity code. The dataset was built to have a balanced number of examples in each of the classes and each document was labelled by hand. The third dataset, referred to as **Questions**, is a bank of questions labelled with a question type and used in [6]. This corpus was selected for its relevance in a classification task. The Questions corpus contains 3500 documents divided over 50 classes.

4.2   Experimental Results

Using those three datasets, we validate the effectiveness of our extension of BNS with various aggregation function. We evaluate the performance of each candidate function using the macro F1 score of the classification of a logistic regression classifier. To validate that EBNS value holds in a multiclass problem, we compare these F1 scores against results obtained with the TD-IDF metric. The aggregation functions are minimum, mean and maximum. Table 1 presents the F1 scores of a classification task using various scaling schemes on the three datasets.

The first observation is that EBNS scaling is overperforming TF-IDF feature scaling scheme in all three one-of-m classification tasks. It suggests that EBNS is a superior scaling metric for classification tasks that are small or medium-sized (either in vocabulary size, number of examples or document length) and is efficient even when there is a large number of classes. The second observation is that the maximum function is the most appropriate to aggregate BNS sub-score vector into a single EBNS score. This result corroborates our intuition, as explained in Sect. 3.2.

The most remarkable difference between the TF-IDF and the EBNS scores is obtained with **Wikipedia**. This can be explained by the uniformity between one text to another in articles from Wikipedia. The smallest improvement comes from **Yellow Pages**.

Table 1. F1 scores on one-of-m classification tasks

Scaling metric	Aggregation function	Wikipedia	Yellow Pages	Questions
TF-IDF	N/A	56%	25%	31%
EBNS	Minimum	75%	24%	37%
EBNS	Mean	92%	23%	33%
EBNS	Maximum	92%	26%	61%

This can be explained by the normalization of words and a small vocabulary size, which does not offer enough discriminating power between classes.

5   Conclusion

In this paper, we presented a modification of Forman's BNS that allows this feature scaling formulation to be extended to one-of-m classification problems. We showed that this methodology is overperforming TF-IDF feature scaling in this context. For future work, it would be interesting to use EBNS with word embeddings to assess classification performance from the experiment in [1] by using BNS word ordering instead of TF-IDF. Another experiment would be to challenge the use of true and false positive rate in (2) by using statistical indicators, such as Kendall's τ (see [5]), that have desirable behaviour in class dependency evaluation.

Acknowledgements. The authors gratefully acknowledge the Natural Sciences and Engineering Research Council of Canada, the Chaire d'actuariat de l'Université Laval and Intact Financial Corporation for financial support, and Véronique Barras-Fugère for her illustrations.

References

1. De Boom, C., Van Canneyt, S., Demeester, T., Dhoedt, B.: Representation learning for very short texts using weighted word embedding aggregation. Pattern Recogn. Lett. **80**, 150–156 (2016)
2. Forman, G.: An extensive empirical study of feature selection metrics for text classification. J. Mach. Learn. Res. **3**, 1289–1305 (2003)
3. Forman, G.: BNS feature scaling: an improved representation over TF-IDF for SVM text classification. In: Proceedings of the 17th ACM Conference on Information and Knowledge Management, pp. 263–270. ACM (2008)
4. Kapoor, A., Dhavale, S.: Control flow graph based multiclass malware detection using bi-normal separation. Defence Sci. J. **66**(2), 138–145 (2016)
5. Kendall, M.G.: Further contributions to the theory of paired comparisons. Biometrics **11**(1), 43–62 (1955)
6. Li, X., Roth, D.: Learning question classifiers: the role of semantic information. Natural Lang. Eng. **12**(3), 229–249 (2006)
7. Salton, G., Buckley, C.: Term-weighting approaches in automatic text retrieval. Inf. Process. Manage. **24**(5), 513–523 (1988)

A Deep Learning Approach for Diagnosing Long QT Syndrome Without Measuring QT Interval

Habib Hajimolahoseini[1,2(✉)], Damian Redfearn[1], and Andrew Krahn[3]

[1] Department of Medicine, Queen's University, Kingston, ON K7L3N6, Canada
[2] Stradigi AI, Montreal, QC H3A1T1, Canada
habibh@stradigi.ai
[3] Division of Cardiology, University of British Columbia,
Vancouver, BC V6E1M7, Canada

Abstract. For decades, ECG segmentation and QT interval measurement have been two fundamental steps in ECG-based diagnosis of the long QT syndrome (LQTS). However, due to the subjective nature of the definition of Q and T wave boundaries and confusion with an adjacent U wave, it suffers from a high degree of inter- and intra-analyst variability. In this paper, without measuring the QT interval and extracting the ECG waves, we propose a convolutional neural network which receives the raw ECG signal, and classifies every heartbeat as Normal or LQTS. The network is trained using a dataset of genotype-positive LQTS, and genotype-negative normal ECGs of family relatives. Experimental results reveal a high accuracy in diagnosing LQTS non-invasively, with a very low computational complexity, guaranteeing the clinical application of the proposed method.

Keywords: Long QT Syndrome · Deep Learning · Classification

1 Introduction

Long QT Syndrome (LQTS) is an arrhythmogenic cardiac disorder, associated with an abnormal ventricular repolarization which results in ventricular arrhythmias. It is clinically characterized by a prolonged QT interval (defined as the time between the start of the Q wave and the end of the T wave shown in Fig. 1) and abnormal T wave morphology. LQTS could result in an abnormally fast heart rhythm which is associated with sudden cardiac death [5]. Therefore, developing an automatic system which is able to predict patients who are at a high risk of mortality is an invaluable asset.

In clinical practice, QT interval is the only standard and universally accepted quantitative measure used for non-invasive diagnosis of LQTS. It is usually measured by an ECG specialist manually, using one of the well-known formulas such as Bazett's formula known as QTc [1]. If the maximum QTc measured in all

© Springer Nature Switzerland AG 2019
M.-J. Meurs and F. Rudzicz (Eds.): Canadian AI 2019, LNAI 11489, pp. 440–445, 2019.
https://doi.org/10.1007/978-3-030-18305-9_42

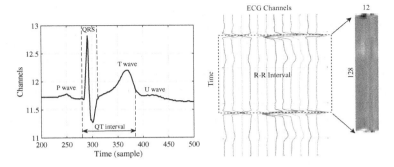

Fig. 1. Left figure shows a typical ECG signal, its corresponding waves and QT interval while the right figure represents a cropped R-R intervals as a 2D image

12 leads of ECG is greater than a threshold (450 ms in males and 470 ms in females), it is typically considered as LQTS [10].

However, there is a wide range of overlap between QTc in normal and LQTS patients. According to the observations reported in [10], the QTc of around 25% of genotype-positive LQTS patients is in the normal range. Hence, even if we ignore the error in measuring the QTc, using the QTc as the only feature for diagnosis LQTS leads to a high rate of false positives and/or false negatives.

A variety of algorithms have been proposed in the literature for automatic QT interval analysis including: threshold-based algorithms [12], hidden Markov models [4], curve fitting [5], wavelet transform [7] and machine learning techniques [3,9]. In general, the existing algorithms for QT analysis consist of the following main steps: (1) preprocessing, (2) ECG segmentation, (3) QT measurement and T wave analysis and (4) LQTS classification.

The common strategy in most of the existing algorithms is to extract the QRS and T waves and measure the QT interval prior to classification. Therefore, the accuracy of the overall system highly depends on the accuracy of the ECG segmentation (second step) and feature measurement (third step). However, because of the subjective nature of most of the measurements in the second and third steps, there is always a high degree of uncertainty and variability.

Furthermore, the existing approaches analyse each of the 12 ECG channels independently, which ignores the correlation between different channels (see [13] for example). In this paper, skipping the ECG segmentation and QT interval measurement steps, we propose a Convolutional Neural Network (CNN) which receives the raw ECG signal and classifies every heartbeat as Normal or LQTS. The network is trained using a dataset of LQTS and normal ECGs. It also considers the correlation between different channels of the ECG signals.

2   Database

The ECG signals were collected from 81 unique patients, using a resting 12-lead ECG. The database includes 45 genotype-positive LQTS patients and 36 genotype-negative normal controls. For every patient, we recorded 10 s of 12-channel ECG signals at a sampling rate of 250 Hz. The genetic screening result for all patients was also available which was used for labeling the ECGs. In order to create a dataset of individual heartbeats, the ECGs were segmented using a QRS peak detector. The peak of the QRS complex, known as the R wave, is the most recognisable feature of the ECG waveform which could be extracted with 100% accuracy [3].

A total number of 10,500 heartbeat cycles from 12 channels are then extracted as R-R intervals. The resulting R-R intervals are then downsampled/upsampled to 128 samples per heartbeat in all 12 channels. A typical 2D R-R interval is depicted in Fig. 1. After repeating this process for all patients, a dataset of 875 distinct labeled heartbeat matrices of dimension 12×128 is created. Almost half of the dataset corresponds to genotype-positive LQTS patients and the other half corresponds to genotype-negative normal controls.

3 Convolutional Neural Network

We propose a CNN which automatically extracts the morphological features of the heartbeats from the raw ECG signals [2]. Here, we consider the prolongation of QT interval as a change in heartbeat morphology. Thus, we do not need to extract the T wave and/or calculate the QT interval as CNN is a powerful tool for automatically extracting the morphological features that are not visible to human eyes. The CNN receives every single heartbeat as a 2D image X with dimension 12×128 at the input layer and delivers one of the two labels $y = [\text{Normal}, \text{LQTS}]$ at the output.

The input images are first normalized by subtracting the mean and dividing by standard deviation. Note that in contrast with the conventional strategies which process the ECG signals in different channels independently, we process every 12-channel heartbeat as a 2D image and hence, the correlation between different channels is also considered in the proposed CNN. This will significantly improve the performance and robustness as we also use the information related to the correlation between different channels.

3.1 Network Architecture and Training

The proposed network consists of four convolutional and two fully connected layers. The number of filters used in the convolutional layers are 32, 32, 64 and 64, respectively. The dimension of all filters is 3×5. A number of 64 neurons is also used in the first fully-connected layer. Dropout with probability of 0.5 is also applied to the third and fourth convolutional and the first fully connected layers in order to avoid over-fitting [11]. Batch normalization is also applied to the output of every convolutional layer [6].

Except from the first convolutional layer, the ReLU function is applied after all other convolutional layers. The average pooling of size 3×5 with stride 2 is also used to reduce the dimensionality. The fully connected and softmax layers generate a probability function at the output which calculates the probability that the input heartbeat belongs to one of the two categories: Normal and LQTS. We use the cross-entropy as the cost function of optimization in the training step as follow:

$$H(X, y) = \frac{1}{3} \sum_{i=1}^{3} \log p(Y = y_i | X) \tag{1}$$

where $p(.)$ represents the probability. The Adam optimizer, which is an efficient gradient-based algorithm, is also employed for optimizing the weights in the training step [8]. The hyperparameters of the CNN is tuned experimentally by trying different values over a certain range.

4   Results and Discussion

In order to evaluate the performance of our algorithm, we use the F_1 score defined as follow:

$$F_1 = 2 \times \frac{\text{Precision} \times \text{Recall}}{\text{Precision} + \text{Recall}} \tag{2}$$

Here we introduce two metrics: Beat-Level accuracy, and Patient-Level accuracy. The Beat-Level accuracy gives the accuracy of LQTS detection for every heartbeat, regardless of the patients. To this end, precision and recall are calculated using detection results for all heartbeats in the test set, altogether. On the other hand, the Patient-Level accuracy, which is more practical clinically, is calculated using the following rule: if more than 50% of the heartbeats corresponded to a single patient are normal, consider that patient as normal, otherwise LQTS.

We exclude 10% of the population as the test set, containing heartbeats from normal and LQTS patients which have not been used in the training step. We select those genotype-positive LQTS patients whose QTc is in the normal range (less than 470 ms) which makes conventional ECG-based diagnosis challenging.

The confusion matrix of the beat-level analysis is depicted in Fig. 2 for all of the ECGs in the test set. The beat-level and patient-level precision and recall as well as the F_1 score are also reported in Table 1. As reported in these results, although there is a small beat-level error in diagnosing the LQTS, the LQTS patients could be diagnosed with 100% accuracy. It implies that although there are some apparently normal heartbeats in the ECG of a given LQTS patient, but since the rate of normal heartbeats is less than 50% for that patient, our system correctly considers that as an LQTS. In fact, the patient-level accuracy is more realistic in clinical applications as the electrophysiologists do not make decision based on a patient's single heartbeat, but by observing the rate of irregular heartbeats in his/her ECG.

In order to evaluate the performance of the proposed model and its ability in generalizing to the ECG signals from new patients, we adopt the cross-validation approach. Therefore, we train our model 50 times and each time a random split

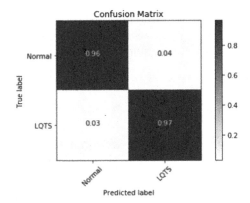

Fig. 2. The confusion matrix of the beat-level analysis for the ECGs in the test set.

Table 1. Beat-level and patient-level precision and recall (%)

	Beat-level	Patient-level
Precision	96.77	100
Recall	96.03	100
F_1 Score	96.40	100

of 10% and 90% as test and train sets is used, respectively. In every trial, the number of normal and LQTS patients in the test set is equal and their heartbeats are not used in the training step. The F_1 score statistics of the cross validation analysis is given in Table 2. As seen in this table, although the F_1 score of the average model is lower than the best model, it generalizes better to the ECG signals of the new patients which may be included in the dataset in the future.

Table 2. F_1 score statistics of the cross-validation analysis (%)

	Mean	STD	Min	Max
F_1 score	93.26	3.19	89.93	100

5 Conclusion

In this study, we proposed a convolutional neural network which is able to diagnose the LQTS from 12-channel ECGs of the patients' heartbeats without measuring the QT interval and segmenting the ECG waves. In contrast with the conventional methods which perform ECG segmentation and QT interval measurement prior to classification, the proposed method analyzes the 12-channel heartbeats as an image, considering the QT prolongation as a change in heart-beat morphology. This makes the proposed method robust to the subjectivity in definition of QT boundaries and confusion between U wave and T wave fluctuations. It also takes into account the correlation between different ECG channels while the conventional methods analyze them independently. The network was trained and tested using a cross-validation technique. The experimental results prove a high performance and low complexity, which are two assets in clinical applications. Our future studies involve collecting ECGs from new LQTS patients and modifying the network design in order to be able to diagnose and classify different types of LQTS in a larger dataset of ECGs.

References

1. Bazett, H.: An analysis of the time-relations of electrocardiograms. Ann. Noninvasive Electrocardiol. **2**(2), 177–194 (1997)
2. Goodfellow, I., Bengio, Y., Courville, A.: Deep Learning. MIT Press, Cambridge (2016). http://www.deeplearningbook.org
3. Hajimolahoseini, H., Hashemi, J., Redfearn, D.: ECG delineation for QT interval analysis using an unsupervised learning method. In: IEEE International Conference on Acoustic, Speech and Signal Processing (2018)
4. Hughes, N.P., Tarassenko, L., Roberts, S.J.: Markov models for automated ECG interval analysis. In: Advances in Neural Information Processing Systems, pp. 611–618 (2004)
5. Immanuel, S., et al.: T-wave morphology can distinguish healthy controls from LQTS patients. Physiol. Meas. **37**(9), 1456 (2016)
6. Ioffe, S., Szegedy, C.: Batch normalization: accelerating deep network training by reducing internal covariate shift. arXiv preprint arXiv:1502.03167 (2015)
7. İşcan, M., Yilmaz, A., Yilmaz, C.: A novel algorithm combining continuous wavelet transform and philips method for QT interval analysis. In: 2016 National Conference on Electrical, Electronics and Biomedical Engineering (ELECO), pp. 507–511. IEEE (2016)
8. Kingma, D.P., Ba, J.: Adam: a method for stochastic optimization. arXiv preprint arXiv:1412.6980 (2014)
9. Maršánová, L., et al.: ECG features and methods for automatic classification of ventricular premature and ischemic heartbeats: a comprehensive experimental study. Sci. Rep. **7**(1), 11239 (2017)
10. Page, A., Aktas, M.K., Soyata, T., Zareba, W., Couderc, J.P.: "QT clock" to improve detection of QT prolongation in Long QT Syndrome patients. Heart Rhythm **13**(1), 190–198 (2016)
11. Srivastava, N., Hinton, G., Krizhevsky, A., Sutskever, I., Salakhutdinov, R.: Dropout: a simple way to prevent neural networks from overfitting. J. Mach. Learn. Res. **15**(1), 1929–1958 (2014)
12. Struijk, J.J., et al.: Classification of the long-QT syndrome based on discriminant analysis of T-wave morphology. Med. Biol. Eng. Comput. **44**(7), 543–549 (2006)
13. Warrick, P., Homsi, M.N.: Cardiac arrhythmia detection from ECG combining convolutional and long short-term memory networks. In: 2017 Computing in Cardiology (CinC), pp. 1–4. IEEE (2017)

Learning Career Progression by Mining Social Media Profiles

Zakaria Soliman[1]([✉]), Philippe Langlais[1], and Ludovic Bourg[2]

[1] Université de Montréal, Montreal, QC H3C 3J7, Canada
{solimanz,felipe}@iro.umontreal.ca
http://rali.iro.umontreal.ca/rali/en/
[2] LittleBIGJob, Montreal, QC H3B 4W5, Canada
ludovic.bourg@lorenzandhamilton.com

Abstract. With the popularity of social media, large amounts of data have given us the possibility to learn and build products to optimize certain areas of our existence. In this work, we focus on exploring methods by which we can model the career trajectory of a given candidate, with the help of data mining techniques applied to professional social media data. We first discuss our efforts to normalizing raw data in order to get good enough data for predictive models to be trained. We then report the experiments we conducted. Results show that we can predict job transitions with 67% accuracy when looking at the 10 top predictions.

Keywords: Data mining · Machine learning ·
Carrier path prediction · Contextual embedding

1 Introduction

Labor flow networks have been an area of interest by researchers in the social sciences and by economists [2–4]. Labour mobility between industries has been shown to have a positive effect on economic indicators [3]. For this work, we gathered a large collection of user profiles, with the motivation of being able to predict a set of plausible recommended positions that a professional can take as his next career move given his working history. However, systems to predict the next job position has received less attention. Building predictive models that are personalized for every individual professional is a hard task because it involves several hard subproblems that need to be solved such as normalizing the large number of job titles, and skill set since these are all defined by individuals. This results in job titles that seemingly have the same function but different names for the role. Recommendation systems have been proposed as well [1,10] which are based on whether or not a user applies or clicks for a recommended job.

In this article, we discuss in Sect. 2 the specificities of the data we collected, and our approach to design a dataset amendable to benchmarking. We report in Sect. 3 the predictive models we implemented, the results they obtained in Sect. 4, and conclude in Sect. 5.

© Springer Nature Switzerland AG 2019
M.-J. Meurs and F. Rudzicz (Eds.): Canadian AI 2019, LNAI 11489, pp. 446–452, 2019.
https://doi.org/10.1007/978-3-030-18305-9_43

2   Dataset

We gathered a very large set of over 9.5 million public user profiles from LinkedIn. We removed the numerous profiles where no job experience was reported, yielding a still substantial set of more than 7.1 million profiles where 40% of users only have one filled out job experience. Since we are primarily interested in modeling career moves, we dropped all user profiles with less than 2 job experiences reported, and focused on profiles written in English[1] to identify the language of a profile. Removing such profiles leaves us with roughly 3 millions ones.

Table 1. Main statistics of our dataset

Total number of user profiles	2 789 111
Total number of unique job titles	3 859 835
Total number of unique job titles used as last job	927 209
Avr. job title length (# of words)	4.5
Longest job title string (# of words)	42
Avr. length of job history	5.2 (positions held)
Shortest job history	2 (positions held)
Longest job history	140 (positions held)

The main characteristics of our dataset are reported in Table 1. One striking figure of Table 1 is that there is slightly less than twice the number of unique job titles than we have user profiles. In other words, there is more job titles than profiles. Suggesting that we may face problems fitting this data into a powerful predictive model we discuss in Sect. 2.1 our approach at resolving the issue.

2.1   Normalizing Job Titles

The distribution of job title names is expectedly Zipfian, which means that most job titles in our dataset appear rarely. In fact, 98.2% of all the job titles appear less than 10 times illustrating the root of the problem. Rare job titles are actually overly specific, therefore reducing the likelihood that they match another job title.

Looking at Fig. 1, we observe that punctuations and conjunctions seem to separate long strings describing a job title into smaller sub-strings of individual entities in order to extract the most relevant information (the more general job title) out of a longer job title containing specificities that are not relevant to us. For example, we split the job title `co instructor and teaching`

---

[1] The `langid` [9] toolkit was used: https://github.com/saffsd/langid.py.

```
co instructor and teaching assistant, executive programs & undergraduate...
accountant, sales and marketing department
senior manager, project management methodology & governance
journalist and travel writer
vice president for marketing department of university art group
project manager, key accounts - human health therapeutics
information technology manager - operations & architecture
```

Fig. 1. Random sampling of rarely seen job title strings. Colored text is the reduced job title we would like to have.

assistant, executive programs & undergraduate programs into the substrings co instructor, teaching assistant, executive programs, and undergraduate programs. Among those, teaching assistant is the most frequent one we keep. Manual inspection of the resulting job titles did not reveal any ill-formed job titles.

2.2   Selecting User Profiles

In an effort to minimize the ratio of number of unique job title strings to the total number of user profiles, we looked at what happens when we discard the least common job titles based on some lower bound on their frequency in user profiles. We ordered the job titles in $\mathcal{J}$ in decreasing order of frequency leading to $\mathbf{j} = [j_1, \ldots, j_{|\mathcal{J}|}]$ with $j_{|\mathcal{J}|}$ being the most recent job title. We denote the set of user profiles that **only** use the k most common job titles in their profile as $N(\mathbf{j}_{1:k})$. We can then define the gain as:

$$\delta(k, k+1) = \frac{N(\mathbf{j}_{1:k+1}) - N(\mathbf{j}_{1:k})}{N(\mathbf{j}_{1:k})}$$

For example, we may look at how many user profiles solely use the 10 most common job titles in their job history (1 166 626), and compare it to how many users exclusively use the 110 most common ones. This leads to a gain of 211.22%. This method resulted in a dataset of 550 job titles with 120 371 user profiles. Considering more job titles yielded small gains only.

3   Experimental Setup and Models

The dataset was randomly split into a train (65%), a validation (15%) and a test (20%) sets prior to any experimentation. All hyperparameter tuning was made on the validation set, and all results presented here are those measured on the test set.

Along with standard classification accuracy, we use the mean percentile rank (MPR), a metric used for recommendation systems [6] as well as for evaluating models for similar tasks [7].

We compared 4 families of models briefly described in the sequel.

Baseline. A simple baseline consisting in always predicting the last job title in the job history. We call it *PreLa* for **Pre**dicting the **la**st seen job title. In the dataset, 34.15% of the profiles have their last experience identical to the previous one.

Naive Bayes. We experimented with the multinomial and bernoulli Naive Bayes variants along with different methods to represent the data features. The input was always a bag-of-words, the difference was in what we considered 'words' (complete job title as a token v.s. individual words as tokens).

N-gram Model. We tested an N-gram language model, which computes an estimation of the transition probability of a sequence of elements, by using the KenLM library [5]. In our case, instead of focusing on a sequence of words in a text, we look at the sequence of job titles.

RNN-based models. We used an LSTM based RNN network as the decoder of our encoder-decoder approach as well as a stand-alone network. The difference is in the initial state h_0 that is the encoder's output in the former case and a zero vector in the latter case. The choice of a recurrent network is due to it's success in sequential modeling [11].

As previously mentioned, we tested variants making use of the job history alone, as well as some incorporating the skills of a user. To do this, we use an encoder that takes as input the skills and outputs a real-valued vector that we interpret as a representation of that user's skills. The networks inputs (skills) were represented by pre-trained, `FastText` [8] embeddings. The output is a real valued vector that is fed into the LSTM decoder.

4   Results

Table 2 shows the MPR on the prediction of the next job title. There is no way to compute this metric on the PreLa baseline since it outputs a single prediction. Figure 2 shows the accuracy of the models when considering the top K predictions, so a model is counted correct if the reference prediction is found in the top K best scored ones.

Table 2 shows that the N-gram models outperform the Naive Bayes model on this metric suggesting that keeping contextual historic information about past job transitions give the models a better understanding of the future outcome. This is also reflected in Fig. 2, where the Naive Bayes model is beaten by every other model.

Looking at Fig. 2, we notice that the LSTM network is lagging behind the other models when we look at predictive accuracy. However, as can be seen in Table 2, when evaluating the models as recommendation engines, both neural models outperform the other model families. It is difficult to clearly understand why that is the case; this would be an interesting area of inquiry for future work. We notice that the addition of the skill set of a user (a feature given to the CNN-LSTM encoder-decoder model) increases the performance of the model as is evident in Fig. 2 as well as in Table 2.

Table 2. Mean percentile rank of the correct job title we are looking for within the model's predictions. A lower value is better.

Models	MPR × 100
Bigram	0.164
Trigram	0.168
Multinomial Naive Bayes	0.213
LSTM	0.148
CNN-LSTM	**0.106**

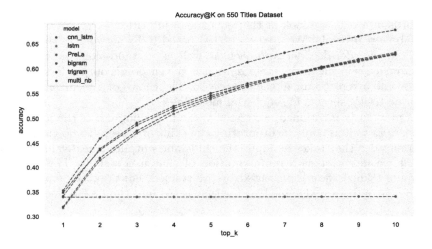

Fig. 2. Prediction accuracy@k for best performing models within each model family.

5   Conclusion

In this work, we have explored various methods to model a given candidate's career progression. We have discussed the difficulties in prepossessing and normalizing the dataset we gathered. A major difficulty for this task is the variation in job title names for the same apparent responsibilities. We compared neural models with classical language models applied to our problem and Naive Bayes methods to find that, surprisingly, the N-gram models are competitive with the neural models, since N-gram models use a smaller historical context (the value of N). This suggests that the positions held at the start of a career don't contribute as much to predict the end of a career. Thus, further motivating the use of a contextual representation of the user profile along with the sequence of experiences a user has had.

Considering that we had a large amount of different job titles to predict from (550 job titles), the models that were trained perform surprisingly well; we were able to get an accuracy at rank 1 of about 35% on the dataset with the CNN-LSTM model for exact job title matches, and we attained 67% accuracy when looking at the top 10 predictions. When looking at the mean percentile rank, the CNN-LSTM gives the correct target within the 6 best scoring prediction labels on average.

Data normalization still remains to be explored more thoroughly in future work. Potential solutions could be a rule based approach or a clustering method that could create clusters of job titles that describe very similar responsibilities and use these clusters as prediction labels. This would allow us to have more coarsely grained target labels we could then predict on these clusters instead of the individual job titles. Pushing this idea a bit further we could also group together the job titles by type like `manager` or `engineer` for instance thus yielding a smaller pool of possible prediction targets.

Additionally, the heuristic presented here introduced a trade off, and we dramatically reduced the size of our dataset. Further investigations on methods to standardize the job titles is an interesting area of inquiry for future work. We also want to benefit the additional information available in user profiles that we discarded here.

References

1. Al-Otaibi, S.T., Ykhlef, M.: A survey of job recommender systems. Int. J. Phys. Sci. **7**(29), 5127–5142 (2012)
2. Bjelland, M., Fallick, B., Haltiwanger, J., McEntarfer, E.: Employer-to-employer flows in the United States: estimates using linked employer-employee data. J. Bus. Econ. Stat. **29**(4), 493–505 (2011)
3. Boschma, R., Eriksson, R.H., Lindgren, U.: Labour market externalities and regional growth in Sweden: the importance of labour mobility between skill-related industries. Reg. Stud. **48**(10), 1669–1690 (2014)
4. Guerrero, O.A., Axtell, R.L.: Employment growth through labor flow networks. PLOS ONE **8**(5), 1–12 (2013). https://doi.org/10.1371/journal.pone.0060808
5. Heafield, K.: KenLM: faster and smaller language model queries. In: Proceedings of the EMNLP 2011 Sixth Workshop on Statistical Machine Translation, Edinburgh, Scotland, United Kingdom, pp. 187–197, July 2011. https://kheafield.com/papers/avenue/kenlm.pdf
6. Hu, Y., Koren, Y., Volinsky, C.: Collaborative filtering for implicit feedback datasets. In: Eighth IEEE International Conference on Data Mining, ICDM 2008, pp. 263–272. IEEE (2008)
7. James, C., Pappalardo, L., Sirbu, A., Simini, F.: Prediction of next career moves from scientific profiles (2018). arXiv preprint: arXiv:1802.04830
8. Joulin, A., Grave, E., Bojanowski, P., Mikolov, T.: Bag of tricks for efficient text classification (2016). arXiv preprint: arXiv:1607.01759
9. Lui, M., Baldwin, T.: Langid.py: an off-the-shelf language identification tool. In: Proceedings of the ACL 2012 System Demonstrations, pp. 25–30. Association for Computational Linguistics (2012)

10. Paparrizos, I., Cambazoglu, B.B., Gionis, A.: Machine learned job recommendation. In: Proceedings of the Fifth ACM Conference on Recommender Systems, pp. 325–328. ACM (2011)
11. Sundermeyer, M., Schlüter, R., Ney, H.: LSTM neural networks for language modeling. In: Thirteenth Annual Conference of the International Speech Communication Association (2012)

Toward a Conceptual Framework for Understanding AI Action and Legal Reaction

Raheena Dahya[1] and Alexis Morris[2]([✉]) [iD]

[1] Barrister & Solicitor, Toronto, ON, Canada
raheena@dahyalaw.com
[2] OCAD University, 100 McCaul St., Toronto, ON, Canada
amorris@faculty.ocadu.ca
https://www2.ocadu.ca/research/acelab

Abstract. Artificial Intelligence (AI), refers to computational components that process, classify, make decisions, or act on information from
data inputs, and in recent years more capable autonomous systems with
realtime decision-making properties have become tenable. In this landscape it becomes imperative to consider the socio-technical implications
of such systems, particularly at the legal level. This work facilitates this
discussion, broadly highlighting the relationship between law and AI, and
proposes a conceptual framework to understand the intersection between
these disciplines. AI designers and legal reasoners are encouraged to
apply this work to identify the connection and constraints involved when
developing AI systems, and the legal responses to these systems.

Keywords: Artificial intelligence · Legal analysis ·
AI design considerations · AI and Law · Legal personality ·
Autonomous decision-making

1 Introduction

Understanding the socio-technical implications of AI systems is imperative, and
may be considered from multiple perspectives, such as the physical, psychological, team, organizational, or political dimension [10]. The essence of this paper
is: (a) to outline the relationship between law and artificial intelligence in general; and (b) to identify the impact of law on the multiplicity of stakeholders,
which can (c) influence the way AI designers think about the ecologies in which
their designs will operate. It culminates in a conceptual tool, the AI and Law
Landscape (AILL), an inter-sectional landscape between the disciplines of AI
design and law, designed for AI system developers, who are not likely to have
legal knowledge; and for legal reasoners, who are likewise not expected to have
knowledge of AI domains.

© Springer Nature Switzerland AG 2019
M.-J. Meurs and F. Rudzicz (Eds.): Canadian AI 2019, LNAI 11489, pp. 453–459, 2019.
https://doi.org/10.1007/978-3-030-18305-9_44

The remainder of this work is presented as follows: Sect. 2 highlights conflicting terminology between the disciplines and indicates the need for a cohesive vocabulary. Section 3 presents the intersectional AI and Law Landscape in a new unified framework. Lastly, Sect. 4 provides a summary and directions for further research.

2   Agency: Towards Compatible Terminology

Each time an AI application is deployed it intersects with various areas of law. Law is a vast subject with a multiplicity of regimes, schemes, and systems nested within the overarching term "legal system". As such, this work is necessarily limited to providing high level and general explanations of certain legal concepts. Further, most legal schemes are rooted in the notion of physical and geographical limits in which a given legal regime applies. However, technological systems have no inherent geographical domain, and questions of jurisdiction can arise.

Agency in the AI Context: Multiple forms of agent can be employed within AI systems [11], e.g., for ambient intelligence [1], where agents are embedded (within the environment), context-aware (able to understand situations from contextual information), personalized (toward specific data sources, perhaps from individuals), adaptive (able to respond to dynamics within environments), and anticipatory (able to predict situational states from preceding patterns of information). Agent architectures to enable these kinds of ambient intelligence can be simple control programs, or complex deliberative systems (e.g., beliefs-desires-intentions architectures or reinforcement learning [6,7]), and these agents can also be organizational, communicating within a complex multi-agent system [11].

Agency in the Legal Context[1]: In contrast, the term agency, in the legal context, is a complex topic and there can be multiple definitions[2] of the word *agent*. For example, an agent may be a (i) separate legal entity which (ii) acts on behalf of another, being the agent's principal, (iii) whereby the agent consents to act on the principal's behalf[3]. In another sense the term agent may be used to describe "...the position of a person who is employed by another to perform duties of a technical or professional nature which he discharges as that other's alter-ego..."[4]. Both definitions demand an exploration as to whether an AI is capable of attaining legal personality, which may be required in order to consent to an employment or agency relationship. It is clear from this discussion that

---

[1] As this work is interdisciplinary, citations have been adapted to the Springer LNCS format; consequently, legal citations have been minimized and legal digest references have been provided as an alternative.

[2] For the purposes of this paper legal constructs and definitions are based on English or Canadian law.

[3] Halsbury's Laws of England, 5th Ed., Vol. 1, Pg. 5.

[4] ibid.

there are linguistic incompatibilities between the disciplines of AI design and law. To solve this incompatibility, for the purposes of this work, unless explicitly stated otherwise, this paper refers to agents as computational agents and not legal agents. Going forward, a cohesive shared vocabulary should be developed.

3   The AI and Law Landscape (AILL): A Conceptual Tool

In this work a conceptual tool is introduced to provide insight into the domains of AI and law and how they intersect in the context of AI design and deployment. There are a variety of perspectives to consider when approaching this discourse, and this tool, Fig. 1, presents a wide vantage point from which to explore AI designs and legal considerations. Iterative dialog between the disciplines is inevitable as the discourse regarding advanced uses of AI progresses. As shown, there are four key components in the dialog cycle section of the AILL, namely (i) stakeholders; (ii) considerations in the AI decision-making process (whether in realtime or at design time); (iii) foreseeable facts and events (which are useful to prepare for events, that may or may not happen, and prompting development of legal safeguards and AI designs); and (iv) considerations in the legal decision-making process. This iterative dialog is represented both as an interaction between all four components, and intense interaction between the AI and legal decision-making processes.

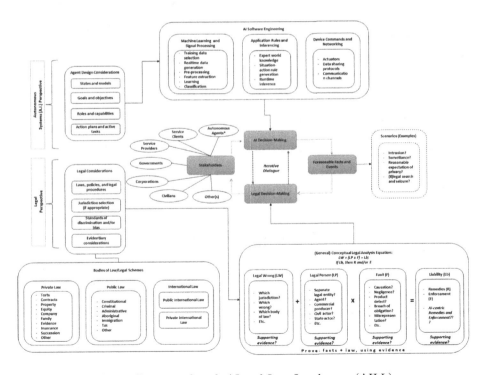

Fig. 1. Conceptual tool: AI and Law Landscape (AILL).

Stakeholders: There are various kinds of stakeholders to consider with respect to AI design and deployments and the legal implications of, and responses to, both. Stakeholders are not limited to these disciplines, as is exemplified in the AILL. While stakeholders may potentially include autonomous agents themselves, it more typically refers to (a) the providers and clients of AI systems; (b) governments and states (who enact and enforce regulatory instruments); (c) corporations who may be involved in the sale of items that rely on AI services; and (d) civilians (who may or may not be direct consumers of AI technologies, and may be directly or indirectly impacted by their use. The list of stakeholders in this framework is far from exhaustive; other stakeholders may for instance include educational institutions or advocacy groups.

AI Decision-Making: The autonomous system perspective as depicted in the AILL, presents a view toward agent design and software engineering for AI-based systems. From an agent system design perspective, this includes considerations of: (a) the kinds of states and model of how those states relate within the agent; (b) the goals and objectives of the agent system; (c) the roles and capabilities which the agent system can perform; and (d) the specific action plans and policies undertaken by the agent toward system goals. These are impacted by world dynamics, and require the system to respond or adapt. Herein the distinction is made between the system learning/being trained from pre-existing data, and the application of expert knowledge entered directly by system designers. These both translate directly into system behaviors which can lead to subsequent events.

Foreseeable Facts and Events: Legal decision-making considers the proverbial "fence" (implementing limits and boundaries), "shield" (creating a protective mechanism), and "sword" (designing and employing assistive tools during legal conflict). The foreseeable facts and events component in the AILL inform legal decision-making with respect to the proverbial "fence" and "shield," prompting robust legal mechanisms to be built in contemplation of foreseeable problem scenarios. The framework highlights, as an example, a suite of scenarios which can impact the way that legal mechanisms are crafted considering this component; and may lead to pre-emptive adaptations by AI designers.

Legal Decision-Making: When considering the implications of AI design and deployment, it is imperative to look to existing laws, policies, and legal procedures in order to both (a) enable AI designers to incorporate legal principles within the development lifecycle; and (b) create suitable legal mechanisms that can respond appropriately to benign, malicious, deliberate, or inadvertent AI action. Additional legal considerations may include: (i) jurisdiction selection, which may be appropriate, for example, in federal states or international dealings; (ii) standards of discrimination and/or bias which might involve referring to legal instruments which outline categories of discrimination[5] with a view toward cross-checking the grounds of discrimination with training datasets and other

---

[5] See, for example, Section 15(1) of the Canadian Charter of Rights and Freedoms (Constitution Act 1982, Enacted as Schedule B to the Canada Act 1982, 1982, c. 11 (U.K.)).

AI design considerations; and (iii) evidentiary considerations which may be at issue per event, whether or not it is a pre- or post- deployment event.

3.1 Conceptual Legal Analysis Equation (CLAE): Applications Within the AILL

The Conceptual Legal Analysis Equation (CLAE) in Fig. 1 presents a general explanation of how legal analysis occurs in the context of proving liability. In adversarial legal systems,[6] the decision-maker has two roles. Where a jury is present they serve as the "trier of fact", whereas a judge serves as the "trier of law." Where no jury is present the judge undertakes both roles. In any legal dispute, facts and law are considered in light of evidence. Evidence is the "...data [...triers of fact] use when resolving factual controversies" [5]. The CLAE provides a method, framed as an equation to explain how legal reasoners identify liability $(LW + (LP * F) = Lb)$, where LW refers to a legal wrong, LP refers to a legal person, F refers to fault, and Lb refers to liability). The corollary to this equation is that, if liability is proved, then legal remedies R will result, either in conjunction with, or potentially followed by, other enforcement mechanisms, E, i.e., (if Lb then R and/or E). For non-legal experts, determining liability can be a challenge, and the CLAE aims to elucidate risk factors in AI decision-making that are legal in nature.

Within the CLAE framework the legal wrong must be selected by first identifying the jurisdiction(s) in which the wrong takes place. Like the internet, the AI agent "is wholly insensitive to geographic distinctions. In almost every case, users of the [AI agent] neither know nor care about the physical location of the [AI agent] resources they access" [2]. This poses a particular difficulty when attempting to identify the appropriate jurisdiction in which to situate the legal wrong. Second, one must identify the bodies of law at play and select, from the bodies identified, the specific legal wrong that has taken place. The next step in the CLAE is identifying a legal person allegedly at fault. Legal personality is not necessarily restricted to human individuals. For example "incorporation brings into existence a new legal person whose rights and obligations may be thought of as analogous to those of a human person" [9]. It is rare to find examples wherein an AI agent has separate legal personality, at present. However, given Saudi Arabia's recent conferral of citizenship to an embodied humanoid AI agent known as Sophia[7], it is arguable that Sophia, as an AI agent, holds some level of, if not full, legal personality. When considering an AI agent's action from a legal perspective, it is worth asking if there are arguments to support the AI agent having legal personality within the jurisdiction. The potential for agents to attain legal personhood represents an avenue for future exploration.

If an AI agent is not capable of holding legal personality, then we must consider which legal person is responsible for its actions. For example, this could

---

[6] For example England and Wales; and the majority of Canada.

[7] There has been some debate [4,8] regarding whether the level of agency that Sophia displays is beyond that of a chatbot, which may bring into question Sophia's AI agency. However, Sophia received widespread media recognition as an AI agent [3].

be a service provider, corporation, or state actor deploying the AI agent on its behalf. Once a legal person has been identified, fault must be established. The CLAE expresses *LP * F* because the fault must be that of the legal person in question. There are tests and criteria that assist legal reasoners in proving fault, using concepts such as causation, negligence, or product defect. Once the legal wrong, and a person to whom fault attaches, has been established, liability can flow and legal remedies can be put into place to rectify the legal wrong. An example of a remedy might be where a liable party pays damages to an injured party. Where remedies are not produced despite a court order requiring a party to produce them, enforcement mechanisms may be put into effect, for example a party in breach of a court order may be incarcerated for being in contempt of a court order. Interesting questions arise: Would remedial action and/or enforcement result in a court order to a re-write of a specific algorithm? Perhaps an order to cease and desist the deployment of an AI agent? These questions provide rich fodder for further discourse.

4 Summary and Future Work

There is a need for analytical frameworks and conceptual tools that allow members in the AI and Law disciplines to understand, interact, and cooperate in mutual endeavours. In this work, a new conceptual tool, the AILL has been designed to assist legal reasoners and AI designers in thinking about AI engineering considerations and legal mechanisms. The AILL remains to be properly situated within the growing body of research on the intersection of AI and law, and this is an avenue for future research; other avenues include: evaluations of scenarios, proof-of-concept AI applications, and present and potential legal frameworks that affect AI systems.

References

1. Aarts, E.: Ambient intelligence: a multimedia perspective. IEEE Multimed. **11**(1), 12–19 (2004)
2. Castel, J.G.: The internet in light of traditional public and private international law principles and rules applied in Canada. Can. Yearb. Int. Law/Annuaire canadien de droit international **39**, 3–67 (2002)
3. Forbes.com: Everything you need to know about Sophia the world's first robot citizen (2018). http://www.forbes.com/sites/zarastone/2017/11/07/everything-you-need-to-know-about-sophia-the-worlds-first-robot-citizen
4. Goertzel, B., Giacomelli, S., Hanson, D., Pennachin, C., Argentieri, M.: Singularitynet: a decentralized, open market and inter-network for AIS (2017)
5. Paciocco, D.M., Stuesser, L.: The Law of Evidence, 7th edn. Irwin Law, Toronto (2015)
6. Rao, A.S.: AgentSpeak(L): BDI agents speak out in a logical computable language. In: Van de Velde, W., Perram, J.W. (eds.) MAAMAW 1996. LNCS, vol. 1038, pp. 42–55. Springer, Heidelberg (1996). https://doi.org/10.1007/BFb0031845
7. Russell, S., Norvig, P.: Artificial intelligence: a modern approach (2009)

8. TheVerge.com: Facebook's head of AI really hates Sophia the robot (and with good reason) (2018). https://www.theverge.com/2018/1/18/16904742/sophia-the-robot-ai-real-fake-yann-lecun-criticism

9. VanDuzer, A.J., Daniels, R.J.: Law of Partnerships and Corporations. The Essentials of Canadian Law, 3rd edn. Irwin Law Incorporated (2009)

10. Vicente, K.: The Human Factor: Revolutionizing the Way People Live with Technology. Routledge, New York (2004)

11. Wooldridge, M., Jennings, N.R.: Intelligent agents: theory and practice. Knowl. Eng. Rev. **10**(2), 115–152 (1995)

Unsupervised Sentiment Analysis of Objective Texts

Qufei Chen[1(✉)] and Marina Sokolova[1,2]

[1] University of Ottawa, Ottawa, ON, Canada
{qchen037, sokolova}@uottawa.ca
[2] Institute for Big Data Analytics, Dalhousie University, Halifax, NS, Canada

Abstract. Unsupervised learning is an emerging approach in sentiment analysis. In this paper, we apply unsupervised word and document embedding algorithms, Word2Vec and Doc2Vec, to medical and scientific text. We use SentiWordNet as the benchmark measures. Our empirical study is done on the Obesity NLP Challenge data set and four Science subgroups from Reuters 20 Newsgroups. Our results show that Word2Vec demonstrates a reliable performance in sentiment analysis of the text, whereas Doc2Vec requires more detailed studies.

1 Introduction

In this study, we use unsupervised embedding algorithms Word2Vec and Doc2Vec in sentiment analysis of medical and scientific texts. In those texts, sentiments are not directly stated, thus making manual annotation unreliable and supervised learning unfeasible.

To assess the unsupervised results, we use SentiWordNet (SWN) [4], an extension of WordNet that groups words together into sets of synonyms (synsets). SWN assigns each synset a positive score pos_w, a negative score neg_w, and the objective score $obj_w = 1 - (pos_w + neg_w) = 1 - sub_w$. We compute SWN scores Sc_T and SWN term ratios R_T as sentiment benchmarks for a data set. We use Welch's t-test to generalize the results.

Word2Vec is an unsupervised two-layer neural network model that produces numeric representations of words (i.e. word embedding) [6]. For each word, its embedding represents the contextual semantics of the word in a text. We employ Word2Vec's Continuous Bag-of-Words model (CBOW) that takes in a context as input and aims to predict a specific word. Doc2Vec is an extension of Word2Vec applied to a document as a whole [5].

We used four Science subgroups (cryptography, electronics, medicine, and space) from Reuters 20 Newsgroups and clinical discharge summaries from Obesity NLP Challenge Dataset. We separated the Obesity NLP Challenge Dataset into 12 subsets, as reported in details in [1], e.g. *'Hypertension and Obesity'* and *'Obesity not Hypertension'*. We removed the stop words, converted capital letters to lower cases and

© Springer Nature Switzerland AG 2019
M.-J. Meurs and F. Rudzicz (Eds.): Canadian AI 2019, LNAI 11489, pp. 460–465, 2019.
https://doi.org/10.1007/978-3-030-18305-9_45

lemmatized the sets. The final Science sets had approx. 460,000 word tokens and 36,800 word types, the final Obesity sets had approx. 804,000 word tokens and 32,100 word types.

2   SentiWordNet Sentiment Evaluation

The Obesity datasets[1]. We use numerical IDs for the data sets: *Obesity - 1, Hypertension - 2, Diabetes - 3, Obesity and Hypertension - 4, Obesity not Hypertension - 5, Hypertension not Obesity - 6, Obesity and Diabetes - 7, Obesity not Diabetes - 8, Diabetes not Obesity - 9, Hypertension and Diabetes - 10, Hypertension not Diabetes - 11*, and *Diabetes not Hypertension - 12.*

The results of SWN analysis, reported in Table 1, show that the majority of terms for each subset are objective (87.7%). Among the datasets, subset 4 contains the most negative overall sentiment (0.026).

Table 1. SWN Sentiment Scores and Ratios of the Obesity datasets

Set ID	Pos Score	Neg Score	Obj Score	Overall (Neg-Pos)	Number Neg	Number Pos	Neg Ratio	Pos Ratio
6	0.052	0.070	0.878	0.018	4322	4447	0.493	0.507
9	0.051	0.071	0.879	0.020	4321	4467	0.492	0.508
12	0.052	0.073	0.876	0.021	2931	3016	0.493	0.507
3	0.051	0.072	0.877	0.021	4941	5114	0.491	0.509
10	0.050	0.071	0.879	0.021	4597	4768	0.491	0.509
2	0.051	0.072	0.878	0.021	4971	5163	0.491	0.509
5	0.051	0.073	0.876	0.022	2636	2708	0.493	0.507
7	0.051	0.072	0.877	0.022	3503	3657	0.489	0.511
1	0.048	0.073	0.879	0.022	4142	4341	0.488	0.512
8	0.049	0.072	0.879	0.023	3006	3122	0.491	0.509
11	0.050	0.074	0.876	0.023	3186	3338	0.488	0.512
4	0.049	0.075	0.877	0.026	3744	3948	0.487	0.513
Avg	**0.050**	**0.072**	**0.877**	**0.022**	**3858.333**	**4007.417**	**0.491**	**0.509**

The SWN ratios (Table 1) demonstrate the results are stable across the sets and the differences are insignificant. For example, subset 5 contains the highest percentage of positive terms (49.3%) along with the lowest percent of negative terms (50.7%), thus making it the most positive subset. In some cases, scores and ratios results support each other. For example, subset 6 is the second most positive subset according to SWN ratios (49.3% positive, 50.7% negative), and also the most positive subset according to SWN scores. Subset 4 is the most negative subset according to SWN ratios (48.7% positive, 51.3% negative) and also the most negative subset according to SWN scores.

---

[1] https://www.i2b2.org/NLP/DataSets/Main.php

We performed an independent two-sample Welch's t-test. It shows a reliable estimation on samples with unequal variances and sizes [3]. Its null hypothesis states that the means of both samples are equal ($\mu1 = \mu2$). Difference in score between the most negative (subset 4) and most positive (subset 6) has a p-value $= 0.061$, indicating a 93.9% chance of being *statistically significant*. We confidently reject the null hypothesis and state that the difference between these subsets is *statistically significant*. The largest p-value $= 0.989$ occurs between subsets 3 and 12, having a 1.1% chance of being statistically significant. The result aligns with the overall SWN sentiments scores. Therefore, for these two subsets, we accept the null hypothesis that the means of both subsets are equal. The SWN ratios too show insignificant differences between subsets.

The Reuters-Science Datasets[2]. For the Reuters-Science texts, the score results (Table 2) show that the majority of terms for each dataset are objective (89.24%). The subset *Crypto* is the most positive (score $= -0.011$), and the subset *Med* is the most negative (score $= -0.005$).

Table 2. SWN Sentiment Scores and Ratios for the Reuters-Science data

Dataset	Pos Score	Neg Score	Obj Score	Overall (Neg-Pos)	Number Pos	Number Neg	Pos Ratio	Neg Ratio
Crypto	0.060	0.049	0.890	-0.011	78341	75301	51.0	49.0
Space	0.049	0.039	0.912	-0.010	74604	71826	50.9	49.1
Electronics	0.056	0.046	0.898	-0.010	45278	43947	50.7	49.3
Med	0.066	0.061	0.872	-0.005	71019	69175	50.7	49.3
Avg	**0.058**	**0.049**	**0.893**	**-0.009**	**67311**	**65062**	**50.8**	**49.2**

The sentiment ratios (Table 2) contain approx. 1% difference between positive and negative terms. Among the data sets, changes in those differences are insignificant: the most positive *Crypto* has the highest percent of positive terms (51.0%) and the lowest percent of negative terms (49.0%), whereas the most negative *Med* has the lowest percent of positive terms (50.7%) and the highest percent of negative terms (49.3%). The Welch's t-test results show that the subsets *Crypto* and *Med* have a p-value of 0.168, meaning that the difference between the subsets have an 83.2% change of being statistically significant. The subsets *Space* and *Crypto* contain the highest p-value (0.757), having only a 24.3% chance of being statistically significant.

3 Unsupervised Sentiment Analysis

Obesity Dataset. We trained a Word2Vec model on all the Obesity subsets, and then computed the cosine distance between each subset. For the Doc2Vec evaluation, we also trained a Doc2Vec model on all the documents in the dataset. We then created an

---

[2] http://www.daviddlewis.com/resources/testcollections/reuters21578/

inferred vector for each of the subsets, and then calculated the cosine similarities between each of the inferred vectors.

The Word2Vec similarities (Table 3) indicate that similarity of the terms in the subsets is very high. The subsets with the highest cosine similarity are subset 10 and subset 3, with a 0.9997 cosine similarity. This result is supported by both the SWN scores and ratios, as these both have the same overall sentiment score of 0.021 and the same SWN ratios (49.1% positive, 50.9% negative). The subsets with the lowest cosine similarities are subset 5 and subset 4, with a 0.9722 cosine similarity. This result is supported by SWN ratios, as subset 5 contains the highest percentage of positive terms (49.3%) and the lowest percent of negative terms (50.7%), and subset 4 contains the lowest percentage of positive terms (48.7%) and the highest percent of negative terms (51.3%).

Table 3. Obesity Word2Vec Cosine Similarity

ID	1	2	3	4	5	6	7	8	9	10	11	12
1	1.0000	0.9987	0.9977	0.9986	0.9831	0.9955	0.9989	0.9966	0.9954	0.9980	0.9947	0.9892
2	0.9987	1.0000	0.9994	0.9972	0.9820	0.9987	0.9981	0.9945	0.9986	0.9996	0.9953	0.9916
3	0.9977	0.9994	1.0000	0.9957	0.9828	0.9987	0.9982	0.9913	0.9995	0.9997	0.9922	0.9942
4	0.9986	0.9972	0.9957	1.0000	0.9722	0.9922	0.9985	0.9936	0.9927	0.9972	0.9912	0.9821
5	0.9831	0.9820	0.9828	0.9722	1.0000	0.9855	0.9788	0.9855	0.9836	0.9791	0.9855	0.9925
6	0.9955	0.9987	0.9987	0.9922	0.9855	1.0000	0.9946	0.9919	0.9995	0.9980	0.9948	0.9949
7	0.9989	0.9981	0.9982	0.9985	0.9788	0.9946	1.0000	0.9917	0.9958	0.9987	0.9899	0.9893
8	0.9966	0.9945	0.9913	0.9936	0.9855	0.9919	0.9917	1.0000	0.9896	0.9915	0.9982	0.9838
9	0.9954	0.9986	0.9995	0.9927	0.9836	0.9995	0.9958	0.9896	1.0000	0.9988	0.9920	0.9955
10	0.9980	0.9996	0.9997	0.9972	0.9791	0.9980	0.9987	0.9915	0.9988	1.0000	0.9920	0.9913
11	0.9947	0.9953	0.9922	0.9912	0.9855	0.9948	0.9899	0.9982	0.9920	0.9920	1.0000	0.9862
12	0.9892	0.9916	0.9942	0.9821	0.9925	0.9949	0.9893	0.9838	0.9955	0.9913	0.9862	1.0000

The results on the Doc2Vec model are more ambiguous. The highest cosine similarity occurs between subset 10 and subset 3 (0.9623), supported by both the SWN scores and the Word2Vec cosine similarity results. At the same time the lowest cosine similarity occurs between subset 12 and subset 8 (0.3546). This result is not strongly supported by SWN scores or ratios.

Reuters-Science Dataset. For the Word2Vec cosine similarity on the Reuters-Science dataset (Table 4), the subsets *Med* and *Electronics* have the highest cosine similarity (0.727). However, both SWN scores and ratios do not support this similarity. The lowest cosine similarity score occurs between *Space* and *Crypto* (0.542), which is not supported by SWN scores or ratios.

Table 4. Reuters-Science Word2Vec Cosine Similarity

	Crypto	Electronics	Med	Space
Crypto	1.0000	0.6896	0.6109	0.5422
Electronics	0.6896	1.0000	0.7274	0.6061
Med	0.6109	0.7274	1.0000	0.6382
Space	0.5422	0.6061	0.6382	1.0000

The results of the Doc2Vec cosine similarities show that the highest cosine similarity occurs between the subsets *Med* and *Electronics* (0.379), which is not supported by the results in SWN. The lowest cosine similarity occurs between the subsets *Crypto* and *Space* (−0.32), which is also not supported by the results of SWN. Note that the Word2Vec model showed the same result for both the highest and lowest cosine similarities.

To enforce differentiation between positive and negative sentiment, we created two lists: **positive_terms:** "love", "good", "excellent", "enjoy", "happiness"; **negative_terms:** " sorry", "sad", "bad", "worst", "anger". All terms in the *positive_terms* list have a SWN positive score of 1 and negative score of 0, and all terms in the *negative_terms* list have a SWN negative score of 1 and a positive score of 0. The terms were *randomly* chosen from SWN terms with positive (negative) scores of 1. We added the two lists as training sentences/documents to our Word2Vec/Doc2Vec models, introducing these terms into the models. We then calculated the cosine similarities of each subset to the lists and evaluated the results with the benchmarks.

Table 5. Obesity Word2Vec Cosine Similarity against Sentiment Lists.

	1	2	3	4	5	6	7	8	9	10	11	12
Pos List	0.635	0.635	0.640	0.626	0.660	0.640	0.637	0.630	0.642	0.636	0.626	0.655
Neg List	0.789	0.789	0.781	0.798	0.749	0.781	0.784	0.795	0.779	0.787	0.793	0.756

Obesity Dataset. The Word2Vec cosine similarity results (Table 5) show that subset 5 has the highest cosine similarity to the list *positive_terms*, indicating that this subset contains the most positive context out of all the subsets. We also see that subset 4 has the highest cosine similarity to the list *negative_terms*, indicating that this subset contains the most negative context out of all the subsets. Both results are strongly supported by the SWN benchmarks.

However, Doc2Vec cosine similarity score results against the *positive_terms* and *negative_terms* lists were not supported by any of the SWN results.

Reuters-Science Dataset. In the Reuters-Science dataset, we see that the subset *Crypto* contains the highest Word2Vec cosine similarity score to the *positive_terms* list, indicating that this subset contains the most positive context out of all the subsets (Table 6). This result is strongly supported by the SWN benchmarks, where *Crypto* contains the most positive overall sentiment score and the most positive overall sentiment ratio. However, the results of the *negative_terms* list is not supported by the SWN benchmarks. The Doc2Vec cosine similarity evaluation again was not supported by the SWN results.

Table 6. Reuters-Science Word2Vec Cosine Similarity against Sentiment Lists

	Crypto	Electronics	Med	Space
Pos List	0.0786	0.0656	0.0339	0.0271
Neg List	−0.2075	−0.1574	−0.1987	−0.0980

4 Discussion and Future Work

The following studies directly relate to ours. In unsupervised setting, clinicians' stance concerning treatment options was studied in [7]. The study focused on rule-based Information Extraction of epistemic terms appearing in articles published in medical journals. Implicit and explicit judgments expressed by radiologists and clinicians in clinical notes were studied in [2]. The authors concluded that a simple count method is not suited to analyze sentiment in clinical narratives. In supervised sentiment analysis, Tang et al. [8] showed that sentiment specific word embeddings can outperform by 10% generic neural network models such as Word2Vec. We, on the other hand, showed that Word2Vec can have a reliable sentiment analysis, by comparing its performance with the SentiWordNet sentiment assessment.

We applied Welch's t-test to generalize the results. Word2Vec reliably predicted sentiments of the Obesity dataset and of its subsets. Word2Vec also showed a reliable performance in predicting the most negative subset from the Obesity dataset and the most positive subset from the Reuters-Science dataset when evaluated against the sentiment lists. A more detailed study of Doc2Vec is left for future work.

References

1. Chen, Q., Sokolova, M.: Word2Vec and Doc2Vec in Unsupervised Sentiment Analysis of Clinical Discharge Summaries. arXiv:1805.00352
2. Deng, Y., Stoehr, M., Denecke, K.: Retrieving attitudes: sentiment analysis from clinical narratives. In: Medical Information Retrieval Workshop at SIGIR 2014 (2014)
3. Derrick, B., Toher, D., White, P.: Why Welchs test is Type I error robust. Quant. Methods Psychol. **12**(1), 30–38 (2016)
4. Esuli, A., Sebastiani, F.: SENTIWORDNET: a publicly available lexical resource for opinion mining. In: LREC 2006, Genova (2006)
5. Le, Q., Mikolov, T.: Distributed representations of sentences and documents. In: 31st ICML (2014)
6. Mikolov, T., Chen, K., Corrado, G., Dean, J.: Efficient estimation of word representations in vector space. CoRR, vol. abs/1301.3781, January 2013
7. Sokolova, M., Ioshikhes, I., Poursepanj, H., MacKenzie, A.: Helping parents to understand rare diseases. In: NLP for Medicine and Biology - RANLP 2013 (2013)
8. Tang, D., Wei, F., Yang, N., Zhou, M., Liu, T., Qin, B.: Learning sentiment-specific word embedding for twitter sentiment classification. In: ACL (2014)

Efficient Sequence Labeling
with Actor-Critic Training

Saeed Najafi[1(✉)], Colin Cherry[2(✉)], and Grzegorz Kondrak[1(✉)]

[1] University of Alberta, Edmonton, Canada
{snajafi,gkondrak}@ualberta.ca
[2] Google, Montreal, Canada
colin.a.cherry@gmail.com

Abstract. Neural approaches to sequence labeling often use a Conditional Random Field (CRF) to model their output dependencies. We set out to establish Recurrent Neural Networks (RNNs) as an efficient alternative to CRFs especially in tasks with large number of output labels. We propose an adjusted actor-critic reinforcement learning algorithm to fine-tune RNN network (AC-RNN). Our comprehensive experiments suggest that AC-RNN efficiently matches the performance of the CRF on NER and CCG tagging, and outperforms it on Machine Transliteration; with an overall faster training time, and smaller memory footprint.

1 Introduction

Conditional Random Fields (CRF) built over a neural feature layer have recently emerged as a best practice for sequence labeling tasks in NLP [6,9,11,13]. Alternatively, one can track the same output dependencies using a Recurrent Neural Network (RNN) over the output sequence in an auto-regressive fashion generating output tokens from left to right, similar to the decoder component in Neural Machine Translation (NMT). Using a decoder RNN instead of a CRF has several potential advantages, such as simplified implementation, tracking longer dependencies, and allowing for larger output vocabularies. However, shifting from the CRF's sequence-level training objective to the decoder RNN's sequence of token-level objectives may lead to suboptimal performance due to exposure bias. Exposure bias stems from the token-level maximum likelihood objective typically used in decoder RNN training, which does not expose the model to its own errors [2]. During training, decoder RNNs are typically conditioned on the gold-standard labels of the previous time steps (a procedure known as Teacher Forcing [4]), while at test time, the model is conditioned on its own predictions, creating a train-test mismatch. As sequence-level models, CRFs are immune to exposure bias.

We set out to establish the decoder RNN as an attractive alternative to the CRF by adopting a simple and effective training strategy to counter the exposure-bias problem, and by providing an experimental comparison to demonstrate that our strategy helps decoder RNN match the CRF in accuracy, and

M.-J. Meurs and F. Rudzicz (Eds.): Canadian AI 2019, LNAI 11489, pp. 466–471, 2019.
https://doi.org/10.1007/978-3-030-18305-9_46

surpass it in flexibility. Our chosen training method is Actor-Critic reinforcement learning [8]. We demonstrate that on Named Entity Recognition (NER) and Combinatory Categorical Grammar (CCG) supertagging, the AC-RNN can match the performance of the CRF, while training more efficiently. To demonstrate its flexibility, we also test our method on machine transliteration, a monotonic transduction problem that straddles the boundary between sequence labeling and full sequence-to-sequence modeling. On NER tagging, we also demonstrate that our reinforcement learning method is significantly better than the related work of scheduled sampling [2].

2   Prior Work

In sequence labeling, several neural methods have recently been shown to outperform earlier systems that use hand-engineered features. For the tasks of POS tagging, chunking and NER, Huang et al. [6] apply a CRF output layer on top of a bi-directional RNN over the source. For the NER task, Lample et al. [9] extend the RNN layer with character-level RNNs that capture information about word prefixes and suffixes. We build upon Lample et al.'s approach, replacing their CRF with a decoder RNN trained with an adjusted Actor-Critic objective. We also apply the AC-RNN to CCG supertagging [3], where the model labels each word in a sentence with one of 1,284 morphosyntactic categories from CCG-bank [5]. Finally, we consider transliteration, where the goal is to convert a word from a source script to a target script on the basis of the word's pronunciation. This task goes beyond sequence labeling by allowing many-to-many monotonic alignments between the source and target symbols.

Other approaches have been proposed to address exposure bias in decoder RNN, especially for NMT. Bengio et al. [2] introduce the notion of scheduled sampling as the decoder RNN is gradually exposed to its own errors, where a sampling probability is annealed at every training epoch so that we use gold-standard labels at the beginning of the training, but while approaching the end, we instead condition the current prediction on the model-generated previous labels. Ranzato et al. [16] apply the REINFORCE algorithm [19] to NMT, to train the network with a reward derived from the BLEU score of each generated sequence. Bahdanau et al. [1] apply the actor-critic algorithm in NMT by applying a reward-reshaping approach to construct intermediate BLEU feedback at each step. Unlike these previous works, we demonstrate that sequence labeling tasks benefit from the binary rewards that are available at each step. In contrast to [1], our method employs a simpler critic architecture, without any schedules to pre-train the critic model.

3   Method

We adopt the actor-critic algorithm [8,12,17] to fine-tune the decoder RNN. In AC training, the decoder RNN first generates a greedy output sequence according to its current model, similar to how it would during testing. We calculate a

---

Algorithm 1. Adjusted Actor-Critic Training

- Input: Source X, Target Y, and n as hyper-parameter
- Greedy decode X using θ to get:
    - the output sequence $\hat{Y} = (\hat{y}_1, ..., \hat{y}_l)$
    - decoder RNN states $D = (d_1, ..., d_l)$
    - context vectors $C = (c_1, ..., c_l)$
- For each output target position t:
    - $r_t = 1$ if $\hat{y}_t = y_t$, 0 otherwise
    - $V_{\theta'}(t) = \text{CriticNetwork}(d_t, c_t, \theta')$
- $loss_\theta = 0$; $loss_{\theta'} = 0$
- For each output target position t:
    - $G_t = \sum_{i=0}^{n-1} [r_{t+i}] + V_{\theta'}(t+n)$
    - $\delta_t = G_t - V_{\theta'}(t)$
    - $a\delta_t = \text{adjust}(y_t, \hat{y}_t, \delta_t) \times \delta_t$
    - $loss_\theta = loss_\theta - a\delta_t \ln p_\theta(\hat{y}_t|X, \hat{y}_{t'<t})$
    - $loss_{\theta'} = loss_{\theta'} + \delta_t \times \delta_t$
- Back-propagate through $loss_\theta$ as normal to update θ
- Perform a semi-gradient step along loss θ' to update θ'

---

sequence-level credit (return) for each prediction by comparing it to the gold-standard. The AC update modifies our RNN to improve credits at each step. This process exposes the decoder to its own errors, alleviating exposure bias. Algorithm 1 provides pseudo code for the training process, which we expand upon in the following paragraphs.

We define the token-level reward r_t as $+1$ if the generated token $\hat{y}_t$ is the same as the gold token y_t, and as 0 otherwise. We compute the sequence-level credit G_t for each decoding step using the multi-step Temporal Difference return [17]:

$$G_t = \sum_{i=0}^{n-1} [r_{t+i}] + V_{\theta'}(t+n)$$

The step count n allows us to control our bias-variance trade-off, with a large n resulting in less bias but higher variance. The critic $V_{\theta'}(t)$ is a regression model that estimates the expected return $E[G_t]$, taking the context vector c_t[1] and the decoder's hidden state d_t as input. It is trained jointly alongside our decoder RNN, using a distinct optimizer (without back-propagating errors through c_t and d_t). With this critic in place, the update for the AC algorithm is defined as

$$\frac{\partial J_{ac}(\theta)}{\partial \theta} = \sum_t \frac{\partial \log(p_\theta(\hat{y}_t|X, \hat{y}_{t'<t}))}{\partial \theta} (\delta_t)$$

where:

$$\delta_t = G_t - V_{\theta'}(t)$$

---

[1] The context vector summarizes the input X for the current time step via soft or hard attention mechanisms [10].

The AC update changes the prediction likelihood proportionally to the advantage δ_t of the token $\hat{y}_t$. Therefore, if $G_t > V_{\theta'}(t)$, the decoder should increase the likelihood. The AC error $\frac{\partial J_{ac}(\theta)}{\partial \theta}$ back-propagates only through the actor's prediction likelihood p_θ.

Adjusted Training: Due to the inevitable regression error of the critic (two-layer feed-forward network), and the fact that it is randomly initialized at the beginning, the advantage δ_t can undesirably become negative for a correctly-selected tag, or positive for a wrongly-selected tag. Optimizing the network according to these invalid advantages would increase the probability of the wrong tags, while decreasing the probability of the true tag. In such cases, to help critic update itself and form better estimates in the next iteration, we clip δ_t to zero by defining the adjusted advantage $a\delta_t$ as $\mathrm{adjust}(y_t, \hat{y}_t, \delta_t) \times \delta_t$ where:

$$\mathrm{adjust}(y_t, \hat{y}_t, \delta_t) = \begin{cases} 0 & \text{if} \quad \hat{y}_t = y_t \quad \& \quad \delta_t < 0 \\ 0 & \text{if} \quad \hat{y}_t \neq y_t \quad \& \quad \delta_t > 0 \\ 1 & otherwise \end{cases}$$

By setting the advantage $a\delta_t$ to 0, the adjust term effectively switches off the entire actor update when the advantage has the wrong polarity. Note that the critic is always updated.

4   Experiments

We comprehensively study the effect of adjusted actor-critic training on RNN, and compare AC-RNN with CRF on three tasks: NER tagging, CCG supertagging, and machine Transliteration. We then compare our technique with Scheduled Sampling on NER. The full experimental setup is discussed in [14,15].

Training Details: Our different models share the same encoder, using the same number of hidden units. The maximum-likelihood training is done with the Adam optimizer [7] with a learning rate of 0.0005. The RL training is done with the mini-batch gradient ascent using a fixed step size of 0.5 for NER & CCG, and 0.1 for Transliteration. The critic is trained with a separate Adam optimizer with the learning rate of 0.0005. We employ a linear-chain first-order undirected graph in the CRF model. As performance varies depending on the random initialization, we train each model 20 times for NER and 5 times for CCG using different random seeds which are the same for all models. We report scores averaged across these runs $\pm$ the standard deviations. Due to time constraints, for the transliteration experiments, we train each model only once.

Evaluation: We compute the standard evaluation metric for each task: entity-level F1-score for NER, tagging accuracy for CCG, and word-level accuracy for transliteration. For the models with decoder RNNs, we report the results achieved using a beam search with a beam of size 10. For the NER and CCG experiments, we conduct the significance tests on the unseen final test sets, using the Student's t-test over random replications at the significance level of 0.05.

4.1   Results

The results of the NER experiments are shown in Table 1. As expected, we observe that by modelling the output dependencies using either an RNN or a CRF, we achieve a significant improvement over the independent prediction of the labels INDP, about 1% F1-score on both English and German datasets. Moreover, AC-RNN significantly outperforms both RNN and CRF on both English and German test sets. These results demonstrate that AC-RNN is successful at overcoming the RNN's exposure bias, and represents a strong alternative to CRF for NER.

Table 1. Average entity-level F1-score for English & German NER on the CoNLL-2003 test sets [18]. Average top-1 accuracy on English CCG supertagging. The word-level transliteration accuracy on the development sets of NEWS-2018 shared task.

Model	NER (En)	NER (De)	CCG	EnCh	EnJa	EnPe	EnTh
INDP	89.77 ± 0.21	72.15 ± 0.57	94.25 ± 0.11	na	na	na	na
CRF	90.80 ± 0.19	73.59 ± 0.36	94.15 ± 0.11	67.6	45.8	75.6	32.2
RNN	90.75 ± 0.23	73.52 ± 0.36	94.28 ± 0.09	70.6	51.6	76.3	39.7
AC-RNN	$\mathbf{90.96 \pm 0.15}$	$\mathbf{73.82 \pm 0.29}$	$\mathbf{94.39 \pm 0.06}$	**72.3**	**52.4**	**77.8**	**41.4**

On CCG supertagging, AC-RNN is significantly better than all other models. We observe that, due to the large output vocabulary size of the task (1284 supertags), CRF is five times slower than AC-RNN during training, while the batched version of its Forward algorithm requires six times more GPU memory during training. The transliteration results show that AC-RNN outperforms CRF (likely due to CRF's inability to predict an output of a different length from its input), as well as RNN (likely due to its exposure bias). The transliteration experiments support our hypothesis that AC-RNN is more generally-applicable than CRF, and the improvements from the adjusted actor-critic training transfer to other tasks.

Scheduled Sampling Comparisons: we compare the adjusted actor-critic objective to Scheduled Sampling [2] (SS-RNN). On German NER, both systems improve over RNN, but AC-RNN (**73.82** ± 0.29) significantly outperforms SS-RNN (73.65 ± 0.29).

5   Conclusion

We have proposed an adjusted actor-critic algorithm to train encoder-decoder RNNs for sequence labeling tasks. Our proposed AC-RNN is specialized to sequence-labeling by taking advantage of the per-position rewards. On NER and CCG supertagging, our system significantly outperforms both RNN and CRF, establishing the AC-RNN as an efficient alternative for sequence labeling.

We have also demonstrated the advantages of the AC-RNN in terms of its flexibility, fast training, and small memory footprint. Our models are available at https://github.com/SaeedNajafi/ac-tagger.

References

1. Bahdanau, D., et al.: An actor-critic algorithm for sequence prediction. In: ICLR (2017)
2. Bengio, S., Vinyals, O., Jaitly, N., Shazeer, N.: Scheduled sampling for sequence prediction with recurrent neural networks. In: NIPS, pp. 1171–1179 (2015)
3. Clark, S., Curran, J.R.: The importance of supertagging for wide-coverage ccg parsing, In: COLING (2004)
4. Goodfellow, I., Bengio, Y., Courville, A.: Deep Learning. MIT Press, Cambridge (2016)
5. Hockenmaier, J., Steedman, M.: CCGbank: a corpus of CCG derivations and dependency structures extracted from the Penn treebank. Comput. Linguist. **33**(3), 355–396 (2007)
6. Huang, Z., Xu, W., Yu, K.: Bidirectional LSTM-CRF models for sequence tagging. CoRR abs/1508.01991 (2015)
7. Kingma, D.P., Ba, J.: Adam: a method for stochastic optimization. In: ICLR (2015)
8. Konda, V.R., Tsitsiklis, J.N.: On actor-critic algorithms. SIAM J. Control Optim. **42**(4), 1143–1166 (2003)
9. Lample, G., Ballesteros, M., Subramanian, S., Kawakami, K., Dyer, C.: Neural architectures for named entity recognition. In: NAACL-HLT, pp. 260–270 (2016)
10. Luong, T., Pham, H., Manning, C.D.: Effective approaches to attention-based neural machine translation. In: EMNLP, pp. 1412–1421 (2015)
11. Ma, X., Hovy, E.: End-to-end sequence labeling via bi-directional LSTM-CNNS-CRF. In: ACL, pp. 1064–1074 (2016)
12. Mnih, V., et al.: Asynchronous methods for deep reinforcement learning. In: ICML, vol. 48, pp. 1928–1937 (2016)
13. Moon, S., Neves, L., Carvalho, V.: Multimodal named entity recognition for short social media posts. In: Proceedings of the 2018 Conference of the North American Chapter of the Association for Computational Linguistics: Human Language Technologies, New Orleans, Louisiana, vol. 1 (Long Papers), pp. 852–860. Association for Computational Linguistics, June 2018. https://doi.org/10.18653/v1/N18-1078. https://www.aclweb.org/anthology/N18-1078
14. Najafi, S.: Sequence labeling and transduction with output-adjusted actor-critic training of RNNs (2018). https://doi.org/10.7939/R39Z90T8B
15. Najafi, S., Cherry, C., Kondrak, G.: Efficient sequence labeling with actor-critic training. CoRR (2018). http://arxiv.org/abs/1810.00428
16. Ranzato, M., Chopra, S., Auli, M., Zaremba, W.: Sequence level training with recurrent neural networks. In: ICLR (2016)
17. Sutton, R.S., Barto, A.G.: Introduction to Reinforcement Learning, 1st edn. MIT Press, Cambridge (1998)
18. Tjong Kim Sang, E.F., De Meulder, F.: Introduction to the CoNLL-2003 shared task: Language-independent named entity recognition. In: CoNLL (2003)
19. Williams, R.J.: Simple statistical gradient-following algorithms for connectionist reinforcement learning. Mach. Learn. **8**(3–4), 229–256 (1992)

Detecting Depression from Voice

Mashrura Tasnim[(✉)] and Eleni Stroulia

Department of Computing Science, University of Alberta, Edmonton, AB, Canada
{mashrura,stroulia}@ualberta.ca

Abstract. In this paper, we present our exploration of different machine-learning algorithms for detecting depression by analyzing the acoustic features of a person's voice. We have conducted our study on benchmark datasets, in order to identify the best framework for the task, in anticipation of deploying it in a future application.

Keywords: Depression · Acoustic features · Classification · Regression

1 Introduction

Depression is the most common psychological disorder affecting more than 300 million people around the globe and is considered as the leading cause of disability worldwide. Current depression diagnostic instruments require active participation of the depressed individuals. But due to lack of awareness and the nature of the disorder itself, a large percentage of population refrain from seeking expert assistance. Recent studies reveal that depression is reflected in behavioral fluctuations of certain day-to-day activities and also in the ways people talk [16]. These findings have motivated a wave of research efforts aimed at developing automated depression-detection methods based on vocal acoustic features. Introduction of Depression Recognition Sub-Challenge (DSC) as a part of Audio/Visual Emotion Challenge (AVEC) since 2013 has accelerated interventions in depression recognition combining different modalities, i.e., audio, video and text features [15]. Different directions of feature engineering, algorithms and contextual information incorporation has been explored in four challenges taken place this far.

In our work, we are interested in developing a practical system that can capture the audio of the users' voice during phone-call conversations and analyze it to detect their depression level. A pre-requisite for such a system is a model capable of detecting evidence of depression from conversational audio. In this work we explored the effectiveness of different machine-learning algorithms for the anticipated depression detection model with the currently available AVEC data sets. The rest of the paper is organized as follows. Section 2 reviews the related research on depression detection based on a subject's vocal biomarkers. We explain the analysis methodology and experimental results in Sect. 3. The paper has been concluded discussing future prospects in Sect. 4.

This work has been partially funded by the GRA Rice Graduate Scholarship in Communications, the AGE-WELL NCE and NSERC.

M.-J. Meurs and F. Rudzicz (Eds.): Canadian AI 2019, LNAI 11489, pp. 472–478, 2019.
https://doi.org/10.1007/978-3-030-18305-9_47

2 Background and Related Work

Much of the work in this area (including our own) has been done on two publicly available benchmark audio datasets: AVEC 2013 and AVEC 2017. The AVEC 2013 audio corpus is a subset of the audio-visual depressive language corpus (AVDLC) [15]. 84 subjects performed two Power Point guided tasks in German while being recorded, resulting in 300 recordings. Each of training, development and test partition consists of 100 recordings. The training and development partitions are labelled with depression score in 21 items BDI-II scale ranging from 0–63, where score greater than 19 is considered to belong in the *"depressed"* class [6]. The mean BDI-II score of the recordings is M = 15.3 with standard deviation (SD) = 12.3. The AVEC 2017 DSC dataset consists of 189 audio recordings of clinical interviews [11]. The recordings are labelled with depression scores of the participants in the 8 items PHQ-8 depression inventory ranging 0–24. Score of 10 or greater is considered as major depression. The average depression severity on the training and development set of the challenge is M = 6.67 (SD = 5.75). The training, development and test data sets contain 107, 35 and 47 audio recordings respectively.

Distinguishing depressed individuals from non-depressed ones is a binary classification task, while determining severity by predicting depression score formulates a regression problem. The binary-classification task has been explored on a variety of datasets in [1,6–8,13] and [3]. Prosodic, spectral, cepstral, glottal, energy related features have been experimented for this purpose. Moore II *et al.* reported superiority of glottal features over prosodic ones [8], while spectral and energy related features were found most effective by Lopez-Otero *et al.* [6]. Low *et al.* reported significant gender dependency in classification accuracy using Teager energy operator (TEO) features [7]. Besides these, covariance structure of Gaussian Mixture Model (GMM) of recorded speech was found informative by Cummins *et al.* [1]. To overcome small sample size in high dimensional feature space, Moore II *et al.* adopted one-feature-at-a-time strategy and Sanchez *et al.* used backward elimination. The highest classification accuracy was reported by Moore II *et al.* (95.6%) on recordings from 15 depressed and 18 control subjects using quadratic discriminant analysis [8], followed by SVM achieving 87.0%, 81.3% and 65.8% accuracy reported in [7,13] and [3] respectively. Linear discriminate analysis (LDA), adopted by Lopez-Otero *et al.*, performed the classification task with 70% accuracy on the AVEC 2013 development dataset.

Besides binary classification, researchers have also endeavoured to determine depression severity by predicting depression score using audio features. AVEC 2013 depression sub-challenge (DSC) dataset has been used in [5,10,17] and [9] to predict BDI-II score. In [17] combination of formant and delta-mel-cepstral features were used to train Gaussian staircase regression system. Their subject-based and subject-independent adaptation achieved root mean square error (RMSE) 8.68 and 7.42 respectively. He & Cao used combination of Median Robust Extended Local Binary Patterns (MRELBP) and AVEC 2013 baseline features (mentioned as hand-crafted features in the literature) with deep-learned features extracted from raw audio and spectrogram images for their proposed deep convolutional neural network (DCNN) architecture [5]. The proposed model obtained RMSE 9.89 and mean absolute error (MAE) 8.19. Özkanca *et al.* compared the performance of their proposed framework using Turkish and German (AVEC 2013) dataset. They applied minimum redundancy maximum relevance (MRMR) feature selection criteria on AVEC 2013 baseline feature set of 2268 features prior to using Support Vector Regressor (SVR) [10]. The best RMSE of 9.42 was reported on this dataset. Morales also applied SVR for comparing depression detection systems on several publicly available depression datasets using prosodic and speech rate related features and documented RMSE = 10.70 (MAE 8.59).

The AVEC 2017 DSC dataset was used in [2,12,14,18] and [4]. In addition to challenge baseline audio features, Sun *et al.* took text topics into account [14], while Gong & Poellabauer considered a more extended set of features, including audio, video and text features. Yang *et al.* extracted deep learned features from spectrograms and Samareh *et al.* added Delta and Delta-Delta coefficients, mean, median, SD, peak-magnitude to RMS ratio to the set of challenge baseline audio features. On the AVEC 2017 dataset best performance was obtained using Deep Convolutional Neural Network (DCNN) and Deep Neural Network (DNN) based audio visual multi-modal depression recognition framework [18], followed by stochastic gradient descent (SGD) regressor [4], random forest [12] and SVM [2]. RMSE 6.32 (MAE 4.40) and 5.45 (MAE 4.32) were reported by [2] and [12] respectively on the development set using audio features exclusively. The challenge baseline RMSE was set 6.74 (MAE 5.36) and 7.78 (MAE 5.72) for development and test set respectively [11].

Considering the fact that conversational audio provides valuable information to assist depression detection, we plan to develop our audio based depression detection system from phone conversation. Here we analyze performance of different classification and regression model for sensing prevalence and severity of depression with a view to finding the best model for future usage.

3   Our Method

We applied four algorithms for the classification and regression tasks using the AVEC 2013 [15] and AVEC 2017 [11] data sets.

Data Pre-processing: The audio recording in these data sets is anywhere between 5 and 50 min long. The AVEC 2013 corpus provides with features extracted from 20s long windows (shifting forward at the rate of one second). For AVEC 2017 dataset we did segmentation based on subjects' voice activity.

In this work we experimented on AVEC 2013 baseline feature set consisting of 2268 audio features. The feature set comprises of 76 low-level descriptors (LLD) features and their statistical, regressional and local minima/maxima related functionals. The LLD features include energy and spectral related, voicing related, delta coefficients of the energy/spectral features, delta coefficients of the voicing related LLDs and voiced/unvoiced durational features. We standardized the features by removing the mean and scaling to unit variance. We applied Principal Component Analysis (PCA) to identify the minimum number of features that is capable to retain 95% of the variance, resulting in 791 features.

Model Training: We have trained the following four algorithms for each of classification and regression task on the processed features from the training partitions of the datasets.

Random Forest: 100 estimator trees with learning rate of 0.1 for both the classification and regression tasks.

Support Vector Machine (SVM): We used the radial basis function (RBF) kernel for SVM.

Gradient Boosting Tree (GBT): For both GBT classifier and regressor, 100 estimator trees were used.

Deep Neural Network: The network consists of three fully connected layers with 512, 256 and 512 neurons respectively. To avoid overfitting 30–50% dropout was added between layers. We trained the model with mini-batch gradient descent with a batch size of 64. Categorical cross-entropy and mean squared error was considered as the loss function for classification and regression respectively. We exploited the best model chosen from 500 epochs. The learning rate of 10^{-4} was found best fit for classification while 10^{-3} did well for regression. Adam optimizer was used for model optimization.

3.1   Experimental Results and Discussion

The first question of interest in this study is *"how effective are the chosen algorithms in distinguishing between depressed and non-depressed individuals?"*. The accuracy, precision and recall of binary classification are summarized in Table 1.

Table 1. "Depressed" and "Not-depressed" classification

Algorithm	AVEC 2013			AVEC 2017 Dev			AVEC 2017 Test		
	Acc.	Prec.	Rec.	Acc.	Prec.	Rec.	Acc.	Prec.	Rec.
SVM	67.78	0.64	0.68	57.92	0.50	0.58	63.81	0.58	0.64
Random forest	64.50	0.51	0.64	60.19	0.52	0.60	69.24	0.58	0.69
GBT	62.26	0.57	0.62	58.40	0.54	0.58	63.58	0.59	0.64
DNN	**72.85**	0.70	0.72	**74.65**	0.49	0.56	**80.11**	0.59	0.64

Based on Table 1, one can see that the deep neural network (DNN) performed best on both data sets. The DNN accuracy for AVEC 2013 dataset is 72.85%, which is a marginal improvement from the accuracy reported by Lopez-Otero *et al.* [6]. The DNN accuracy is higher with the AVEC 2017 data set, which

Table 2. "Level of depression" regression

Algorithm	AVEC 2013		AVEC 2017 Dev		AVEC 2017 Test	
	RMSE	MAE	RMSE	MAE	RMSE	MAE
SVM	10.55	**7.93**	7.50	6.11	6.44	5.37
Random forest	**9.75**	8.21	**6.60**	**5.55**	**6.17**	**5.22**
GBT	14.60	10.38	6.63	5.49	6.26	5.26
DNN	10.74	8.75	8.07	6.67	6.55	5.33
Baseline [11, 15]	11.52	8.93	6.74	5.36	7.78	5.72

may likely be attributed to the fact that the AVEC 2013 data set is smaller and may not be sufficient to train the DNN. The DNN achieved 74.65% and 80.11% accuracy respectively on the development and test partition of the AVEC 2017 dataset. The low recall values (0.56 on development set and 0.64 on test set) indicate that a significant portion of depressed cases has been misclassified. This is a very undesirable phenomenon when considering to apply this method to support real-world diagnosis. One possible reason of this outcome is the imbalanced proportion of class samples in the training set (30 depressed vs 77 not depressed). In the future, we plan to solve this issue by applying synthetic over-sampling technique on the minority class.

The second question of interest is *"how effective are the chosen algorithms in assessing an individual's level of depression?"*. Results of the regression task are summarized in Table 2. The random forest algorithm performed best on both datasets, outperforming the baseline models.

It is important to note, however, that DCNN reported in [5] performs marginally better than our random forest regressor on the AVEC 2013 dataset (RMSE 8.89). As the classes in BDI-II depression scale are 5 to 8 points apart (<14: minimal, 14–19: mild, 20–28: moderate, >28: severe), the current results indicate that there is high possibility of misclassification, implying room for further improvement.

For the AVEC 2017 dataset, most of the existing models use additional data beyond the voice audio, i.e., including video and text features into their process. As the motivation of our work is to find a reliable model to detect depression prevalence and severity from phone-call conversation, we only took audio features into account. In person-invariant unimodal (audio only) scenario results of our random forest model are consistent with results reported in [2] and [12] on the development set. For the test set the only unimodal result is available in conference baseline [11] (RMSE = 7.78, MAE = 5.72) which is outperformed by our proposed model (RMSE = 6.17, MAE = 5.22). Still the range of score on PHQ-8 scale is 0–24 where score higher than 9 indicate major depression, therefore more accurate model will increase reliability of our envisioned system.

4   Conclusions and Future Work

In this work, detection of prevalence and severity of depression from acoustic features of conversational speech in two languages has been explored using different classification and regression algorithms. We have found that the deep neural network performs best in binary classification while random forest gives competitive results for the regression task. In the future, we will consider synthetically balancing the classes as our next measure for performance improvement. We also envision to incorporate our findings in a real-time depression detection application with a view to ensure emotional support when necessary.

References

1. Cummins, N., Epps, J., Sethu, V., Breakspear, M., Goecke, R.: Modeling spectral variability for the classification of depressed speech. In: Interspeech, pp. 857–861 (2013)
2. Dham, S., Sharma, A., Dhall, A.: Depression scale recognition from audio, visual and text analysis. arXiv preprint arXiv:1709.05865 (2017)
3. Fraser, K.C., Rudzicz, F., Hirst, G.: Detecting late-life depression in alzheimer's disease through analysis of speech and language. In: Proceedings of the Third Workshop on Computational Lingusitics and Clinical Psychology, pp. 1–11 (2016)
4. Gong, Y., Poellabauer, C.: Topic modeling based multi-modal depression detection. In: Proceedings of the 7th Annual Workshop on Audio/Visual Emotion Challenge, pp. 69–76. ACM (2017)
5. He, L., Cao, C.: Automated depression analysis using convolutional neuralnetworks from speech. J. Biomed. Inform. **83**, 103–111 (2018)
6. Lopez-Otero, P., Docio-Fernandez, L., Garcia-Mateo, C.: A study of acoustic features for the classification of depressed speech. In: 2014 37th International Convention on Information and Communication Technology, Electronics and Microelectronics (MIPRO), pp. 1331–1335. IEEE (2014)
7. Low, L.S.A., Maddage, N.C., Lech, M., Sheeber, L.B., Allen, N.B.: Detection of clinical depression in adolescents' speech during family interactions. IEEE Trans. Biomed. Eng. **58**(3), 574–586 (2011)
8. Moore II, E., Clements, M.A., Peifer, J.W., Weisser, L.: Critical analysis of the impact of glottal features in the classification of clinical depression in speech. IEEE Trans. Biomed. Eng. **55**(1), 96–107 (2008)
9. Morales, M.R.: Multimodal depression detection: an investigation of features and fusion techniques for automated systems (2018)
10. Özkanca, Y., Demiroglu, C., Besirli, A., Celik, S.: Multi-lingual depression-level assessment from conversational speech using acoustic and text features. In: Proceedings of Interspeech 2018, pp. 3398–3402 (2018)
11. Ringeval, F., et al.: AVEC 2017: Real-life depression, and affect recognition workshop and challenge. In: Proceedings of the 7th Annual Workshop on Audio/Visual Emotion Challenge, pp. 3–9. ACM (2017)
12. Samareh, A., Jin, Y., Wang, Z., Chang, X., Huang, S.: Predicting depression severity by multi-modal feature engineering and fusion. In: Thirty-Second AAAI Conference on Artificial Intelligence (2018)

13. Sanchez, M.H., Vergyri, D., Ferrer, L., Richey, C., Garcia, P., Knoth, B., Jarrold, W.: Using prosodic and spectral features in detecting depression in elderly males. In: Twelfth Annual Conference of the International Speech Communication Association (2011)
14. Sun, B., et al.: A random forest regression method with selected-text feature for depression assessment. In: Proceedings of the 7th Annual Workshop on Audio/Visual Emotion Challenge, pp. 61–68. ACM (2017)
15. Valstar, M., et al.: AVEC 2013: the continuous audio/visual emotion and depression recognition challenge. In: Proceedings of the 3rd ACM International Workshop on Audio/Visual Emotion Challenge, pp. 3–10. ACM (2013)
16. Wang, R., et al.: StudentLife: assessing mental health, academic performance and behavioral trends of college students using smartphones. In: Proceedings of the 2014 ACM International Joint Conference on Pervasive and Ubiquitous Computing, pp. 3–14. ACM (2014)
17. Williamson, J.R., Quatieri, T.F., Helfer, B.S., Horwitz, R., Yu, B., Mehta, D.D.: Vocal biomarkers of depression based on motor incoordination. In: Proceedings of the 3rd ACM International Workshop on Audio/Visual Emotion Challenge, pp. 41–48. ACM (2013)
18. Yang, L., Sahli, H., Xia, X., Pei, E., Oveneke, M.C., Jiang, D.: Hybrid depression classification and estimation from audio video and text information. In: Proceedings of the 7th Annual Workshop on Audio/Visual Emotion Challenge, pp. 45–51. ACM (2017)

A Machine Learning Method
for Nipple-Areola Complex Localization
for Chest Masculinization Surgery

Mohammad Ghodratigohar[1], Kevin Cheung[2], Natalie Baddour[1],
and Hussein Al Osman[1(✉)]

[1] Faculty of Engineering, University of Ottawa, Ottawa, ON, Canada
{mghod021, nbaddour, hussein.alosman}@uottawa.ca
[2] Division of Plastic Surgery, Children's Hospital of Eastern Ontario,
Ottawa, ON, Canada
kcheung@cheo.on.ca

Abstract. Appropriately positioning the Nipple-Areola Complex (NAC) during chest masculinization surgery is a principle determinant of the aesthetic success of the procedure. Nonetheless, today, this positioning process relies on the subjective judgement of the surgeon. Therefore, this paper proposes a novel machine learning solution that leverages Artificial Neural Networks (ANNs) for estimating the NAC location on the chest wall. A dataset composed of 173 pictures of male subjects of various ages and body types was used. The ANN was fed a set of features inputs based on distance ratios between features of the upper body that are common between both biological sexes (e.g. umbilicus, anterior axillary fold, suprasternal notch). Using the proposed ANN regressive model, we achieved a Root Mean Square Error (RMSE) of 0.0617 for the ratio of distances from the suprasternal notch to the center between the NACs, and from the latter point to the umbilicus. Furthermore, an RMSE of 0.0560 for the ratio of the distances between the NACs and from the anterior axillary fold to the umbilicus was obtained. Our results demonstrate that machine learning can be used to support the surgeon in the localization of the NAC for chest masculinization surgery.

Keywords: Machine learning · Neural network regression ·
Chest masculinization surgery · Transgender surgery

1 Introduction

The Nipple-Areola Complex (NAC) is a prominent aesthetic feature of the chest wall for both biological sexes. Female-to-male transgender patients may undergo chest masculinization surgery to achieve a male appearing chest. Often this surgery requires removal of the NAC and reconstruction with placement in a location appropriate for the male chest. In addition to gender reassignment [1, 2], corrective chest surgeries due to trauma or tumors, massive weight loss, or high-grade gynecomastia, can result in the need for NAC repositioning on the newly reshaped thoracic wall [3–5]. The reconstruction of the masculine characteristic shape often presents a challenge to the plastic

© Springer Nature Switzerland AG 2019
M.-J. Meurs and F. Rudzicz (Eds.): Canadian AI 2019, LNAI 11489, pp. 479–485, 2019.
https://doi.org/10.1007/978-3-030-18305-9_48

surgeon during preoperative planning [1]. Therefore, there is a growing interest in investigating methods for male breast configuration associated with various corrective or cosmetic chest surgeries [6, 7]. For instance, Beer et al. [8] measure the thoracic circumference and the length of the sternal notch to localize the horizontal and vertical position of the NAC. Based on the golden number Phi, Bishara et al. [9] localize the horizontal and vertical position of the NAC by measuring the distances between the umbilicus-anterior axillary fold apex and umbilicus-suprasternal notch. To the best of our knowledge, none of the NAC positioning methods are automated. Moreover, they are based on a small sample population. These techniques require the surgeon to perform the measurements and computations during pre-operative planning. Due to the labor involved and the possibility of miscalculation, these schemes can become clinically impractical [10]. Therefore, in this paper we propose an image-based method to automatically localize the appropriate location of the NAC by leveraging artificial intelligence techniques.

2   Method

2.1   Landmarks

Since individual anatomic features affect the position of the NAC [11], we estimate the position in the image in relation to key landmarks in the upper body, namely the Suprasternal Notch (SN), Anterior Axillary Fold (AX), Nipple (N), Umbilicus (U), and Center of Inter-Nipple Distance (CN), which falls on the sagittal plane. Figure 1 depicts the landmarks used to localize the NAC. Since we are using images, distances are measured in pixels. We use the ratios of distances between some of the landmarks to create a set of features to feed to the ANN. Likewise, the ANN outputs ratios of distances between landmarks that enable the calculation of the coordinates of the NAC.

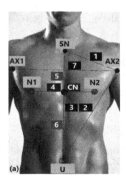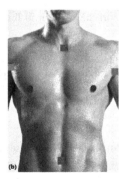

Fig. 1. (a): Anatomical landmarks and lines for NACs positioning on the thoracic wall: Suprasternal Notch (SN), Anterior Axillary Fold (AX), Nipple (N), Umbilicus (U), and Center of Inter-Nipple Distance (CN). (b): Localizing NACs positions of the given image as indicated by the red filled circles (the other colored indicators depict the landmarks used for the ratios calculation) (Color figure online)

The subset of landmarks chosen to determine inputs to the ANN are such that they remain unaltered by the operative procedure, as they are distant from the surgical site. All the landmarks are used for training the ANN. Hence, for a chest masculinization procedure, the calculated position of the NAC corresponds to an estimate of their location on male subjects with similar upper body characteristics to those of the patient.

2.2   Dataset and Ratios Calculation

We collected 173 random images of the male upper body (similar to Fig. 1) from the web. The images depict male subjects that are standing straight and facing the camera. We recorded the coordinates of the following landmarks for each image: SN, AX, N, U, and CN. Table 1 describes all the distances used to calculate the ratios at the input and output of the ANN. The number associated with each distance corresponds to the one depicted in Fig. 1.

Table 1. Distances for ratio calculations.

Distance No.	Landmarks
1	Suprasternal notch to left anterior axillary fold (SN_AX2)
2	Umbilicus to left anterior axillary fold (U_AX2)
3	Suprasternal notch to umbilicus (SN_U)
4	Right nipple to left nipple (N1_N2)
5	Suprasternal notch to center of inter nipples (SN_CN)
6	Center of inter-nipples to umbilicus (CN_U)
7	Right anterior axillary fold to left anterior axillary fold (AX1_AX2)

Using the distances of Table 1, Table 2 depicts some of the calculated ratios that have been used in the literature [8, 9, 11, 12]. Ratios A, B, C, D, and E serve as inputs to the ANN and ratios G and H are estimated by the ANN.

Table 2. Ratios inputted or outputted by the ANN.

Ratio	Related lines
A	SN_AX2/U_AX2
B	U_AX2/SN_U
C	SN_AX2/SN_U
D	SN_AX2/AX1_AX2
E	U_AX2/AX1_AX2
F	SN_U/AX1_AX2
G	SN_CN/CN_U
H	N1_N2/U_AX2

2.3 Neural Network Regression Model

The structure of the proposed ANN is shown in Fig. 2. The ANN takes 6 inputs (ratios A, B, C, D, E, and F) and outputs ratios G and H. These two latter ratios can be used to determine the vertical and horizontal position of NAC, respectively (Sect. 2.4). The number and size of the hidden layers was determined through a trial-and-error process. To evaluate the model within each epoch of training, we utilized Mean Squared Error (MSE) as the loss function. Since this is a regression model, we used a rectifier activation function for all layers. By holding out 20% of the data for final evaluation, the Root Mean Square Error (RMSE) was adopted as the evaluation metric to assess the performance of the model.

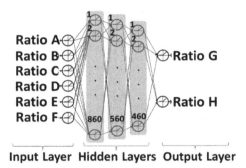

Fig. 2. The structure of neural network regression model

2.4 NAC Localization

After estimating the ratios G and H, we calculate the x and y coordinates of the NAC (X_{NAC}, Y_{NAC}) using the procedure described below. The X_{NAC} and Y_{NAC} are measured in pixels, with the origin located at the top left corner of the image.

To estimate Y_{NAC}:

1. Using Eqs. (1) and (2), we calculate SN_CN and CN_U:

$$CN\_U = \frac{SN\_U}{1+G} \tag{1}$$

$$SN\_CN = G \times CN\_U \tag{2}$$

2. The horizontal position of CN (X_{CN}) is taken as approximately equal to the average horizontal positions of SN (X_{SN}) and U (X_U), we estimate the vertical position of CN (Y_{CN}) using Eq. (4). First, we write

$$SN\_CN = \sqrt{(X_{SN} - X_{CN})^2 + (Y_{SN} - Y_{CN})^2} \tag{3}$$

which can be rearranged to give

$$Y_{CN}^2 - 2Y_{SN}Y_{CN} + X_{SN}(X_{SN} - 2X_{CN}) + X_{CN}^2 + Y_{SN}^2 - (SN\_CN)^2 = 0 \qquad (4)$$

3. Solving the above quadratic Eq. (4) in terms of Y_{CN} produces two values. One of the values places CN above the SN and the other below it, hence we discard the former and retain the latter. Given that CN is aligned vertically with both NAC, we adopt Y_{CN} as the value of Y_{NAC}.

To estimate X_{NAC}:

1. We establish the relationship between ratio H and the distance between CN and NAC (CN_NAC) using Eq. (5):

$$CN\_NAC = \frac{H \times U\_AX2}{2} \qquad (5)$$

2. Given that we know X_{CN}, CN-NAC, $Y_{CN} = Y_{NAC}$, we define Eq. (6), then rearrange and simplify to obtain Eq. (7):

$$CN\_NAC = \sqrt{(X_{CN} - X_{NAC})^2 + (Y_{CN} - Y_{NAC})^2} \qquad (6)$$

$$X_{NAC}^2 - 2X_{CN}X_{NAC} + X_{CN}^2 - (CN\_NAC)^2 = 0 \qquad (7)$$

3. Solving the quadratic Eq. (7) in term of X_{NAC} and by replacing CN_NAC into its value from Eq. (5), we obtain two solutions. The first solution corresponds to X_{NAC} of the left NAC and the second corresponds to X_{NAC} of the right NAC.

For example, Fig. 1(b) depicts the estimated positions of both NACs for the given image as indicated by the red filled circles.

3 Results

We randomly split the dataset into a training (80% of the data) and testing (20% of the data) sub-sets. Our goal was to assess whether we can accurately estimate ratios G and H, which are used to calculate the NAC coordinates. After training, we executed the model on the testing dataset (Fig. 3).

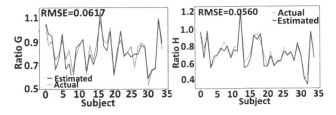

Fig. 3. Comparison of actual and estimated data for ratio G and H

Using the procedure described in Sect. 2.4 and the output of the ANN, we estimated the vertical and horizontal coordinates of the NAC for the images in the testing subset. Figure 4 shows the actual and estimated NAC coordinates for each subject in the testing subset (scaled by the size of the image to account for the various image-dimensions in the dataset). Furthermore, the distance between the actual and estimated NAC is scaled by dividing by AX1-AX2, we obtain an RMSE of 0.0254 for the left and 0.0314 for the right NAC horizontal position and 0.0383 for the vertical position.

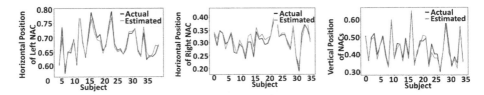

Fig. 4. Comparison of actual and estimated vertical and horizontal positions of the NACs

4   Discussion and Conclusion

The distance between the subject and the camera differed from one image to another. Our reliance on ratios to localize the NACs resolves this issue. Nonetheless, the collected dataset was limited in size due to the lack of appropriate publically available images on the web. The images were taken under varying lighting conditions and resolution. The body of the subjects in some of the images was slightly tilted to the right or left. All these factors can further introduce inaccuracies into the estimates. Hence, the collection of a larger data set with consistent environmental parameters can lead to further ameliorating the results. In this paper, we proposed and evaluated a machine learning method for male NAC positioning. The proposed method is based on an ANN regression model trained with ratios of distances between suprasternal notch, anterior axillary folds and umbilicus. In our experimental results, we trained the regression model on 80% of our 173 subjects dataset and measured the RMSE for two desired ratios that lead to the estimation of the horizontal and vertical positions of the NAC. When we applied the testing subset (remaining 20% of the dataset) to the model, we achieved an RMSE of 0.0617 for the ratio of distances of SN - CN and U - CN that enables the localization of vertical NAC position. Furthermore, we obtained an RMSE of 0.0560 for the ratio of the distances between N1-N2 and AX2-U, which in turn leads to an estimated horizontal NAC position.

References

1. Lindsay, W.R.N.: Creation of a male chest in female transsexuals. Ann. Plast. Surg. **3**(1), 39–46 (1979)
2. Hage, J.J., Bloem, J.J.: Chest wall contouring for female-to-male transsexuals: Amsterdam experience. Ann. Plast. Surg. **34**(1), 59–66 (1995)

3. Altintas, M.A., Vogt, P.M.: Postbariatric plastic surgery. Der Chirurg; Zeitschrift fur alle Gebiete der operativen Medizen **84**, 527–540 (2013)
4. Hage, J.J., van Kesteren, P.J.: Chest-wall contouring in female-to-male transsexuals: basic considerations and review of the literature. Plast. Reconstr. Surg. **96**, 386–391 (1995)
5. Kornstein, A.N., Cinelli, P.B.: Inferior pedicle reduction technique for larger forms of gynecomastia. Aesthet. Plast. Surg. **16**, 331–335 (1992)
6. Mett, T.R., et al.: Optimal positioning of the nipple–areola complex in men using the Mohrenheim-Estimated-Tangential-Tracking-Line (METT-Line): an intuitive approach. Aesthet. Plast. Surg. **41**(6), 1295–1302 (2017)
7. Murphy, T.P., Ehrlichman, R.J., Seckel, B.R.: Nipple placement in simple mastectomy with free nipple grafting for severe gynecomastia. Plast. Reconstr. Surg. **94**(6), 818–823 (1994)
8. Beer, G.M., Budi, S., Seifert, B., Morgenthaler, W., Infanger, M., Meyer, V.E.: Configuration and localization of the nipple-areola complex in men. Plast. Reconstr. Surg. **108**(7), 1947–1952 (2001)
9. Atiyeh, B.S., Dibo, S.A., Chafic, A.H.E.: Vertical and horizontal coordinates of the nipple-areola complex position in males. Ann. Plast. Surg. **63**(5), 499–502 (2009)
10. Kääriäinen, M., Salonen, K., Helminen, M., Karhunen-Enckell, U.: Chest-wall contouring surgery in female-to-male transgender patients: a one-center retrospective analysis of applied surgical techniques and results. Scand. J. Surg. **106**, 74–79 (2016)
11. Kasai, S., et al.: An anatomic study of nipple position and areola size in Asian men. Aesthet. Surg. J. **35**(2), NP20–NP27 (2015)
12. Shulman, O., Badani, E., Wolf, Y., Hauben, D.J.: Appropriate location of the nipple-areola complex in males. Plast. Reconstr. Surg. **108**(2), 348–351 (2001)

A Generic Evolutionary Algorithm for Efficient Multi-Robot Task Allocations

Muhammad Usman Arif[(⊠)]

Institute of Business Administration, Karachi, Pakistan
musman@iba.edu.pk

Abstract. Task allocation in multi-robot teams is conventionally carried out using customized algorithms against individual distributions due to their NP-hard nature. The expanding range of autonomous multi-robot operations demands for a generic allocation scheme capable of working across a variety of problem distributions. This paper presents an intelligently crafted, novel, evolutionary algorithm based task allocation scheme capable of working across a range of multi-robot problem distributions. Qualitative analysis against exact optimal solutions and a state of the art auction based scheme verify the capabilities of the proposed algorithm.

Keywords: Multi-Robot Task Allocation · Plan formulation · Evolutionary algorithm · Optimization

1 Introduction

Multi-Robot Task Allocation (MRTA) as a research domain handles allocations of tasks amongst a team of robots for efficient completion of the job at hand. A 3 axes based taxonomy [1] classifies different MRTA problems on the basis of (a) Robot Type: Single Task (ST) Robot or Multi-Task (MT) Robot (b) Task Type: Single Robot (SR) Tasks against Multi-Robot (MR) Tasks and (c) Allocation type: Instantaneous Allocations (IA) vs. Time Extended Allocations (TA). The NP-hard nature of the majority of these task distributions [1] encourages customized solution approaches. These solutions though perform well within the designed scenario but struggle to perform against even minor changes in the problem structure [2]. Increased utilization of multi-robot teams and their expanding range of operations demands generic solution approaches capable of autonomously guiding the team through a large range of scenarios.

The need for such solutions can be best understood with the help of an example. Figure 1 provides four frequently faced scenarios a team of fire-fighting robots may encounter. Consider allocating a low number of small magnitude fires amongst a team of robots, Fig. 1(a), if all the fires require services from a single fire truck it is an instance of the ST-SR-IA distribution. This can escalate into an ST-SR-TA distribution if the number of fires exceeds the number of resources available, Fig. 1(b). The fires might expand into larger magnitude ones, demanding more resources for completion. This becomes an ST-MR-IA distribution if the total number of resources required is

© Springer Nature Switzerland AG 2019
M.-J. Meurs and F. Rudzicz (Eds.): Canadian AI 2019, LNAI 11489, pp. 486–491, 2019.
https://doi.org/10.1007/978-3-030-18305-9_49

less than the number available, Fig. 1(c), and an ST-MR-TA problem type, Fig. 1(d), if the total count of tasks gets more than the available resources. Hence, one distribution can fast escalate into another without prior notice. This paper presents a novel task allocation scheme capable of seamlessly switching between different task distributions while efficiently allocating tasks amongst robots. It must be highlighted that in the theme of this paper different distributions of MRTA are considered as the dynamics present in the problem and the EA is designed to be able to adjust against them.

(a)   (b)   (c)   (d)

Fig. 1. Frequently faced Multi-Robot Task Allocation distributions

2   Literature Review and Problem Formulation

On an organizational paradigm, the solution approaches for MRTA can be classified into two broad categories, Centralized and Distributed. Centralized schemes use traditional analytical solutions such as Branch and Bound and Linear Programming for optimal allocation of tasks among the robots. These schemes have limited scalability against the NP-hard distributions of MRTA [3]. Further, they require certain mathematical conditions to be satisfied during problem formulation for successful implementation; making the system rigid towards a variety of distributions.

Distributed working model, on the other hand, scales well to larger team and problem sizes but does not guarantee optimal allocations. Behavior-based solution and auction-based schemes are the more famous forms of distributed allocation algorithms [3] with the auction-based schemes having better performance efficiencies. However, deterioration in solution quality has been observed for different auction schemes against minor environmental changes such as robot starting locations and task distribution in the physical space [2].

There has been little effort to provide schemes capable of working across a large range of distributions. Heuristics due to their low problem formulation requirements and close to optimal results are the perfect candidate for a generic solution towards a wide range of complex problem sets. A generic and novel EA based solution is presented in [4] to efficiently allocate tasks against two complex task distributions among a heterogeneous team of robots. This paper expands the previous implementation to four possible task distributions with a team of heterogeneous robots without any changes to the EA structure.

2.1   Problem Formulation

A Standard MRTA problem comprises of a team of robots $R = \{r_1, r_2, \ldots r_n\}$ and a group of tasks, $T = \{t_1, t_2, \ldots, t_m\}$. The objective is to find an allocation of tasks for the robots in the team $f : R \rightarrow T$ while optimizing some known performance measure, $P : R \times T \rightarrow \mathbb{R}^+$. A task $t_i \in T$ is represented as a tuple $t_i = \{(x_i, y_i, q_i) : x_i, y_i, q_i \in \mathbb{N}\}$ where x_i and y_i are the coordinates of the task on the map, and q_i gives the robot requirement of the task. In the current representation, if $\Sigma m_i q_i \leq n$ and $(\forall t_i \in T : q_i = 1)$ the problem is an ST-SR-IA problem. The problem can be optimally solved using the classic Hungarian algorithm [1]. The second distribution to be considered is ST-MR-IA where $\Sigma m_i q_i \leq n$ with $(\forall t_i \in T : q_i \geq 1)$. Due to the NP-hard nature only smaller cases can be solved using exact solution approaches such as the Hungarian algorithm. The ST-SR-TA distribution is where $m \geq n$ with $(\forall t_i \in T : q_i = 1)$. The problem is similar to the NP-hard Multiple-Traveling Salesman Problem [5] where a time-extended schedule of optimal task execution for each robot needs to be formed.

The final task distribution handled by this paper is ST-MR-TA where $m \geq n$ and $(\forall t_i \in T : q_i \geq 1)$. Robot coalitions need to be formed in a time extended manner to solve this type of problems. This paper considers loosely coupled MR tasks, which do not require simultaneous execution by the robot coalitions. With this assumption, the scheme can allocate all instances for a single MR task to the same robot. To avoid such behavior a constraint is added not allowing a single robot more than one instance of an MR task. Irrespective of these timing and allocation constraints ST-MR-TA distributions are NP-hard and often employ approximation based solution for plan generation.

3   Proposed Solution

The proposed EA is equipped with a versatile chromosome design, customized generational operators, and fitness evaluation mechanism to achieve seamless transitions between the targeted distributions. A two-part chromosome representation is used by the EA. The first part of the chromosome, g_{1i} where $1 \leq i \leq m$, uses a decimal point notation to represent a permutation of task ids. Any SR task is represented as a single instance using its id such as 10.1, and an MR task is represented using multiple instances, against its task id such as 10.1, 10.2, 10.3 for task 10 which needs 3 robots for completion. The second part of the chromosome, $g_{2j} \in 0,1$ for the IA cases and $g_{2j} \in \mathbb{N}$ (set notation, N with double line) for the TA cases with $\Sigma g_{2j} = \Sigma m_i q_i$.

An Ordered Crossover operator (ORX) and two mutation operators, namely swap and inverse mutation are used for the first part of the chromosome. The ORX is selected due to the permutation of tasks present in the first part of the chromosome. Creep mutation is the only operator applied to the second part of the chromosome. Creep mutation works by subtracting a small number from a randomly selected gene value and adds the same to another randomly selected gene thus altering the tour length of the two the robots involved while keeping Σg_{2j} intact.

For fitness evaluation a distance based function is used initially, this function minimizes the total distance traveled by the team for mission completion. A time-based fitness function is later introduced which minimizes the time taken by the team to accomplish the operation. Since robot executions proceed in parallel, the most time taken by any member of the team is deemed as the total team time. Other EA parameters included a fixed population size of 100 chromosomes with crossover and mutation rates of 1.0 and 0.1. The number of allowed generations was the only parameter which varied for different cases. The EA was allowed 500 generations for the ST-SR-IA & ST-MR-IA distributions, 1000 generations for ST-SR-TA and 2000 generations for the ST-MR-TA distribution.

As already discussed, no robot was allowed to attempt more than one instance of an MR task in the ST-MR-TA distribution. Such solutions were penalized by the EA using an adaptive penalty function. The penalty value started from 5 units for every violation made and was increased or decreased by 15% every 10 generations based on the validity of the last 10 best solutions.

3.1 Validation Schemes

Exact solutions using the Hungarian algorithm were used for the ST-SR-IA and the ST-MR-IA task distributions. An integer linear program was solved for the ST-SR-TA task distributions using the CPLEX solver. Also, an auction-based scheme was utilized for the ST-SR-TA and ST-MR-TA distributions. The auction-based scheme employed was Sequential Single Item Auction (SSI) [6], a popular auction scheme with performance guarantees in a fully known centralized environment. In SSI robots submit bids against each unallocated task in every round. The auctioneer makes only one allocation per round, allocating the task with the lowest bid to the respective robot. Thus for m tasks, m auction rounds take place.

4 Experiments and Results

For all the distributions considered, experiments were conducted against 10–50 randomly generated tasks with a step size of 5, meaning that cases of 10, 15, 20, 25, 30, 35, 40, 45, and 50 tasks were handled. 3 random cases were generated against each task count and each one of them was solved 5 times by the EA for an average response. Hence, for a 50 task case, 3 instances were generated, each of which was solved 5 times using the EA to observe the average behavior.

Heterogeneous teams were considered for the IA cases where each robot's performance against every individual task varied between 25%–100% efficiency. Teams of 3 and 5 homogeneous robots were considered for the ST-SR-TA and ST-MR-TA distributions respectively. The heterogeneous team performance of the EA for these cases has already been explored in our previous work [4]. Figure 2 provides the performance curves for the EA and the Hungarian algorithm against ST-SR-IA and ST-MR-IA task distributions. For both the distributions EA was able to almost exactly replicate all the optimal allocations irrespective of the problem size.

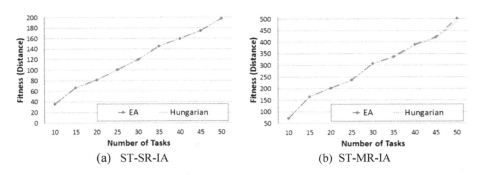

(a) ST-SR-IA   (b) ST-MR-IA

Fig. 2. Performance curves EA vs. the Hungarian Algorithm for IA based task distributions

Figure 3(a) provides the performance curves of EA, and CPLEX for a distance based and EA, CPLEX and SSI for a time-based fitness function against the ST-SR-TA distribution. The EA performed well with both the fitness functions. The distance-based fitness function provided uneven load distributions amongst the robots for all allocation strategies. A robot tour plot in Fig. 3(b) for a 50 task case highlights the use of only two out of the three robots. The time-based fitness function provided better load distribution amongst the robots as can be seen in Fig. 3(c). The change in fitness function held CPLEX from providing optimal solutions beyond 20 tasks cases.

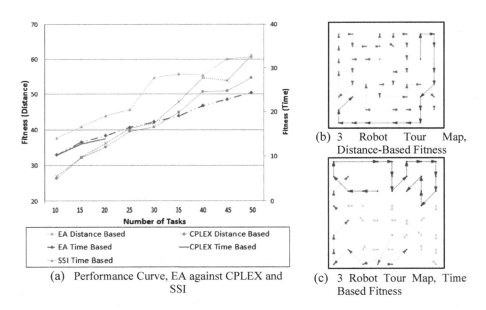

(a) Performance Curve, EA against CPLEX and SSI

(b) 3 Robot Tour Map, Distance-Based Fitness

(c) 3 Robot Tour Map, Time Based Fitness

Fig. 3. ST-SR-TA, performance curve and robot tour plot

A similar performance gap was observed between SSI and EA for the ST-MR-TA distribution as shown in Fig. 4(a). Figure 4(b) gives the convergence graph of the EA for one of the 50 task cases. Lack of convergence highlights the need of more generations to achieve quality solutions for the larger problem sizes.

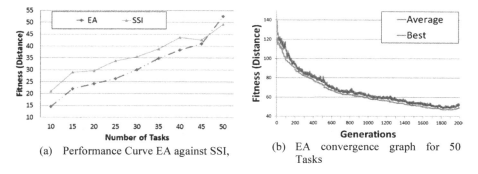

(a) Performance Curve EA against SSI,

(b) EA convergence graph for 50 Tasks

Fig. 4. ST-MR-TA, performance curve and EA convergence graph

5   Conclusion

This paper presented a generic and novel task allocation scheme capable of working across a variety of MRTA distribution. The proposed EA was able to provide high-quality solutions against all the task distributions considered. Efficient results were achieved across four task distributions while keeping the EA structure completely static. Future work will explore the EAs capability to adjust to other dynamics in the system such as robot failures and new task arrivals.

References

1. Gerkey, B.P., Matarić, M.J.: A formal analysis and taxonomy of task allocation in multi-robot systems. Int. J. Robot. Res. **23**, 939–954 (2004)
2. Sen, S.D., Adams, J.A.: An influence diagram based multi-criteria decision making framework for multirobot coalition formation. Auton. Agent. Multi-Agent Syst. **29**, 1061–1090 (2015)
3. Gerkey, B.P., Mataric, M.J.: Multi-robot task allocation: analyzing the complexity and optimality of key architectures. In: IEEE – ICRA, vol. 3, pp. 3862–3868 (2003)
4. Arif, M.U., Haider, S.: A flexible evolutionary algorithm for task allocation in multi-robot team. In: Nguyen, N.T., Pimenidis, E., Khan, Z., Trawiński, B. (eds.) ICCCI 2018. LNCS (LNAI), vol. 11056, pp. 89–99. Springer, Cham (2018). https://doi.org/10.1007/978-3-319-98446-9_9
5. Arif, M.U., Haider, S.: An evolutionary traveling salesman approach for multi-robot task allocation. In: 9th International Conference on Agents and Artificial Intelligence, pp. 567–574 (2017)
6. Koenig, S., et al.: The power of sequential single-item auctions for agent coordination. In: Proceedings of the National Conference on Artificial Intelligence. p. 1625. Menlo Park, CA; Cambridge, MA; London; AAAI Press; MIT Press; 1999 (2006)

Neural Prediction of Patient Needs in an Ovarian Cancer Online Discussion Forum

Hyeju Jang[1(✉)], Young Ji Lee[2], Giuseppe Carenini[1], Raymond Ng[1], Grace Campbell[2], and Kendall Ho[3]

[1] Department of Computer Science, University of British Columbia, Vancouver, BC, Canada
{hyejuj,carenini,rng}@cs.ubc.ca
[2] School of Nursing, University of Pittsburgh, Pittsburgh, PA, USA
{leeyoung,gbc3}@pitt.edu
[3] Department of Emergency Medicine, University of British Columbia, Vancouver, BC, Canada
kendall.ho@ubc.ca

Abstract. Social media is an important source to learn the concerns and needs of patients and caregivers in home care settings. However, manually identifying their needs can be labor-intensive and time-consuming. In this paper, we address the problem of need detection, automatically identifying patient needs in text. We explore both neural and traditional machine learning approaches, and evaluate them on a newly annotated dataset in an ovarian cancer discussion forum. We discuss issues and challenges of this novel task.

Keywords: Patient need prediction · Social media · Text classification

1 Introduction

Many patients with serious diseases and conditions often visit social media to connect with others in the same situation. They support each other by sharing experiences and knowledge that involve medical, psychological, and even social aspects of the disease. Hence, social media has been recognized as an important source to learn the concerns and needs of patients and caregivers in home care settings [6,9,12]. Understanding their perspectives through social media may facilitate effective interventions and provide proper care, thus improving the quality of life of patients and their caregivers.

This paper focuses on identifying patient needs in social media data. We detect patient needs of four types including physical needs, emotional needs, social needs, and health information needs. We formulate need detection as a text classification problem, and investigate neural network approaches. Unlike traditional machine learning models such as logistic regression, end-to-end neural

© Springer Nature Switzerland AG 2019
M.-J. Meurs and F. Rudzicz (Eds.): Canadian AI 2019, LNAI 11489, pp. 492–497, 2019.
https://doi.org/10.1007/978-3-030-18305-9_50

network models usually do not require feature engineering because the models can learn high-level features from data in a non-linear manner by exploiting distributional representations of words or sentences. Additionally, unlike Bag-of-Words (BoW) models, a recurrent neural network (RNN) can take into account the order of words and sentences. Specifically, we apply Hierarchical Attention Networks (HANs) [11] for our task. We choose to use HAN because of its attention mechanism that helps interpretability, as well as its hierarchical RNN structure that considers sequences and document structures.

Our Contributions: (1) We build a labeled dataset for patient need detection in social media.[1] To the best of our knowledge, this is the first attempt to annotate large patient-generated data with different types of needs. (2) We select a suitable and interpretable neural model that predicts patient needs in text, and discuss the results. We reveal that the difficulty of patient need detection varies considerably across need types.

2   Related Work

Prior work has investigated computational approaches to predicting patient needs in social media. [1] asked 184 social media users about their needs to use social media as well as their online behaviors based on demographics, reading behavior, posting behavior, and perceived roles in social networks. Then, the authors built a model to predict patient needs based on the behavior patterns obtained from the survey, using traditional classification algorithms. [4] annotated each online forum post with whether the post was responded by a moderator or by only patients to identify posts that need moderators' help. Based on the annotated data, they classified texts using BoW, sentiment analysis, and thread length features. [10] also tried traditional classification algorithms to identify clinically related sentences (e.g., a medical question, a symptom, or a treatment) assuming that those sentences indicate patient needs for clinical resources. Our work is aligned with [4,10] in that we detect needs on patient-generated online forum data. However, our work is different from the prior work in that (1) we annotate data with more detailed and diverse types of patient needs, and (2) we use end-to-end neural models that leverage document structures.

3   Creating Labeled Dataset

Our data was obtained from the American Cancer Society: the Cancer Survivors Network online peer-support forum.[2] We collected user posts from the ovarian cancer discussion board, posted from January 2006 to March 2016. The collected posts include post IDs, thread titles, thread IDs, poster IDs, and post contents.

---

[1] This dataset will be made public after publication for those who get both IRB approval and permission from the American Cancer Society.

[2] http://csn.cancer.org.

To build a hand-coded dataset, we selected posts based on the assumption that initial posts of users would more likely expose needs about why they visited the online forum. First, we picked the initial posts of all the users. Then, among these posts, we excluded the ones that were replies to other posts. This process resulted in a dataset of 853 posts in total.

Table 1. Types of patient needs, adapted from [3,5,7], and annotation statistics.

No.	Need type	Definition	n	κ
N1	Physical	Experience of physical symptoms such as fatigue, urinary symptoms, pain, hot flushes, etc.	100	.88
N2	Emotional	Experience of psychological/emotional symptoms such as anxiety, depression, worry, despair, fear, etc.	136	.83
N3	Family-related	Experience of fears/concerns for the family, dysfunctional relationships, etc.	46	.85
N4	Social	Looking for people with similar experience, Experience of reduced social support, social isolation, loneliness, etc.	319	.87
N5	Intimacy	Experience of difficulties with self-image and/or body-image, reduced libido, sexual dysfunction, compromised intimacy with partner, fertility, etc.	8	.42
N6	Practical	Situations of transportation, out-of-hours access to healthcare, financial support, etc.	32	.87
N7	Daily living	Experience of restriction in daily living tasks such as exercise, housekeeping, etc.	2	.57
N8	Spiritual	Existential concerns such as fear of death, death and dying, fears regarding afterlife, etc.	7	.50
N9	Health information	Experience of a lack of information, uncertainty of follow-up care, lack of information in relation to treatment and diagnosis, etc.	455	.93
N10	Patient-clinician communication	Quality of communication between patients and healthcare professionals, satisfaction with care, shared decision-making, etc.	18	.85
N11	Cognitive	Experience of cognitive impairments, memory loss, etc.	3	.75
N12	Etc.	Any concerns/issues do not fit into 11 categories	105	.49

We annotated our dataset according to the 12 types of needs as shown in Table 1, which was adapted from the medical literature [3,5,7] by medical domain experts in our team. We employed one undergraduate student and one faculty member in the Nursing field to annotate each post with the need types. First, they were asked to annotate who wrote each post, between *patient* and *caregiver*, based on the information available within the post. Then, they answered whether each type of the need is shown in the post, given the definitions of the need types. A post could be annotated with multiple need types. We computed inter-agreement between the two annotators (last column in Table 1), and the final annotations of disagreed instances between the two were decided by consulting the third annotator, a nurse scientist who has 20 years of clinical experience.

The statistics of the annotated dataset are also shown in Table 1. Overall 1,231 needs were annotated in 853 posts. More specifically, 321 posts are assigned more than one need type, and a post is assigned 1.4 need types on average. For our experiments, we focus on physical needs, emotional needs, social needs, and health information needs where the number of need occurrences is not too small ($n \geq 100$).

4 Experiments

Using the annotated dataset, we build neural network models for need detection. We formulate need detection as a binary classification problem – for each need, we build a computational model that decides whether a post expresses the given need or not.

4.1 Hierarchical Attention Network

We adopt Hierarchical Attention Networks (HANs) [11] as our neural model. HANs use an RNN that incorporates a hierarchical structure of document with attention mechanism. This model progressively builds a document vector by aggregating important words into a sentence vector and then aggregating important sentence vectors into a document vector.

In spite of our small data, we choose to use HAN for the following reasons. First, HAN is an RNN, a sequence model that can take into account the order of words and sentences, information that is arguably critical to recognize complex needs. Second, the document representation of HAN considers a hierarchical structure (words - sentences - documents) rather than looking at a document as a flat sequence of words. Therefore, we expect the HAN model to better capture weak clues for patient needs in text by leveraging such structural information. Finally, HAN uses an attention mechanism which supports a sophisticated interpretation of prediction results; something really important in health care applications.

4.2 Evaluation

We evaluate HAN compared to a simple unsupervised baseline and traditional BoW models. The simple unsupervised baseline uses the definition of needs in Table 1. If a post contains at least one content word from a definition of a need type, it is classified positive for that need, and negative otherwise. Note that we only use unique content words for the definition of a need type, i.e., content words not shared with any other definition. We also employ the standard BoW model as our baseline. BoW represents a post as the occurrence of words within the post ignoring any information about the position or order of words. For the BoW model, we use logistic regression with L1 regularization.[3]

---

[3] We tried other algorithms such as Support Vector Machine and Random Forest, but logistic regression outperformed others.

Experiment Settings: The parameters used for our neural models are as follows: We set the word embedding dimension to be 100, initialized with the pre-trained embeddings from GloVE [8]. The GRU dimension is set to be 100. We use Adagrad [2] for training with default settings of learning rate 0.01, epsilon 1e-07, and decay 0.0. We use a mini-batch size of 64. We conduct 10-fold cross validation ten times for each need type to reduce variance in the data.

Table 2. Performance of the different models on the need detection task. (**Models**) Simple: Simple unsupervised model, LR: Logistic regression with L1 regularizer, HAN: Hierarchical attention network, (**Metrics**) κ: Cohen's kappa between gold standard and prediction, F1: weighted F1 score on need-exist/need-not-exist, P-T: precision on need-exist, R-T: recall on need-exist, P-F: precision on need-not-exist, R-F: recall on need-not-exist, A: accuracy, *: statistically significant improvement over LR ($p < .05$).

Model	κ	F1	P-T	R-T	P-F	R-F	A
Physical needs							
Simple	.36	.81	**.67**	.44	**.94**	.80	.78
LR	.37	**.87**	.35	.43	.91	**.96**	**.88**
HAN	**.39**	**.87**	.38	**.45**	.91	**.96**	.87
Social needs							
Simple	.10	.58	.18	.26	.66	**.91**	.64
LR	**.25**	**.66**	**.46**	**.50**	**.72**	.78	**.66**
HAN	.20	.63	.45	.47	71	.75	.64

Model	κ	F1	P-T	R-T	P-F	R-F	A
Emotional needs							
Simple	.21	.80	.31	**.79**	.86	.93	.82
LR	.20	.79	.24	.29	.86	.93	.82
HAN*	**.34**	**.83**	**.36**	.43	**.88**	**.94**	**.84**
Health information needs							
Simple	.06	.52	.47	.51	.49	.58	.52
LR	.40	.70	.73	.72	.69	.67	.70
HAN*	**.46**	**.73**	**.75**	**.75**	**.72**	**.71**	**.73**

Experiment Results: Our experiment results are presented in Table 2. Remarkably, the classification performance varies drastically depending on the need type. Overall, the physical needs and the emotional needs show better performance than the other two need types. We also see that the neural network model significantly outperform the logistic regression model, showing that the information the neural models capture is effective for predicting needs. Additionally, it suggests that the attention module in HAN is able to correctly focus on the most predictive words and sentences in a document. However, this is not the case for the social needs. Another interesting observation is that the precision and recall scores on need-exist are lower for the physical and emotional needs than for the social and health information needs although the overall F1 scores show the opposite pattern. This may be due to the fact that the latter datasets are more balanced.

5   Conclusion

We built a labeled dataset for different types of patient needs, by annotating data from an ovarian cancer online discussion forum. We showed that using HAN outperforms baselines such as BoW despite the small data size, suggesting the effectiveness of considering the document hierarchical structure based on word/sentence sequences and of the attention mechanism for this task. For future work, we plan to improve the performance of HAN, especially exploring how to incorporate medical domain knowledge. Also, incorporating sentiment analysis could be beneficial for certain need types. Additionally, formulating the task as a multi-task problem could be a direction to improve the performance of each need type by learning them jointly. Finally, pursuing an unsupervised learning approach would be another direction as well.

Acknowledgement. The Titan Xp used for this research was donated by the NVIDIA Corporation.

References

1. Choi, M.J., et al.: Toward predicting social support needs in online health social networks. J. Med. Internet Res. **19**(8) (2017)
2. Duchi, J., Hazan, E., Singer, Y.: Adaptive subgradient methods for online learning and stochastic optimization. J. Mach. Learn. Res. **12**(Jul), 2121–2159 (2011)
3. Fitch, M.: Supportive care framework. Can. Oncol. Nurs. J./Revue canadienne de soins infirmiers en oncologie **18**(1), 6–14 (2008)
4. Huh, J., Yetisgen-Yildiz, M., Pratt, W.: Text classification for assisting moderators in online health communities. J. Biomed. Inf. **46**(6), 998–1005 (2013)
5. Hui, D.: Definition of supportive care: does the semantic matter? Curr. Opin. Oncol. **26**(4), 372–379 (2014)
6. Moorhead, S.A., Hazlett, D.E., Harrison, L., Carroll, J.K., Irwin, A., Hoving,C.: A new dimension of health care: systematic review of the uses, benefits, and limitations of social media for health communication. J. Med. Internet Res. **15**(4) (2013)
7. Paterson, C., Robertson, A., Smith, A., Nabi, G.: Identifying the unmet supportive care needs of men living with and beyond prostate cancer: a systematic review. Eur. J. Oncol. Nurs. **19**(4), 405–418 (2015)
8. Pennington, J., Socher, R., Manning, C.D.: Glove: global vectors for word representation. In: Empirical Methods in Natural Language Processing (EMNLP), pp. 1532–1543 (2014). http://www.aclweb.org/anthology/D14-1162
9. Smailhodzic, E., Hooijsma, W., Boonstra, A., Langley, D.J.: Social media use in healthcare: a systematic review of effects on patients and on their relationship with healthcare professionals. BMC Health Serv. Res. **16**(1), 442 (2016)
10. VanDam, C., Kanthawala, S., Pratt, W., Chai, J., Huh, J.: Detecting clinically related content in online patient posts. J. Biomed. Inform. **75**, 96–106 (2017)
11. Yang, Z., Yang, D., Dyer, C., He, X., Smola, A., Hovy, E.: Hierarchical attention networks for document classification. In: Proceedings of the 2016 Conference of the North American Chapter of the Association for Computational Linguistics: Human Language Technologies, pp. 1480–1489 (2016)
12. Zhang, S., O'Carroll Bantum, E., Owen, J., Bakken, S., Elhadad, N.: Online cancer communities as informatics intervention for social support: conceptualization, characterization, and impact. J. Am. Med. Inform. Assoc. **24**(2), 451–459 (2017)

Fully End-To-End Super-Resolved Bone Age Estimation

Mohammed Gasmallah$^{(\boxtimes)}$, Farhana Zulkernine, Francois Rivest,
Parvin Mousavi, and Alireza Sedghi

School of Computing, Queen's University, Kingston, ON K7L3N6, Canada
11mhg@queensu.ca

Abstract. With the release of large-scale bone age assessment datasets and competitions looking at solving the problem of bone age estimation, there has been a large boom of machine learning in medical imaging which has attempted to solve this problem. Although many of these approaches use convolutional neural networks, they often include some specialized form of preprocessing which is often lengthy. We propose using a subpixel convolution layer in addition to an attention mechanism similar to those developed by Luong et al. in order to overcome some of the implicit problems with assuming particular placement and orientation of radiographs due to forced preprocessing.

Keywords: Image processing · Bone age estimation ·
Convolutional neural networks

1 Introduction

In radiology, bone age estimation is useful for a variety of reasons. Bone age is an indicator of the skeletal and biological maturity of a person and can often be different from the chronological age of an individual [7]. Bone age estimation is often requested for diagnosing pediatric diseases which indicates the maturity of a child's skeletal structure [1,7]. Other measures have far too much variation in development to be used as established techniques for skeletal maturity [1].

Bone age estimation using radiographs is invaluable for pediatricians and orthopedic surgeons [1]. The most employed methods for bone age estimation are the Greulich and Pyle and Tanner-Whitehouse (TW2) atlases [1,3,7,8]. A radiologists spends approximately thirty minutes per patient, comparing the radiograph to reference bone ages and estimating the age of the patient based off these references [3,4]. This can be quite a time-consuming task and the accuracy and efficiency of the process is mainly determined by the experience of the reviewer in question [4].

© Springer Nature Switzerland AG 2019
M.-J. Meurs and F. Rudzicz (Eds.): Canadian AI 2019, LNAI 11489, pp. 498–504, 2019.
https://doi.org/10.1007/978-3-030-18305-9_51

Computerized methods and computer-assisted automated systems can help radiologists save precious time and increase accuracy at estimating bone age [2,4]. The Radiological Society of North America (RSNA) has released a dataset and held a competition to assess and evaluate machine learning and automated bone age estimation methods [2]. In many of these methods, artificial neural networks are used to preprocess the dataset and transform the radiograph into some standardized form to create a better and augmented training dataset [2]. For the same reason, we explored the technique of super-resolution by adding a subpixel convolution layer before the feature extractor in a deep convolutional network based on the work of Shi et al. [9]. Our model thus super resolves the input by extracting higher resolution information before passing it to feature extraction phase.

The rest of the paper is organized as follows. Section 2 discusses the methods and our proposed neural network model including the dataset and training and evaluation protocol. Section 3 presents the results and some additional images. Finally Sect. 4 concludes the paper and discusses some future work.

2   Methods and Proposed Neural Network Model

We describe the dataset used to train the model and our proposed augmented neural network model, in this section. We discuss in detail our implementation and training/evaluating protocol as well as reasoning for our design choices.

Dataset: The dataset that was posted on Kaggle is from the Radiological Society of North America's pediatric bone age assessment challenge in 2018 [6]. The dataset contains 12611 images each labelled with the gender and bone age of the patient estimated by six reviewers. The ground truth was also provided in the dataset [2].

In order to make sure the network does not predict based on the non-uniform distribution of the dataset (see Fig. 1) and train a good model, we resample the dataset based on bone age and sex. Additionally, we separate an initial 15% of the dataset to use for testing and another 25% for validation. The remainder is the training set and is augmented using rotation, horizontal/vertical flips and small affine transformations. This allows the network to learn the features that are necessary to evaluate the bone age regardless of rotation and other affine transformations of the image. It also allows the network to generalize better to radiographs which it has not seen yet. Finally, the images are normalized between 0–1 by dividing all pixel values by 255 [2].

In order to make the regression smaller and more contained, we decide not to regress over the actual bone age of the patients. Instead, we regress over the z-score distributions of the patients' bone age using Eq. 1. This keeps the values that the network must regress much smaller and allows for generally smaller gradients to flow through the network.

$$z_n = \frac{x - \mu}{2\sigma} \tag{1}$$

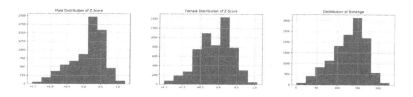

(a) Male Distribution   (b) Female Distribution   (c) Total Distribution

Fig. 1. Distribution of bone age by gender and stand-alone of gender.

where x is the bone age, μ is the mean of the bone ages, σ is the standard deviation of the bone ages and z is the z-score.

Model Architecture: As our base model, we use the network architecture that won the bone age estimation contest. We add on three blocks of convolutional neural network layers that lead to a subpixel phase shift layer. The subpixel phase shift layer is presented in Shi et al.'s paper and transforms depth wise information into space wise information [9]. The input image is then resized to match the size of the output of the subpixel phase shift layer and concatenated in order to feed into a feature extractor (such as VGG16 or Resnetv2 50). The output of the feature extractor is passed on to an attention module composed of a series of layers similar to those found in [5]. This accentuates particular areas of the feature extractor's output. The sex of the patient is also processed and then concatenated with the output of the feature extractor and attention module. Finally, two dense layers attempt to regress the z-score using this final output. The network architecture can be seen in Fig. 2.

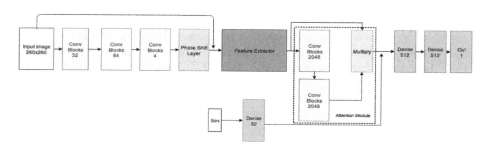

Fig. 2. Basic architecture used for bone age estimation. Feature extractor is replaced with either VGG or Resnetv2 50.

Variants for the purpose of testing include a network without the phase shift layer and any preceding convolutions, and two networks which either use the VGG16 feature extractor or the Resnetv2 50 feature extractor. Imagenet trained weights were loaded and the feature extractor was not trained. The final layers of the feature extractors were fed into our regressor and the attention module as illustrated in Fig. 2.

Training and Evaluation: After we separate the dataset into the training, validation and testing sets, we began training the network using the ADAM optimizer. The ADAM optimizer is an optimizer which computes adaptive learning rates for differing parameters. The network is trained using a learning rate reduction scheduler, an early stopping mechanism to avoid overfitting and we only save the network which performs best on the validation set. The input dimension for the images is set to 260×260 and the batch size is set to 2 as any value higher causes certain variant networks to fail as they cannot fit in the memory of one graphics card. During training, the network is trained using a root mean squared error (RMSE) loss (see Eq. 2), however, a custom metric of mean average error is calculated so that we can directly compare performance when we relate it to the differences in months between the predicted value and the label.

$$RMSE = \sqrt{\frac{\sum_{i=1}^{n}(P_i - O_i)^2}{n}} \tag{2}$$

3 Results

Our results focus on the performance of the network using subpixel layers versus those that do not. In order to evaluate the networks, we decided to use the testing set to report the mean average deviation (MAD) and root mean squared error (RMSE) in both years and months. The networks reported in Table 1 have only been trained for a max of 25 epochs. The state of the art performance from the RSNA 2018 challenge was an ensemble network which was trained for 300 epochs on this problem [2]. It should be noted that the state of the art was trained on the same dataset as our network, but uses an ensemble method in order to outperform other methods. Additionally the images in Fig. 3 provide insight on each networks' validation as it trains.

Table 1. Metrics of results comparing MAD score against RMSE in years and months.

Networks	Mean average deviance (months)	RMSE (years)	RMSE (months)
RSNA state of the art [2]	6.12	-	-
Resnetv2 50 **with subpixel layers**	36.31 ± 22.30	3.58	43.02
Resnetv2 50 with no subpixel layers	36.63 ± 23.00	3.60	43.26
VGG16 **with subpixel layers**	16.48 ± 13.46	1.77	21.28
VGG16 with no subpixel layers	17.93 ± 15.94	2.00	23.99

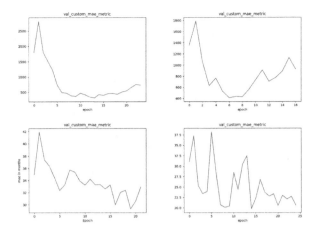

Fig. 3. Validation mean average error. Top left is VGG with super resolution layers, top right is VGG without the super resolution layers. Bottom left is Resnet with super resolution layers and bottom right is Resnet without the super resolution layers.

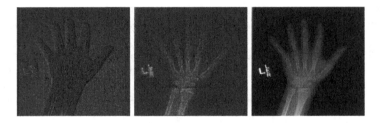

Fig. 4. The image on the far right is the input to the network, left is the super-resolution output for the network with Resnet and the middle is the super-resolution output for the network with VGG16 (both have brightness adjustments in order to be able to view the hand better).

4   Future Works and Conclusion

Overall the use of the subpixel layers before the feature extraction network lead to inconclusive results regarding an increase in performance of the network. Although the networks with the subpixel layers do take longer to train and converge, they also tend to be much smoother when they reach convergence (see Fig. 3). We have provided an example of the output of the super-resolution layer as well as the input in Fig. 4. What is interesting to note is that the super-resolved hand includes higher activations in specific areas of the bone. Particularly, certain bone edges seem to be highly valued. This is the type of information we expect that the network is learning and propagating through the super-resolution layers which is similar to those described in [7].

Although the networks we have described do not perform as well as many of the other networks as in [2, 8], the purpose of this endeavour was to view whether the subpixel convolution layers as introduced by [9] would be useful for the purposes of extracting and super-resolving information that the network deems important for the purposes of bone age estimation. In this endeavour, we believe our results are inconclusive. Although all networks are trained the same way and the subpixel convolution layers have lower MAD scores, and are more consistent, they do not perform significantly better (see Table 1). Regardless, the subpixel convolution layers are very useful and show that networks for bone age estimation do not require more complex preprocessing steps but instead require important and useful layers that can aid or skip the preprocessing steps altogether and a more varied data augmentation phase.

There is still much to be done in this field. Bone age estimation is still an open problem in machine learning, although networks often perform better than radiologists and often more consistent [2,8] they are less trusted and not as well understood. Additionally, the problem is quite simplistic and allows for a variety of different approaches to be taken. This makes it a useful problem to evaluate new and novel layers and how they perform in practical scenarios.

Future Works: There is still much to be done with the implementation of the subpixel convolutional layers. For one, we attempted early on to develop a multi-stage training step which required freezing particular layers of the network in order to focus on training either the subpixel layers or the regression layers. This proved to be unsuccessful, but we believe that modifying the training it may be possible to get the multi-stage training step to work. This changes the framing of the problem and intuitively, we believe that this will increase the network performance. Finally, looking into whether training from scratch or retraining the feature extractor may be beneficial.

References

1. Gilsanz, V., Ratib, O.: Hand Bone Age A Digital Atlas of Skeletal Maturity, 1st edn. Springer, Los Angeles (2005). https://doi.org/10.1007/978-3-642-23762-1
2. Halabi, S.S., Prevedello, L.M., Kalpathy-cramer, J., Mamonov, A.B.: The RSNA pediatric bone age machine learning challenge, pp. 1–6 (2018). https://doi.org/10.1148/radiol.2018180736
3. Iglovikov, V.I., Rakhlin, A., Kalinin, A.A., Shvets, A.A.: Paediatric bone age assessment using deep convolutional neural networks. In: Stoyanov, D., et al. (eds.) DLMIA/ML-CDS -2018. LNCS, vol. 11045, pp. 300–308. Springer, Cham (2018). https://doi.org/10.1007/978-3-030-00889-5_34
4. Kim, J.R., et al.: Computerized bone age estimation using deep learning-based program: Evaluation of the accuracy and efficiency. Am. J. Roentgenol. **209**(6), 1374–1380 (2017). https://doi.org/10.2214/AJR.17.18224
5. Luong, M.T., Pham, H., Manning, C.D.: Effective approaches to attention-based neural machine translation. In: Proceedings of the 2015 Conference on Empirical Methods in Natural Language Processing, pp. 1412–1421 (2015). https://doi.org/10.18653/v1/D15-1166, https://re-work.co/blog/deep-learning-ilya-sutskever-google-openai
6. Mader, K.: Kaggle RSNA Bone Age Dataset (2018). https://www.kaggle.com/kmader/rsna-bone-age

7. Manzoor Mughal, A., Hassan, N., Ahmed, A.: Bone age assessment methods: a critical review **30**(1), 211–215 (2014). https://doi.org/10.12669/pjms.301.4295
8. Mutasa, S., Chang, P.D., Ruzal-Shapiro, C., Ayyala, R.: MABAL: a novel deep-learning architecture for machine-assisted bone age labeling. J. Digit. Imaging 1–7 (2018). https://doi.org/10.1007/s10278-018-0053-3
9. Shi, W., et al.: Real-time single image and video super-resolution using an efficient sub-pixel convolutional neural. Network (2016). https://doi.org/10.1109/CVPR.2016.207

Name2Vec: Personal Names Embeddings

Jeremy Foxcroft, Adrian d'Alessandro, and Luiza Antonie$^{(\boxtimes)}$

School of Computer Science, University of Guelph, Guelph, Canada
{jfoxcrof,dalessaa,lantonie}@uoguelph.ca

Abstract. Predicting if two names refer to the same entity is an important task for many domains, such as information retrieval, record linkage and data integration. In this paper, we propose to create name-embeddings by employing a Doc2Vec methodology, where each name is viewed as a document and each letter in the name is considered a word. Our hypothesis is that representing names as documents, with letters as words, will help capture the internal structure of names and relationships among letters. We present and discuss an experimental study where we explore the effect of various parameters, and we assess the stability of the models built for the embedding of names. Our results show that the new proposed method can predict with high accuracy when a pair of names matches.

1 Introduction

The motivation behind this research comes from record linkage, where name similarities are crucial features in determining if two records refer to the same entity or not. Record linkage is the process of matching data records that refer to the same entity across different data sources, typically without the aid of a shared unique identifier. The resulting linked data can provide a wealth of new information and shed fresh light on previously unanswerable questions. Record linkage is often used to identify records that refer to the same person, bringing data collected at various times and by various systems together to form a more complete picture. Historical research [2], medical research [6], and business intelligence are just a few of the many areas that can benefit from data integration.

At the core of record linkage is the notion of how similar two data records are. A data record representing a person generally contains a name. These names are potentially valuable identifiers, however discrepancies between different data sources must be accounted for. Inconsistencies may be introduced through the use of nicknames (as explored in [3]), data entry typos, errors in optical character recognition during record digitization, common misspellings, or differing transliteration approaches. Determining the likelihood that two names refer to the same individual is therefore of great value. Different similarity measures are often useful in different situations; there is no one-size-fits-all approach. This has led to the diverse array of approaches that exist today, and the wide range of applications continues to incentivize increasingly advanced and accurate techniques. There have been many studies evaluating and comparing the effectiveness of existing string comparison techniques [4,5].

© Springer Nature Switzerland AG 2019
M.-J. Meurs and F. Rudzicz (Eds.): Canadian AI 2019, LNAI 11489, pp. 505–510, 2019.
https://doi.org/10.1007/978-3-030-18305-9_52

In this paper, we propose a novel method to calculate name similarities. We train a model to transform names to a vector space. Then we employ the model to embed names into this vector space and we use the vector representation when we calculate the cosine distance based name similarity.

2 Building Name-Embeddings

Word embedding is the process of mapping a vocabulary to vectors of real numbers. A set of one-hot vectors is a simple way to represent each word in a vocabulary, but the dimensionality of such vectors would need to be the size of the vocabulary. Word embeddings are used to map words to a continuous vector space with a much lower dimension. There are a large number of techniques that can be used to generate this mapping, many of which utilize neural networks. Word2Vec [8] in particular is one of these methods that has been used extensively. Doc2Vec [7] builds on Word2Vec, and is used to map each document in a corpus to a vector representation. Where Word2Vec trains the vector for each word by examining which words appear nearby, Doc2Vec is more concerned with the internal structure of documents, and how the individual words in a document form topics. The vectors that emerge from successful embeddings are not easily understood, but have a number of valuable properties. Words or documents with similar contextual significance will tend to have similar embeddings. It has also been found that an element-wise addition of two vectors produces an embedding that is similar to both of the source vectors [9]. These traits reinforce the idea that the contextual significance of words and documents is captured by the embeddings they are mapped to. Embeddings of words, documents, and entire records have been used successfully in record linkage applications [10, 12].

To generate the name-embeddings (i.e., mapping names to a vector space representation), we employ a Doc2Vec methodology, where each name is viewed as a document and each letter in the name is considered a word. The vocabulary was composed of the 26 lower-case English letters. Gensim [11] is used for the Doc2Vec implementation. During model generation, there are a number of parameters that can significantly affect the model that is produced. We investigate the three main parameters: **epochs** (the number of iterations over the corpus during the training of the model), **vector_size** (the dimensionality of the feature vectors) and **window** (the maximum distance between the current and predicted word within a document).

To test the effect that each of these parameters have on the resulting model, a grid search was performed. The following values were considered for each of the parameters: epochs $\in \{1, 5, 10, 20, 40, 80, 160, 320, 640\}$, vector_size $\in \{5, 10, 15, 20, 25, 30, 40, 50, 70, 100, 130, 180, 210, 240, 270, 300\}$ and window $\in \{1, 2, 3, 4, 5, 6\}$. Parameter sets were generated by taking the cross product of the above sets. Altogether this resulted in 864 unique combinations of parameters to test.

Vector size is typically set in the range of 100 to 300, however given the size of the documents (i.e., the length of the names), it seemed appropriate to also test smaller sizes. Window size is typically set to a value around 5, but the small document sizes also justified testing smaller values for this parameter. The other consideration for the model generation was the stability of the results of a particular set of parameters. To measure this, 5 models were generated with each unique combination of parameters, for a total of 4320 models. The implementation of our method can be found in the associated GitHub repository [1].

3   Results

The data used in this paper was collected by Ancestry.com [13]. In this paper, two of the datasets from that work are used: the *Records* dataset and the *Surnames* dataset. The *Records* dataset consists of 25,000 surname pairs. The second dataset is the *Surnames* dataset. This dataset represents the "universe" in which all last names reside. It contains the 250,000 most commonly occurring surnames in Ancestry.com data. The *Surnames* dataset is used to train the embedding models. For a more detailed description of this data and how it was collected, see [13].

3.1   Name2Vec Models

To test the name-embedding models that are generated, the name pairs in the *Records* dataset is used. For each of the 25,000 pairs of names in this dataset, a cosine distance of their embeddings is calculated. All cosine scores are in the range $[0, 2]$[1]. To provide a comparison, 25,000 random name pairs were also generated from the *Surname* dataset. These random name pairs were cross checked with the *Records* dataset to ensure they were not present (in either order, as the records dataset contains directed name pairs). A high quality model should be able to differentiate as many of the matching name pairs as possible from the random name pairs. To measure the effectiveness of each model, the cosine scores of the matching names and the random names were sorted into 50 buckets, where each bin represents 2% of the range between 0 and 2. For a good model, the cosine distance distribution for the matching name pairs should be skewed toward 0, while the random name pairs one should be uniformly distributed around some high value. Where the two histograms overlap, there is ambiguity whether these name pairs represent matches or not. The cut-off point was defined as the leftmost bucket where more random name pairs occurred than matching name pairs. Each model was awarded a score that represents the number of the matching name pairs with cosine scores lower than the cut-off point. The maximum score is 25,000, and the perfect model would have zero overlap between the two distributions. Of the 4320 models, scores ranged from 0 to 24563 (i.e., the worst model could differentiate none of the matching name pairs, and the best model could differentiate 98.252% of the matching name pairs). The histograms for the best and worst performing models are shown in Figs. 1 and 2.

---

[1] This is due to the fact that some of the vectors could have negative values as well.

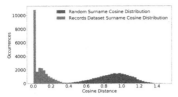

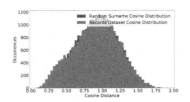

Fig. 1. Cosine distance histogram of the best performing model. The model was generated with epochs $= 640$, vector_size $= 30$, window $= 2$.

Fig. 2. Cosine distance histogram of one of the worst performing models. The model was generated with epochs $= 1$, vector_size $= 15$, window $= 4$.

3.2 Parameter Significance and Consistency of Model

The number of iterations each model is trained for is the most significant factor that determined the performance of the model. As shown in Fig. 3, average score consistently increases with the number of iterations up to a ceiling. The lowest score achieved by a model trained for 640 iterations was 21,742. Consequently, the impact of other parameters will be considered within the context of epochs $= 640$. The effect of vector size on average score is shown in Fig. 4. Doc2Vec normally performs best with a vector size of at least 100, however for this application, setting it higher than 20 made very little difference to the quality of the model. Window size was the least significant of the three parameters, making no discernible difference to the quality of the model for the range in which it was set. This is shown in Fig. 5.

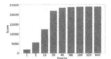
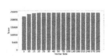
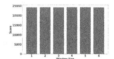

Fig. 3. Average score broken down by epochs.

Fig. 4. Average score by vector size when epochs $= 640$.

Fig. 5. Average score by window size when epochs $= 640$.

Five models were generated with each set of parameters, allowing for analysis of how consistently different parameters perform. Consistency in model quality is shown to be tied to average model quality. This is illustrated in Fig. 6. For parameters that did not perform well, score variation was high. Amongst the top 100 models, parameter sets consistently produced models that scored within 12 points of each other. Amongst the top 50 models, parameter sets scored within 6 points of each other. This demonstrates that given good parameters, model generation is highly stable.

3.3   The Outcome of the Best Performing Model

The best performing model was generated with epochs = 640, vector_size = 30, and window = 2. It is interesting to inspect some of the cosine scores of name pairs under this model. This model had a cosine similarity cut-off of 0.4 (i.e., a cosine similarity of less than 0.4 indicates that two names match), as determined by the method laid out in Sect. 3.1. This cut-off point is clearly visualized in Fig. 1 as the point where the two histograms overlap. Table 1 shows the 10 names with the highest and lowest scores from the matching set. The names in the first two columns of Table 1 are unsurprising very similar, often differing by a single swap of two adjacent characters. Names in columns 4 and 5 in Table 1 on the other hand are often radically different, despite appearing as a matching pair in the *Records* dataset. Names in the rightmost columns are perhaps the most interesting, as they represent the ambiguous cases. It is interesting to note that the model is able to identify two names containing an identical substring as quite similar, even if one name has a lengthy prefix (e.g., knatchbullhugessen and hugessen). There is also a noticeable relationship between the co-occurrence counter in the records dataset and cosine distance. Names that co-occur with higher frequency tend to have more similar embeddings (Fig. 7).

Table 1. Example name pairs for various similarities

Most similar			Most dissimilar			Around cut-off threshold		
Name A	Name B	Cosine	Name A	Name B	Cosine	Name A	Name B	Cosine
frieberg	freiberg	0.0019	berkeley	martiau	1.1999	knatchbullhugessen	hugessen	0.3982
frieberg	freiberg	0.0019	gitton	channdeler	1.2175	shawnee	spawnell	0.3991
cuffel	cuffle	0.0019	desreux	ancien	1.2476	neuser	neiser	0.3993
watowa	wotawa	0.0021	sitzberger	dahmen	1.2522	sorley	sorlie	0.3994
ziesel	zeisel	0.0021	mansion	cherlot	1.2568	stanier	stanyer	0.3997
dziuk	dzuik	0.0022	lelong	dumats	1.2914	fedarb	fedash	0.4004
peevey	peevy	0.0022	jorisse	blanchan	1.2954	altemose	altimore	0.4013
zeitz	zietz	0.0023	declarke	northos	1.3278	smithdorrien	dorrien	0.4018
zietz	zeitz	0.0023	mcmurrough	leinster	1.3293	fitzjohn	fitzgeoffrey	0.4034
gieg	geig	0.0025	deurcant	willemson	1.3386	craiger	crazer	0.4049

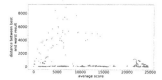

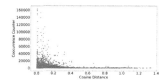

Fig. 6. Maximum difference in score for each parameter set plotted against average score.

Fig. 7. Cosine distance of each Ancestry.com records name pair plotted against its co-occurrence counter.

4   Conclusions

In this paper, a new approach to name comparison is presented. The name-embeddings generated were used to predict whether a pair of names refers to the same entity or not. An experimental study was conducted to evaluate the performance and consistency of the models produced. The effect of three core parameters was investigated, along with the stability of generating a model with specific parameters. Our results show that the new proposed method can predict with high accuracy when a pair of names matches.

References

1. Name2Vec implementation and results (2019). https://github.com/foxcroftjn/CanAI-Name2Vec
2. Antonie, L., Inwood, K., Lizotte, D.J., Ross, J.A.: Tracking people over time in 19th century Canada for longitudinal analysis. Mach. Learn. **95**(1), 129–146 (2014)
3. Carvalho, V.R., Kiran, Y., Borthwick, A.: The Intelius nickname collection: quantitative analyses from billions of public records. In: Proceedings of the 2012 Conference of the North American Chapter of the Association for Computational Linguistics: Human Language Technologies, pp. 607–610 (2012)
4. Christen, P.: A comparison of personal name matching: techniques and practical issues. In: Proceedings of IEEE International Conference on Data Mining - Workshops, pp. 290–294 (2006)
5. Cohen, W., Ravikumar, P., Fienberg, S.: A comparison of string metrics for matching names and records. In: KDD Workshop on Data Cleaning and Object Consolidation, vol. 3, pp. 73–78 (2003)
6. Jaro, M.A.: Probabilistic linkage of large public health data files. Stat. Med. **14**(5–7), 491–498 (1995)
7. Le, Q., Mikolov, T.: Distributed representations of sentences and documents. In: International Conference on Machine Learning, pp. 1188–1196 (2014)
8. Mikolov, T., Chen, K., Corrado, G., Dean, J.: Efficient estimation of word representations in vector space. arXiv preprint arXiv:1301.3781 (2013)
9. Mikolov, T., Sutskever, I., Chen, K., Corrado, G.S., Dean, J.: Distributed representations of words and phrases and their compositionality. In: Advances in Neural Information Processing Systems, pp. 3111–3119 (2013)
10. Müller, M.-C.: Semantic author name disambiguation with word embeddings. In: Kamps, J., Tsakonas, G., Manolopoulos, Y., Iliadis, L., Karydis, I. (eds.) TPDL 2017. LNCS, vol. 10450, pp. 300–311. Springer, Cham (2017). https://doi.org/10.1007/978-3-319-67008-9_24
11. Řehůřek, R., Sojka, P.: Software framework for topic modelling with large corpora. In: Proceedings of the LREC 2010 Workshop on New Challenges for NLP Frameworks, pp. 45–50, May 2010
12. Sim, A., Borthwick, A.: Record2Vec: unsupervised representation learning for structured records. In: IEEE International Conference on Data Mining, ICDM 2018, Singapore, 17–20 November 2018, pp. 1236–1241 (2018)
13. Sukharev, J., Zhukov, L., Popescul, A.: Parallel corpus approach for name matching in record linkage. In: Proceedings of IEEE International Conference on Data Mining, ICDM, pp. 995–1000 (2014)

Genome-Wide Canonical Correlation Analysis-Based Computational Methods for Mining Information from Microbiome and Gene Expression Data

Rayhan Shikder[1,2], Pourang Irani[1], and Pingzhao Hu[1,2(✉)]

[1] Department of Computer Science, University of Manitoba, Winnipeg, Canada
shikderr@myumanitoba.ca,
pourang.irani@cs.umanitoba.ca,
pingzhao.hu@umanitoba.ca
[2] Department of Biochemistry and Medical Genetics,
University of Manitoba, Winnipeg, Canada

Abstract. Multi-omics datasets are very high-dimensional in nature and have relatively fewer number of samples compared to the number of features. Canonical correlation analysis (CCA)-based methods are commonly used for reducing the dimensions of such multi-view (multi-omics) datasets to test the associations among the features from different views and to make them suitable for downstream analyses (classification, clustering etc.). However, most of the CCA approaches suffer from lack of interpretability and result in poor performance in the downstream analyses. Presently, there is no well-explored comparison study for CCA methods with application to multi-omics datasets (such as microbiome and gene expression datasets). In this study, we address this gap by providing a detail comparison study of three popular CCA approaches: regularized canonical correlation analysis (RCC), deep canonical correlation analysis (DCCA), and sparse canonical correlation analysis (SCCA) using a multi-omics dataset consisting of microbiome and gene expression profiles. We evaluated the methods in terms of the total correlation score, and the classification performance. We found that the SCCA provides reasonable correlation scores in the reduced space, enables interpretability, and also provides the best classification performance among the three methods.

Keywords: Canonical correlation analysis (CCA) · RCC · DCCA · SCCA · Comparison study · Multi-omics data · Microbiome and gene expression data

1 Introduction

Datasets comprising of multiple feature sets from different omics sources (e.g., genomics, proteomics, microbiomics etc.) measured on the same subjects are known as multi-omics (or multi-view) data. Integrated study of the multi-omics data has the potential to reveal more information about a disease as it may tell us about the individual associations, interactions among the factors and the flow of information from cause of the disease to consequences [1]. However, most of the omics datasets are very

© Springer Nature Switzerland AG 2019
M.-J. Meurs and F. Rudzicz (Eds.): Canadian AI 2019, LNAI 11489, pp. 511–517, 2019.
https://doi.org/10.1007/978-3-030-18305-9_53

high dimensional in nature and combining them usually results in a unique representation where the numbers of features are very large (e.g., tens of thousands) compared to the number of available samples (e.g., hundreds). These large number of features create challenges in applying most of the statistical methods. Moreover, a large subset of these features may represent redundant or irrelevant information. Therefore, prior to learning any objective functions or finding associations among the omics datasets, the feature sets need to be reduced to a lower dimensional subspace.

Most often, researchers want to investigate the relationships between two omics datasets. Canonical correlation analysis (CCA) - based approaches, which finds the linear combinations of features from two datasets and tries to maximize the correlation between them, are common ways to find such relationships [2]. In addition, CCA also reduces the dimensionality of the original high-dimensional omics datasets, making it suitable for fusion and downstream predictive analysis. However, in a setting where the numbers of features outnumber the number of samples, the basic version of the CCA is not effective. To deal with this situation, regularized versions of the canonical correlation analysis (regularized canonical correlation analysis or RCCA) have been developed [3, 4]. To learn non-linear combinations of the features while calculating the correlations, deep neural network based parametric version of the CCA (named as deep canonical correlation analysis (DCCA)) has been proposed too [5, 6]. In biological applications, researchers also seek to trace the original features that correspond to the resulting correlations, which is hard to achieve with either RCC or DCCA. Hence, sparse versions of the canonical correlation analysis (SCCA) methods have been developed [7–9]. However, there exist no study which highlighted the comparison of the approaches with application to multi-omics datasets specially datasets consisting of microbiome and gene expression profiles.

Our contribution in this paper includes a detailed comparison of the canonical correlation methods (RCC, SCCA, and DCCA) in terms of correlation score and classification performance with applications to a multi-omics dataset consisting of microbiome and gene expression profiles. To the best of our knowledge, this study is the first to investigate the CCA approaches for microbiome and gene expression data together.

2   Preliminaries

2.1   Canonical Correlation Analysis (CCA)

Having two datasets X_1 and X_2 with $(n \times p_1)$ and $(n \times p_2)$ dimensions measured on the same subject $i = 1, 2 \ldots, n$, CCA finds linear combinations of the features from the two datasets which are maximally correlated [2]. In other words, CCA finds the linear projections $w_1^T X_1$ and $w_2^T X_2$ which have a maximum correlation between them, where w_1 and w_2 are the canonical coefficients. Let $\sum_{11}$, and $\sum_{22}$ be the covariances of X_1 and X_2, and $\sum_{12}$ be the cross-covariance between the features of the datasets, then the objective of the CCA method is to maximize the following:

$$corr\left(w_1^T X_1, w_2^T X_2\right) \quad or, \quad \left(\frac{w_1^T \sum_{12} w_2}{\sqrt{w_1^T \sum_{11} w_1 w_2^T \sum_{22} w_2}}\right) \tag{1}$$

2.2   Regularized Canonical Correlation (RCC) Analysis

When the number of features (p_1 or p_2) become larger than the total number of samples (n), the basic version of the CCA doesn't work as the first n canonical variates possess larger values while the rest of the canonical covariates becomes zero [10]. To deal with this, regularization parameters (λ_1 and λ_2) can be added with the covariance matrices in the following manner (I_{p_1} and I_{p_2} are identity matrices) [3, 4].

$$\sum_{11}' = \sum_{11} + \lambda_1 I_{p_1} \tag{2}$$

$$\sum_{22}' = \sum_{22} + \lambda_2 I_{p_2} \tag{3}$$

2.3   Deep Canonical Correlation Analysis (DCCA)

DCCA finds complex nonlinear projections of the input features which are maximally correlated [5]. DCCA is a deep neural network (DNN)-based approach, where two densely connected networks (Network 1 and Network 2 in Fig. 1) are separately trained on two views of the dataset. These two networks learn nonlinear feature combinations and use a correlation maximization objective function.

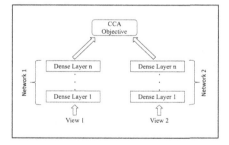

Fig. 1. Schematic diagram of the deep canonical correlation analysis (DCCA) method

2.4   Sparse Canonical Correlation Analysis (SCCA)

For datasets with large number of features, the interpretation of linear combinations become impracticable. In such cases, considering a sparse subset of the features is a viable approach [8, 9]. In this case, the objective function to be maximized takes the following form:

$$corr\left(w_1^T X_1, w_2^T X_2\right)$$

$$where \ ||w_1||^2 \leq 1, ||w_2||^2 \leq 1, P_1(w_1) \leq c_1, \ and \ P_2(w_2) \leq c_2 \qquad (4)$$

Here, P_1 and P_2 are called penalty functions or sparse CCA criterion. These penalty functions are chosen in a way to provide sparse feature combinations and also to make the CCA deal with situations where the feature sets are large compared to the number of samples. P_1 and P_2 can be lasso or fused lasso penalty functions. The parameters c_1, c_2 are used to control the level of penalization.

3   Experiments and Results

3.1   Dataset

We considered a multi-omics dataset consisting of two views: gene expression and microbiome profiles [11]. There are 184 samples with 4 disease subtypes. The gene expression profiles of the data consist of 20,253 features, each of which represents the level of expression for a particular gene. The microbiome profiles consist of 7,000 features which are sparse, discrete in nature and represent the count of an operational taxonomic unit (OTU) in the sample.

3.2   Preprocessing and Hyperparameter Tuning

All the features with no variation (such as zero and constant values) across all of the 184 samples were removed from the dataset. The remaining dataset contained 20,251 gene features and 5,443 OTU features which were normalized afterwards. The 184 samples were divided into train (147) and test (37) groups in a stratified manner. We used R package: CCA for the RCC [10], python implementation from [12] for DCCA, and the PMA package in R [13] for SCCA. We have also considered a supervised version of the SCCA approach (we call this SCCA(S)), where the learned feature projections are also correlated with the output labels. For all the methods, we tuned the hyperparameters to their appropriate values using the training set.

3.3   Total Correlation Scores

After tuning the hyperparameters, we performed the canonical correlation analyses for different output dimensions and learned the canonical coefficients (w_1 and w_2). We multiplied these coefficients with the original dataset to generate the projections.

Total correlation scores were calculated (using the *linear_cca* method provided in [12]) from the projections. From Fig. 2, we can see that RCC provides better correlation scores. On the other hand, SCCA, SCCA(S), and DCCA provide almost similar correlation scores. With the increasing number of output dimensions, the correlation scores become almost the same for all of the approaches when the output dimension surpasses the number of test samples. For SCCA, the sparsity nature may correspond to the compromise in the total correlation score. As deep neural network (DNN)-based approaches are always data hungry, the fewer number of samples is the main reason behind the relatively lower correlation scores of the DCCA method.

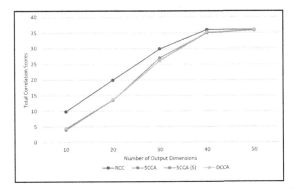

Fig. 2. Total correlation scores for different canonical correlation approaches.

3.4   Classification Performance

We performed binary classifications using support vector machine (SVM) on the projected data from the CCA methods. The classification labels were converted into binary by keeping one class in one group and all the others in another. Hyperparameters (kernels, C, sigma, gamma etc.) of the SVM method were adjusted.

Table 1. Binary-class classification results using SVM on the output projections from different CCA methods. Evaluation metrics are accuracy and area under the ROC curve (AUC).

Dimensions	Metrics	RCC	SCCA	SCCA (S)	DCCA
10	Accuracy	67.56%	72.97%	72.97%	67.56%
	AUC	0.5	0.6	0.6	0.5
20	Accuracy	67.56%	75.67%	75.67%	67.56%
	AUC	0.5	0.71	0.67	0.5
30	Accuracy	70.27%	**75.67%**	**75.67%**	67.56%
	AUC	0.54	**0.756**	**0.67**	0.5
40	Accuracy	70.27%	**78.37%**	**78.37%**	67.56%
	AUC	0.54	**0.69**	**0.69**	0.5
50	Accuracy	67.56%	70.27	70.27	67.56%
	AUC	0.5	0.63	0.6	0.5

From Table 1, we can see that DCCA provides the worst classification performances. The smaller sample size of the dataset may be the reason behind this performance loss. The RCC method's performance is also poor which is easily observed with the low AUC values. Although multi-omics datasets are very high-dimensional, only a handful of these dimensions are actually responsible for a particular phenotype. Therefore, incorporating all the input features may be the reason for the poor classification performance of RCC. Finally, it is visible that the SCCA methods provide relatively better classification performances. The sparse nature of these methods may be the main reason behind this. However, it is surprising that the supervised version of the SCCA didn't provide any better results than the unsupervised one.

4 Conclusion

In this study, we found that sparse canonical correlation analysis provides interpretable correlation scores and better performance in downstream analysis. The regularized canonical correlation analysis, although provides good correlation scores, lacks interpretability and provides poor classification performance. On the other hand, the deep canonical correlation analysis provides moderate correlation scores but lacks interpretability and suffers from poor performance in classification. Therefore, it is advised not to use DCCA with high-dimensional multi-omics datasets having fewer numbers of samples. In future, we will run the comparisons using a larger dataset. We hypothesize that incorporating the outcome variables in the DCCA approach may aid in better performance in the downstream analyses, which we will investigate in future.

Acknowledgements. This work was supported in part by Natural Sciences and Engineering Research Council of Canada, Manitoba Health Research Council and University of Manitoba.

References

1. Hasin, Y., Seldin, M., Lusis, A.: Multi-omics approaches to disease Genome Biol. **18**(1), 83 (2017)
2. Hotelling, H.: Relations between two sets of variates. Biometrika **28**, 321–377 (1936)
3. Vinod, H.D.: Canonical ridge and econometrics of joint production. J. Econom. **4**(2), 147–166 (1976)
4. Leurgans, S.E., Moyeed, R.A., Silverman, B.W.: Canonical correlation analysis when the data are curves. J. R. Stat. Soc. Ser. B. **55**(3), 725–740 (1993)
5. Andrew, G., Arora, R., Bilmes, J.A., Livescu, K.: Deep canonical correlation analysis. In: ICML (2013)
6. Wang, W., Arora, R., Livescu, K., Bilmes, J.: On deep multi-view representation learning. In: International Conference on Machine Learning, pp. 1083–1092 (2015)
7. Hardoon, D.R., Shawe-Taylor, J.: Sparse canonical correlation analysis. Mach. Learn. **83**(3), 331–353 (2011)
8. Parkhomenko, E., Tritchler, D., Beyene, J.: Sparse canonical correlation analysis with application to genomic data integration. Stat. Appl. Genet. Mol. Biol. **8**(1), 1–34 (2009)
9. Witten, D.M., Tibshirani, R., Hastie, T.: A penalized matrix decomposition, with applications to sparse principal components and canonical correlation analysis. Biostatistics **10**(3), 515–534 (2009)
10. Gonzalez, I., Déjean, S., Martin, P., Baccini, A.: CCA: an R package to extend canonical correlation analysis. J. Stat. Softw. **23**(12), 1–14 (2008)

11. Morgan, X.C., et al.: Associations between host gene expression, the mucosal microbiome, and clinical outcome in the pelvic pouch of patients with inflammatory bowel disease. Genome Biol. **16**(1), 67 (2015)
12. Noroozi, V.: VahidooX/DeepCCA. https://github.com/VahidooX/DeepCCA
13. Witten, D., Tibshirani, R., Gross, S., Narasimhan, B., Witten, M.D.: Package 'pma'. Genet. Mol. Biol. **8**, 28 (2013)

Semantic Roles: Towards Rhetorical Moves in Writing About Experimental Procedures

Mohammed Alliheedi[1,3](✉) and Robert E. Mercer[2,3]

[1] Al Baha University, Al Bahah, Saudi Arabia
[2] The University of Western Ontario, London, Canada
`mercer@csd.uwo.ca`
[3] University of Waterloo, Waterloo, Canada
`malliheedi@uwaterloo.ca`

Abstract. Scholarly writing in the experimental biomedical sciences follows the IMRaD (Introduction, Methods, Results, and Discussion) structure. Many Biomedical Natural Language Processing tasks take advantage of this structure. The task of interest in this paper is the identification of semantic roles of procedural verbs as a first step toward identifying *rhetorical moves*, text segments that are rhetorical and perform specific communicative goals, in the Methods section. Based on a descriptive taxonomy of rhetorical moves structured around IMRaD, the foundational linguistic knowledge needed for a computationally feasible model of the rhetorical moves is described: *semantic roles*. Using the observation that the structure of scholarly writing in the laboratory-based experimental sciences closely follows the laboratory procedures, we focus on the procedural verbs in the Methods section. Our goal is to provide FrameNet and VerbNet-like information for the specialized domain of biochemistry. This paper presents the semantic roles required to achieve this goal.

1 Introduction

Scientists must routinely review the scholarly literature in their fields to keep abreast of current advances and to retrieve information relevant to their research. However, this undertaking is becoming more difficult as the volume of scientific literature is immense and rapidly increasing. The types of tasks currently handled by Biomedical Natural Language Processing (BioNLP) systems have generally been aimed at extracting very specific and limited information. Recently, new and more challenging information extraction tasks have been introduced (e.g., [5,6]).

Various studies have used recurrent patterns of text organization called *rhetorical moves* (i.e., text segments that are rhetorical and perform specific communicative goals). Swales' CARS model targets the Introduction section[1] of

---

[1] Experimental articles in the biomedical sciences normally organized in the IMRaD style: Introduction, Methods, Results, and Discussion.

© Springer Nature Switzerland AG 2019
M.-J. Meurs and F. Rudzicz (Eds.): Canadian AI 2019, LNAI 11489, pp. 518–524, 2019.
https://doi.org/10.1007/978-3-030-18305-9_54

scientific articles. Teufel's interests are concentrated on rhetorical moves associated with defining the research space and suggesting the knowledge claims for computational linguistics and chemistry articles. Kanoksilapatham adds to these works by providing the first comprehensive set of rhetorical moves for complete biochemistry articles [7].

Our research goal is to provide a computational model for Kanoksilapatham's descriptive rhetorical move taxonomy. Initially, our focus is on the Methods section of the taxonomy since this provides a description of the procedures followed in the experiment. Because the experimental process is procedural, the moves tend to follow the verbs describing the steps in the experimental process. When a scientist describes her/his method in the writing, it contains a list of experimental steps which are described by verbs (actions). These verbs evoke (initiate) the rhetorical moves in the writing. To understand the moves, we need information about the semantic roles associated with these procedural verbs. Two well known databases containing semantic role information, Framenet [1] and Verbnet [12], do not provide the information appropriate for the verbs found in this scientific domain. So, our purpose in this paper, in the spirit of these two databases, is to introduce the semantic roles that we are proposing for this domain, some of which are the same as those normally found and some which are new and we suggest are required for this domain.

The structure of the document will be as follows: First, an overview of some theoretical and computational approaches to argumentation are presented in Sect. 2. Then, our proposed approach to argumentation analysis is described in Sect. 3. Next, a description of our annotation scheme is given in Sect. 4. Finally, the future work and a conclusion of this paper is given in Sect. 6.

2   Related Work

Swales [13] proposed the Create-A-Research-Space (CARS) model that uses intuition about the argumentative structure of scientific research articles. Kanoksilapatham [7] advanced Swales' approach to move analysis by developing a framework that combines his original CARS model with the use of Biber's multidimensional analysis [2] to enrich the model with additional information about linguistic characteristics. Although Kanoksilapatham provides an extension to the Swales' move analysis study and attempted validation of these moves in biochemistry articles, she only provides a descriptive analysis about rhetorical moves without defining an explicit method for analyzing and recognizing these moves in texts.

Argumentative Zoning (AZ) was developed by Teufel and Moens [14] to categorize sentences based on their contextual information (e.g., determining authorship of knowledge claims). The AZ scheme was later modified to suit the characteristics of biology articles [10]. Furthermore, Teufel [15] proposed a revised version of AZ to include more new categorizes for annotating scientific articles such as chemistry.

We believe that developing a formal representational framework based on verb semantics in procedural scientific discourse will enable a more in-depth analysis in a computationally feasible manner.

3   Our Proposed Approach: Rhetorical Moves Mirror Scientific Experimental Procedures

We aim to develop a formal knowledge representation based on procedural verbs as a method for rhetorical move analysis. We have introduced the notion of moves [13] [7] in Section 2. We also hypothesize that the Method sections in biochemistry articles contain moves which can be correlated with the author's experimental procedures. A key aspect of our hypothesis is that development of a frame-based knowledge representation can be based on the semantics of the verbs associated with these procedures. This representation can provide detailed knowledge for understanding these rhetorical moves. In other words, we propose that a *procedurally rhetorical verb-centric frame semantics* can be used to obtain a deeper analysis of sentence meaning in a computationally feasible manner.

Fillmore [4] introduced the notion of frame semantics as a theory of meaning. Following Fillmore's theory of frame semantics, FrameNet [1] was developed to create an online lexical resource for English. The computational linguistic community has been attracted to the concept of the frame semantics and developed computational resources using this concept, such as VerbNet [12], an on-line verb lexicon for English and PropBank [11], an annotated corpus with basic semantic propositions.

Following the notion of frame semantics, we propose to build a knowledge representation framework to analyze verbs in a procedurally-oriented genre. Our concept of verb-centric frame semantics is intended to address this gap by developing a computationally feasible knowledge representation, based on semantic roles, that will enable the analysis of rhetorical moves.

4   Experimental Event Annotation Scheme

We have developed a new annotation scheme for identifying the structured representation of knowledge in a set of sentences describing the experimental procedures in the Method sections of biochemical articles. An experimental event (as opposed to "bio-events", [16]) is concerned with processes and procedures that are used to investigate biological events. The experimental event is also concerned with the recognition of the biochemist's reasoning of standard biochemical procedures such as using certain instruments or specific biological materials.

Our experimental event scheme was inspired by the annotation scheme for bio-events occurring in biomedical articles [17]. We based our experimental event scheme on the inventory of semantic roles in VerbNet [12] and modified and added new semantic roles to define our scheme. Our experimental event scheme includes: *Theme, Patient, Predicate, Agent, Location, Goal*, etc.

We have extended the VerbNet definition of the semantic role *Instrument* from simply describing "an object or force that comes in contact with an object and causes some change in them" [12] to include a variety of subcategories that correspond to various types of biological and man-made instruments that are

Table 1. Semantic roles in the annotation scheme of our experimental event

Semantic role	Definition
Agent	Generally a human or an animate subject
Patient	Participants that have undergone a process
Predicate	A word that initiates the frame
Theme	Participants in a location or undergoing a change of location
Goal	Identifies a thing toward which an action is directed or a place to which something moves
Factitive	A referent that results from the action or state identified by a verb
Location	The physical place where the experiments took place
Protocol-Detail:	
Time	Identifies the time or a duration of an experimental process
Temperature	Identifies the temperature of an experimental process
Condition	Identifies the condition of how an experimental process is performed
Repetition	Identifies the number of times an experimental process is repeated
Buffer	Identifies the buffer that was used in an experimental process
Cofactor	Identifies the cofactor that was used in an experimental process
Instrument:	
Change	Describes objects (or forces) that come in contact with an object and cause some change
Measure	Describes an object or protocol that can measure another object(s)
Observe	Describes an object which can be used to observe another object(s)
Maintain	Describes an object or protocol which can be used to maintain the state of object(s)
Catalyst	Describes an object that can be used as a catalytic "facilitator" for an experimental event to occur
Reference	Refers to a method or protocol that is being used
Mathematical	Describes a mathematical or computational instrument

used in a biochemistry laboratory. We have also proposed a new semantic role *protocol detail* that identifies certain types of information about experimental processes. The complete set of semantic roles and their definitions in our experimental event scheme is presented in Table 1.

5 Annotation

We have created a data set consisting of 3499 articles between the years of 2013 to 2015 from the top nine journals in biochemistry including *Cell* and *Genome Research*. We have hired ten annotators with a variety of backgrounds (Biochemistry, Bioinformatics, Biology) and different academic levels ranging from Bachelor to PhD degree. Each article is labeled by two annotators using the GATE tool[2]. The labeling is done on a verb basis rather than a full-sentence basis. In other words, each sentence with more than one verb is divided into mini text spans (e.g., clauses) then annotators identity the verbs and label all associated semantic roles for that verb. The annotators are able to decide which constituent is a semantic role. To date, 36 articles have been annotated by two annotators.

We measured the inter-annotator agreement between the two annotations of identical articles using the κ-score [3]. Then, we measured the κ-score after the adjudication step which was done by one of the authors. The adjudication step's main goal is to resolve any disagreement in annotations [11]. We have also measured the kappa score for different configurations of the data set as shown in Table 2. "Original" means the annotation that was given by the annotators, while "Theme = patient and all instrument" indicates theme and patient were combined as one role and all instrument subcategories were considered as one. "Protocol detail" indicates that all protocol detail subcategories were combined as one role. "Adjudicated" means that the disagreements in the original annotations were resolved and any missing semantic roles were added. All of the κ-scores in Table 2 are rated substantial according to [8,9]. The results are very promising.

Table 2. Inter annotator agreement κ-score

Configuration	Kappa score
Original	61.3%
Theme = patient and all instrument	68.9%
Protocol detail	71.6%
Adjudicated	93.6%

---

[2] https://gate.ac.uk/.

6   Conclusion

In this paper, we have presented the semantic roles that we have suggested to be necessary for this scientific domain and which will be used in our annotation scheme. This Experimental Event Scheme, which is based on the proposed semantic roles is the first step towards developing an automated rhetorical moves analysis. We have developed a small set of frames for some verbs (e.g., "wash") based on the manually analyzed data. Once the complete data set is annotated, we will develop a larger set of frames for the most frequent procedural verbs in biochemistry. In future work, we also aim to extend the VerbNet project [12] by providing syntactic and semantic frames for procedural verbs (e.g.,"biotinylated", "annealed", and "carboxymethylated").

References

1. Baker, C.F., Fillmore, C.J., Lowe, J.B.: The Berkeley FrameNet project. In: Proceedings of the 36th Annual Meeting of the Association for Computational Linguistics and 17th International Conference on Computational Linguistics, Stroudsburg, PA, USA, vol. 1, pp. 86–90 (1998)
2. Biber, D.: Variation Across Speech and Writing. Cambridge University Press, Cambridge (1988)
3. Cohen, J.: A coefficient of agreement for nominal scales. Educ. Psychol. Meas. **20**(1), 37–46 (1960)
4. Fillmore, C.J.: Frame semantics and the nature of language*. Ann. N. Y. Acad. Sci. **280**(1), 20–32 (1976)
5. Green, N.: Towards creation of a corpus for argumentation mining the biomedical genetics research literature. In: Proceedings of the First Workshop on Argumentation Mining, Baltimore, Maryland, pp. 11–18 (2014)
6. Green, N.L.: Identifying argumentation schemes in genetics research articles. In: Proceedings of the Second Workshop on Argumentation mining, North American Conference of the Association for Computational Linguistics (NAACL), Denver, CO, USA, pp. 12–21 (2015)
7. Kanoksilapatham, B.: A corpus-based investigation of scientific research articles: Linking move analysis with multidimensional analysis. Ph.D. thesis, Georgetown University, Washington, USA (2003). http://202.28.199.34/multim/3107368.pdf
8. Landis, J.R., Koch, G.G.: The measurement of observer agreement for categorical data. Biometrics **33**, 159–174 (1977)
9. McHugh, M.L.: Interrater reliability: the kappa statistic. Biochem. Med. Biochem. Med. **22**(3), 276–282 (2012)
10. Mizuta, Y., Korhonen, A., Mullen, T., Collier, N.: Zone analysis in biology articles as a basis for information extraction. Int. J. Med. Inform. **75**(6), 468–487 (2006)
11. Palmer, M., Gildea, D., Kingsbury, P.: The proposition bank: an annotated corpus of semantic roles. Comput. Linguist. **31**(1), 71–106 (2005)
12. Schuler, K.K.: VerbNet: a broad-coverage, comprehensive verb lexicon, Doctoral dissertation, University of Pennsylvania, Philadelphia, PA, USA (2005). http://dl.acm.org/citation.cfm?id=1104493
13. Swales, J.: Genre Analysis: English in Academic and Research Settings. Cambridge University Press, Cambridge (1990)

14. Teufel, S., Carletta, J., Moens, M.: An annotation scheme for discourse-level argumentation in research articles. In: Proceedings of the Ninth Conference on European Chapter of the Association for Computational Linguistics, Stroudsburg, PA, USA, pp. 110–117 (1999)
15. Teufel, S.: The Structure of Scientific Articles: Applications to Citation Indexing and Summarization. Center for the Study of Language and Information (2010)
16. Thompson, P., et al.: Building a bio-event annotated corpus for the acquisition of semantic frames from biomedical corpora. In: Proceedings of the Sixth International Language Resources and Evaluation (LREC), European Language Resources Association (ELRA), Marrakech, Morocco (2008)
17. Thompson, P., Nawaz, R., McNaught, J., Ananiadou, S.: Enriching a biomedical event corpus with meta-knowledge annotation. BMC Bioinform. **12**, 393 (2011)

Compression Improves Image Classification Accuracy

Nnamdi Ozah[✉] and Antonina Kolokolova

Memorial University of Newfoundland, St. John's, Canada
{nwo347,kol}@mun.ca

Abstract. We study the relationship between the accuracy of image classification and the level of image compression. Specifically, we look at how various levels of JPEG and SVD compression affect the score of the correct answer in Inception-v3, a TensorFlow-based image classifier trained on the ImageNet database.

Surprisingly, the compression seems to improve the ability of Inception-v3 to recognize images, with the best performance seen at fairly high degrees of compression for most images tested (with half achieving maximal score at JPEG quality under 15, corresponding to more than tenfold reduction in file size). The same behaviour holds for images compressed using the singular value decomposition (SVD) method. This phenomenon suggests that even significant compression can be beneficial rather than detrimental to image classification accuracy, in particular for convolutional neural networks. Understanding when and why compression helps, and which compression algorithm and compression ratio are optimal for any given image remains an open problem.

Keywords: Compression · Image classification · JPEG · SVD · Convolutional neural networks · ImageNet

1 Introduction

Convolutional deep neural networks have been amazingly successful in image classification tasks. Starting with the famous AlexNet [5], they now nearly outperform humans in classification accuracy on tasks such as ImageNet competitions [9]. Yet it is not quite clear what makes recognizing an object in a specific image easier or harder, and, even more importantly from the practical point of view, what preprocessing can be done to improve the accuracy.

It is natural to assume that the more detail there is in the image, the easier it is to classify it correctly. In the image quality studies the "ground truth" is the original, unprocessed image, and any modification that results in information loss produces an inherently inferior result. And indeed, it seems that the studies of the effect of image compression on classification accuracy tend to assume that

Supported by an NSERC Discovery grant.

M.-J. Meurs and F. Rudzicz (Eds.): Canadian AI 2019, LNAI 11489, pp. 525–530, 2019.
https://doi.org/10.1007/978-3-030-18305-9_55

compression degrades accuracy, and study just how much degradation there is. Here we are proposing an alternative view of compression: that of a feature selection mechanism, which can enhance the classification accuracy.

Surprisingly, in our experiments we found that some degree of compression often helps rather than hinders recognition. For each of the 200 images we considered, including 21 from the standard compression literature (SIPI and McGill datasets), there is a compression level for which an ImageNet-trained convolutional neural network (Inception-v3) gives higher score than for the full quality images. Though predictably the images at the lowest quality levels were unrecognizable to both our eyes and Inception-v3, once Inception-v3 was able to identify the images it gave the highest confidence scores when an image was compressed to a fairly low quality. For 102/200 of the images the highest score given by Inception-v3 for a correct label was at quality 15 or below (with 100 being full quality and the range from 1 to 100). The caveat is that for different images different degrees of compression achieve the best classification accuracy, which makes the effect disappear if the accuracy scores are aggregated over different images. The same phenomenon seems to hold for SVD compression, even though SVD is a global (spectral) technique, whereas JPEG works on blocks of pixels.

So far we do not have a conclusive answer to why compression helps, and what determines the best level of compression for a given image. We think that it could be the result of an implicit dimensionality reduction produced by compression, which makes it easier to select features relevant to the image classification task. After all, recognizing a pencil drawing of a familiar face seems easier, not harder than recognizing that face on a photograph.

Viewing compression as preprocessing aimed at improving accuracy has a number of potential applications. First, this suggests that compressing images even before they are sent to a classifier, could be a good strategy: not only would this improve bandwidth and speed of transmitting images to the classifier, it might help with the classification task itself. Second, classifier scores give a more objective metric for evaluating quality of images compressed using different techniques. And third, we hope that understanding how compression affects image recognition might help shed the light on conditions allowing such classifiers to achieve their best performance, and lead to improved techniques for image classification.

1.1   Related Work

There has been a number of papers that considered accuracy of image classification for compressed images. However, it seems that with a few exceptions lossy compression is usually considered a type of degradation [6,8], with the focus on how much compression can be tolerated without significant decrease in performance, rather than looking at compression itself as a preprocessing technique.

One of the first studies of the effect of compression on the automatic image classification is the 1995 paper by Paola and Schowengerdt [7], where they examined the effect of JPEG compression on image classification using minimum-distance, maximum-likelihood, and a 3-layer neural net with backpropagation as classifiers. The conclusion they draw is that compression does not significantly hinder classification even for compression beyond 10:1; our observation from the same data is that accuracy can increase with compression for both the minimum-distance and neural network classifiers. Similarly, representative graphs in Bourlai, Kittler and Messer's work on smart card face verification [1] show that in multiple cases the lowest error rate is when an image is compressed to quality level 20 or below, and the lowest HTER has been achieved for images compressed to below quality level 100. Additionally, several papers on hyperspectral data classification mention improvement with JPEG compression, including Garcia-Vilchez et al. [4] for SVM classification, and Zabala et al. [12] unsupervised classification from Landsat images.

Compression is known to help for classification of adversarially perturbed images. In [3], it was shown that for images with small adversarial perturbation JPEG compression to quality 75 was able to recover the correct classification (albeit with slightly lower confidence scores). Interestingly, adding noise with the same statistics as JPEG compression did not change the effect of adversarial perturbation; there is something specific to JPEG compression that helps recover the image. A subsequent paper by Das et al. [2] explicitly presents JPEG compression as a preprocessing method for removing adversarial perturbations.

Other papers do state that accuracy degrades with compression. Roy et al. [8] found that for several CNNs the accuracy degrades with the increase in JPEG compression ratio. However, their graphs aggregate over a large number of images, hiding this phenomenon. Other papers such as [6,11] use confusion matrix (difference between pixel labels in original vs. compressed image), as a quality metric, which inherently sets the original as the ground truth.

In this paper, we present experimental results suggesting that similar improvement with JPEG and SVD compression holds for the ImageNet challenge-style image classification tasks, with a convolutional neural net such as Inception-v3 as a classifier. We leave as an open problem to understand why such disparate compression methods as JPEG and SVD give improved results, and, more important practically, how to predict the level of compression producing best classification result. Finding an image quality metric correlated with the classifier score is another interesting, though likely challenging, question.

2   Our Results

We tested standard images in the image compression literature, obtained from SIPI and McGill databases, plus photos we took ourselves, for the total of 200 images[1]. From SIPI and McGill database we have selected 21 images with the labels in the training set of Inception-v3. Each image was compressed to each available JPEG quality level (1–100), additionally, this experiment was repeated with grayscale versions of the images. Five SIPI images were also compressed using SVD, generating images at all possible compression levels, then saving

---

[1] The data is available at https://github.com/ozahstonemun/CompressionClassification.

them in JPEG format. Finally, as a classifier we used Inception-v3, pre-trained on 2014 ImageNet data.

Inception-v3 (GoogLeNet) is a TensorFlow-based image classifier, which is claimed to achieve upwards of 95% accuracy on image recognition and classification [10]. The training set for Inception-v3 is the ImageNet database, more specifically images from the ImageNet Large Scale Visual Recognition Challenge from 2014. There are 999 categories in this set, 400 of which are animals; there are no textures and only 3 humans, limiting us to only 21 images from SIPI and McGill datasets; we augmented that with images we took ourselves (mostly vegetable labels). Some images Inception-v3 was not able to recognize at all, but for ones that it did, an average score (across all qualities) was above 95.

To our surprise, the dependence between the score of Inception-v3 and the level of compression appears to be far from linear. Instead, there tends to be a sharp increase from unrecognizable to a peak or near-peak score at a fairly low quality level, followed by worsening and subsequent fluctuations of the score, which improves again, though not necessarily to the previous maximum. Most strikingly, 102 out of 200 images had the maximum score at compression level below 15. Every image achieved the best score when compressed to some degree, with all but 4 having better score at compression levels below 100%. Moreover, for half of the images the difference between the best score and the score at full quality was at least 2.5 points (out of 100), with the maximum above 14 points.

Figure 1 shows representative graphs of Inception-v3 scores vs. JPEG quality. Baboon, Clock, Pepper and Warplane images are from SIPI dataset, and Camel and Monarch butterfly are from McGill dataset. The most pronounced example is the Clock image: there, after spiking above 60 at qualities 7 to 11, with the maximum score 76.74 at quality 8, it only rises above 60 once at quality 32, ending with a meager 41.97 for the full quality image (with the second label

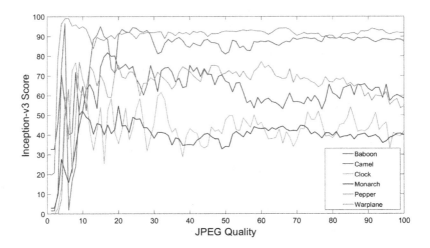

Fig. 1. Inception-v3 score vs. JPEG quality

choice, "stopwatch", at score only 19.25). Note that four of the six images have highest scores at qualities below 10; however, if we were to average the scores across images, we would not see the peaks, as other images have very low scores at the same quality levels.

One might be wondering whether this effect is specific to JPEG compression. However, we found a similar effect with SVD. There, the compression is achieved by computing the singular value decomposition of the image matrix, and disregarding all but k principal eigenvectors. Even though this method seems to be quite different from block-oriented JPEG compression, the same phenomenon of the best score at a low quality seems to arise in SVD compression.

3   Conclusion

In this paper, we looked at the effect of image compression on the performance of a CNN-based image classifier, Inception-v3. To our surprise, we found that compressed images were recognized with better scores by Inception-v3 than uncompressed images, with more than half of maximal scores achieved below quality 15. Moreover, this phenomenon persisted with SVD rather than JPEG algorithm used for compression, suggesting an implicit feature selection. One may alternatively suggest that this is an artifact of possible Inception-v3 classifier bias towards lower-quality images, as a result of being trained on Flickr-sourced images; however, [7] data suggests that even when the training set was not compressed there might be an increase in performance with image compression (see Table 6 in [7]). Overall, our experiments suggest that compression might be useful not only for improving transmission speed and reducing processing time and resources, but also for improving classifier accuracy.

We are working on comparing compression with other standard image perturbations such as noise. Our preliminary results suggest that the effect of (Gaussian) noise and effect of JPEG compression are incomparable, with noise usually hindering recognition, and compression not acting as a denoising mechanism.

For the future work, it would be interesting to determine what causes this phenomenon. To start, it would be good to see if it persists when the classifier is trained on high-quality images, though it would be computationally intensive to train it even if we could find a suitably large high-quality image dataset. One also wonders whether compression plays a role in how animals and humans process visual information: it is believed that the retina compresses the signal received from rods and cones in the eye, sending its compressed representation into the brain. And using compression at the appropriate (possibly domain-dependent) level could have the potential to improve speed and performance in applications.

We believe that improvement of classifier accuracy with compression warrants further investigation, both for practical applications and theoretical insights.

Acknowledgments. We are very grateful to anonymous referees for useful suggestions and literature pointers, and to Jennifer Listgarten, Minglun Gong and Valentine Kabanets for helpful discussions.

References

1. Bourlai, T., Kittler, J., Messer, K.: JPEG compression effects on a smart card face verification system. In: MVA, pp. 426–429 (2005)
2. Das, N., et al.: Keeping the bad guys out: protecting and vaccinating deep learning with JPEG compression. arXiv preprint arXiv:1705.02900 (2017)
3. Dziugaite, G.K., Ghahramani, Z., Roy, D.M.: A study of the effect of JPG compression on adversarial images. arXiv preprint arXiv:1608.00853 (2016)
4. Garcia-Vilchez, F., et al.: On the impact of lossy compression on hyperspectral image classification and unmixing. IEEE Geosci. Remote Sens. Lett. **8**(2), 253–257 (2011)
5. Krizhevsky, A., Sutskever, I., Hinton, G.E.: ImageNet classification with deep convolutional neural networks. In: Advances in Neural Information Processing Systems, pp. 1097–1105 (2012)
6. Lau, W.L., Li, Z.L., Lam, K.W.K.: Effects of JPEG compression on image classification. Int. J. Remote Sens. **24**(7), 1535–1544 (2003). https://doi.org/10.1080/01431160210142842
7. Paola, J.D., Schowengerdt, R.A.: The effect of lossy image compression on image classification. In: Geoscience and Remote Sensing Symposium, IGARSS 1995. Quantitative Remote Sensing for Science and Applications, International. vol. 1, pp. 118–120. IEEE (1995)
8. Roy, P., Ghosh, S., Bhattacharya, S., Pal, U.: Effects of degradations on deep neural network architectures. arXiv preprint arXiv:1807.10108 (2018)
9. Russakovsky, O., et al.: ImageNet large scale visual recognition challenge. Int. J. Comput. Vision (IJCV) **115**(3), 211–252 (2015). https://doi.org/10.1007/s11263-015-0816-y
10. Szegedy, C., Vanhoucke, V., Ioffe, S., Shlens, J., Wojna, Z.: Rethinking the inception architecture for computer vision. In: The IEEE Conference on Computer Vision and Pattern Recognition (CVPR), June 2016
11. Xia, Y., Li, Z., Chen, Z., Yang, D.: Quantitative analysis on lossy compression in remote sensing image classification. In: Visual Information Processing and Communication VI, vol. 9410, p. 94100K. International Society for Optics and Photonics (2015)
12. Zabala, A., Pons, X., Díaz-Delgado, R., García, F., Aulí-Llinàs, F., Serra-Sagristà, J.: Effects of JPEG and JPEG 2000 lossy compression on remote sensing image classification for mapping crops and forest areas. In: IEEE International Conference on Geoscience and Remote Sensing Symposium, IGARSS 2006, pp. 790–793. IEEE (2006)

Predicting Commentaries on a Financial Report with Recurrent Neural Networks

Karim El Mokhtari(✉)(iD), John Maidens(iD), and Ayse Bener(iD)

Data Science Laboratory, Ryerson University, Toronto, Canada
elmkarim@ryerson.ca
https://www.ryerson.ca/dsl

Abstract. Aim: The paper aims to automatically generate commentaries on financial reports. Background: Analysing and commenting financial reports is critical to evaluate the performance of a company so that management may change course to meet the targets. Generating commentaries is a task that relies on the expertise of analysts. Methodology: We propose an encoder-decoder architecture based on Recurrent Neural Networks (RNN) that are trained on both financial reports and commentaries. This architecture learns to generate those commentaries from the detected patterns on data. The proposed model is assessed on both synthetic and real data. We compare different neural network combinations on both encoder and decoder, namely GRU, LSTM and one layer neural networks. Results: The accuracy of the generated commentaries is evaluated using BLEU, ROUGE and METEOR scores and probability of commentary generation. The results show that a combination of one layer neural network and an LSTM as encoder and decoder respectively provides a higher accuracy. Conclusion: We observe that the LSTM highly depends on long term memory particularly in learning from real commentaries.

Keywords: NLP · Recurrent Neural Networks · LSTM · GRU

1 Introduction

Companies generate financial reports for both internal and external purposes. Financial results show the past, current and future performance of a company. Financial reports are generated from the daily sales transactions, inventories, cash flows, supplier transactions, etc. They are critical to evaluate the liquidity, profitability, and capital adequacy of the company. Companies make budgets to set targets for sales, revenues, expenses, assets and liabilities on an annual basis. Depending on the company policy these targets are compared against the actual numbers in different frequencies (i.e. weekly, monthly, quarterly, etc.) and if necessary adjustments are made to targets or some actions are taken in various departments to meet these targets.

Financial analysts within the company periodically study the variances between the actuals and the budget numbers at many levels and they provide

© Springer Nature Switzerland AG 2019
M.-J. Meurs and F. Rudzicz (Eds.): Canadian AI 2019, LNAI 11489, pp. 531–542, 2019.
https://doi.org/10.1007/978-3-030-18305-9_56

management with insights. This task often requires access to many datasets related to different areas (finance, logistics, planning, inventory, etc.) to explain the variance. The analyst creates a summary report that relates each variance with a short commentary that serves as a baseline for top management to take immediate actions.

In this paper we propose an approach for generating commentaries by learning from analysts reports. The proposed algorithm is trained on a dataset of variance and the corresponding commentaries created by analysts. The use of commentaries written in English has many advantages. First, it provides a clear understanding of the actual circumstances of the business in the form of a limited set of sentences and fosters immediate actions especially in a rapidly-changing market. Second, it accelerates the analysis by providing a quick feedback for the learned patterns. Finally, in the long-term, it reduces the cost and time of the analysis by involving the human analyst only in ambiguous cases.

The business considered in this paper is a consumer goods company making and selling hundreds of brands for customers worldwide. The products are grouped into brands, and the brands are then grouped into categories.

The monthly report generated by the company's Enterprise Resource Planning (ERP) system includes the sales details in millions of dollars for each product and customer. Table 1 shows a typical report where the variance is the difference between the forecast and the actual results per product and customer.

Table 1. Sample from the monthly report generated by the company's ERP. Numbers are in Millions of CAD

Customer	Product category	Product brand	Product name	Forecast	Actual	Variance
Customer 1	Category 1	Brand 1	Product 1	2.5	2.4	−0.1
Customer 1	Category 1	Brand 1	Product 2	0.5	0.7	+0.2
...	...	...	...	...	...	...
Customer 32	Category 35	Brand 52	Product 15	0.8	0.75	−0.05
...	...	...	...	...	...	...

The monthly report created by the analyst is shown on Table 2. The sales figures are summed up by brand and compared with the Forecast. The analyst writes his comments to explain the observed variance.

Table 2. Sample from the monthly commentaries report generated by the expert. Numbers are in Millions of CAD

Brand	Forecast	Actual	Variance	Commentary
Brand 1	4.75	5.25	+0.50	The variance is driven by Customer 1 and Customer 2, caused by over delivery
...	...	...	...	...

Generating commentaries is a complex process and requires an extensive experience and meticulous data analysis. This process is time consuming and entails high costs for the company. It needs multi-disciplinary experts and data sources of different natures: Point of Sale (POS) transactions, inventory, manufacturing and marketing to name a few. One downside of this process is its reliance on the expert that may not be consistent over time or might occasionally miss some critical information. Such errors could incur important losses to the company and miss significant opportunities for growth. Our aim is to build a learning based model that is able to understand patterns in data residing in different silos of the organisation and generate commentaries that are as close as possible to the ones created by the experts.

This paper is organised as follows. The second section describes the state of the art in caption generation, the third section describes the learning model. The experimental part is presented in the fourth section and deals with synthetic and real data. The conclusion summarises the findings and presents the prospects.

2   Related Work

Commentary generation is a process where numerical data is encoded and transformed into natural language. This concept is similar in its nature to many problems found in the literature such as caption generation for images and language translation. Many of these methods rely on Recurrent Neural Networks that were applied to sequence to sequence models in language translation [1,2]. We are motivated by this encoder-decoder structure as it translates data between different representations through a context vector linking both ends of the architecture. In our case, we encode financial data into vector representation then we translate it to language form by the decoder.

This same concept has been applied to create captions for images [3]. In this context, retrieval-based methods use neural networks to map image and text into a vector representation [4] or use similarity metrics applied on predefined image features [5]. Mao et al. [6] use a language model that relies on a recurrent neural network instead of a feed-forward based model. Vinyals et al. [7] use Long Short Term Memory (LSTM) networks to generate captions for an image from a pre-trained Convolutional Neural Network (CNN). Donahue et al. [8] apply a similar structure based on LSTM to caption a video sequence. Karpathy and Li [9] follow a different approach in captioning an image that consists of learning through a multimodal embedding space in a model that uses a bidirectional RNN over sentences and a R-CNN over image regions. There is a comparative study that applies different captioning methods namely Visual-Back [11], Guiding-LSTM [12], ShowTell [7] and DeepSemantic [9] to describe car images [10].

3   Learning Model

An overview of the method is illustrated in Fig. 2. For each month, data is aggregated by brand and customer resulting in the pivot table V shown on

Fig. 1. In this matrix denoted V, an element $v(i,j)$ is the variance of Brand i for Customer j for the given month. More specifically, an element $v(i,j)$ is the difference between the sum of the forecast and the sum of the actual of all products belonging to Brand i and delivered to Customer j. The row vector $V^{(i)}$ contains the values of the variance of all customers for Brand i. As shown in Fig. 1, the expert may write a comment like the one created for brand 2.

During training, matrix V is parsed row by row and each row vector $V^{(i)}$ along with the corresponding analyst commentary $C^{(i)}$ is fed into the learning model. The model learns to predict the next word of the commentary jointly from the current word and the variance vector of the Brand i.

	Customer 1	Customer 2	...	Customer j	...		Commentaries
Brand 1	+0.10	+0.11	...	-0.01	...		No comment
Brand 2	-0.13	-0.09	...	+0.27	...		Promotion did well for Customer j
...							...
Brand i	+0.02	+0.03	...	+0.14	...	← $V^{(i)}$	No comment
...							...

Fig. 1. Matrix V generated for one month. An element of V denoted $v(i,j)$ - highlighted on the matrix - is the variance of Brand i for Customer j. The row vector $V^{(i)}$ is the variance of all customers for brand i. (Color figure online)

We propose to use an encoder-decoder architecture similar to [10] but with both encoder and decoder defined as LSTM [13] or GRU [3]. The LSTM network is an RNN that is designed to handle vanishing and exploding gradient issues through the use of forget, input and output gates. The GRU network is a simplified version of the LSTM that uses only two gates to solve the vanishing gradient problem called update gate and reset gate. We also propose to use a simple neural network as encoder.

The commentary for brand i noted $C^{(i)}$ is a sequence of n words such that:

$$C^{(i)} = \{w_0^{(i)}, w_1^{(i)}, \ldots, w_{n+1}^{(i)}\} \qquad (1)$$

$w_0^{(i)}$ and $w_{n+1}^{(i)}$ are special tokens that mark respectively the START and the END of the commentary. The ideal commentary is the one that maximises the probability $p(C^{(i)}|V^{(i)})$ for a given variance vector $V^{(i)}$. This probability is defined with a joint probability as follows:

$$\log p(C^{(i)}|V^{(i)}) = \sum_{k=2}^{n} \log p(w_k^{(i)}|V^{(i)}, w_1^{(i)}, \ldots, w_{k-1}^{(i)}) \qquad (2)$$

The encoder RNN converts the variance vector $V^{(i)}$ into a fixed-size context vector serving as an initial value for the decoder. The chain nature of Eq. 2 is modelled by the decoder RNN. This RNN maps the current word and the

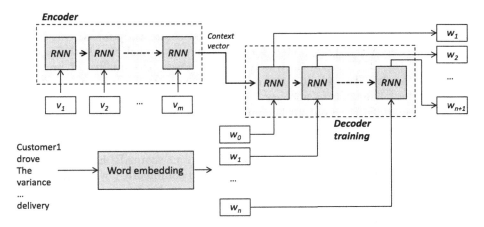

Fig. 2. Learning model architecture: the model consists in an encoder and decoder both implemented by a Recurrent Neural Network depicted unrolled

previous cell state that includes all past information learned up to the current step to the output vector. This vector is applied to a softmax classifier [14] that calculates the probability of each word in the dictionary. The objective function is defined as:

$$\theta = \arg\max_{\theta} \sum_{(C^{(i)}, V^{(i)})} \log p(C^{(i)}|V^{(i)}; \theta) \tag{3}$$

where θ is the model parameter that contains all RNN parameters and word embedding. The objective function can be reformulated as a loss function $L(C, V^{(i)}|\theta)$ to be used in neural network training:

$$L(C^{(i)}, V^{(i)}|\theta) = -\sum_{(C^{(i)}, V^{(i)})} \sum_{k=1}^{n} \log p_k(w_k^{(i)}|V^{(i)}; \theta) \tag{4}$$

During training, the pair variance vector for brand i and the corresponding commentary $(V^{(i)}, C^{(i)})$ is fed to the model. The words of the commentary $C^{(i)}$ are applied to the decoder RNN input at each time step. We use the START token as first input to predict the first word, the commentary words are then applied sequentially to the decoder until the END token is reached. We use the RMSprop optimizer [15] to accelerate the learning. This method divides the learning rate for each weight in the model by a running average of the magnitudes of recent gradient for that weight thus allowing a fast convergence. This optimizer is usually a good choice for recurrent neural networks [16].

4   Experiment

The experimental part includes model testing on synthetic and real data. The real dataset is proprietary and is provided by our research partner. In both experiments, we use the following evaluation metrics: BLEU (BiLingual Evaluation Understudy) [17], ROUGE (Recall-Oriented Understudy for Gisting Evaluation) [18] and METEOR (Metric for Evaluation of Translation with Explicit ORdering) [19]. BLEU score is widely used to assess the quality of machine translation models. It evaluates the quality of text which has been translated from a natural language to another by comparing the machine's output with that of a human. BLEU's output is a number between 0 and 1 that indicates the similarity between the reference and generated text, with values closer to one representing more similar texts. Clarity of grammatical correctness are not taken into account. For each commentary, we compute four scores BLEU-1, BLEU-2, BLEU-3 and BLEU-4 that compare the generated commentary with the human created one. BLEU-1 score considers only the precision at an unigram level, while BLEU-4 considers all four-grams.

While BLEU is a precision-based score used in the machine translation community, ROUGE is a recall-based from the summarization community. It compares the overlapping n-grams, words sequences and word pairs. In our evaluation, we use the ROUGE-L version. ROUGE metric penalises short sentences as it relies on recall [20]. METEOR is used in machine translation and calculates the harmonic mean of precision and recall of unigram matches between the reference and the generated sentences. It also applies synonyms and paraphrase matching for implicit word matching. We use the package nlg-eval in [21] to calculate ROUGE and METEOR scores.

We also assess the model in terms of predicting when a commentary needs to be generated. In this context, we want to evaluate whether the model has learned to detect specific patterns on the variance vector $V^{(i)}$ that triggers the commentary generation. This is a binary classification problem where a positive output stands for a nonempty commentary and a negative output for an empty commentary. We use accuracy and F1-score [22] as evaluation metrics. The accuracy is the number of correct predictions to the total number of input samples. In our case, it refers to the ratio of the generated commentaries to the total number of non empty commentaries or the commentary generation probability. F1-score is the harmonic mean of the precision and recall [22] defined by Eq. 5. Precision and Recall are respectively defined by $TP/(TP + FP)$ and $TP/(TP + FN)$, where TP is the number of True Positives and FN the number of False Negatives.

$$F1 = 2.\frac{Precision.Recall}{Precision + Recall} \tag{5}$$

In both experiments, we compare different networks for both encoder and decoder, namely GRU and LSTM. We also test a one layer neural network as an encoder denoted NN that receives $V^{(i)}$ and encodes it into an initial value for the encoder.

4.1   Results with Synthetic Data

In the synthetic case, we consider that the analyst generates a commentary whenever one or many values in the variance vector $V^{(i)}$ are over a threshold that we define here as ± 0.5. Therefore, we initialise the values of $V^{(i)}$ according to a uniform distribution between -0.5 and $+0.5$, then we choose randomly one to three customers and bring their variance outside the range $[-0.5, +0.5]$. A graphic visualisation of $V^{(i)}$ is shown on Fig. 3. We consider three types of commentaries depending on the number of customers with a high variance. Some samples do not includes any high variance, in such case, en empty commentary is generated.

We train the model by applying at each time step the synthetic variance vector $V^{(i)}$ for brand i to the encoder and the corresponding synthetic commentary $C^{(i)}$ to the decoder. Commentaries are tokenized and applied to a word embedding neural network. We generate 23562 commentaries and we run the simulations for 100 epochs with a batch size of 20.

Table 3. BLEU, ROUGE, METEOR scores with synthetic data

Enc./Dec.	BLEU-1	BLEU-2	BLEU-3	BLEU-4	ROUGE	METEOR
GRU/GRU	0.769	0.626	0.563	0.438	0.814	0.367
NN/GRU	0.546	0.405	0.370	0.273	0.552	0.255
NN/LSTM	0.526	0.390	0.360	0.269	0.558	0.249

Table 3 shows that the highest BLEU, ROUGE and METEOR scores are obtained with the architecture GRU/GRU. The measured BLEU-1 score for this architecture is 0.769 and shows that almost 77% of the words generated in the commentaries were identical to the synthetic ones, while 43.8% of the four-grams were identical. The ROUGE score shows 81.4% of similarity by measuring the recall over the longest common subsequences between reference and predicted sentences. METEOR is also the highest for the GRU/GRU architecture but it is lower than BLEU and ROUGE scores as it takes into account the order of the matched unigrams of the predicted commentary with respect to the human generated one.

Table 4 shows the architecture NN/GRU (NN as encoder and GRU as decoder) provides the best prediction regarding generating a commentary according to the rules above mentioned. Indeed, in predicting that a commentary is needed for a given variance vector, the accuracy of this architecture was as high as 0.863 with a F1-score of 0.906. This score shows a higher rate of true positives compared to the two other architectures.

Figure 4 shows samples of reference and predicted commentaries using the GRU/GRU architecture. In general, the model is able to detect when the variance is associated with one, two or three customers although sometimes the customer numbers are not always correctly predicted.

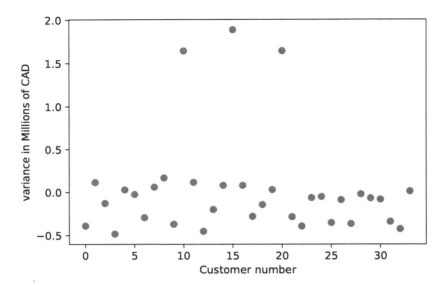

Fig. 3. Sample of a variance vector $V^{(i)}$ for brand i where three customers have a high variance between actual and forecast

Table 4. Commentary generation metrics with synthetic data

Enc./Dec.	Accuracy	F1-score
GRU/GRU	0.769	0.438
NN/GRU	0.863	0.906
NN/LSTM	0.838	0.889

Reference1: | <start> | driven | only | by | customer | 6 | <end> |

Prediction1: | <start> | driven | only | by | customer | **7** | <end> |

Reference2: | <start> | both | customers | 6 | and | 28 | drive | the | variance | <end> |

Prediction2: | <start> | both | customers | 6 | and | **14** | drive | the | variance | <end> |

Reference3: | <start> | customers | 0 | , | 14 | and | 28 | are | behind | the | observed | variance | <end> |

Prediction3: | <start> | customers | 0 | , | **17** | and | 28 | are | behind | the | observed | variance | <end> |

Fig. 4. Examples of reference and predicted commentaries in the experiment with synthetic data

4.2 Results with Real Data

We test the model with real data where patterns are more complex than in the synthetic case and the content of a commentary depends on the analyst experience. The dataset consists of 1330 commentaries. We apply stemming to reduce

Table 5. BLEU, ROUGE, METEOR scores with real data

Enc./Dec.	BLEU-1	BLEU-2	BLEU-3	BLEU-4	ROUGE	METEOR
GRU/GRU	0.172	0.076	0.067	0.041	0.470	0.122
NN/GRU	0.367	0.279	0.283	0.234	0.594	0.237
NN/LSTM	0.399	0.306	0.310	0.258	0.615	0.252

Table 6. Commentary generation metrics with real data

Enc./Dec.	Accuracy	F1-score
GRU/GRU	0.685	0.605
NN/GRU	0.828	0.816
NN/LSTM	0.844	0.830

the dictionary size before tokenizing the comments. As in the synthetic case, we compare different architecture and measure their performance in terms of BLEU, ROUGE and METEOR scores and commentary generation probability.

The results on Tables 5 and 6 show that the NN/LSTM architecture performs best. Indeed, the BLEU-1 score with this architecture is the higher reaching 0.399. This means that almost 40% of the predicted words are identical to the original commentaries. The BLEU-4 score is also the highest and reaches 0.258. Thus, 25.8% of the four-grams in the generated commentaries are identical to the original commentaries. ROUGE and METEOR scores are also the highest. This performance may be explained by the greater dependency of the LSTM on long term memory which is more needed in real commentaries than on synthetic ones.

Regarding the commentaries generation, the NN/LSTM was 84.4% accurate in predicting that a commentary needs to be generated. An F1-score of 83% indicates a higher rate of true positives compared to the other architectures.

Figure 5 shows samples of the predicted commentaries compared with the real ones. The first predicted sample is different from the source, while in the second one, the model was able to predict the customer name but unable to completely predict the commentary. In the last sample, the prediction and reference correspond exactly.

By comparing synthetic and real results, we notice the higher performance of the GRU decoder over the LSTM decoder with synthetic data. This can be interpreted by the similarity between commentaries generated synthetically. Indeed, we created four different classes of commentaries where the only difference within the same class was the customers number. On the other hand, in the real case, the similarity is low between commentaries and learning relies more on long term relation between words in the same commentary. This may explain the higher performance of the LSTM over the GRU network. In fact, the LSTM uses a more complex structure with four gates which allows a better dependency with the past compared to the GRU network.

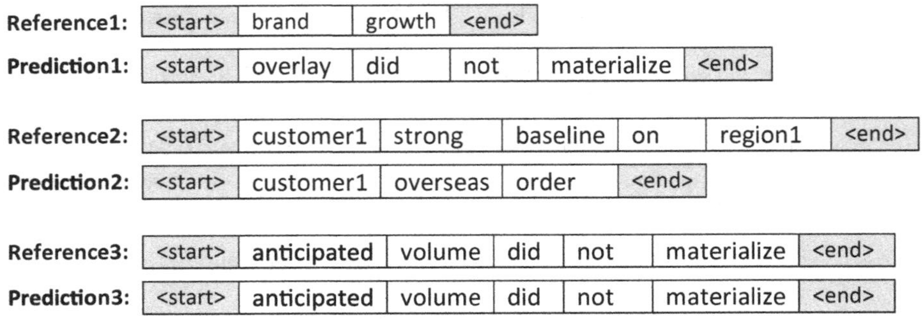

Reference1:	<start>	brand	growth	<end>		
Prediction1:	<start>	overlay	did	not	materialize	<end>

Reference2:	<start>	customer1	strong	baseline	on	region1	<end>
Prediction2:	<start>	customer1	overseas	order	<end>		

Reference3:	<start>	anticipated	volume	did	not	materialize	<end>
Prediction3:	<start>	anticipated	volume	did	not	materialize	<end>

Fig. 5. Examples of reference and predicted commentaries in the experiment with real data. As the dataset is the property of our research partner, some words may not be identical to the real dataset and real customer names were replaced by customer 1, 2 and 3.

5 Threats to Validity

In this paper, we note the following threats to validity:

- In the real dataset, in a high number of cases, the analyst decides not to generate any commentary. This results in an imbalanced dataset that would create a wrongly high accuracy. To mitigate the impact of this fact, we combined undersampling of empty commentaries and over sampling of non empty commentaries. In addition, we used F1-score as a second evaluation metric.
- Over-fitting is a problem that was noted during training on both synthetic and real data. We mitigated this problem by using random validation sets and by changing batch sizes.
- A highly diversified vocabulary used in the commentaries impacts negatively the accuracy and convergence of the algorithm. We reduced the impact of such a problem by stemming words and replacing all number by a unique token. The prediction of numbers is out of the scope of this work.

6 Conclusion

In this paper, we propose an encoder-decoder architecture that aims at generating commentaries from a dataset reporting the variance between forecast and actual for a consumer good company. The commentaries are originally created by financial analysts depending on their observation of the variance dataset. We train the architecture on synthetic and real commentaries and we compare different variants of the architecture. We assess the performance of the model using BLEU, ROUGE and METEOR scores to compare the accuracy of the generated commentaries. We also assess the probability of generating a commentary by learning from the analyst. We show that an LSTM decoder associated with a one layer neural network encoder performs better than a GRU network in the real case. The GRU, on the other hand is more accurate in the synthetic case. This difference may be explained by the high variance among real commentaries that needs a long term dependency in the learning process. The more complex four-gate structure of the LSTM allows a better long-term dependency, and provides a better accuracy in commentary generation. The prospects of this work include comparing the encoder-decoder model with an attention mechanism. In such model, the interaction between both entities does not rely only on the context vector but extends the interaction to learning the dependency between each word and all financial data. Such model may allow to learn more complex patterns between brands and between variances over time.

Acknowledgement. This work is supported by a grant from Smart Computing For Innovation (SOSCIP) consortium, Toronto, Canada.

References

1. Sutskever, I., Vinyals, O., Le, Q.V.V.: Sequence to sequence learning with neural networks. In: NIPS, pp. 3104–3112 (2014)
2. Bahdanau, D., Cho, K., Bengio, Y.: Neural machine translation by jointly learning to align and translate. arXiv:1409.0473 (2014)
3. Cho, K., et al.: Learning phrase representations using RNN encoder-decoder for statistical machine translation. In: Conference on Empirical Methods in Natural Language Processing (EMNLP), pp. 1724–1734 (2014)
4. Farhadi, A., et al.: Every picture tells a story: generating sentences from images. In: Daniilidis, K., Maragos, P., Paragios, N. (eds.) ECCV 2010. LNCS, vol. 6314, pp. 15–29. Springer, Heidelberg (2010). https://doi.org/10.1007/978-3-642-15561-1_2
5. Hodosh, M., Young, P., Hockenmaier, J.: Framing image description as a ranking task: data, models and evaluation metrics. JAIR **47**, 853–899 (2013)
6. Mao, J., Xu, W., Yang, Y., Wang, J., Yuille, A.: Deep captioning with multimodal recurrent neural networks (M-RNN). arXiv:1412.6632 (2014)
7. Vinyals, O., Toshev, A., Bengio, S., Erhan, D.: Show and tell, a neural image caption generator. arXiv:1411.4555 (2014)
8. Donahue, J., et al.: Long-term recurrent convolutional networks for visual recognition and description. arXiv:1411.4389v2 (2014)
9. Karpathy, A., Li, F.-F.: Deep visual-semantic alignments for generating image descriptions. arXiv:1412.2306 (2014)
10. Chen, L., He, Y., Fan, L.: Let the robot tell: describe car image with natural language via LSTM. Pattern Recogn. Lett. **98**, 75–82 (2017)
11. Fang, H., et al.: From captions to visual concepts and back. In: Proceedings of the IEEE Conference on Computer Vision Pattern Recognition, pp. 1473–1482 (2015)
12. Jia, X., Gavves, E., Fernando, B., Tuytelaars, T.: Guiding long-short term memory for image caption generation. In: Proceedings of the ICCV (2015)
13. Hochreiter, S., Schmidhuber, J.: Long short-term memory. Neural Comput. **9**(8), 1735–1780 (1997)
14. Bridle, J.S.: Probabilistic interpretation of feedforward classification network outputs with relationships to statistical pattern recognition. In: Soulié, F.F., Hérault, J. (eds.) Neurocomputing: Algorithms, Architectures and Applications, pp. 227–236. Springer, Heidelberg (1990). https://doi.org/10.1007/978-3-642-76153-9_28

15. Tieleman, T., Hinton, G.: Lecture 6.5-RMSprop: divide the gradient by a running average of its recent magnitude. COURSERA 4(2), 26–31 (2012)
16. Keras Documentation. https://keras.io/optimizers/. Accessed 28 Jan 2019
17. Papineni, K., Roukos, S., Ward, T., Zhu, W.J.: BLEU: a method for automatic evaluation of machine translation. In: ACL (2002)
18. Lin, C.Y.: Rouge: a package for automatic evaluation of summaries. In: Text Summarization Branches Out: Proceedings of the ACL 2004 Workshop, pp. 74–81 (2004)
19. Elliott, D., Keller, F.: Image description using visual dependency representations. In: EMNLP, pp. 1292–1302. ACL (2013)
20. Kilickaya, M., Erdem, A., Ikizler-Cinbis, N., Erdem, E.: Re-evaluating automatic metrics for image captioning. In: Proceedings of EACL 2017, pp. 199–209 (2017)
21. Sharma, S., El Asri, L., Schulz, H., Zumer, J.: Relevance of unsupervised metrics in task-oriented dialogue for evaluating natural language generation. CoRR, vol. abs/1706.09799 (2017). http://arxiv.org/abs/1706.09799
22. Dangeti, P.: Statistics for Machine Learning, 1st edn. Packt Publishing Ltd., Birmingham (2017)

Applications of Feature Selection Techniques on Large Biomedical Datasets

Nicolas Ewen[1]([⊠]), Tamer Abdou[1,2], and Ayse Bener[1]

[1] Data Science Laboratory, Ryerson University, Toronto, ON M5B 2K3, Canada
{nicolas.ewen,tamer.abdou,ayse.bener}@ryerson.ca
[2] Faculty of Science, Arish University, North Sinai 45516, Egypt

Abstract. The main goal of this paper is to determine the best feature selection algorithm to use on large biomedical datasets. Feature Selection shows a potential role in analyzing large biomedical datasets. Four different feature selection techniques have been employed on large biomedical datasets. These techniques were Information Gain, Chi-Squared, Markov Blanket Discovery, and Recursive Feature Elimination. We measured the efficiency of the selection, the stability of the algorithms, and the quality of the features chosen. Of the four techniques used, the Information Gain and Chi-Squared filters were the most efficient and stable. Both Markov Blanket Discovery and Recursive Feature Elimination took significantly longer to select features, and were less stable. The features selected by Recursive Feature Elimination were of the highest quality, followed by Information Gain and Chi-Squared, and Markov Blanket Discovery placed last. For the purpose of education (e.g. those in the biomedical field teaching data techniques), we recommend Information Gain or Chi-Squared filter. For the purpose of research or analyzing, we recommend one of the filters or Recursive Feature Elimination, depending on the situation. We do not recommend the use of Markov Blanket discovery for the situations used in this trial, keeping in mind that the experiments were not exhaustive.

Keywords: Feature selection · Bio-medical · Large dataset

1 Introduction

In this paper we consider four techniques used for feature selection on large biomedical datasets. We analyzed the performance of Information Gain and Chi-Squared filters, as well as Markov Blanket discovery and Recursive Feature elimination on a number of large biomedical datasets from the UCI dataset repository. We then used various measures to compare the selection techniques and drew conclusions about when they might be useful. When considering feature selection on large datasets, we found that the best technique to use depends

The work in this paper is conducted in the Capstone Project Course of the Certificate in Data Analytics, Big Data, and Predictive Analytics at Ryerson University.

© Springer Nature Switzerland AG 2019
M.-J. Meurs and F. Rudzicz (Eds.): Canadian AI 2019, LNAI 11489, pp. 543–548, 2019.
https://doi.org/10.1007/978-3-030-18305-9_57

on the situation. For example, the needs for education purposes would be different from those of research purposes. For example, when instructing those in the biomedical field, higher emphasis would be put on understandability and speed. Even among researchers, the needs can be different. Some will want an algorithm that produces the best predictors by some measure, while others will want an algorithm that has high feature stability [8].

2   Background and Related Work

Information Gain and Chi-Squared based methods are filters, and thus quick to perform on large datasets [7]. Markov Blanket discovery is a feature selection technique used in the biomedical field [1]. A static Bayesian Network is good for modelling probabilistic dependencies between attributes without a time component [9]. Using this, we can infer causality for a dependant variable [9]. A Markov Blanket consists of the attributes that the target is conditionally dependant on (parents), the attributes that are conditionally dependant on the target (children), and the children's other parents. Markov Blankets are accurate and useful for feature selection [5]. We used an algorithm called IAMB (Incremental Association Markov Blanket) to find the Markov Blanket of the target attribute [11]. Recursive Feature Elimination is another feature selection technique. Features are ranked and eliminated. Different sized subsets are tested to pick the best one. Resampling is done to avoid selection bias [3].

3   Datasets

3.1   Gene Expression Cancer RNA-Seq Data Set

This dataset is from the Pan-cancer project [12], available through the UCI datasets. Each row is data from one patient. The dataset has one column listing the type of cancerous tumour (of 5 possible) that data was taken from. The rest of the columns are readings of the RNA from the cells in those tumours.

3.2   Arcene Data Set

The Arcene dataset originally from the NIPS 2003 challenge on feature selection [6]. This dataset uses mass spectrometry readings from a number of patients with different cancers, as well as patients without cancer. The dataset was simplified to be used with binary classification: those with, and those without cancer (Table 1).

Table 1. Datasets

Dataset	Instances	Features	Domain
Arcene	200	10,000	Mass spectrometry
RNA	801	20,531	Gene expression

4   Methods

4.1   Pre-selection

The first thing we did to the datasets was removing zero variance attributes. This initial step reduced the features without any loss. We then proceeded with feature selection. Four different methods of feature selection were performed on each dataset. Various results were then collected from them and compared.

4.2   Feature Selection Methods

The first method of feature selection we used was **Information Gain**. The second method used was a **Chi-Squared** based filter. The third method we used was to take the **Markov Blanket** of the class labels. We used this method because using Bayesian Networks has been a successful method of feature selection in the biomedical field [10]. We found the Markov Blanket using the IAMB algorithm. This was done to save the time of building the full network [11]. The fourth method we used was recursive **Feature Elimination (RFE)** with a resampling wrapper to avoid overfitting.

4.3   Evaluation

After feature selection was applied, we compared the selected subsets using a variety of measures: the time each algorithm took to select features; the average Kappa from three predictors we built using each subset; and the average Kappa standard deviation. We measured the time from when the selection algorithm was given the dataset to when it finished running. For each selected subset, we performed normalization, and then created three predictors: A K-nearest-neighbours, a random forest, and a regression. For each kind of predictor, we performed 10 iterations of 10-fold cross-validation. We averaged out all the scores of all predictors to give a score for each feature selection method. Then, we compared the stability of the selection methods on each dataset. To do this, we split the datasets into four subsets, each with three quarters of the data points. We then ran each selection method on each of the subsets, and compared the features they selected with those selected by the same method performed on the other subsets to calculate the stability of each selection method. The measure for stability we used was the Adjusted Stability Measure [8]. Our aim was to be able to categorize each of the selection methods for reference on when someone might want to use them. Speed of selecting, quality of the predictors produced, and relative simplicity of the algorithm would be relevant in a teaching environment. In a medical environment, stability of feature selection and quality of predictors produced would be relevant, for example [8].

5   Results

5.1   Efficiency

Since the datasets we chose to work with were large, the speed of the feature
selection algorithm was important [2]. Speed is relevant whenever time is limited,
whether that be for instruction purposes, or in cases when only a quick overview
is necessary, or for streaming data that must be analyzed quickly. Of the four
selection algorithms we compared, the filter methods were the fastest, finishing many times faster. The Markov Blanket algorithm and Recursive Feature
Elimination algorithm took significantly longer. See Fig. 1.

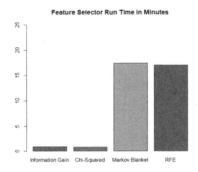
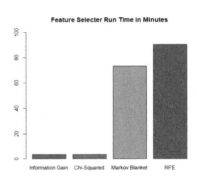

(a) Speed in minutes on Arcene dataset   (b) Speed in minutes on RNA dataset

Fig. 1. Efficiency

5.2   Stability

Feature stability is a measure of how much a feature selection algorithm chooses
the same features from datasets with slight variation [8]. Feature stability is
important in a couple of cases. One is when researchers want to narrow down
their feature set in order to closely examine a few particular features. For example, this might be important to cancer researchers. The other is when analysis
is being done on streaming data, or data that is recorded over time. If feature
selection is being done on streaming data using an algorithm with high stability,
then a significant change in the features selected would imply a change in the
data that could be explored. The Information Gain, and Chi-Squared algorithms
had the highest stability of the four, as shown in Fig. 2.

5.3   Quality of Predictors

To get a measure on the quality of features that were selected, we measured
the Cohen's Kappa of the predictors built from the chosen features. Cohen's
Kappa is a measure of how much more accurate than a random guess a predictor
performed [4]. RFE performed the best. Information Gain and Chi-Squared also
performed quite well, and better than the Markov Blanket (Fig. 3).

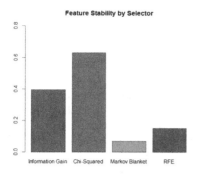

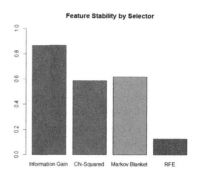

(a) Stability for Arcene dataset

(b) Stability for RNA dataset

Fig. 2. Stability

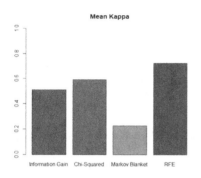

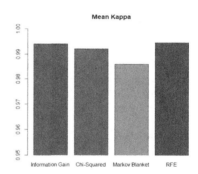

(a) Kappa for the Arcene dataset

(b) Kappa for the RNA dataset

Fig. 3. Kappa

6 Threats to Validity

Internal Threats: The feature selection methods we used were not optimized. Furthermore, the four techniques do not cover all methods of feature selection. Likewise, the three predictors built for each one do not cover all types of predictors, and were not optimized. Performance evaluation was only done with Kappa. External Threats: This study was only performed on biomedical datasets with large numbers of features. Datasets of different sizes and datasets from different fields may produce different results on the same methods.

7 Conclusion

We identified three different areas with different feature selection requirements on large datasets. For early instruction purposes we think that the Information Gain and Chi-Squared filters are the best algorithms of the four. They are efficient on large datasets, simple to understand, and produce predictors that perform fairly well. For research, we identified two areas with different needs. When all that matters is the quality of the predictors, RFE outperformed the other three. When the speed of the selection process or stability of the features selected is of higher priority than having the best predictor results, Information Gain and Chi-squared performed better than the others. As this was not a thorough analysis of the topic, for further research we propose adding more selection methods to our trials, as well as a variety of different sized large bio-medical datasets.

References

1. Aliferis, C.F., Tsamardinos, I., Statnikov, A.: Large-scale feature selection using Markov blanket induction for the prediction of protein-drug binding. Technical report (2002)
2. Alzubi, R., Ramzan, N., Alzoubi, H., Amira, A.: A hybrid feature selection method for complex diseases SNPs. IEEE Access **6**, 1292–1301 (2018)
3. Ambroise, C., McLachlan, G.J.: Selection bias in gene extraction on the basis of microarray gene-expression data. Proc. Nat. Acad. Sci. U.S.A. **99**(10), 6562–6566 (2002)
4. Ben-David, A.: About the relationship between ROC curves and Cohen's kappa. Eng. Appl. Artif. Intell. **21**(6), 874–882 (2008)
5. Fu, S., Desmarais, M.C.: Markov blanket based feature selection: a review of past decade. In: Proceedings of the World Congress on Engineering 2010, vol. I, pp. 321–328 (2010)
6. Guyon, I.: Design of experiments of the NIPS 2003 variable selection benchmark. In: NIPS 2003 Workshop on Feature Extraction, July 2003
7. Lazar, C., et al.: A survey on filter techniques for feature selection in gene expression microarray analysis. IEEE/ACM Trans. Comput. Biol. Bioinf. **9**(4), 1106–1119 (2012)
8. Lustgarten, J.L., Gopalakrishnan, V., Visweswaran, S.: Measuring stability of feature selection in biomedical datasets. In: AMIA Annual Symposium Proceedings, vol. 2009, no. 3, pp. 406–410 (2009)
9. Nagarajan, R., Scutari, M., Lèbre, S.: Bayesian Networks in R. Springer, New York (2013)
10. Sachs, K., Perez, O., Pe'er, D., Lauffenburger, D.A., Nolan, G.P.: Causal protein-signaling networks derived from multiparameter single-cell data. Science (New York N.Y.) **308**(5721), 523–529 (2005)
11. Tsamardinos, I., Aliferis, C., Statnikov, A., Statnikov, E.: Algorithms for large scale Markov blanket discovery. In: Flairs (i), pp. 376–381 (2003)
12. Weinstein, J.N., et al.: The cancer genome atlas pan-cancer analysis project. Nat. Genet. **45**(10), 1113–1120 (2013)

DeepAnom: An Ensemble Deep Framework for Anomaly Detection in System Processes

Okwudili M. Ezeme$^{(\boxtimes)}$ ⓘ, Michael Lescisin, Qusay H. Mahmoud ⓘ, and Akramul Azim ⓘ

Department of Electrical, Computer and Software Engineering,
University of Ontario Institute of Technology, Oshawa, ON L1G 0C5, Canada
{mellitus.ezeme,michael.lescisin,qusay.mahmoud,akramul.azim}@uoit.ca

Abstract. Model checking and verification using Kripke structures and computational tree logic* (CTL*) use abstractions from the process to create the state-transition graphs that verify the model behavior. This scheme of profiling the behavior of a process means that the depth of the model behavior that can be synthesized correlates with the level of the model abstraction. Therefore, for complex processes, this approach does not produce a fine-grained behavioral model and does not capture the execution time interactions amongst processes, hardware, and the kernel because of state explosion problems. Hence, in this paper, we introduce *DeepAnom:* an ensemble deep framework for anomaly detection in system processes. *DeepAnom* targets anomalies in both time-driven and event-driven processes. We test the model with dataset generated from autonomous aerial vehicle application, and the results confirm our hypothesis that *DeepAnom* presents a deeper view of the system processes and can therefore capture anomalies of various scenarios.

Keywords: Anomaly detection · Machine learning · Context modeling

1 Introduction

Model checking and verification using Kripke structures and CTL* [2] use abstractions from the model/process/application to create the state-transition graphs that verify the model behavior. This scheme of profiling the behavior of a process relies on different layers of abstraction and this means that the depth of the model behavior that can be synthesized correlates with the level of model abstraction. However, because the symbolic model behavior verifications like CTL* requires that there exists a finite number of states, deep and broad level of abstraction may result in state explosion problems. Hence, the often resort to Kripke structures that have finite and manageable states to represent model behavior. Therefore, for complex processes, this approach does not return a fine-grained behavioral model and does not capture in-depth execution time

© Springer Nature Switzerland AG 2019
M.-J. Meurs and F. Rudzicz (Eds.): Canadian AI 2019, LNAI 11489, pp. 549–555, 2019.
https://doi.org/10.1007/978-3-030-18305-9_58

interactions amongst processes, hardware, and the kernel. However, the salient difference between *DeepAnom* and the symbolic model checking schemes is that the former uses *boolean* probabilities to create the state relationships while the later utilizes *non-binary* probabilities in capturing the state interactions. The implication of the use of non-binary probabilities is that a previously unseen state can be handled without resort to generating an entirely new model as is obtainable with the use of boolean probabilities. For example, assuming that Φ is a transformative function in (1), the use of boolean coefficients makes no difference when the order of the system calls are changed. However, with non-binary probabilities, the linear combinations will result in different output values based on the ordering of the events.

$$y_t = \Phi\left(a \times y_{t-1} + b \times y_{t-2} + c \times y_{t-3}\right) \tag{1}$$

Hence, in this paper, we introduce an ensemble framework of models that uses RNN to profile the behavior of time-driven and event-driven processes based on system call information. Therefore, our contributions in this paper are as follows; **(a)** design and development of a unique ensemble architecture for profiling system processes or applications. **(b)** creation of fine-grained features from system calls to capture the broad representation of a system process behavior. **(c)** development of a simulation platform for generating system calls of real applications under different operation scenarios.

The paper is organized as follows: Sect. 2 highlights the related anomaly detection work concerning trace analysis while Sect. 3 presents our model architecture and the technical details of the work. Section 4 discusses our experimental setup and results while we conclude the work in Sect. 5 and give further research directions.

2   Related Work

Detecting deviations or attacks have been largely classified as *intrusion* or *anomaly* detection [1]. While intrusion detection techniques rely on the use of *signatures* to detect a misuse behavior by storing signatures of discovered anomalies, anomaly detection models construct profiles using known standard features and labels any deviation from the constructed profile as anomalous. The obvious limitation of this signature-based approach is that *zero-day* vulnerabilities cannot be detected as it only searches for observed signatures, i.e. *it emphasizes memorization over generalization*. On the other hand, the anomaly-based approaches target both known and unknown anomalies and can detect *zero-day* vulnerability. Because *DeepAnom* centers on anomaly detection approach, we highlight mostly the works in this area. In [4], the authors used kernel events to build an offline anomaly detection model using some vector space model concepts, agglomerative clustering and imputation techniques. Reference [3] used deep LSTM model augmented with a workflow model constructed from system logs to create anomaly detection models.

In [5,6], the authors performed context modeling and anomaly detection using system call IDs but other properties of the events are skipped.

Authors of [7] have an anomaly model built using system call frequency distribution as well as clustering techniques to detect when processes deviate from their standard profiles.

3   DeepAnom: The Ensemble Framework

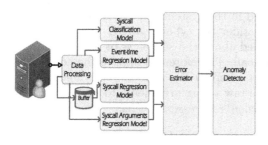

Fig. 1. The ensemble framework

The high-level overview is that we stream system call traces from the instrumented kernel and we feed the streams to the model via the *Data Processing* block where the system call properties: *system call ID, arguments*, and the *timestamp* are parsed. The *syscall regression model* requires a buffer because $F_{TD}/F_{ED} \geq 1$ where F_{TD} and F_{ED} are the operational frequencies of the *Time-Driven* and *Event-Driven*

(syscall regression model) models respectively. In summary, while the *syscall classification model, event-time regression model*, and the *syscall arguments regression model* target anomalous sequence calls with *local temporal profile*, the *syscall regression model* targets anomalous system call sequences that has *long temporal profile*.

3.1   Data Processing Module

The system call traces contain several properties, but the features of interest to us are *timestamp, system call ID* and *system call arguments*. Given *timestamp* as **t**, system call string name as **k** and system call arguments as **a**, we process these properties to yield a multivariate feature space that we feed to our models to help determine the presence and type of anomlay in the process execution.

Given an observation of system call samples $\{s_1, s_2, s_3, \ldots, s_n\}$ where **n** is the total number of the observed traces, then the function β of (2) defines the time window between the observation of successive system calls. Equation (2) ensures that we detect illegal actions which affects the timing operation of the process.

$$\beta : (t_i, t_{i+1}) \longmapsto t_{i+1} - t_i \text{ where } i \in \mathbb{N} \tag{2}$$

We use the function defined in (3) to encode the system call strings into their corresponding Linux tables for a total of **330** unique system calls in the X_64 platform.

$$S : k \longmapsto w; \text{ where } w = \{w \in \mathbb{N} \mid 1 \leq w \leq 330\} \tag{3}$$

The system call argument is alphanumeric and special character strings. System call arguments are strings, and in order to maintain a relatively few vocabularies, we encode the string characters using **ASCII** values. This encoding provides us with a range of unique **256** classes. After encoding the string values, we compute the *frequency distribution* and *relative frequency distribution* for the **256** classes in each system call argument. Our aim is not to check how consistent each *ASCII* value is across the system call but to find the steepness of the distribution of the characters when we sort them. Checking for consistency of each *ASCII* value seen during learning amounts to assuming that other characters not seen during learning cannot be used in the system call arguments, and this is unrealistic. Therefore, we do not consider the type of *ASCII* values present in the argument; we are rather concerned about the *steepness* of the sorted distribution of the characters per system call. Therefore, we impose the condition of (4) on the counted string arguments where n has a maximum value of 256 and (5) provides the mapping from the raw string arguments to the $SARFD$ feature which is fed to the *syscall arguments regression model*.

$$SARFD_i \geq SARFD_{i+1}, \ldots, \geq SARFD_n \text{ where } n \in \mathbb{N} \qquad (4)$$

$$SARFD : L \longmapsto M \text{ where} \qquad (5)$$
$$L = \{l \in \mathbb{N} \mid 1 \leq l \leq 256\} \qquad (6)$$
$$M = \{m \in \mathbb{R} \mid 0 \leq m \leq 1\} \qquad (7)$$
$$\sum_{i=1}^{256} SARFD_i = 1 \qquad (8)$$

3.2 Error Estimator and Anomaly Detector

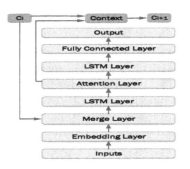

Fig. 2. The ensemble RNN model

Figure 2 is our deep recursive *auto-encoder* used for the sequence reconstruction. The operation of the stacked layers are intuitive and easy to follow and we are skipping other details because of space constraint.

Given $x_i \in \mathbb{R}$ which serves as the input, the target during learning is a shifted version $x_{i+w} \in \mathbb{R}$ where w is the lookahead window. Therefore, the goal is to replicate the sequence at the output by creating $x' \in \mathbb{R}$ where $x_k \equiv x'_k$. Hence, when we have $f : x \longmapsto x'$ given x as the ground truths, the classification error $d = \varnothing(x \equiv x')$ is the *boolean* difference in the lookahead window. The regression error estimator of the *regression models* of Fig. 1 is computed as the square of the $d = \varnothing(x - x')$ where x and x' are the ground truth and predicted value respectively.

Given the sample μ and variance σ^2 from the validation data, we employ the *Bienaymé-Chebyshev* inequality given in (9) to compute the threshold.

$$P\left(|X - \mu| \geq \varUpsilon\right) \leq \frac{\sigma^2}{\varUpsilon^2} \qquad (9)$$

We are interested in the range where the observation error $d - \mu > 1$. Therefore, we interchange $\varUpsilon$ in (9) with $|d - \mu|$ to derive (10) which computes the decreasing probability for increasing deviation $|d - \mu|$ where $d > \mu$.

$$P\left(d > \mu\right) = P\left(|X - \mu| \geq |d - \mu|\right) = \frac{\sigma^2}{(d - \mu)^2} \qquad (10)$$

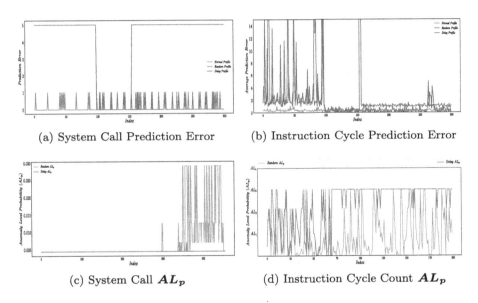

(a) System Call Prediction Error   (b) Instruction Cycle Prediction Error

(c) System Call AL_p (d) Instruction Cycle Count AL_p

Fig. 3. Prediction error and anomaly level probability AL_p

4   Experiments and Analysis

The UAV controller application has four states and these are explained as follows: **(a)** *Take Off*: Apply the Z-axis throttle until we attain cruising height.**(b)** *Cruising*: Adjust Z-axis throttle to counteract gravity, and set the X-axis and Y-axis throttles to move at a uniform velocity to the destination.**(c)** *Landing*: Turn off X-axis and Y-axis throttles, and lower the Z-axis throttle to ensure a smooth landing of the UAV. **(d)** *Landed*: The UAV reaches the ground and the instrumentation script exits.

We have the *normal profile (no anomalies), delay (contains timing anomalies),* and *random (contains sequence ordering and burst anomalies)* as the three modes we simulated. **100964**, **436615**, and **434310** represent the total number of events logged for the *delay, normal,* and *random* respectively. Figures 3a and b are the predictions errors for each of the profile. As can be seen, the *normal* profile generalizes well and forms our baseline for the anomaly detector. We have quantized the anomaly probabilities AL_p to emphasize the severity levels and these are shown in Figs. 3c and d. $\mathbf{0.75 \leq AL_4 < 1}$ while $\mathbf{0.5 \leq AL_3 < 0.75}$. Also, $\mathbf{0.25 \leq AL_2 < 0.5}$ and the worst anomaly level is $\mathbf{AL_1 < 0.25}$. From Fig. 3, increasing error values in Figs. 3a and b have decreasing AL_p in Figs. 3c and d. This behavior is in tandem with the operation of (10) and this confirms our hypothesis that deeper insight and wider scope of anomalies can be detected by exploring hitherto unused properties as input features.

5   Conclusion

In this work, we propose a deep ensemble framework called *DeepAnom* for detecting anomalies in system processes using system calls. We present how to extract more useful features from a system call to broaden the scope of anomalies that can be detected by the model. The results of Fig. 3 confirm our hypothesis that exploring the hitherto unused properties of the system call can lead to a more effective anomaly detection framework. Our future work will be exploring the *cause* and *effect* relationship between layers.

Acknowledgment. This research was funded in part by PTDF Nigeria and the Natural Sciences and Engineering Research Council of Canada (NSERC).

References

1. Chandola, V., Banerjee, A., Kumar, V.: Anomaly detection: a survey. ACM Comput. Surv. (CSUR) **41**(3), 15 (2009)
2. Clarke, E.M., Klieber, W., Nováček, M., Zuliani, P.: Model checking and the state explosion problem. In: Meyer, B., Nordio, M. (eds.) LASER 2011. LNCS, vol. 7682, pp. 1–30. Springer, Heidelberg (2012). https://doi.org/10.1007/978-3-642-35746-6_1
3. Du, M., Li, F., Zheng, G., Srikumar, V.: Deeplog: anomaly detection and diagnosis from system logs through deep learning. In: Proceedings of the 2017 ACM SIGSAC Conference on Computer and Communications Security, pp. 1285–1298. ACM (2017)
4. Ezeme, M., Azim, A., Mahmoud, Q.H.: An imputation-based augmented anomaly detection from large traces of operating system events. In: Proceedings of the Fourth IEEE/ACM International Conference on Big Data Computing, Applications and Technologies, pp. 43–52. BDCAT 2017, ACM, New York (2017). http://doi.acm.org/10.1145/3148055.3148076
5. Ezeme, M.O., Mahmoud, Q.H., Azim, A.: Hierarchical attention-based anomaly detection model for embedded operating systems. In: 2018 IEEE 24th International Conference on Embedded and Real-Time Computing Systems and Applications (RTCSA), pp. 225–231. IEEE (2018)

6. Ezeme, O.M., Mahmoud, Q.H., Azim, A.: Dream: deep recursive attentive model for anomaly detection in kernel events. IEEE Access **7**, 18860–18870 (2019). https://doi.org/10.1109/ACCESS.2019.2897122
7. Yoon, M.K., Mohan, S., Choi, J., Christodorescu, M., Sha, L.: Learning execution contexts from system call distribution for anomaly detection in smart embedded system. In: Proceedings of the Second International Conference on Internet-of-Things Design and Implementation, pp. 191–196. ACM (2017)

Predicting Sparse Clients' Actions with CPOPT-Net in the Banking Environment

Jeremy Charlier[1]([✉]), Radu State[1], and Jean Hilger[2]

[1] University of Luxembourg, 1855 Luxembourg, Luxembourg
{jeremy.charlier,radu.state}@uni.lu
[2] BCEE, Avenue de la liberte, 1930 Luxembourg, Luxembourg
j.hilger@bcee.lu

Abstract. The digital revolution of the banking system with evolving European regulations have pushed the major banking actors to innovate by a newly use of their clients' digital information. Given highly sparse client activities, we propose CPOPT-Net, an algorithm that combines the CP canonical tensor decomposition, a multidimensional matrix decomposition that factorizes a tensor as the sum of rank-one tensors, and neural networks. CPOPT-Net removes efficiently sparse information with a gradient-based resolution while relying on neural networks for time series predictions. Our experiments show that CPOPT-Net is capable to perform accurate predictions of the clients' actions in the context of personalized recommendation. CPOPT-Net is the first algorithm to use non-linear conjugate gradient tensor resolution with neural networks to propose predictions of financial activities on a public data set.

Keywords: Tensor decomposition · Personalized recommendation · Neural networks

1 Motivation

The modern banking environment is experiencing its own digital revolution. Strong regulatory directives are now applicable, especially in Europe with the Revised Payment Directive, PSD2, or with the General Data Protection Regulation, GDPR. Consequently, financial actors are now exploring the latest progress in data analytics and machine learning to leverage their clients' information in the context of personalized financial recommendation and client's action predictions. Recommender engines usually rely on second order matrix factorization since their accuracy has been proved in various publications [1–3]. However, matrix factorization are limited to the unique modeling of *clients* × *products*. Therefore, tensor factorization have skyrocketed for the past few years [4–6].

© Springer Nature Switzerland AG 2019
M.-J. Meurs and F. Rudzicz (Eds.): Canadian AI 2019, LNAI 11489, pp. 556–562, 2019.
https://doi.org/10.1007/978-3-030-18305-9_59

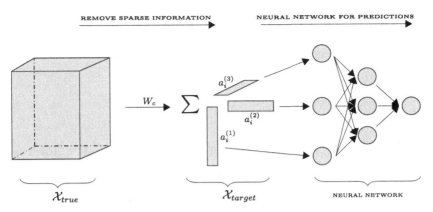

Fig. 1. In CPOPT-Net, the function W_c between the original tensor $\mathcal{X}_{true}$ and the decomposed tensor $\mathcal{X}_{target}$ is minimized. Then, the latent factor vectors $\mathbf{a}^{(1)}, \mathbf{a}^{(2)}, \mathbf{a}^{(3)}$ of each order are sent as input to the neural network. Following the neural network training, CPOPT-Net is able to predict the financial activities of the bank's clients.

Various tensor factorization, or tensor decomposition, exist for different applications [7,8]. However, the CP decomposition [9,10] is the most frequently used. Two of the most popular resolution algorithms, the Alternating Least Square (ALS) [9,10] and the non-negative ALS [11], offer a relatively simple mathematical framework explaining its success for the new generation of recommender engines [12–14]. In this paper, we use the gradient-based resolution for the CP decomposition [15] to address the predictions of clients' financial activities based on time, clients' ID and transactions type. The method, illustrated in Fig. 1, reduces the sparsity of the information while a neural network performs the predictions of events. We outline three contributions of our paper:

- We use the CP decomposition for separate modeling of each order of the data set. Since one client can have several financial activities simultaneously, we include the independent modeling of clients and financial transactions.
- We build upon non-linear conjugate gradient resolution for the CP decomposition, CPOPT [15]. We show CPOPT applied on a financial data set leads to small numerical errors while achieving reasonable computational time.
- Finally, we combine CPOPT with neural network leading to CPOPT-Net. A compressed dense data set, inherited from CP, is used as an optimized input for the neural network to predict the financial activities of the clients.

The remaining of the paper is organized as follows. Section 2 describes the CP tensor decomposition with its gradient-based resolution applied to third order financial predictions with neural network. Then, we highlight the experimental results in Sect. 3 and we conclude by emphasizing pointers to future work.

2   CPOPT-Net and Third Order Financial Predictions

In the CP tensor decomposition [9,10], the tensor $\mathcal{X} \in \mathbb{R}^{I_1 \times I_2 \times I_3 \times ... \times I_N}$ is described as the sum of the rank-one tensors

$$\mathcal{X} = \sum_{r=1}^{R} \mathbf{a}_r^{(1)} \circ \mathbf{a}_r^{(2)} \circ \mathbf{a}_r^{(3)} \circ ... \circ \mathbf{a}_r^{(N)} \tag{1}$$

where $\mathbf{a}_r^{(1)}, \mathbf{a}_r^{(2)}, \mathbf{a}_r^{(3)}, ..., \mathbf{a}_r^{(N)}$ are vectors of size $\mathbb{R}^{I_1}, \mathbb{R}^{I_2}, \mathbb{R}^{I_3}, ..., \mathbb{R}^{I_N}$. Each vector $a_r^{(n)}$ with $n \in \{1, 2, ..., N\}$ refers to one order and one rank of the tensor $\mathcal{X}$. We point out to [7] for further information. We use the Nonlinear Conjugate Gradient (NCG) method proposed in [15], CPOPT, with the strong Wolfe line search as it appears to be more stable in our case. Let $\mathcal{X}_{true}$ a real-valued N-order tensor of size $I_1 \times I_2 \times ... \times I_N$. Given R, the objective is to find a factorization

$$\mathcal{X}_{true} \approx \mathcal{X}_{target} = \sum_{r=1}^{R} \mathbf{a}_r^{(1)} \circ ... \circ \mathbf{a}_r^{(N)} \tag{2}$$

with the *factors* $\mathbf{a}_r^{(1)}, ..., \mathbf{a}_r^{(N)}$ initially randomized. Therefore, we denote by $\mathcal{X}_{target}$ the target tensor composed of the *factor vectors* $\mathbf{a}_r^{(1)}, ..., \mathbf{a}_r^{(N)}$.
The objective minimization function is denoted by $W_c(\mathcal{X}_{true}, \mathcal{X}_{target})$.

$$W_c(\mathcal{X}_{true}, \mathcal{X}_{target}) = \min f(\mathcal{X}_{true}, \mathcal{X}_{target}) = \frac{1}{2}||\mathcal{X}_{true} - \mathcal{X}_{target}||^2 \tag{3}$$

The values of the factor vectors can be stacked in a parameter vector $\mathbf{x}$.

$$\mathbf{x} = [\mathbf{a}_1^{(1)} \cdots \mathbf{a}_R^{(1)} \cdots \mathbf{a}_1^{(N)} \cdots \mathbf{a}_R^{(N)}]^T \tag{4}$$

Therefore, we can rewrite the objective function (3) as three summands.

$$W_c(\mathbf{x}) = W_c(\mathcal{X}_{true}, \mathcal{X}_{target}) = \frac{1}{2}||\mathcal{X}_{true}||^2 - \langle \mathcal{X}_{true}, \mathcal{X}_{target} \rangle + \frac{1}{2}||\mathcal{X}_{target}||^2 \tag{5}$$

From (5), we deduce the gradient function of the CP decomposition involved in the minimization process according to the factor vectors $\mathbf{a}_1^{(1)}, ..., \mathbf{a}_R^{(N)}$. We refer to [15] for more details about the gradient computation. Therefore, CPOPT-Net achieves a NCG resolution of the objective function $W_c(\mathcal{X}_{true}, \mathcal{X}_{target})$. Sparse information contained in $\mathcal{X}_{true}$ are removed in the factor vectors $\mathbf{a}^{(1)}, ..., \mathbf{a}^{(N)}$ of $\mathcal{X}_{target}$. Then, the factor vectors are sent as optimized inputs to the neural network. Through the training of the data set to learn the function $g(.) : \mathbb{R}^3 \to \mathbb{R}^1$, the neural network is able to predict the financial activities of the bank's clients. The implementation of CPOPT-Net is summed up in Algorithm 1.

Algorithm 1. CPOPT-Net for third order financial predictions

Data: tensor $\mathcal{X} \in \mathbb{R}^{I \times J \times K}$, rank R
Result: time series containing financial activities predictions, $\mathbf{y} \in \mathbb{R}^1$

1 /* $\mathrm{A} = \mathbf{a}^{(1)}, \mathrm{B} = \mathbf{a}^{(2)}, \mathrm{C} = \mathbf{a}^{(3)}$ */
2 **begin**
3   random initialization $\mathbf{A} \in \mathbb{R}^{I \times R}$, $\mathbf{B} \in \mathbb{R}^{J \times R}$, $\mathbf{C} \in \mathbb{R}^{K \times R}$
4      $\mathbf{x}_0 \leftarrow$ flatten$(\mathbf{A}, \mathbf{B}, \mathbf{C})$ as described in (4)
5      $\nabla W_{c_0} = \frac{\partial}{\partial x_i} W_c(\mathbf{x}_0) \leftarrow$ gradient of 5 at $\mathbf{x}_0$
6      $\alpha_0 \leftarrow \underset{\alpha}{\operatorname{argmin}} f(x_0 - \alpha \nabla W_{c_0})$
7      $x_1 = x_0 - \alpha_0 \nabla W_{c_0}$
8      $n = 0$
9   **repeat**
10         $\nabla W_{c_n} = \frac{\partial}{\partial x_i} W_c(\mathbf{x}_n) \leftarrow$ gradient of 5 at $\mathbf{x}_n$
11         $\beta_n^{HS} \leftarrow \dfrac{\nabla W_{c_n}^T (-\nabla W_{c_n} + \nabla W_{c_{n-1}})}{s_{n-1}^T(-\nabla W_{c_n} + \nabla W_{c_{n-1}})}$
12         $s_n \leftarrow -\nabla W_{c_n} + \beta_n^{HS} s_{n-1}$
13         $\alpha_n \leftarrow \underset{\alpha}{\operatorname{argmin}} f(x_n - \alpha \nabla W_{c_n})$
14   update $\mathbf{x}_{n+1} = \mathbf{x}_n - \alpha_n \nabla W_{c_n}$
15         $n = n + 1$
16   **until** *maximum number of iterations or stopping criteria*
17     $\mathbf{A}, \mathbf{B}, \mathbf{C} \leftarrow$ unflatten$(\mathbf{x}_n)$
18   send $\mathbf{A}, \mathbf{B}, \mathbf{C}$ to the input of the NN
19   training of the NN to learn the function $g(.) : \mathbb{R}^3 \rightarrow \mathbb{R}^1$
20     $\mathbf{y} \in \mathbb{R}^1 \leftarrow$ NN prediction of financial activities
21   **return** $y \in \mathbb{R}^1$

---

3 Predictions of Clients' Actions for Banking Recommendation

Data Availability and Experimental Setup. In 2016, the Santander bank released an anonymized public dataset containing financial activities from its clients[1]. The file contains activities of 2.5 millions of clients classified in 22 transactions labels for a 16 months period between 28 January 2015 and 28 April 2016. We choose the 200 clients having the most frequent financial activities since regular activities are more interesting for the prediction modeling. All the information is gathered in the tensor $\mathcal{X}_{true}$ of size $200 \times 22 \times 16$. We define the tensor rank equal to 25. We use the Adam solver with the default parameters $\beta_1 = 0.5, \beta_2 = 0.999$ for the training of the neural network[2].

Results and Discussions on CPOPT-Net. We test CPOPT-Net using three different type of neural networks: Multi-Layer Perceptron (MLP), Convolutional Neural Network (CNN) and Long-Short Term Memory (LSTM) network. Additionally, we cross-validate the performance of the neural networks with a Decision Tree (DT). The models have been trained on one year period from 28 January 2015 until 28 January 2016. Then, the activities for the next three months are

---

[1] The data set is available at https://www.kaggle.com/c/santander-product-recommendation.

[2] The code is available at https://github.com/dagrate/cpoptnet.

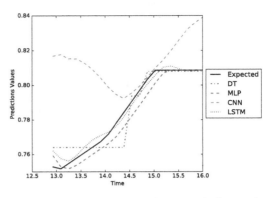

Fig. 2. Three months prediction of the evolution of the personal savings of one client. We can observe the difference of CPOPT-Net depending on the neural network chosen for the predictions.

Table 1. Residual errors of the objective function W_c between CPOPT-Net resolution and ALS resolution at convergence (the smaller, the better). Both methods have similar computation time.

	CPOPT-Net	CP-ALS
W_c Error	**10.099**	15.896

Table 2. Latent predictions errors on personal savings. LSTM achieves superior performance.

Error measure	DT	MLP	CNN	LSTM
MAE	0.044	0.004	0.282	**0.002**
Jaccard dist	0.053	0.027	0.348	**0.003**
cosine sim	0.967	0.953	0.966	**0.969**
RMSE	0.047	0.031	0.354	**0.003**

Table 3. Aggregated predictions errors on all transactions. LSTM achieves superior performance.

Error measure	DT	MLP	CNN	LSTM
MAE	0.029	0.027	0.272	**0.014**
Jaccard dist	0.034	0.032	0.290	**0.018**
cosine sim	0.827	0.909	0.880	**0.965**
RMSE	0.033	0.030	0.290	**0.017**

predicted with a rolling time window of one month. First, the Table 1 highlight the lower numerical error obtained with the CPOPT resolution in comparison to the ALS resolution. Then, the Fig. 2 shows that the LSTM models the most accurately the future personal savings activities followed by the MLP, the DT, and finally the CNN. The CNN fails visually to predict accurately the savings activity in comparison to the other three methods, while the LSTM seems to achieve the most accurate predictions. We highlight this preliminary conclusion for Fig. 2 in Table 2 by reporting four metrics: the Mean Absolute Error (MAE), the Jaccard distance, the cosine similarity and the Root Mean Square Error (RMSE). In Table 3, we show the aggregated metrics among all transaction predictions. In all the experiments, the LSTM network predicts the activities the most accurately, followed by the MLP, the DT and the CNN.

4   Conclusion

Building upon the CP tensor decomposition, the non-linear conjugate gradient resolution and the neural networks, we propose CPOPT-Net, a predictive method for the banking industry in which the sparsity of the financial transactions is removed before performing the predictions on future clients' transactions. We conducted experiments on a public data set highlighting the prediction differences depending on the neural network involved in CPOPT-Net. Due to the recurrent activities of most of the financial transactions, we underlined the best results were found when CPOPT-Net was used with LSTM. Future work will concentrate on a limited memory resolution for a usage on very large data sets. Furthermore, the personal financial recommendation will be assessed on smaller time frame discretization, weekly or daily, with other financial transactions. It will offer a larger choice of financial product recommendations depending on the clients' mid-term and long-term interests.

References

1. Brand, M.: Fast online svd revisions for lightweight recommender systems. In: Proceedings of the 2003 SIAM International Conference on Data Mining, pp. 37–46. SIAM (2003)
2. Ghazanfar, M.A., Prugel, A.: The advantage of careful imputation sources insparse data-environment of recommender systems: generating improved svd-basedrecommendations. Informatica **37**(1) (2013)
3. kumar Bokde, D., Girase, S., Mukhopadhyay, D.: Role of matrix factorization model in collaborative filtering algorithm: A survey. CoRR, abs/1503.07475 (2015)
4. Lian, D., Zhang, Z., Ge, Y., Zhang, F., Yuan, N.J., Xie, X.: Regularized content-aware tensor factorization meets temporal-aware location recommendation. In: 2016 IEEE 16th International Conference on Data Mining (ICDM), pp. 1029–1034. IEEE (2016)
5. Zhao, S., Lyu, M.R., King, I.: Aggregated temporal tensor factorization model for point-of-interest recommendation. In: Hirose, A., Ozawa, S., Doya, K., Ikeda, K., Lee, M., Liu, D. (eds.) ICONIP 2016. LNCS, vol. 9949, pp. 450–458. Springer, Cham (2016). https://doi.org/10.1007/978-3-319-46675-0_49
6. Song, T., Peng, Z., Wang, S., Fu, W., Hong, X., Yu, P.S.: Review-based cross-domain recommendation through joint tensor factorization. In: Candan, S., Chen, L., Pedersen, T.B., Chang, L., Hua, W. (eds.) DASFAA 2017. LNCS, vol. 10177, pp. 525–540. Springer, Cham (2017). https://doi.org/10.1007/978-3-319-55753-3_33
7. Kolda, T.G., Bader, B.W.: Tensor decompositions and applications. SIAM Rev. **51**(3), 455–500 (2009)
8. Acar, E., Kolda, T.G., Dunlavy, D.M.: All-at-once optimization for coupled matrix and tensor factorizations. arXiv preprint arXiv:1105.3422 (2011)
9. Harshman, R.A.: Foundations of the parafac procedure: Models and conditions for an explanatory multimodal factor analysis (1970)
10. Carroll, J.D., Chang, J.J.: Analysis of individual differences in multidimensional scaling via an n-way generalization of decomposition. Psychometrika **35**(3), 283–319 (1970)

11. Welling, M., Weber, M.: Positive tensor factorization. Pattern Recogn. Lett. **22**(12), 1255–1261 (2001)
12. Ge, H., Caverlee, J., Lu, H.: Taper: a contextual tensor-based approach for personalized expert recommendation. In: Proceedings of the 10th ACM Conference on Recommender Systems, pp. 261–268. ACM (2016)
13. Almutairi, F.M., Sidiropoulos, N.D., Karypis, G.: Context-aware recommendation-based learning analytics using tensor and coupled matrix factorization. IEEE J. Sel. Top. Sig. Process. **11**(5), 729–741 (2017)
14. Cai, G., Gu, W.: Heterogeneous context-aware recommendation algorithm with semi-supervised tensor factorization. In: Yin, H., et al. (eds.) IDEAL 2017. LNCS, vol. 10585, pp. 232–241. Springer, Cham (2017). https://doi.org/10.1007/978-3-319-68935-7_26
15. Acar, E., Dunlavy, D.M., Kolda, T.G.: A scalable optimization approach for fitting canonical tensor decompositions. J. Chemometr. **25**(2), 67–86 (2011)

Contextual Generation of Word Embeddings for Out of Vocabulary Words in Downstream Tasks

Nicolas Garneau[✉], Jean-Samuel Leboeuf, and Luc Lamontagne

Département d'informatique et de génie logiciel, Université Laval,
Quebec City, QC, Canada
{nicolas.garneau.1,jean-samuel.leboeuf.1}@ulaval.ca,
luc.lamontagne@ift.ulaval.ca

Abstract. Over the past few years, the use of pre-trained word embeddings to solve natural language processing tasks has considerably improved performances on every end. However, even though these embeddings are trained on gigantic corpora, the vocabulary is fixed and thus numerous out of vocabulary words appear in specific downstream tasks. Recent studies proposed models able to generate embeddings for out of vocabulary words given its morphology and its context. These models assume that we have sufficient textual data in hand to train them. In contrast, we specifically tackle the case where such data is not available anymore and we rely only on pre-trained embeddings. As a solution, we introduce a model that predicts meaningful embeddings from the spelling of a word as well as from the context in which it appears for a downstream task without the need of pre-training on a given corpus. We thoroughly test our model on a joint tagging task on three different languages. Results show that our model helps consistently on all languages, outperforms other ways of handling out of vocabulary words and can be integrated into any neural model to predict out of vocabulary words.

Keywords: Natural language processing · Sequence labeling · Out of vocabulary words · Contextual word embeddings

1 Introduction

Distributed word embeddings such as Word2Vec [1] and Polyglot [2] are able to capture fine-grained semantic and syntactic features for each word within a large body of text. The use of these embeddings has improved performance of many models within natural language processing (NLP) tasks such as part of speech tagging (POS) and morphosyntactic attribute tagging (MORPH).

Several of these embeddings databases have been trained on a huge corpus such as Wikipedia[1] and thus have a fixed size vocabulary on a given domain.

---

[1] https://dumps.wikimedia.org.

© Springer Nature Switzerland AG 2019
M.-J. Meurs and F. Rudzicz (Eds.): Canadian AI 2019, LNAI 11489, pp. 563–569, 2019.
https://doi.org/10.1007/978-3-030-18305-9_60

When working on a downstream task, such as POS tagging, we are exposed to words that are in our corpus but not within the embeddings database. For these specific out of vocabulary (OOV) words, it is common to either assign them random embeddings or to map all of them to a unique "unknown" embedding.

We here propose a novel model to handle OOV words, where we distinguish ourselves from previous work with the following contributions:

- By taking into account not only the characters of the word, but also the words of its surrounding context, each prediction our model makes is unique and context dependent;
- Our model does not need a separate pre-training step as others since it learns on-the-fly how to generate meaningful embeddings for a given task;

Our model essentially consists of three recurrent neural networks trained on the left context of the target word, on its right context and on its characters. As a baseline, we use random embeddings as well as Pinter's Mimick model [3]. We evaluate the predicted embeddings extrinsically on the tasks of POS and MORPH tagging on a subset of the Universal Dependencies (1.4) corpus for a fair and meaningful comparison of our contribution.

2   Related Work

Handling OOV words is often an underestimated problem within common NLP tasks. Due to the lack of proper ways to handle OOV words, researchers often resort to simply assign random embeddings to unknown words, hoping that their model will generalize well nonetheless. As a solution, Goldberg [4] proposed the *Word Dropout* approach, which was shown to be better than using a special vector dedicated to unknown words (Kiperwasser et al. [5]).

Bojanowski et al. [6] propose to overcome this problem by modeling rare words as a bag of n-grams of characters. While this representation may capture morphology phenomena such as suffixes and prefixes, it fails to characterize properly some meanings of a word. Ling et al. [7] proposed a similar approach where they compositionally model word embeddings by using the characters of the word.

Pinter et al. [3] use the same intuition as Ling et al. [7], but they learn unknown word representations as a first step. Their proposed model, Mimick, is a recurrent neural network trained on the characters using pre-trained embeddings as target values. They then use the pre-trained model to improve the accuracy of POS tagging as well as MORPH tagging on the Universal Dependencies (1.4) corpus [8]. As it relies solely on the morphology of a word, their efficient model can be used with any NLP tasks to predict OOV words beforehand. A variant of Mimick has also recently been proposed by Zhao et al. [9]. In a similar way to [6], it decomposes words into "bags of substrings" used to predict an embedding.

Recently, Peters et al. [10] proposed a deep recurrent language model to generate embeddings using the context in which a word appears. While this architecture looks similar to ours, it relies on a large body of text and a pre-trained language model. In our setting, we do not assume that we have access to a large corpus for a target language and this is what makes our model interesting.

In this perspective, we base our work on a model that generates word representations using characters embeddings, just as [3], and take it further by including the context in which unknown words may appear in the training sentences of the task at hand. Our work is also related to Garneau et al. [11] which uses an LSTM to capture the semantic representation of a target word given its characters and the context in which it appears.

3   Contextually Generating Word Embeddings

Pinter et al. [3] proposed an approach to generate OOV word embeddings compositionally by learning a function that maps characters of a word to its distributed representation. This simple yet powerful model did improve the performances of their POS and MORPH tagging models in a low-resource setting.

We follow their intuition but we do not assume that there is a generative wordform-based protocol to create the word embeddings provided by Polyglot. Instead, we design our model so that it learns to generate embeddings for a specific task, regardless of the pre-trained embeddings we have in hand. In addition to the model presented by [3] that mimics the spelling of a given word, we use bidirectional LSTMs to model the context in which this specific word appears.

The architecture of our model is illustrated in Fig. 1. The first step is to translate every words and characters into their proper embedding. Then, three

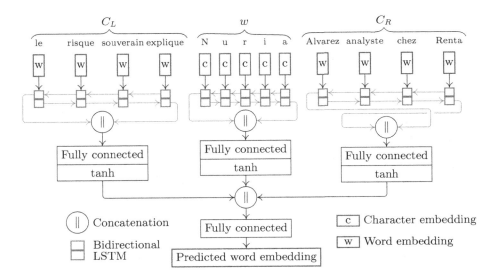

Fig. 1. Architecture of our proposed model. The example is drawn from the French part of the Universal Dependencies dataset, where *Nuria* is an actual OOV word.

bidirectional LSTMs respectively takes as input the left context C_L, the characters that compose the target word w and the right context C_R. For the next step, we only consider the last hidden state in both directions of the LSTMs.

Next, we transform the concatenated hidden states with three respective fully connected layers followed by a hyperbolic tangent as activation function. We choose to use the tanh function which showed faster convergence than sigmoid and ReLU in our tests. Afterward, we concatenate again these state and we compress them into a valid embedding with a last fully connected layer.

4   Joint Prediction of Part-of-Speech Tags and Morphosyntactic Attributes

Dataset. We train our model on three different languages, French, English, and Spanish, available in the Universal Dependencies dataset (1.4) [8]. For each of these languages, we jointly predict part of speech tags as well as the morphosyntactic attributes for each word in order to evaluate the contribution of our embedding generator model.

Each of these datasets is split in a training, validation and test sets that we call $\mathcal{D}$, $\mathcal{V}$ and $\mathcal{T}$, respectively. The French corpus contains 356,419, 38,758 and 7,020 tokens in $\mathcal{D}$, $\mathcal{V}$ and $\mathcal{T}$, respectively. The English corpus contrains 204,586, 25,148 and 25,096 tokens in $\mathcal{D}$, $\mathcal{V}$ and $\mathcal{T}$, respectively. The Spanish corpus contains 382,436, 41,198 and 7,953 tokens in $\mathcal{D}$, $\mathcal{V}$ and $\mathcal{T}$, respectively. As a collection of pre-trained embeddings, we use Polyglot because it offers low dimensional embeddings for several languages, but also because it allows us to compare our model to Pinter's Mimick model [3].

OOV Embedding Model. Section 3 describes generically the OOV embedding model but does not include the specificities of it, which are documented here. We set the character embeddings dimension to 20, like [3]. We set the hidden state of all LSTMs to a dimension of 128 for the sake of comparison with the Mimick model. The two first fully connected layers hence take as input 2×128 components and output 64 components, 64 being the dimension of the word embeddings. The last fully connected layer is then 64×64. All other weights are initialized at random, sampled from a Kaiming normal distribution [12]. We considered short and long context windows size n from 5 to 41 words but our experiments showed that larger windows yielded better results.

Tagging Task Model. Our tagging model is a sequence labeling neural network identical to the new proposed by Pinter [3], at the exception of the OOV handling part. It takes as input a sequence of word embeddings and of their characters embeddings that is fed to a OOV handling module that predicts embeddings if needed. The characters embeddings are passed through a char2tag model that generates an internal representations of the words that are then concatenated to the word embeddings. These outputs are then passed through two stacked bidirectional LSTMs before being fed to a task specific fully connected layer. Finally, they are converted into probabilities with the help of a softmax.

A label is then chosen by taking the argmax of this distribution. We use the same model as [3] to perform the joint prediction of part-of-speech tags and morphosyntactic attributes so we can compare our model with this previous architecture.

Our model has been trained using the Adam optimizer, a batch size of 32 and a learning rate of 0.005 with a learning rate scheduler dividing it by 2 when no improvements are made for more than 10 epochs. The hidden state dimension of the LSTMs is 128. We use the validation set to do early stopping when there is no more improvement on the validation loss for at least 20 epochs. We measure the performances of our models for POS tagging accuracy and MORPH F1-score.

5 Results and Discussion

To properly measure the impact of our proposed model, we ran our experimentations with two other schemes that handle OOV words differently. The first one assigns a random embedding to every OOV in the train, valid and test set. Note here that the embeddings for the OOV words in the training set are fine-tuned while training the network for the tagging task. The other scheme uses the embeddings generated by Mimick for 2 of the languages (English and Spanish). Mimick generates the embeddings for all the OOV words in the dataset as a first step and then these embeddings are used within the task's neural network. We can see from the results shown in Table 1 that our model performs better than the naive baseline and the Mimick model on all tested tasks.

Table 1. Comparison of the different models using random embeddings for the OOV words, Mimick's and those generated from our model. We can see that our model outperforms all approaches on three languages of the Universal Dependencies dataset (1.4). Occ. stands for the number of occurrences in the test set.

	Vocab	Occ.	POS (accuracy)			MORPH (F1-score)		
			Random	Mimick	Ours	Random	Mimick	Ours
French	45,388	7,020	96.43	-	**96.93**	97.12	-	**97.63**
English	23,020	25,096	94.35	94.60	**95.06**	96.33	96.55	**96.78**
Spanish	49,792	7,953	94.63	95.45	**95.57**	97.10	97.43	**97.90**

Since our model aims to help tagging tasks by handling OOV words, it is relevant to compute the effective accuracy on the POS tagging as well as the F1-Score of the Morphosyntactic attributes for the OOV words only. Table 2 presents the performances of our model and the baselines on the OOV only. We can see that for every languages, we have a gain from 0.6% to 3% on the POS accuracy and 3,44 to 4,81 on the Morphosyntactic attributes F1-Score, which is a significant margin.

Table 2. Comparison of the different models on the effective accuracy and F1-Score of the OOV words of the test set. We clearly see the advantage of our model compared to the two other approaches. The percentage of OOV words in the vocabulary in presented in parenthesis.

	OOV	Occ.	POS (accuracy)			MORPH (F1-score)		
			Random	Mimick	Ours	Random	Mimick	Ours
French	14,459 (32%)	672	91.52	-	**94.34**	90.36	-	**94.93**
English	6,022 (26%)	1,470	72.79	78.23	**80.48**	82.90	87.37	**90.90**
Spanish	14,475 (29%)	383	75.98	81.46	**81.98**	73.32	78.54	**81.34**

6   Conclusion

In this paper, we showed that using not only the characters of a word but also the context in which it appears contributes to predicting useful embeddings for downstream NLP tasks. We introduced a model that efficiently handles OOV words and can be integrated and trained along with the neural network of the sequence labeling task. In our future works, we want to add an attention mechanism within our model so that it may learn to pay attention to either the characters, the left or the right context. We also plan to run our experiments on the new Universal Dependencies dataset (2.3) recently released comprising more than 70 languages. Hopefully this experiment will confirm that our model expands and performs on every possible language.

References

1. Mikolov, T., Sutskever, I., Chen, K., Corrado, G.S., Dean, J.: Distributed representations of words and phrases and their compositionality. In: Advances in Neural Information Processing Systems, pp. 3111–3119 (2013)
2. Al-Rfou, R., Perozzi, B., Skiena, S.: Polyglot: distributed word representations for multilingual NLP. arXiv preprint arXiv:1307.1662 (2013)
3. Pinter, Y., Guthrie, R., Eisenstein, J.: Mimicking word embeddings using subword RNNs. arXiv preprint arXiv:1707.06961 (2017)
4. Goldberg, Y.: Neural network methods for natural language processing, vol. 10, pp. 83–97 (2017)
5. Kiperwasser, E., Goldberg, Y.: Simple and accurate dependency parsing using bidirectional LSTM feature representations. arXiv preprint arXiv:1603.04351 (2016)
6. Bojanowski, P., Grave, E., Joulin, A., Mikolov, T.: Enriching word vectors with subword information. arXiv preprint arXiv:1607.04606 (2016)
7. Ling, W., et al.: Finding function in form: compositional character models for open vocabulary word representation. arXiv preprint arXiv:1508.02096 (2015)
8. De Marneffe, M.C., et al.: Universal Stanford dependencies: a cross-linguistic typology. In: LREC, vol. 14, pp. 4585–4592 (2014)
9. Zhao, J., Mudgal, S., Liang, Y.: Generalizing word embeddings using bag of subwords. In: EMNLP (2018)

10. Peters, M.E., et al.: Deep contextualized word representations. In: NAACL-HLT (2018)
11. Garneau, N., Leboeuf, J.S., Lamontagne, L.: Predicting and interpreting embeddings for out of vocabulary words in downstream tasks. In: Proceedings of the 2018 EMNLP Workshop BlackboxNLP: Analyzing and Interpreting Neural Networks for NLP, pp. 331–333 (2018)
12. He, K., Zhang, X., Ren, S., Sun, J.: Delving deep into rectifiers: surpassing human-level performance on imagenet classification. In: Proceedings of the IEEE International Conference on Computer Vision, pp. 1026–1034 (2015)

Using a Deep CNN for Automatic Classification of Sleep Spindles: A Preliminary Study

Francesco Usai[✉][iD] and Thomas Trappenberg

Dalhousie University, 1355 Oxford Street, Halifax, NS B3H 4R2, Canada
francesco.usai@dal.ca

Abstract. In this work we applied a deep convolutional neural network to a binary classification task of clinical relevance, namely detecting sleep spindles. Specifically, we studied the conditions that are conducive of successful training on small data, emphasizing how the number of processing layers and the relative proportion of the two classes of events affect performance. We demonstrate that, in contrast with our expectations, the number of processing layers did not influence performance. Instead, the relative proportion of events affected the speed of learning but did not affect accuracy. This ceases to be the case when one class represents less than 30% of the total events, wherein training does not lead to improvement above the chance level. Overall, this preliminary study provides a picture of the dynamics that characterize training on small data, while providing further insights to explore the potential of automatic detection of sleep spindles based on deep learning.

Keywords: Convolutional Neural Network · EEG · Sleep spindles · Small datasets

1 Introduction

Sleep spindles are neurophysiological events occurring during the Non-Rapid Eye Movement (NREM) sleep stage 2 (N2) and present a characteristic waxing and waning shape whose duration ranges between 0.3 and 1.5 s [1,2] in the frequency range 11–16 Hz. Sleep spindles are used as a biomarker of specific neurodegenerative disorders and are considered a neural event of interest for understanding the mechanisms underlying sleep-mediated memory consolidation. The gold standard to identify sleep spindles is currently based on human expert analysis of the polysomnogram [2], a task that is very labour intensive and prone to considerable inter-rater variability. This motivates the significant efforts made so far to develop automated methods for the detection of sleep spindles, although a recent study [1] has shown that most of the proposed algorithms have not yet reached the level of reliability attained by humans.

In this regard, Deep Learning (DL) has emerged as one of the most promising solutions, although as of now its success has been largely dependent on

© Springer Nature Switzerland AG 2019
M.-J. Meurs and F. Rudzicz (Eds.): Canadian AI 2019, LNAI 11489, pp. 570–575, 2019.
https://doi.org/10.1007/978-3-030-18305-9_61

the availability of massive amounts of data. In this study, we aimed to explore
the potential of DL-based models applied to a classification task involving the
detection of sleep spindles. In particular, we were interested in exploring the conditions under which successful learning might occur when using a small dataset
for training. As mentioned before, learning from small data typically represents a
significant challenge for deep networks, and therefore we wanted to characterize
which factors facilitate learning when this is the only available option. For this
reason, we focused on investigating the role of two factors: depth of the network
(i.e. number of processing layers); and the relative proportion of events from
different classes. We made this choice because we wanted to verify whether what
we know about learning dynamics with big datasets - namely that increasing
the depth of the network tend to improve classification performance, and that
learning is faster and better when the datasets are balanced rather than when
they are heavily skewed towards a subset of classes - applies also to small ones.
Given that sleep spindles are relatively rare events in overnight EEG recordings,
we wanted to assess the minimum amount of spindle events - relative to non-
spindles - that is required for deep networks to function. To this end, the goals
of this study were to:

(1) Assess whether a CNN can learn to classify sleep-spindles when applied to
a relatively small dataset.
(2) Characterize under which conditions (i.e. depth of the network and relative
proportion between events of the different classes) learning does occur.

We hypothesized that, in case networks demonstrate to learn the classification
task to some extent, adding extra layers would improve performance. Also, in
line with what mentioned above, we hypothesized that learning would be faster
and more effective as the relative proportion of events from different classes gets
more balanced.

2 Methods

2.1 EEG Pre-processing

In this study we used EEG data from two datasets publicly available online,
MASS (the SS2 subset of cohort 1) [4] and the DREAMS dataset [5]. The SS2
dataset contains data from 19 subjects recorded at 256 Hz during an overnight
session. In line with previous works using this dataset [6], we excluded data
from subjects 4, 8, 15 and 16 due to excessive noise. In addition to it, we also
excluded data from subject 12 due to problems recovering the annotations made
by human experts. The DREAMS dataset consists of EEG data recorded from
8 subjects at different times of the night (30 min from each subject). Since the
recordings were performed at different sampling rates - subject 3 at 50 Hz, subject 1 at 100 Hz, the others at 200 Hz - we excluded subject3 and resampled the
other files to 128 Hz. We then pre-processed both MASS and DREAMS EEG
signals by applying a band-pass filter with cut-off frequencies at 0.1 and 30 Hz.

The signals from each subject were then normalized by subtracting their mean and dividing by their standard deviation. All analyses were performed using Matlab R2018b (MathWorks). Visual scorings from the experts were used to identify which parts of the signals were identified as spindles, which we labelled accordingly (9311 spindles for MASS and 306 for the DREAMS dataset). The remaining activity was labelled with a unique label since this study was based on a binary classification task. Afterwards, the continuous EEG signal was divided into 221 samples long epochs - equivalent to 1.73 s. We made this choice to have time windows that were long enough to include events whose duration generally varies within the range of 0.3–1.5 s [2]. Starting from the first 221 samples of the EEG signal, each epoch was created by shifting the previous one ahead of 64 samples (0.5 s). Each epoch was then assigned to the class of spindles if it contained at least 64 consecutive samples that were labelled as spindle by the human expert. This resulted in 25648 and 1822 spindles for the MASS and DREAMS datasets, respectively. As described below, we used data from the MASS dataset to conduct the preliminary search for the optimal hyperparameters and number of layers. Since the MASS dataset is heavily unbalanced towards non-spindles events - with spindle events representing only 3.4% of the entire dataset - we decided to subsample (at random, without replacement) this class in order to have different ratios between the two classes. To this end, we gradually shifted the balance from 50/50 to the original ratio through decrements of 10% in the proportion of spindles. This way we could test which level of balance the CNN can tolerate in the training phase.

2.2   Training and Validation on MASS and DREAMS Datasets

The first part of this study involved training the CNN on the MASS dataset. Specifically, we tested network architectures with 2 and 3 layers, and searched for optimal hyperparameters to be used in the test phase, which was later conducted using the DREAMS dataset.

The validation phase consisted of 14-fold cross-validation, wherein each fold contained data from a specific subject. Training was done using Tensorflow, using Adam optimizer to minimize cross entropy. Weights were initialized by taking random values from a normal distribution with mean and standard deviation equal to 0 and 0.1, respectively. Networks were trained on different versions of the dataset (differing for the ratio between spindles and non-spindles) and using every combination of hyperparameters such as learning rate (4 levels, from 0.0005 increasing of one order of magnitude until 0.5), dropout (5 levels, from 0.5 increasing of 0.1 until 1), and size of the batch (3 levels, from 256 increasing of a factor of 2 until 1024) with number of epochs set at 30. The combination of parameters that yielded best performance was 0.005 for learning rate, 0.9 for dropout level and 256 for batch size. To compare performance obtained by different architectures and training sets we used the F1-score defined in [1]. Briefly, the F1-score reflects the ability to minimize both the amount of false positives and false negatives, two conditions that are often hard to attain together, especially when using unbalanced datasets.

$$F1 = \frac{2TP}{2TP + FN + FP} \tag{1}$$

After performing cross-validation on the MASS datasets, all models were trained on the entire MASS dataset with the optimal hyperparameters, and then tested on a balanced version of the DREAMS dataset (i.e. with a 50/50 partition between the two classes of events).

3 Results

Results from the validation and test phase are summarized in the upper and bottom halves of Table 1, respectively. Here, performance for each metric is expressed as the Mean and Standard Deviation (within brackets) of the various metrics across the 14 folds. The top half of the table presents results of networks embedding 2 layers, while the bottom half contains results of the network embedding 3 layers. Results of networks trained with different partitions are presented in distinct rows (see column sp/nsp). We excluded results obtained using 20/80, 10/90 and original partitions since networks did not show any learning.

Broadly, learning did not occur when one class accounted for less than 30% of the total amount of events; when the least represented class had more than 30% of events the relative proportion did not affect performance, except for the network with 3 layers which did not learn successfully when trained on a dataset with a 30/70 partition. This trend is observed both when using a network with 2 and 3 convolutional layers, which attained similar performance. Results of the test phase conducted using the DREAMS dataset (see bottom half of Table 1) showed a slight drop in performance on all the metrics that we have considered

Table 1. Validation and test results

		sp/nsp	Metrics			
			Accuracy	*Precision*	*Recall*	*F1*
MASS	2 L	50–50	0.85 (0.07)	0.84 (0.12)	0.88 (0.06)	0.85 (0.06)
		40–60	0.84 (0.08)	0.86 (0.10)	0.91 (0.06)	0.88 (0.05)
		30–70	0.86 (0.06)	0.91 (0.07)	0.90 (0.07)	0.90 (0.03)
	3 L	50–50	0.86 (0.07)	0.86 (0.12)	0.86 (0.08)	0.85 (0.05)
		40–60	0.84 (0.07)	0.88 (0.10)	0.88 (0.09)	0.87 (0.05)
		30–70	0.81 (0.09)	0.85 (0.12)	0.91 (0.12)	0.86 (0.06)
DREAMS	2 L	50–50	0.78	0.75	0.86	0.80
		40–60	0.78	0.73	0.89	0.80
		30–70	0.72	0.66	0.94	0.77
	3 L	50–50	0.78	0.84	0.70	0.76
		40–60	0.78	0.73	0.89	0.80
		30–70	0.5	–	0	–

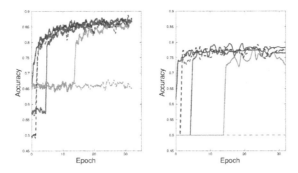

Fig. 1. Example of profiles of training and test curves over 30 epochs obtained from
one particular fold. The figure on the left shows the profile of accuracy on the training
set (MASS dataset), while the figure on the right shows the profile of accuracy on
the test set (DREAMS dataset). Note that, due to different proportions of spindles,
accuracy curves have different baselines. (Color figure online)

when compared to the validation phase. To investigate learning dynamics we
looked at the training curves (see Fig. 1). Here, solid and dashed lines are associated with the dynamics of networks embedding 2 and 3 layers, respectively,
while the color codes for the type of dataset used to train the networks (blue,
red and green are associated with the datasets were spindle events accounted
for 50%, 40% and 30% of the entire dataset, respectively). Despite the fact that
most curves have reached the plateau after 10 epochs, networks get to this stage
at different speed. Specifically, as we hypothesized, networks trained on more
balanced datasets reached the plateau faster, although in contrast with what we
expected the networks embedding 3 layers did not always learn faster and better
than those with 2 layers - networks with 2 convolutional layers learned faster
when employing the 50/50 partition, while the opposite was true for the 40/60
partition.

4   Discussion

This work aimed to assess under which conditions a deep CNN could perform
best at detecting sleep spindles when trained on a relatively small dataset. In
general, we observed similar performance when comparing networks featuring 2
and 3 layers. One potential explanation for this result is that, given the size of the
extra convolutional layer (1×3) the information encoded in it might have been
redundant with that in the fully connected layer (1×2). Interestingly, results
were comparable to those obtained by other algorithms that have been recently
proposed [3,6], and far superior to those obtained by other algorithms which have
been systematically tested on a recent study [1] where F-scores did not exceed
0.53. We also showed that the networks could tolerate a certain amount of unbal-
ance between the two classes of events, wherein one of the classes accounted for
less than 30% of all events. These performances generalized to another independent EEG sleep dataset, although there was a drop of about 8% in most

metrics, including F1 scores. Despite the promising results, the reach of our conclusion is severely limited by the fact that, due to the exploratory nature of this study, we did not go as far as conducting a direct comparison of performance with other proposed algorithms using the same data. This is clearly a major limitation that should be addressed in future investigations. This is one of the aspects that could be improved in this study, which was preliminary in nature. Future studies should improve upon these promising results by testing whether networks can be trained to also estimate the length and frequency content of the spindles, which are crucial features when aiming to use sleep data to support diagnosis of neurodegenerative/neurodevelopmental disorders or to study learning processes mediated by sleep. Alternatively, these features can be analysed using traditional methods (i.e. bandpass filtering and threshold on energy) during a post-classification phase.

5   Conclusions

In this study we have demonstrated the feasibility of using a deep CNN to detect sleep spindles even when using a relatively small dataset, showing that the number of processing layers is irrelevant under the particular set of conditions we employed to set and that the network cannot tolerate a level of unbalancement between the two classes of events wherein one class represents less than 30% of the total events. All in all, this study provides insights to further explore the potential of automatic detection of sleep spindles, and analysis of neuroimaging data more in general, based on deep learning.

References

1. Warby, S.C., et al.: Sleep-spindle detection: crowdsourcing and evaluating performance of experts, non-experts and automated methods. Nat. Methods **11**(4), 385–392 (2014)
2. Iber, C., Ancoli-Israel, S., Chesson, A., Quan, S.F.: AASM Manual for the Scoring of Sleep and Associated Events, 1st edn. American Academy of Sleep Medicine, Darien (2007)
3. Chambon, S., Thorey, V., Arnal, P.J., Mignot, E., Gramfort, A.: A Deep Learning Architecture to Detect Events in EEG Signals During Sleep (2018). arXiv:1807.05981v1
4. O'Reilly, C., Gosselin, N., Carrier, J., Nielsen, T.: Montreal archive of sleep studies: an open-access resource for instrument benchmarking and exploratory research. J. Sleep Res. **23**, 628–635 (2014)
5. The DREAMS Sleep Spindle Project. http://www.tcts.fpms.ac.be/devuyst/DataBaseSpindles/
6. Ray, L.B., et al.: Expert and crowd-sourced validation of an individualized sleep spindle detection method employing complex demodulation and individualized normalization. Front. Hum. Neurosci. **9**(507) (2015). https://doi.org/10.3389/fnhum.2015.00507

Graduate Student Symposium Papers

Principal Sample Analysis
for Data Ranking

Benyamin Ghojogh$^{(\boxtimes)}$ (ID)

Machine Learning Lab, Department of Electrical and Computer Engineering,
University of Waterloo, Waterloo, ON, Canada
bghojogh@uwaterloo.ca

Abstract. Because of the ever growing amounts of data, challenges
have appeared for storage and processing, making data reduction still
an important field of study. Numerosity reduction or prototype selection is one of the primary methods of data reduction. In this paper,
we propose some possible improvements for Principal Sample Analysis
(PSA) which is a numerosity reduction algorithm. The improvements are
PSA in Hilbert space, improving its time complexity using anchor points,
sample size estimation using PAC learning, and PSA for regression and
clustering tasks.

Keywords: Principal sample analysis · Data reduction ·
Data ranking · Numerosity reduction · Prototype selection

1 Introduction

Nowadays, because of the growing amount of data, data reduction has become
crucial. Data reduction is useful for (1) better storage efficiency, (2) improving
time of computation, (3) better representation of data or discrimination of classes
(4) removing outliers, and (5) even better recognition performance. Data reduction falls into two research areas, i.e., dimensionality reduction and numerosity
reduction which reduce the dimensionality and the sample size of data, respectively. The selected samples in numerosity reduction are also called *instances* or
prototypes. Numerosity reduction itself divides into prototype selection [1] and
prototype generation [2] where the prototypes are selected in the former and
selected or generated as new prototypes in the latter. In this paper, we propose
several possible improvements for our previously proposed method in numerosity
reduction, Principal Sample Analysis (PSA) [3], which ranks the prototypes and
can be categorized as a prototype selection method. These improvements might
be also possible to be combined to make an efficient PSA algorithm.

B. Ghojogh—Supervised by Dr. Mark Crowley and Dr. Fakhri Karray, Department of
Electrical and Computer Engineering, {mcrowley, karray}@uwaterloo.ca.

M.-J. Meurs and F. Rudzicz (Eds.): Canadian AI 2019, LNAI 11489, pp. 579–583, 2019.
https://doi.org/10.1007/978-3-030-18305-9_62

2   Our Prior Work: Principal Sample Analysis

The PSA [3] is a method for data ranking and numerosity reduction. This method mostly deals with scatter of data. After some pre-processing which might be necessary (see [3] for more details), It includes three main stages: (1) First, it finds the most important samples in each class. These important samples are named *major samples* and the set containing them is the *major set*. The rest of samples of the class which are not major are called *minor samples*. (2) After finding the major set in every class, it ranks the major samples in the set. (3) Then, it ranks the minor samples of every class. Finally, in every class, the major and minor samples are sorted based on their ranks while the minor samples are ranked after the major ones. Finding the major sets includes calculation of regression score, variance score, between-scatter score, and within-scatter score where fitting a parallel hyperplane to data, spread of data, discrimination of classes, and similarity of samples of the same class are considered, respectively. The second stage, i.e., ranking major samples, contains between- and within-scatter scores where again discrimination of classes, and similarity of samples of the same class are noticed. The between- and within-scatter scores in ranking minor samples has similar intuitions; however, it considers major samples as representatives of the classes. Here, we suggest some improvements for PSA.

3   Improving Principal Sample Analysis

3.1   Principal Sample Analysis in Hilbert Space

The PSA uses within and between scatters as in Fisher Discriminant Analysis (FDA). Inspired by $\boldsymbol{w}^\top \boldsymbol{S} \boldsymbol{w}$ in Fisher criterion [4], where $\boldsymbol{w}$ is the Fisher direction in the Hilbert space $\mathcal{H}$, and according to the theory of Reproducing Kernel Hilbert Space (RKHS), any solution $\boldsymbol{w} \in \mathcal{H}$ must lie in the span of all the training vectors mapped to $\mathcal{H}$, i.e., $\boldsymbol{\Phi}(\boldsymbol{X}) = [\boldsymbol{\phi}(\boldsymbol{x}_1), \ldots, \boldsymbol{\phi}(\boldsymbol{x}_n)] \in \mathbb{R}^{p \times n}$ ($p \gg d$). Therefore, $\mathbb{R}^p \ni \boldsymbol{w} = \sum_{\ell=1}^{n} \alpha_\ell \boldsymbol{\phi}(\boldsymbol{x}_\ell) = \boldsymbol{\Phi}(\boldsymbol{X}) \boldsymbol{\alpha}$ where n is the size of training dataset. Suppose the scatter in the Hilbert space is in the form: $\boldsymbol{S} = \sum_i \sum_j \left(\boldsymbol{\phi}(\boldsymbol{x}_i) - \boldsymbol{\phi}(\boldsymbol{x}_j) \right) \left(\boldsymbol{\phi}(\boldsymbol{x}_i) - \boldsymbol{\phi}(\boldsymbol{x}_j) \right)^\top$. Then, we have: $\boldsymbol{w}^\top \boldsymbol{S} \boldsymbol{w} = \sum_\ell \alpha_\ell^2 \sum_i \sum_j \left(\boldsymbol{K}(\boldsymbol{x}_\ell, \boldsymbol{x}_i) - \boldsymbol{K}(\boldsymbol{x}_\ell, \boldsymbol{x}_j) \right) \left(\boldsymbol{K}(\boldsymbol{x}_\ell, \boldsymbol{x}_i) - \boldsymbol{K}(\boldsymbol{x}_\ell, \boldsymbol{x}_j) \right)^\top := \boldsymbol{\alpha}^\top \boldsymbol{S}_k \boldsymbol{\alpha}$, where $\boldsymbol{K}(\boldsymbol{x}_\ell, \boldsymbol{x}_i) = \boldsymbol{\phi}(\boldsymbol{x}_\ell)^\top \boldsymbol{\phi}(\boldsymbol{x}_i)$ is the (ℓ, i)-th element of the kernel matrix $\boldsymbol{K}$, and $\boldsymbol{S}_k$ is:

$$\boldsymbol{S}_k := \sum_i \sum_j \left(\boldsymbol{K}(\boldsymbol{X}, \boldsymbol{x}_i) - \boldsymbol{K}(\boldsymbol{X}, \boldsymbol{x}_j) \right) \left(\boldsymbol{K}(\boldsymbol{X}, \boldsymbol{x}_i) - \boldsymbol{K}(\boldsymbol{X}, \boldsymbol{x}_j) \right)^\top. \quad (1)$$

The scatters in PSA can be rewritten as or similar to Eq. (1) to have the kernel PSA. The variance score and between scatter score of finding major sets in [3], which include mean of a set of samples in the scatter, should be in the form $\boldsymbol{S}_k = \sum_i \left(\boldsymbol{K}(\boldsymbol{X}, \boldsymbol{x}_i) - \boldsymbol{M} \right) \left(\boldsymbol{K}(\boldsymbol{X}, \boldsymbol{x}_i) - \boldsymbol{M} \right)^\top$ where $\boldsymbol{M} := (1/|\mathcal{S}|) \sum_{j=1}^{|\mathcal{S}|} \boldsymbol{K}(\boldsymbol{X}, \boldsymbol{x}_j)$ and $\mathcal{S}$ is the set of samples.

3.2   Fast Principal Sample Analysis Using Anchor Points

Let i_r, d, c, n_M^c, n^c, and n denote the number of RANSAC iterations, the dimensionality of data, the number of classes, the cardinality of major set, the number of samples in a class, and the total number of samples in PSA, respectively.

Proposition 1. *The time complexity of PSA algorithm is $\mathcal{O}(i_r dc + nc + (n_M^c)^2 c)$ if the sorting algorithm is an oracle with $\mathcal{O}(1)$ run-time.*

Proof. In finding major set for each class, the regression score, the variance score, and the between- and within-scatter scores take $\mathcal{O}(i_r d)$, $\mathcal{O}(n_M^c)$, $\mathcal{O}(n^c c) \in \mathcal{O}(n)$, and $\mathcal{O}((n_M^c)^2)$, respectively. In ranking major samples for each class, the between- and within-scatter scores consume $\mathcal{O}(n^c c) \in \mathcal{O}(n)$, and $\mathcal{O}(n^c)$. In ranking minor samples for each class, we have $\mathcal{O}(n_M^c c) \in \mathcal{O}(n)$, and $\mathcal{O}(n^c)$. Hence, the overall time complexity is $\mathcal{O}(i_r dc + nc + (n_M^c)^2 c)$.

Two bottlenecks of the time complexity arise from the regression score and the scatter calculations. For improving the time of regression score, we can randomly choose k dimensions in every RANSAC iteration rather than the whole $d - 1$ dimensions. Note that $k \ll d$ is a constant. This improves the time of regression score from $\mathcal{O}(i_r dc)$ to $\mathcal{O}(i_r c)$. On the other hand, improving the time complexity of scatters can be done by introducing *anchor points* (or *landmarks*). We define the anchor points in an ellipsoid as the points located in $\boldsymbol{\mu}$, $\boldsymbol{\mu} \pm \boldsymbol{\sigma}$, and $\boldsymbol{\mu} \pm 2\boldsymbol{\sigma}$ where $\boldsymbol{\mu}$ and $\boldsymbol{\sigma}$ are the centroid and spread of ellipsoid, respectively. The data can be assumed as mixture of Gaussians or ellipsoids with anchor points. To find the anchor points, first a mixture of Gaussians is fitted on every class (please see our tutorial [5]) and each fitted Gaussian is translated to origin for processing. Every Gaussian distribution has an ellipsoid contour in $\mathbb{R}^d$, formulated as $\sum_{i=1}^{d}(x_i^2/a_i^2) = t$ where $\boldsymbol{x} = [x_1, \ldots, x_d]^\top$ is a point in the class subset having maximum likelihood in the Gaussian, a_i is the i-th largest eigenvalue in applying principal component analysis to the subset, and the length of the i-th dimension axis in the aligned ellipsoid is $2a_i\sqrt{t}$. As the Gaussian is centered, the distribution of t is $\chi_{(df)}^2$ where degree of freedom is $df = d$. Based on the desired confidence level and the df, the proper value for t is found from χ^2 table. Then, by re-translating the ellipsoid to $\boldsymbol{\mu}$ and according to the angle of the eigenvector having the eigenvalue a_i, the i-th dimension of the anchor points are found. Finally in PSA, the scatters use only the anchor points of a class instead of the whole samples of it. In scatters of ranking minor samples, the major samples are replaced with the nearest anchor points of their class. As the number of anchor points is a small constant, it greatly improves the time complexity.

3.3   Sample Size Estimation Using PAC Learning

There exist two hyperparameters in PSA which are the sample size of the major set and the cardinality of principal samples. Cross validation is one way to tune the hyperparameters. Moreover, one of the two parameters can be determined using Probably Approximately Correct (PAC) learning [6]. The aim of PAC learning is to put an upper-bound on the true (test) error, denoted by **Err**, given the training error shown by **err**. However, PAC learning requires the size of the hypothesis set H which contains the hypotheses for estimating the classifier. With the help of Vapnik-Chervonenkis (VC) dimension, we can have a bound on the true error without knowing $|H|$.

The *dichotomy* of a set is a partition of the set into disjoint subsets. *Shattering* a set means for every dichotomy of the set, there exists at least a hypothesis classifying all the data correctly. The *VC dimension* of H is the size of the largest finite subset of data shattered by H [6]. The bound on the true error for having $\mathbb{P}(\mathbf{Err} \leq \mathbf{err} + \varepsilon) = 1 - \eta$ is:

$$\varepsilon = \sqrt{\frac{1}{n'}\Big[\mathrm{VC}_H\big(\ln(\frac{2n'}{\mathrm{VC}_H}) + 1\big) - \ln(\frac{\eta}{4})\Big]}, \qquad (2)$$

where $\eta \in [0,1]$, VC_H is the VC dimension of H, and $n' \gg \mathrm{VC}_H$ is the desired sample size of training set (see Sects. 3.3 to 3.7 in [6]). Note that VC_H for any d-dimensional hyperplane as a classifier is $d+1$. Considering an upper-bound on VC_H and the amount of error tolerance (values for η and ε), we can determine n' from Eq. (2). The obtained n' can be used for estimating an appropriate size for major set or principal samples cardinality.

3.4 Principal Sample Analysis for Regression and Clustering

Most of the instance ranking and selection methods, such as PSA, are proposed for classification tasks [1,2]. However, numerosity reduction can be useful for regression and clustering tasks as well. Applying slight changes to PSA makes it useful for these tasks. In regression and clustering, there exists merely one set or class of data; hence, the between scatter scores in [3] should be set to 1 because we do not have several classes to compute their between scatters. Moreover, calculation of scatters in [3] can be simplified to iterate over only one existing class which is the whole dataset. For regression case, the regression score can also be simplified because the labels of regression are now available enabling us to omit the loop over the $(d-1)$ dimensions in [3]. Therefore, PSA can be used for regression and clustering after these changes.

References

1. Garcia, S., Derrac, J., Cano, J., Herrera, F.: Prototype selection for nearest neighbor classification: taxonomy and empirical study. IEEE Trans. Pattern Anal. Mach. Intell. **34**(3), 417–435 (2012)
2. Triguero, I., Derrac, J., Garcia, S., Herrera, F.: A taxonomy and experimental study on prototype generation for nearest neighbor classification. IEEE Trans. Syst. Man Cybern. Part C **42**(1), 86–100 (2012)
3. Ghojogh, B., Crowley, M.: Principal sample analysis for data reduction. In: 2018 IEEE International Conference on Big Knowledge (ICBK), pp. 350–357. IEEE (2018)

4. Mika, S., Ratsch, G., Weston, J., Scholkopf, B., Mullers, K.R.: Fisher discriminant analysis with kernels. In: Proceedings of the 1999 IEEE signal processing society workshop on Neural networks for signal processing IX, pp. 41–48. IEEE (1999)
5. Ghojogh, B., Ghojogh, A., Crowley, M., Karray, F.: Fitting a mixture distribution to data: tutorial. arXiv preprint arXiv:1901.06708 (2019)
6. Vapnik, V.: The Nature of Statistical Learning Theory. Statistics for Engineering and Information Science. Springer, Heidelberg (2000). https://doi.org/10.1007/978-1-4757-3264-1

Discrete-Event Systems for Modelling Decision-Making in Human Motor Control

Richard Hugh Moulton$^{(\boxtimes)}$

Department of Electrical and Computer Engineering, Queen's University,
Kingston, ON, Canada
richard.moulton@queensu.ca

Abstract. Artificial intelligence, control theory and neuroscience have a long history of interplay. An example is human motor control: optimal feedback control describes low-level motor functions and reinforcement learning explains high-level decision-making, but where the two meet is not as well understood. Here I formulate the human motor decision-making problem, describe how discrete-event systems could model it and lay out future research paths to fill in this gap in the literature.

Keywords: Online control · Decision-making · Limited lookahead · Reinforcement learning · Human motor control

1 Introduction

The role of decision-making in human motor control can be considered using artificial intelligence, control theory and computational neuroscience. Although optimal feedback control (OFC) describes low-level human motor control and reinforcement learning (RL) can explain high level decision-making, there is a gap in our understanding when motor control meets decision-making.

Decision-making is a discrete process: a possible option is taken or not. This naturally suggests discrete-event systems (DES) as a paradigm. DES for hierarchical, online and optimal control are of particular interest because these capture specific aspects of human motor decision-making.

My hypothesis is that "decision-making in human motor control for reaching-based tasks can be modelled using online control for dynamic DES". In the remainder of this paper I formulate the human motor decision-making problem, summarize three approaches to DES control and identify lines of inquiry.

R. H. Moulton—I acknowledge that Queen's University is situated on traditional Anishinaabe and Haudenosaunee Territory. I am thankful for the guidance I have received from my supervisors, Dr. Karen Rudie and Dr. Stephen Scott. This research is supported by the Dean's Graduate Research Assistant Award.

© Springer Nature Switzerland AG 2019
M.-J. Meurs and F. Rudzicz (Eds.): Canadian AI 2019, LNAI 11489, pp. 584–587, 2019.
https://doi.org/10.1007/978-3-030-18305-9_63

2   Problem Definition

Scott's taxonomy of sensory feedback processing for motor actions leads to a hierarchical view of the human motor system (Fig. 1) [6]. The continuous motor commands to a goal are captured by OFC, [7], but how that goal is selected is less understood. Put another way, although the control of a reach is understood, the selection of its target is less clear.

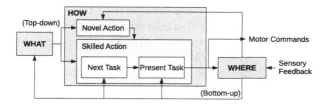

Fig. 1. Functional divisions of the human motor control system, adapted from Scott [6]. Defining a motor action's goal and defining the subsequent motor action's goal take place in the "Present Task" and "Next Task" blocks respectively.

The link between motor control and decision-making has been shown using statistical decision theory to explain movement under different conditions [9]. Although skilled motor control results from a series of decisions and although both have been well-studied individually, Gallivan et al. highlight that a gap exists in the literature for decision-making and motor control together [3].

3   Discrete-Event System Models for Decision-Making

DES models processes as automata and their behaviour as formal languages containing actions that are either controllable or uncontrollable and observable or unobservable. A supervisor enacts control by constraining the process's language to a set of desirable behaviours [10, Chap. 3], resulting in computationally tractable models and provably correct supervisors.

3.1   Hierarchical Control

Control problems can be decomposed into hierarchies when functional divisions in its task lead to top-down/bottom-up processes, e.g. human motor control [6]. Hierarchical control for DES results in models with multiple layers. Each layer has a different view of the environment with higher layers exhibiting broader temporal horizons or deeper logical dependencies than lower layers [10, Sect. 5.1].

Specific questions to investigate for hierarchical control are, "how many layers are needed to capture human motor decision-making?" and "what is the minimum information required by each level to effect motor control?"

3.2   Online and Optimal Control

Another important part of the human motor decision-making problem is the dynamic environment. Not only is the OFC policy being executed in real-time, but so is sensory integration, optimal state estimation and goal estimation [6].

The DES literature includes approaches for large state-spaces, complex possible behaviours and time-varying components where supervisor synthesis is impractical. In these cases an online DES controller is synthesized instead to implement a limited lookahead policy (LLP) that will guide decision-making [1].

Online controllers such as these can also be used to optimize some utility or cost function. Grigorov and Rudie demonstrated that the size of the lookahead window used by a model greatly impacts the utility of its decisions and, critically, that larger windows are not necessarily better [4]. This work provides a concrete framework for combining the online nature of human motor decision-making and the utility inherent in the different options considered.

Specific questions to address include "how are sensory feedback and the efference copy incorporated into the decision-making process?" and "what environmental variables impact the size of lookahead window used by humans?"

3.3   Learning the Model

Although the primary goal of this research is a DES model of human motor decision-making for reaching-based tasks, there is a deeper question that must be asked as well. For any model that we propose, can the human brain learn it?

RL captures how past experiences can be used to shape future outcomes; its role in human decision-making as well as a possible neural basis have both been subjects of investigation [5]. In the field of DES, RL has been used to synthesize supervisors for optimal control in online environments [8].

4   Experimental Design

Thought has also been given to experimental designs for which DES models of human behaviour could make testable predictions. Diamond et al.'s recent work

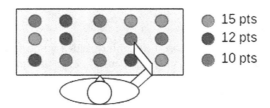

Fig. 2. Diamond et al.'s foraging task, figure adapted [2]. Targets are assigned a value based on size or colour; participants are then instructed to maximize the value harvested within a limited amount of time [2].

used a reaching-based foraging task to isolate the decision-making/motor control link (Fig. 2) [2]. This task is an attractive starting point because it restricts the range of movement to reaching, provides easy-to-interpret discrete goals, and permits a wide range of possible variations.

Analysis will be required to determine which values for the independent variables will be the most illustrative; one possible path is to run simulations with the developed DES models to pinpoint potential critical points for decision-making.

5   Conclusion

The human motor decision-making problem represents a fusion of motor control and decision-making and its solution will require contributions from multiple fields. This paper formalized the problem of interest and identified goal definition for present and upcoming actions as excellent candidates for modelling by DES.

This paper also summarized concepts from DES control that fit naturally with the human motor decision-making problem: hierarchical, online and optimal control. Each of these suggests specific research questions to ask about the larger problem and whose answers would contribute to the literature.

In parallel to this problem, the question of how any model could be (continually) learned was also identified, leading to the idea of incorporating RL into a holistic framework for human motor decision-making.

References

1. Chung, S.L., Lafortune, S., Lin, F.: Limited lookahead policies in supervisory control of discrete event systems. IEEE Trans. Autom. Control **37**(12), 1921–1935 (1992)
2. Diamond, J.S., Wolpert, D.M., Flanagan, J.R.: Rapid target foraging with reach or gaze: the hand looks further ahead than the eye. PLoS Computat. Biol. **13**(7), 1–23 (2017)
3. Gallivan, J.P., Chapman, C.S., Wolpert, D.M., Flanagan, J.R.: Decision-making in sensorimotor control. Nat. Rev. Neurosci. **19**(9), 519–534 (2018)
4. Grigorov, L., Rudie, K.: Near-optimal online control of dynamic discrete-event systems. Discrete Event Dyn. Syst. **16**(4), 419–449 (2006)
5. Lee, D., Seo, H., Jung, M.W.: Neural Basis of Reinforcement Learning and Decision Making. Ann. Rev. Neurosci. **35**(1), 287–308 (2012)
6. Scott, S.H.: A functional taxonomy of bottom-up sensory feedback processing for motor actions. Trends Neurosci. **39**(8), 512–526 (2016)
7. Todorov, E., Jordan, M.I.: Optimal feedback control as a theory of motor coordination. Nat. Neurosci. **5**(11), 1226–1235 (2002)
8. Umemoto, H., Yamasaki, T.: Optimal LLP supervisor for discrete event systems based on reinforcement learning. In: IEEE International Conference on Systems, Man, and Cybernetics, pp. 545–550. IEEE, Kowloon, October 2015
9. Wolpert, D.M., Landy, M.S.: Motor control is decision-making. Curr. Opin. Neurobiol. **22**(6), 996–1003 (2012)
10. Wonham, W.M., Cai, K.: Supervisory Control of Discrete-Event Systems (2017). http://www.control.utoronto.ca/DES/Research.html

Event Prediction in Social Graphs Using 1-Dimensional Convolutional Neural Network

Bonaventure C. Molokwu[(✉)]

School of Computer Science, University of Windsor,
401 Sunset Avenue, Windsor, Ontario N9B-3P4, Canada
molokwub@uwindsor.ca

Abstract. Social network graphs and structures possess implicit knowl-edge embedded about their respective nodes and edges which may be exploited, using effective and efficient methods, for relative event pre-diction upon these network structures. Thus, understanding the intrin-sic patterns of relationship among spatial social actors as well as their respective properties are very crucial factors to be taken into consider-ation with respect to event prediction in social network graphs. Gen-erally, event prediction problems are considered to be NP-Complete. This research work proposes an original approach (Graph-ConvNet) for predicting events in social network structures using a one-dimensional convolutional neural network (1D-ConvNet) model. In this regard, two distinct methodologies have been proposed herein with each having its individual characteristics and advantages. The first methodology intro-duces a pre-convolution layer that involves reframing the input social net-work graph to a two-dimensional adjacency matrix. Thereafter, feature-extraction operations are applied to reduce the linear dimensionality (across the $x-axis$) of the input matrix before it is introduced to a repet-itive series of non-linear convolution and pooling operations. The second methodology operates on a joint input comprising the edge list (E) of the social graph, and its associated feature space matrix. With regard to the observations and findings from experiments thus far: the first method is suitable for relatively smaller network graphs ($nodes \leq 30,000$); while the second method is a good fit for much larger network graphs ($nodes > 30,000$). Training and evaluation of these proposed approaches have been done on datasets (compiled: November, 2017) extracted from real social network communities with respect to 3 European countries where each dataset comprises an average of 280,000 edges and 48,000 nodes.

Keywords: Graph-ConvNet · Pre-convolution · Feature extraction

Supported by International Business Machines (IBM) and Professor Ziad Kobti.

M.-J. Meurs and F. Rudzicz (Eds.): Canadian AI 2019, LNAI 11489, pp. 588–592, 2019.
https://doi.org/10.1007/978-3-030-18305-9_64

1 Introduction

A social network consists of finite set(s) of actors, and the relationship(s) defined between these actors [5]. Analyzing and learning intrinsic knowledge from communities, which are comprised of social actors, using a given set of standards still remains a significant research problem in social network analysis (SNA). Furthermore, an event prediction problem can be expressed as a Satisfiability problem [2] such that an event is said to exist if the variables governing the event's formal definition reduces it to *true*. Accordingly, the Cook-Levin Theorem [1] has proven that Satisfiability problem is NP-Complete. The proposed methodologies herein are based on a neural network architecture assembled using deep layers of stacked processing units comprising convolutional neural network (ConvNet) units and multilayer perceptron (MLP) units. This architecture makes it feasible to develop and train neural network models to be capable of learning the nonlinear distributed representations enmeshed in the graph structures [3].

2 Problem Statement

Social network graphs are characterized by their complex size and dynamic nature; and this makes it relatively challenging and difficult to develop effective machine learning (ML) as well as deep learning (DL) models which can be trained to predict events over a given network graph with respect to its constituent vertices (or actors) and edges (or relationships).

3 Proposed Methodologies

Fundamentally, these methodologies focus on harnessing the immense power of ConvNets in a bid to extract and learn the inherent knowledge embedded in social network graphs. In this regard, we frame the event prediction task as a classification problem where every row of the input matrix is associated with a corresponding output event label.

3.1 Method 1: Using the Proposed Pre-convolution Layer

This is essentially a feature extraction layer which is sandwiched between the input layer and the convolutional learning layer(s) of the network model. Firstly, from the input graph, we derive the adjacency matrix, $A[i, j]$:

$$G(V, E) \vdash A[i, j] \tag{1}$$

Secondly, we generate an identity matrix, $I[i, j]$, which is of same dimension as the adjacency matrix, A. The aim of the identity matrix is to help enforce the properties/attributes of each node with respect to the adjacency matrix.

$$B[i, j] = (A + I)[i, j] \tag{2}$$

Thirdly, we compute the degree of each node as a diagonal matrix, $C[i,j]$, and multiply its inverse with $B[i,j]$; such that:

$$D[i,j] = B[i,j].C^{-1}[i,j] = (B.C^{-1})[i,j] \qquad (3)$$

Thereafter, a linear (aggregate) dimensionality reduction operation is performed across the j columns of the resultant matrix, $D[i,j]$, viz:

$$E[i,k] = \sum_{m=0}^{k} \sum_{p=m*q}^{p+q} (D[i,p]) : k,p,q < j \qquad (4)$$

where i and j denote the rows and columns of the respective matrix; k and q are hyperparameters which denote the number of iterations required to traverse the columns of the matrix and the step-increment per iterative pass respectively. Hence, the operations within the pre-convolution layer are expressed in Algorithm 1. The output, $E[i,k]$, is fed as input to the convolutional learning layers.

Algorithm 1. Pre-Convolution Layer Operation

Input: Adjacency matrix, $A[i,j]$ **Output:** Reduced matrix, $E[i,k]$

---

local:
$B[i,j], D[i,j]$
$C[i,j] \leftarrow compute\_node\_degree()$
$I[i,j] \leftarrow compute\_identity\_matrix()$

$B[i,j] \leftarrow (A+I)[i,j]$
$D[i,j] \leftarrow B[i,j].C^{-1}[i,j]$
Initialize: $q \leftarrow step(int)$ // Hyperparameter
Initialize: $temp() \leftarrow list()$
Initialize: $k \leftarrow j/q$
Initialize: $z \leftarrow 0$
for $t \leftarrow 0$ **to** k **do**
    $z \leftarrow t + q$
    Append: $temp() \leftarrow sum(D[i,t..z], axis = cols)$
    $t \leftarrow z$ // Update for-loop() index
end for
$E[i,k] \leftarrow transpose(temp())$

return $E[i,k]$

3.2   Method 2: Using the Edge List and Feature Space

Practically, this is a feature selection layer that preprocesses the constituent features of the input space, X, before it is applied to the convolutional learning layers. Given a social network, SN, defined by the expression 6 below; this approach redefines the input space, with respect to SN, such that every input vector, $X[i, j]$, is defined by the function expression 5, viz:

$$X[i, j] = f(E[i, g], W[i, r], C[i], D[j]) \tag{5}$$

where i and j denote the horizontal and vertical vectors respectively of the input space, X. The graph's edge list, $E[i, g]$, is a subspace which is defined by the sequence, $E[i, g] := \{(u_i, v_g)...(u_{i+n}, v_{g+n})\}$; such that every tuple (u_i, v_g) denotes a connection/link between pairs of vertices in $\{V\}$. Thus, $\forall (u_i, v_g) \in \{V : v_0, v_2, ..., v_{n-1}\}$. Also, $W[i, r]$ is a two-dimensional array comprising the connection weights associated with each graph edge, $E[i, g]$. The source degree, $C[i]$, per source node, u_i; and the target degree, $D[j]$, per target node, v_g, are derived via the function: $G(V, E) \vdash C[i], D[j]$.

$$SN = \begin{cases} (G, f_V, f_E) \\ G : V, E; & \text{graph on a set of vertices, } V, \text{ and a set of edges, } E \\ f_V : V \to M; & \text{vertices' metadata function} \\ f_E : E \to N; & \text{edges' metadata function} \end{cases} \tag{6}$$

4   Results and Conclusion

Table 1 shows the experiment results using Method 1 and Method 2 respectively.

Table 1. Average performance of the proposed methods over dataset

Model	Validation acc.(%)	Training time(s)	Training acc.(%)
Method 1/Method 2	96.70/90.47	1205.77/218.84	93.58/90.39

Training of the neural model follows a supervised learning function, $f : X \to Y$, where Y denote the set of event labels. Intrinsic patterns and knowledge are extracted and learnt by the neural network model during the convolutional learning phase, and every sequence of learnt pattern is classified with respect to a corresponding event label in the output layer of the network model. Currently, I am validating my propositions with regard to expanding the experimentation scope to include more social network datasets as well as baseline models/methods.

Acknowledgements. This research was supported by International Business Machines (IBM) via the provision of a high performance IBM Power System S822LC Linux Server. Also, I want to acknowledge my doctoral advisor, Professor Ziad Kobti, for his guidance and funding thus far. Consecutively, credit goes to Stanford University [4] for making their compiled datasets readily available to us on research grounds.

References

1. Cook, S.A.: The complexity of theorem-proving procedures. In: Proceedings of the Third Annual ACM Symposium on Theory of Computing, vol. 1, pp. 151–158 (1971)
2. Maaren, H.V., Biere, A., Walsh, T. (eds.): Handbook of Satisfiability. IOS Press, Amsterdam (2009)
3. Goodfellow, I., Bengio, Y., Courville, A. (eds.): Deep Learning. MIT Press, Cambridge (2017)
4. Leskovec, J., Krevl, A.: SNAP datasets: Stanford large network dataset collection, June 2014. http://snap.stanford.edu/data
5. Scott, J. (ed.): Social Network Analysis. SAGE Publications Ltd., Newbury Park (2017)

A Framework for Determining Effective Team Members Using Evolutionary Computation in Dynamic Social Networks

Kalyani Selvarajah[(✉)]

School of Computer Science, University of Windsor, Windsor, ON, Canada
selva111@uwindsor.ca

Abstract. The team formation problem (TFP) concerns the process of bringing the experts together from Social Networks (SN) as teams in a collaborative working environment for a productive outcome. It was proven to be NP-hard problem. Our findings on a static SN using Evolutionary Computations (EC) achieved a significant improvement than State-of-Art methods on different datasets such as DBLP and Palliative care network. Since complexity and dynamics are challenging properties of real-world SN, our current research focuses on these properties in discovering new individuals for the teams. The process of detecting suitable members for teams is typically a real-time application of link prediction. Although different methods have been proposed to enhance the performance of link prediction, these methods need significant improvement in accuracy. Moreover, we examine the changes in attributes over time between individuals of the SN, especially on the co-authorship network. We introduce a time-varying score function, to evaluate the active researcher, that uses the number of new collaborations and number of frequent collaborations with existing connections. Moreover, we incorporate the shortest distance between any two individuals and introduces a score function to evaluate the skill similarity between any two individuals to form an effective team. We introduce Link prediction as a multi-objective optimization problem for optimizing three objectives, score of active researchers, skill similarity and shortest distance. We solved this problem by applying the NSGA-II and MOCA frameworks.

Keywords: Team Formation · Social Networks · Link prediction · Evolutionary Computation · Multi-objective optimization

1 Introduction

Today, it is essential to have some collective thoughts and creative ideas for productive results in various disciplines such as business, academic and healthcare. As expectation for specialization in every field increases, there is a need for experts in specific skills. Bringing these experts together as a team in a collaborative working environment is significant to have a productive outcome. The

© Springer Nature Switzerland AG 2019
M.-J. Meurs and F. Rudzicz (Eds.): Canadian AI 2019, LNAI 11489, pp. 593–596, 2019.
https://doi.org/10.1007/978-3-030-18305-9_65

links connecting these individuals and their attributes collectively is a complex social system or social network (SN).

SNs range from online social interactions including Facebook, and Linkedin to human interactions such as co-authorship and healthcare. Such networks can be represented as a graph with a collection of nodes connected by links. Nodes are individual actors or people in the network, whereas links represent relationships between those actors. Complexity and dynamics are essential properties of real-world SNs since they evolve quickly over time through the appearance or disappearance of new links and nodes. In this complex social system, forming a group of experts while maximizing their social compatibility is a great challenge.

Furthermore, finding suitable individuals for teams in a SN has been extensively studied [3,4]. Majority of these studies considered the snapshot of the entire network and evaluated the potential future collaborations in the next time period and paid less attention to the temporal changes of the networks. Significant efforts have been made to explain the evolution of networks during the past decades. However, such research is yet to achieve the desired results, leaving the door open for further advances in the field. Especially, the formation of teams has been proven to be NP-hard problem [4] that remains open to be addressed in the research community.

2   Progress to Date

We believe that EC is the next level in the progress of AI and produces significant changes in the results when examining the challenging properties such as complexity and dynamics of real-world SNs. To examine the potentials of Evolutionary Algorithms (EA) which mimic biological processes to solve complex problems, we tested the TFP particularly with the knowledge-based population algorithms such as Cultural Algorithms (CA) [5] on a snapshot of SN to predict the effective team of experts in the future.

Our proposed model optimized social compatibles as well as other significant parameters; (a) Communication cost: defined based on the past frequent collaboration of experts. (b) Contact cost: defined based on the availability to work on projects. (c) Geographical distance: defined based on geographical proximity. It's important in health care services. (d) Productivity: determined based on the past performance of the expert.

Consequently, our proposed models achieved a significant improvement than State-of-Art methods concerning accuracy on different datasets such as DBLP and Palliative care network. Our findings have been already presented [6–8], published [6,8] in some conferences and accepted in currently proceeding conferences ANT 2019- "Identifying a Team of Experts in Social Networks using a Cultural Algorithm", EVOApplication 2019 - "A Cultural Algorithm for Determining Similarity Values Between Users in Recommender Systems" and FLAIRS 2019- "Productive and Profitable Cluster Hire". Addition to this, as part of multidisciplinary health application teams in analyzing the effects, we have made a significant contribution. Our work has taken the interest of Hospice and the

Windsor Essex Compassion Care Community in partnership with the Faculty of Nursing. We implemented a simulator in order to generate customized palliative care and visually represent the overall structure of it as a weighted social graph which has been accepted in InMed-2019, "A Palliative Care Simulator and Visualization Framework".

From the findings of our researches, we conclude that considering a snapshot of a network at a certain time to predict the effective teams to collaborate in the future is not utilizing entire information of the network. Therefore, we decided to consider changes in attributes of individuals and links between individuals through a time span would be a potential contribution to further research. The next section will elaborate the proposed solution and approach.

3   Proposed Research Problem

As said in the introduction, the complexity and dynamics are essential properties of real-world SNs. Thus, our hypothesis is that solving TFP on dynamic SNs using EC would be advantageous to enhance the accuracy and the time complexity while handling dynamic skill sets for successful completion of task.

In reality, individuals search for the fruitful collaborations in order to work on challenging projects and complete successfully. The process of detecting suitable members for teams in a SN is typically a real-time application of link prediction. A dynamic network is a sequence of network snapshots within a time interval and evolving. The size of the network can occasionally shrink or expand as the network evolves. This work focus on only undirected graphs. Given a series of snapshots $\{\mathcal{G}_1, \mathcal{G}_2, \ldots, \mathcal{G}_{t-1}\}$ of an evolving graph $\mathcal{G}_T = \langle \mathcal{V}_T, \mathcal{E}_T \rangle$, where the edge $e = (u, v) \in \mathcal{E}_{t'}$ represents a link between $u \in \mathcal{V}_{t'}$ and $v \in \mathcal{V}_{t'}$ at a particular time t'. The dynamic or temporal link prediction approaches attempt to predict the most probable links in the next time step $\mathcal{G}_t$. The list of graphs $\{\mathcal{G}_1, \mathcal{G}_2, \ldots, \mathcal{G}_{t-1}\}$ corresponding to a list of symmetric adjacency matrices $\{A_1, A_2, \ldots, A_{t-1}\}$. The adjacency matrix A_T of $\mathcal{G}_T$ is a $\mathcal{N} \times \mathcal{N}$ matrix where each element $A_T(i, j)$ takes 1 if the nodes $v_i \in \mathcal{V}$ and $v_j \in \mathcal{V}$, are connected at least once within time period T and takes 0 if they are not. Given a sequence of t snapshots $\{A_1, A_2, \ldots, A_{t-1}\}$, the goal is to predict the adjacency matrix A_t at future time t. We are currently working on the following major contributions;

1. Active researchers are more popular among both existing and new researchers, who believe that active researchers always update their research with current trends as well as being open to new ideas. To evaluate the popularity of any experts, we consider two factors on the temporal network. (a) The score for building new connections (b) The score of the increased number of collaborations with existing connected experts. We introduce a new score function to incorporate the impact of the time stamps and the gap between the current time and considered timespan.
2. The smaller the distance between any two authors is higher the chances of a future collaboration. The shortest distance is as another objective function.

3. We propose a score function to evaluate the similarity of experts' skills. The score will measure how two individuals can be similar in terms of their knowledge to work as a team in the same project.
4. We define the link prediction problem as a multi-objective optimization problem for link prediction. We designed our model with Non-dominated Sorting Genetic Algorithm (NSGA-II) [2] and compare with Multi-objective Cultural Algorithms (MOCA) [1]. We are currently trying on different possibilities of designing a new improved MOCA.

4   Conclusions

The Team Formation in Social Networks tackles the problem of finding a team of experts from a SN to complete given tasks successfully. We examined a couple of static datasets using the evolutionary computational method. Since our findings attained considerable improvements than State-of-Art methods, we are currently examining dynamic SNs. In future, we are planning to bring the concept of dynamic skill sets with the skill similarity score function as well as the concept of knowledge utilization where the individual with a specific percentage of knowledge can work on a project that requires equal or less percentage of the skills compare to experts' skills.

Acknowledgement. I would like to express my very great appreciation to Dr. Ziad Kobti for his valuable and constructive suggestions during the planning and development of this research work. I would also like to thank Dr. Mehdi Kargar, for his advice and assistance in keeping my progress on schedule.

References

1. Best, C., Che, X., Reynolds, R.G., Liu, D.: Multi-objective cultural algorithms. In: IEEE Congress on Evolutionary Computation, pp. 1–9. IEEE (2010)
2. Deb, K., Pratap, A., Agarwal, S., Meyarivan, T.: A fast and elitist multiobjective genetic algorithm: NSGA-II. IEEE Trans. Evol. Comput. **6**, 182–197 (2002)
3. Juárez, J., Brizuela, C.A.: A multi-objective formulation of the team formation problem in social networks: preliminary results. In: Proceedings of the Genetic and Evolutionary Computation Conference, pp. 261–268. ACM (2018)
4. Lappas, T., Liu, K., Terzi, E.: Finding a team of experts in social networks. In: Proceedings of the 15th ACM SIGKDD International Conference on Knowledge Discovery and Data Mining, pp. 467–476. ACM (2009)
5. Reynolds, R.G.: An introduction to cultural algorithms. In: Proceedings of the Third Annual Conference on Evolutionary Programming. World Scientific (1994)
6. Selvarajah, K., Bhullar, A., Kobti, Z., Kargar, M.: WSCAN-TFP: weighted scan clustering algorithm for team formation problem in social network. In: The Thirty-First International Flairs Conference (2018)
7. Selvarajah, K., Zadeh, P.M., Kargar, M., Kobti, Z.: A knowledge-based computational algorithm for discovering a team of experts in social networks (2017)
8. Selvarajah, K., Zadeh, P.M., Kobti, Z., Kargar, M., Ishraque, M.T., Pfaff, K.: Team formation in community-based palliative care. In: 2018 Innovations in Intelligent Systems and Applications (INISTA), pp. 1–7. IEEE (2018)

Generating Accurate Virtual Examples for Lifelong Machine Learning

Sazia Mahfuz$^{(\boxtimes)}$

Queen's University, Kingston, Canada
`sazia.mahfuz@queensu.ca`

Abstract. Lifelong machine learning (LML) is an area of machine learning research concerned with human-like persistent and cumulative nature of learning. LML system's objective is consolidating new information into an existing machine learning model without catastrophically disrupting the prior information. Our research addresses this LML retention problem for creating a knowledge consolidation network through task rehearsal without retaining the prior task's training examples. We discovered that the training data reconstruction error from a trained Restricted Boltzmann Machine can be successfully used to generate accurate virtual examples from the reconstructed set of a uniform random set of examples given to the trained model. We also defined a measure for comparing the probability distributions of two datasets given to a trained network model based on their reconstruction mean square errors.

Keywords: Artificial intelligence · Machine learning ·
Lifelong machine learning · Restricted Boltzmann Machine

1 Introduction and Background

Humans learn new knowledge as they grow older while retaining prior knowledge. Similarly, LML systems can retain prior tasks' knowledge over a long time while integrating, or consolidating new task's knowledge periodically [1]. Just like living beings, *consolidation* enables the integration of new information into the existing learning system. The main challenge in consolidation is *catastrophic forgetting* and overcoming the *stability-plasticity dilemma*.

The process of retaining the old information and yet being able to integrate the new information is known as the *stability-plasticity dilemma*. A neural network learning mechanism capable of stability and plasticity can rehearse examples of prior knowledge to maintain functional stability while slowly changing its representation to accommodate new knowledge [2].

Catastrophic forgetting can be defined as the disruption or loss of the prior training information while integrating new information to a trained model [3]. The challenge is to reduce the affect of catastrophic forgetting, which can be done by the rehearsal of a subset of the old information forcing the learning system to retain the structure of the old information [3]. One approach is to store a set of examples from the training set for each task. But the space

© Springer Nature Switzerland AG 2019
M.-J. Meurs and F. Rudzicz (Eds.): Canadian AI 2019, LNAI 11489, pp. 597–600, 2019.
https://doi.org/10.1007/978-3-030-18305-9_66

complexity of this approach will grow linearly with the number of tasks, each time lengthening the training time for a new task. An alternate approach is to use the concept of sweep pseudorehearsal discussed by Robins [4]; where examples are created by passing randomly created input vector through the learning system, the generated outputs are recorded for that particular input vector, and these are randomly included in the training session of the new task. From these reconstructed examples if we select only those examples which adhere to the probability distribution of the training data, then those selected examples are referred to as virtual examples (VEs) [5] in our research.

The focus of this research is to develop a method by which to generate VEs of prior tasks from an existing neural network model such that the example(s) adhere to the probability distribution of those prior tasks. Our approach was to investigate feasible approaches for generating accurate VEs, and then evaluate the selected methods based on the VEs' based on their adherence to the prior task distribution. The research aimed to generate VEs adhering to the input variables' distributions.

2   Related Work

Research by Kirkpatrick et al. [6] presented a novel algorithm, elastic weight consolidation (EWC) by decreasing the weight plasticity which avoided the catastrophic forgetting of old training information during the integration of new information. He et al. [7] proposed a variant of the backpropagation algorithm, "conceptor-aided backprop", where conceptors were used to protect the gradients from degradation of prior trained tasks. Compared with the above approaches, our method used the Robins' pseudorehearsal approach to handle catastrophic forgetting.

3   Generating VEs Based on a Trained Restricted Boltzmann Machine (RBM) Reconstruction Error

An accurately trained Restricted Boltzmann Machine (RBM) reconstructs the training data with a low error. This observation led us to investigate that after one oscillation, which is just passing the data to the trained model and recording the reconstruction, of feeding a uniform random set of examples into the model, the examples that are closer to the training data distribution are more accurately reconstructed than the other examples. This finding suggested that we can select the VEs based on the reconstruction error.

Because all of the generated examples from the uniform random set of examples do not adhere to the training data distribution after one oscillation, an approach was taken to select the examples such that a tolerance level was satisfied. Mean Squared Error (MSE) for the training data reconstruction was used as the initial tolerance level. If the sum of squared error for the uniform random input example and its corresponding reconstruction fell below the defined tolerance value, then the example was considered a VE.

Selecting the Tolerance Level: For one, two, four-dimensional input data, the tolerance level had been selected using the MSE between the training data and its corresponding reconstruction.

Success Criterion: To evaluate the success of the adherence of the virtual example distribution to the training data distribution, that is to evaluate the accuracy of the VEs, we defined a measure using the reconstruction error from a trained model, called the *Autoencoder-based Divergence Measure*. We considered unsupervised autoencoder approach to measure the difference between two probability distributions. The idea was that if our model had been accurately trained, we could measure the relative degree of similarity between an example set and the original training set's probability distributions. We defined the measure called the Autoencoder-based Divergence Measure (ADM) as follows: Let, MSE_{TRN} = the MSE of the training data on the RBM model and MSE_{TST} = the MSE of the test data given to the trained RBM model, then

$$ADM = \frac{MSE_{TST}}{MSE_{TRN}} \tag{1}$$

We verified that, "$0 < ADM <= 1$" signifies same probability distribution as the training data;
"$1 < ADM < 2$" signifies similar or partial space of the training data;
"$ADM \geq 2$" signifies increasingly different probability distribution than the training data.

Experiment - Four-dimensional Data: The four-dimensional synthetic input data had 1000 examples. The input data was selected in such a way that there are various numbers of distinct regions in each of the dimensions within the range of 0 and 1.

For this experiment, the tolerance level measured by MSE was 0.000395, which was the MSE between the training data and its corresponding reconstruction after one oscillation. Thus 80 virtual examples were selected from the reconstructed set of 5000 uniform random examples passed to the model trained on the training examples $x_{1(1T)}$, $x_{2(1T)}$, $x_{3(1T)}$, $x_{4(1T)}$.

We calculated the value of ADM as ADM $= \frac{MSE_{VE}}{MSE_{TRN}} = \frac{0.000228}{0.000395} = 0.5772$. This value signified that the virtual examples are from the same probability distribution as the training data. In Fig. 1, the blue probability density functions represent the training data. The red probability density functions represent the virtual examples selected from the reconstructed set of 5000 uniform random set of examples after one oscillation using the trained model.

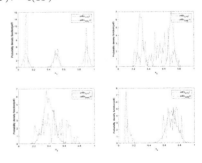

Fig. 1. The probability density funcitons for the training examples and the selected virtual examples

Discussion: The following discusses the merits of the measure based on the experimental results.

For one, two, four-dimensional data, tolerance level measured by MSE had been successful in generating virtual examples validated by the results of the autoencoder-based divergence measure. The success criterion resulted that the selected examples adhere to the training data distribution. So we can consider these examples as VEs for our purpose, and we can use this measure for generating accurate VEs.

4   Conclusion

The important findings of this research are as follows:

1. We investigated approaches for generating VEs from the reconstruction of a uniform random set of examples passed through a generative RBM model. Empirically, we found that we can generate accurate VEs validated by autoencoder-based divergence measure from uniform random examples based on the reconstruction error measured by MSE.
2. We defined and successfully verified a metric called autoencoder-based divergence measure for comparing the probability distributions of two given datasets using the reconstruction MSE from the trained RBM model.

Acknowledgement. I would like to express my sincerest gratitude towards my supervisor Dr. Daniel L. Silver for his consistent, constant support and patience throughout my graduate studies and his valuable comments on writing this summary.

References

1. Silver, D.L., Yang, Q., Li, L.: Lifelong machine learning systems: beyond learning algorithms. In: Lifelong Machine Learning: Papers from the 2013 AAAI Spring Symposium, pp. 49–55. AAAI (2013)
2. Silver, D.L., Mason, G., Eljabu, L.: Consolidation using sweep task rehearsal: overcoming the stability-plasticity problem. In: Barbosa, D., Milios, E. (eds.) CANADIAN AI 2015. LNCS (LNAI), vol. 9091, pp. 307–322. Springer, Cham (2015). https://doi.org/10.1007/978-3-319-18356-5_27
3. Robins, A.: Catastrophic forgetting, rehearsal, and pseudorehearsal. Conn. Sci. **7**(2), 123–146 (1995)
4. Robins, A.: Consolidation in neural network and in the sleeping brain. Conn. Sci. **8**(2), 259–276 (1996)
5. Silver, D.L., Poirier, R., Currie, D.: Inductive transfer with context-sensitive neural networks. Mach. Learn. **73**(3), 313–336 (2008)
6. Kirkpatrick, J., et al.: Overcoming catastrophic forgetting in neural networks. Computing Research Repository abs/1612.00796 (2017)
7. He, X., Jaeger, H.: Overcoming catastrophic interference using conceptor-aided backpropagation. In: International Conference on Learning Representations (2018)

Towards a Novel Data Representation for Classifying Acoustic Signals

Mark Thomas[(✉)]

Dalhousie University Faculty of Computer Science, Halifax, NS B3H 4R2, Canada
markthomas@dal.ca

Abstract. In this paper, we evaluate a novel data representation of acoustic signals that builds upon the traditional spectrogram representation through interpolation. The novel representation is used in training a deep Convolutional Neural Network for the task of marine mammal species classification. The resulting classifier is compared in terms of performance to several other classifiers trained on traditional spectrograms.

Keywords: Deep learning · Convolutional Neural Networks · Classification · Signal processing · Bioacoustics

1 Introduction

There exists a growing collection of research that use computer *vision* technologies like Convolutional Neural Networks (CNNs) on problems which are *auditory* in nature. Such auditory tasks include: speech recognition [1,3] and musical information retrieval [2,5]. In the majority of these cases, the framing of an auditory task is transformed into a visual one through some variation of a Fourier transform. The result is a set of acoustic data represented visually using a spectrogram, and this data can then be used to train a CNN. Transforming acoustic data into "images" will always lead to information-loss, in part due to the necessary step of choosing parameter values for the Fourier transform. In most cases, the parameters in question are the windowing function, window size, and window overlap of a short-time Fourier transform (STFT).

While advances in computer vision have led to significant improvements in performance and indeed state-of-the-art results in many auditory problems, we contend that most of these advances are a byproduct of better performing neural architectures. To the best of our knowledge, very little research has been brought forward which presents a better performing visual representation of an acoustic signal.

The following individuals from Jasco Applied Sciences are thanked for their continued support of this project: Bruce Martin, Katie Kowarski, and Briand Gaudet. Additional thanks to Stan Matwin from Dalhousie University. Collaboration between researchers at Jasco Applied Sciences and Dalhousie University was made possible through an NSERC Engage Grant.

© Springer Nature Switzerland AG 2019
M.-J. Meurs and F. Rudzicz (Eds.): Canadian AI 2019, LNAI 11489, pp. 601–604, 2019.
https://doi.org/10.1007/978-3-030-18305-9_67

In this work, we present a novel representation of acoustic signals that builds upon the commonly used spectrogram representation by way of interpolating and stacking multiple spectrograms produced using distinct STFT parameters. The proposed representation is particularly effective for an ongoing research project aimed at identifying whale species in acoustic recordings as the interpretability of the data is sensitive to the parameters used by the STFT. We show that the novel representation improves the performance of a CNN for the task of marine mammal species classification.

2   Novel Acoustic Representation

Many of the classification tasks cited above train on data sets comprised of single channel inputs in the form of spectrograms. In these scenarios, researchers are faced with selecting a single set of parameters to provide to the STFT, and in some cases, the choice of STFT parameters is not obvious. For example, when it comes to identifying whale species in acoustic data, marine biology experts will typically generate multiple spectrograms using various sets of parameters. The main reason for doing this is that some species tend to make prolonged low-frequency vocalizations with small bandwidth, while other species make short vocalizations with large bandwidth. Depending on the set of parameters used to generate the spectrogram, one can easily misclassify a vocalization as a different species or even ambient noise.

In order to exploit the strategy used by marine biology experts, we propose a novel representation of acoustic data by generating multiple spectrograms using various parameter values.

The equation of the discrete-time STFT can be expressed as:

$$X(n, \omega) = \sum_{m=-\infty}^{\infty} x[m]w[m - n]e^{-j\omega m}, \tag{1}$$

where w is the windowing function and $x[m]$ represents the acoustic signal at time m. The parameters being referred to throughout this paper include the choice of function w as well as the length and overlap of the window.

Once the spectrograms have been produced, they can be interpolated using a simple linear spline (i.e., Eq. 2) such that the dimensions of the matrices containing each spectrogram are the same. Finally the set of spectrograms can be stacked to form a multi-channel input; imitating the idea of RGB channels in an image.

$$\omega = \omega_i + \frac{\omega_{i+1} - \omega_i}{n_{i+1} - n_i}(n - n_i), \tag{2}$$

for some point (n, ω) between (n_i, ω_i) and (n_{i+1}, ω_{i+1}).

3   Dataset

The data used to train the CNN consists of roughly 28,000 human annotated vocalizations pertaining to three species of whales: blue, fin, and sei whales. Due to the varying populations of each species as well as the rate that each species produces vocalizations, the data set is highly unbalanced in favour of fin whales at a ratio of 6:1. The set of acoustic recordings were collected by Jasco Applied Sciences during the summer and fall months of 2015 and 2016 in the areas surrounding the Scotian Shelf and Roseway Basin; along the coast of Nova Scotia, Canada. The original data was sampled at several different rates, however, downsampling to 8 kHz was conducted prior to generating the spectrograms.

Three spectrograms were created for each of the annotations using increasing window lengths (i.e., 256, 2048, and 16,384 frames). All three spectrograms used the Hann window function as well as an overlap of 1/4 the window length. Copies of the three spectrograms were interpolated to fit within a grid of height 512 and width 128, using the smallest resolution in frequency and time of the non-interpolated spectrograms.

The final data set was split into training, validation, and test sets using a ratio of 90/5/5, respectively. A large proportion of data–in comparison to the standard 70/30 ratio used in most machine learning projects–was allocated to the training set in order to satisfy the CNNs hunger for training data.

4   Experimental Results

We evaluate the performance of a deep CNN architecture commonly used for computer vision: ResNet-50 [4], trained on the three-channel version of the novel representation described above and compare these results to the un-interpolated single channel spectrograms used in the three-channel version's creation. The CNN was trained using the Adam optimization routine [6] on a cross-entropy loss function with an exponentially decaying learning rate. The results summarized below depict the median values of ten training runs for each CNN (Table 1) and (Fig. 1).

Table 1. Final evaluation results of the test set. The results of the classifiers trained on the 1-channel spectrograms are listed by their STFT window lengths (in frames).

	3-channels	1-channel		
		256	2048	16384
Accuracy	0.90974	0.80229	0.89327	0.88252
Precision	0.88257	0.68671	0.83691	0.81191
Recall	0.80706	0.48522	0.76841	0.73223
F-1 score	0.83329	0.52779	0.79825	0.75311

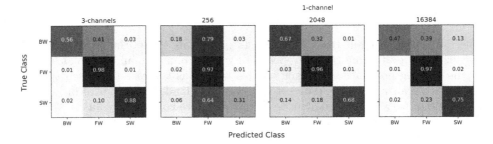

Fig. 1. Normalized confusion matrices of the test set. The three whale species are denoted as follows: blue (BW), sei (SW), and fin (FW).

5 Conclusions and Future Work

The results displayed in Sect. 4 outline that the choice of parameters to the STFT have a significant impact of the ability of a CNN to generalize accurately to the test set. Specifically, the classifier trained on the novel representation outperformed the remaining three classifiers in terms of accuracy, precision, and recall.

The work outlined in this paper is preliminary and further work is being carried out on an ongoing basis as part of the authors thesis. Finally, while the results above are specific to the task of marine mammal species classification, further work is being conducted which uses the novel representation for performing acoustic scene classification using several benchmark data sets.

References

1. Abdel-Hamid, O., Mohamed, A., Jiang, H., Deng, L., Penn, G., Yu, D.: Convolutional neural networks for speech recognition. IEEE/ACM Trans. Audio Speech Lang. Process. **22**(10), 1533–1545 (2014)
2. Choi, K., Fazekas, G., Sandler, M., Cho, K.: Convolutional recurrent neural networks for music classification. In: 2017 IEEE International Conference on Acoustics, Speech and Signal Processing (ICASSP), pp. 2392–2396. IEEE (2017)
3. Deng, L., et al.: Recent advances in deep learning for speech research at Microsoft. In: ICASSP, vol. 26, p. 64 (2013)
4. He, K., Zhang, X., Ren, S., Sun, J.: Deep residual learning for image recognition. In: Proceedings of the IEEE Conference on Computer Vision and Pattern Recognition, pp. 770–778 (2016)
5. Humphrey, E.J., Bello, J.P.: Rethinking automatic chord recognition with convolutional neural networks. In: 2012 11th International Conference on Machine Learning and Applications (ICMLA), vol. 2, pp. 357–362. IEEE (2012)
6. Kingma, D.P., Ba, J.L.: Adam: a method for stochastic optimization. In: Proceedings of the 3rd International Conference on Learning Representations (2014)

Safe Policy Learning with Constrained Return Variance

Arushi Jain[1,2](✉)

[1] McGill University, Montreal, Canada
arushi.jain@mail.mcgill.ca
[2] Mila, Montreal, Canada

Abstract. It is desirable for a safety-critical application that the agent performs in a reliable and repeatable manner which conventional setting in reinforcement learning (RL) often fails to provide. In this work, we derive a novel algorithm to learn a safe hierarchical policy by constraining the direct estimate of the variance in the return in the Option-Critic framework [1]. We first present the novel theorem of safe control in the policy gradient methods and then extend the derivation to the Option-Critic framework.

Keywords: Safety · Policy gradient · Option-Critic

1 Introduction

RL agents learn to solve a task by optimizing the observed return in a conventional setting. While this approach produces the highest return in expectation, it does not provide any constraints on the distribution of the return, making it a vulnerable strategy for the risk-sensitive domains. Safety in AI systems can be defined in several ways - safe exploration, reward hacking, etc. Our notion of safety emphasizes on minimizing the erratic or harmful behavior of an agent - by introducing the constraints on the variance in the return. The variance in the return reflects the uncertainty in the value function which makes an agent behave inconsistently. Therefore, the unsafe states which exhibit harmful or abrupt behavior would have a higher variance in the return.

[3, 4, 6–8] used the estimate of the variance in λ-return by the indirect second-order moment method or directly estimated the cost-to-go returns with the updates provided after completing the entire trajectory. [5] came up with a direct estimation of the variance in the λ-return using a Bellman operator in the policy evaluation methods. [2] used the variance in the temporal difference (TD) error to identify the controllable states in the option-critic framework.

In our preliminary work, we first came up with a Bellman operator to directly estimate the variance in the return given a state-action pair and learn a safe policy in control setting for actor-critic methods. Taking inspiration from this work, we extended the Bellman operator of variance to option-critic setting and introduce a safe hierarchical policy learning approach.

© Springer Nature Switzerland AG 2019
M.-J. Meurs and F. Rudzicz (Eds.): Canadian AI 2019, LNAI 11489, pp. 605–608, 2019.
https://doi.org/10.1007/978-3-030-18305-9_68

2   Background

In a Markov Decision Process (MDP), an agent interacts with the environment in discrete time steps t, where the agent takes an action $a \in A$, transitions from state S_t to state S_{t+1}, and receives an immediate reward R_{t+1} from the environment. The expected reward is $R(S_t, A_t) = \sum_{r \in \mathbb{R}} r \sum_{s'} P(s', r | S_t, A_t)$ where $R : S \times A \to \mathbb{R}$. The environment dynamics is modeled by $P(S_{t+1} | S_t, A_t)$, where $P : S \times A \times S \to [0, 1]$. A stochastic policy $\pi(A_t | S_t)$ determines the probability of taking an action in a given state. The MDP is represented by a tuple $\langle S, A, P, R, \gamma \rangle$, where $\gamma \in [0, 1]$ is a discount factor.

3   Safe Actor-Critic (Safe AC)

Let the policy be given by $\pi_\theta(a|s)$, where θ is the parameter of the policy. Extending the work by [5], the Bellman of the variance in the return given a state-action pair $\sigma_\pi(s, a)$ is similarly given by:

$$\sigma_\pi(s, a) = \mathbb{E}_\pi \left[\delta_t^2 + \bar{\gamma} \sigma_\pi(S_{t+1}, A_{t+1}) \big| S_t = s, A_t = a. \right] \tag{1}$$

where $\bar{\gamma} = \gamma^2 \lambda^2$, λ is the trace-decay parameter and δ_t is the one-step TD error. The proof for the above equation is left because of the space limitation. The new safe objective function now becomes:

$$J_d(\theta) = \mathbb{E}_{d,\pi} \left[\sum_a \pi_\theta(a|s) \big(Q_\pi(s, a) - \psi \sigma_\pi(s, a) \big). \right] \tag{2}$$

where ψ is the penalty coefficient. Here we aim to maximize the mean of the return along with minimizing the variance in the return in order to learn consistently behaving policy. Following the policy gradient theorem, the update for the gradient of the new safe objective function is:

$$\nabla_\theta J_{d,\pi}(\theta) = \mathbb{E}_\pi \left[\nabla_\theta \log \pi_\theta(A_t | S_t) \{ Q_\pi(S_t, A_t) - \psi \sigma_\pi(S_t, A_t) \} \right] \tag{3}$$

4   Safe Option-Critic (Safe OC)

Keeping similar notions to Option-Critic Architecture [1], an option $w \in W$ is defined as a tuple (I_w, π_w, β_w); where I_w contains the initial states set where an option can start, π_w is the option policy defining a distribution over action space and β_w determines the termination probability of an option in a state. The policy over the options is denoted by $\mu(w|s)$. Let $\Theta = [\theta, \nu, \kappa]$ be the parameters of intra-option policy, termination condition and policy over options respectively.

Let us consider $Z_t = (S_t, W_t)$ as an augmented state space, a space of state-option pair. Similar to the (1), the variance given a state-option-action is denoted by:

$$\sigma_{\pi,\mu}(z, a) = \mathbb{E}_{\pi,\mu}[\delta_t^2 + \bar{\gamma} \sigma_{\pi,\mu}(Z_{t+1}, A_{t+1}) | Z_t = z, A_t = a] \tag{4}$$

The safe objective for the option-critic is similar to (2) where state is replaced with augmented states space and ψ_z represents the regularizer for the variance. The updates for the parameters are shown below where blue color highlights the change from [1] due to the safe objective function. The intra-option gradient is:

$$\nabla_\theta J(\Theta) = \mathbb{E}_{d,\Theta}\left[\sum_a \nabla_\theta \pi_\theta(a|z)\big(Q_{\pi,\mu}(z,a) - \psi_z\, \sigma_{\pi,\mu}(z,a)\big)\right] \qquad (5)$$

The termination function gradient update is given by:

$$\nabla_\nu J(\Theta) = \mathbb{E}_{d,\Theta}[-\nabla_\nu \beta_\nu(s',w)\big(A_{\pi,\mu,Q}(s',w) - \psi_z\, A_{\pi,\mu,\sigma}(s',w)\big)] \qquad (6)$$

where $A_{\pi,\mu,\sigma}(s',w) = \sigma_\Theta(s',w) - \sigma_\Theta(s')$ is the *advantage function* for the variance similar to the value function. The update for the policy over options is provided by:

$$\nabla_\kappa J(\Theta) = \mathbb{E}_{d,\Theta}[\beta_\nu(s',w)\sum_{w'}\nabla_\kappa \mu_\kappa(w'|s')\big(Q_\Theta(s',w') - \psi_{s',w'}\, \sigma_\Theta(s',w')\big)] \qquad (7)$$

5   Experiments

We first performed the experiments in Safe Actor-Critic to see the effect of the safety and the experiments in Option-Critic are left as a future work. We added an unsafe frozen region (F) in one of the hallway in the discrete tabular four rooms (FR) environment [1] which has a normal reward distribution from $\mathcal{N}(\mu = 0, \sigma = 8)$. A different action from the one intended by the policy is taken with 0.2 probability. The agent can be initialized from anywhere in the state space and the reward of 50 is received when the agent reaches the goal state denoted by G in the Fig. 1 The reward for all the other states is kept 0. In expectation, the reward for both the hallways is 0.

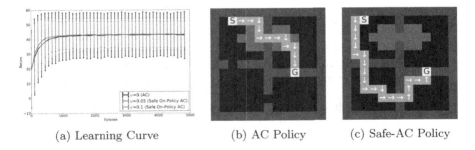

(a) Learning Curve   (b) AC Policy   (c) Safe-AC Policy

Fig. 1. Safe AC in FR domain: (a) Averaged performance over 50 trials where the vertical bands depict the standard deviation demonstrating that safe methods ($\psi > 0$) have lower variation in the return. Sampled policy using (b) baseline actor-critic, (c) safe actor-critic method. The unsafe region is depicted by the red color. (Color figure online)

Safe AC: Figure 1 depicts that using the safe framework (red plot), the variation in the return decreases highlighting that the agent reduces the visits to the variable reward region. The sampled policy from both the baseline and the safe method shows that agent takes a round about path to reach the goal state to avoid the frozen region. The risk-averse policy would generally exhibit faster convergence speed due to decrease in visits to inconsistent regions. If the learning curve is extended over a period of time, the risk-neutral would achieve higher or at par mean performance compared to the risk-averse as the penalty term is not part of the baseline.

6   Conclusion and Future Work

In this work, we presented a generic safe policy learning framework which learns a consistently behaving policy by constraining the direct estimate of the variance in the return. We first presented the safe policy gradient style update in the primitive action space and then extended this framework to the hierarchical option-critic format. Our approach provides an incentive to the agent which minimize the visits to the inconsistently behaving regions. This framework provides the capability to overcome the variability introduced by the environment dynamics.

The future step is to experiment with the safe option-critic framework using the non-linear function approximation in problems like Mujoco and ALE domains. The other step is to explore in the direction of variable value of ψ_z, $\forall z \in Z$ such that each options can have a different value of ψ, the importance factor for learning a safe policy.

References

1. Bacon, P.L., Harb, J., Precup, D.: The option-critic architecture. In: AAAI, pp. 1726–1734 (2017)
2. Jain, A., Khetarpal, K., Precup, D.: Safe option-critic: Learning safety in the option-critic architecture. arXiv preprint arXiv:1807.08060 (2018)
3. Prashanth, L., Ghavamzadeh, M.: Actor-critic algorithms for risk-sensitive MDPs. In: Advances in Neural Information Processing Systems, pp. 252–260 (2013)
4. Sato, M., Kimura, H., Kobayashi, S.: TD algorithm for the variance of return and mean-variance reinforcement learning. Trans. Jpn. Soc. Artif. Intell. **16**(3), 353–362 (2001)
5. Sherstan, C., et al.: Directly estimating the variance of the λ-return using temporal-difference methods. arXiv preprint arXiv:1801.08287 (2018)
6. Tamar, A., Di Castro, D., Mannor, S.: Policy gradients with variance related risk criteria. In: Proceedings of the Twenty-ninth International Conference on Machine Learning, pp. 387–396 (2012)
7. Tamar, A., Di Castro, D., Mannor, S.: Learning the variance of the reward-to-go. J. Mach. Learn. Res. **17**(13), 1–36 (2016)
8. Tamar, A., Xu, H., Mannor, S.: Scaling up robust MDPs by reinforcement learning. arXiv preprint arXiv:1306.6189 (2013)

Artificial Intelligence-Based Latency Estimation for Distributed Systems

Shady A. Mohammed$^{(\boxtimes)}$

School of Electrical Engineering and Computer Science, University of Ottawa,
Ottawa, ON, Canada
smoha191@uottawa.ca

Abstract. Network latency is an important metric specially for distributed systems. Depending on the system size, network latency can be either explicitly measured or predicted. However, prediction methods suffer from several drawbacks which lead to poor performance. The goal of this study is to demonstrate a novel method of network latency estimation which will be considered a valuable addition to the existing works due to its accuracy and efficiency. A number of machine learning techniques such as conventional linear regression, convolutional neural network, and support vector machine are used to predict the value to the end-to-end latency between any given pair of nodes. Two datasets: Ubique and iConnect-Ubisoft are used for training and testing the machine learning algorithms.

Keywords: Machine learning · Network delay measurement ·
Deep learning

1 Introduction

Ever since increasing the value of network latency due to the evolution of large-scale distributed applications such as multiplayer online games and peer-to-peer networks. Moreover, the necessity of predicting rather than explicitly measuring it have led the researchers to focus their efforts into accurate and precise prediction systems.

Some efforts have been done to predict network latency such as Network Coordinates Systems (NCS) [1]. NCS estimate the RTT between any pair of network nodes even without any prior contact between them. These systems explicitly conduct a few sets of actual latency measurements and predict the rest. NCS have two main categories: Euclidean Distance Model e.g. Vivaldi [1], and Matrix Factorization Model e.g. phoenix [2]. Euclidean distance Model maps network hosts to a set of finite-dimensional coordinates and the network latency is the distance between any pair of nodes. Equation 1 shows the calculation of

The work presented in this paper is financially sponsored by the Canadian NSERC, and Ubique Networks Inc.

M.-J. Meurs and F. Rudzicz (Eds.): Canadian AI 2019, LNAI 11489, pp. 609–612, 2019.
https://doi.org/10.1007/978-3-030-18305-9_69

the distance between hosts N and M. On the other hand, matrix factorization model, it model network latencies as NxN matrix such that closer hosts will have nearly identical rows or a linear combination of other rows.

$$E_{N,M} = ||N - M|| = \sqrt{(N_x - M_x)^2 - (N_y - M_y)^2 + ...} \qquad (1)$$

Where N and M are arbitrary network nodes, and x and y represent the coordinates dimensions.

Although NCS succeeded in predicting the network latency, they suffer form serious drawbacks. The Euclidean Distance-based model does not take into account the network asymmetry [3]. As for matrix factorization model, it does not consider the constant propagation delay between hosts due to the geographical distances between them. Besides, both models require a non-trivial time to converge.

To surmount the drawbacks mentioned earlier, I propose a new AI-based system for latency estimation. The system estimates the latency between any pair of network nodes without any explicit latency measurements and without any convergence time. IP addresses of any pair of points are the inputs given to the system to estimate the latency between them. The AI-based system is trained and tested using two datasets provided by iConnect-Ubisoft [4] and Ubique [5]. Note that Sects. 2 and 3 below present the summary of my design ideas and results from [6].

2   Methodology

iConnect-Ubisoft dataset is split into three parts and used for training (60%), validation (10%), and testing (30%). On the other hand, Ubique dataset is used only for testing to ensure model generalization. Figure 1 shows the main building blocks of the proposed system, which works in the following steps:

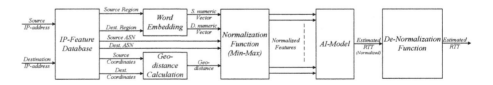

Fig. 1. Considered model of the proposed deep learning system.

Step 1. Feature Extraction: IP addresses alone are not enough to train the system since they expire after some time. IP addresses features are used instead. A RESTful API is developed to contact with the IP-features database provided by KeyCDN, which gives many features of an IP address, including geo-location, autonomous system, DNS, region, and VPN.

Step 2. Data pre-processing: the extracted data from the RESTful API are pre-processed to be ready for the training phase. The main functions used in the pre-processing phase are word embedding, geo-distance function, and normalization. Word embedding encodes strings found in some features such as VPN and ASN into unique codes for each value making these feature understandable by the learning algorithms. Instead of having redundant features, geo-distance is used to extract and condense the information from the geolocation features, host/destinations' longitude and latitude. Finally, normalization function takes place to put all the features in between 0 and 1 in order to speed the training phase [7].

Step 3. Linear regression, support vector machine, and convolutional neural network (CNN) models are trained by the processed data. The convolutional neural network specification is shown in Fig. 2.

Step 4. The AI output is denormalized in order to get the estimated delay.

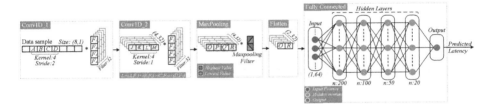

Fig. 2. CNN architecture used for latency estimation.

3   Results To-Date

The three proposed algorithms are tested using the two datasets mentioned earlier. Table 1 shows the performance summary for the three AI models in root mean square error (RMSE). The table shows that the CNN algorithm outperforms the other two models. Furthermore, Fig. 3 depicts the cumulative distributed function of the relative error (RE). RE is the metric that is widely used by researchers in this field to evaluate the performance of the estimated delay beside the ninetieth percentile relative error. The figure shows that CNN outperforms the other two algorithms. CNN's NPRE is 0.591 and 0.575 for Ubique and iConnect-Ubisoft respectively.

4   Future Work

The thesis has made significant progress in dataset preparations and machine learning technique selection. The next step is to compare between the AI-system proposed and the state-of-art. In addition, new datasets is being created to

Table 1. Performance summary for AI approaches in terms of RMSE.

Linear regression				Support vector machine			
Ubisoft			Ubique	Ubisoft			Ubique
Training	Validation	Testing	Testing	Training	Validation	Testing	Testing
0.0949	0.0972	0.1050	0.0901	0.0705	0.0697	0.0720	0.0853
Convolutional neural network							
Ubisoft						Ubique	
Training		Validation		Testing		Testing	
0.0553		0.0543		0.0575		0.0554	

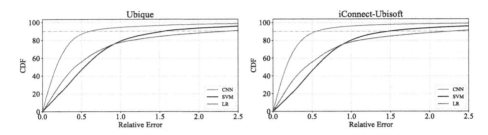

Fig. 3. Cumulative distribution function (CDF) of relative errors (RE).

introduce a new feature to the dataset that will pave the path for determining latency by taking into consideration the state of the network i.e. light or heavy traffic. Finally, I would like to extend the AI-system and predict other network resources such as the bandwidth.

References

1. Dabek, F., Cox, R., Kaashoek, F., Morris, R.: Vivaldi: a decentralized network coordinate system. In: ACM SIGCOMM CCR, pp. 15–26 (2004)
2. Chen, Y., et al.: Phoenix: a weight-based network coordinate system using matrix factorization. IEEE TNSM **8**(4), 334–347 (2011)
3. Cheng, J., Liu, Y., Ye, Q., Du, H., Vasilakos, A.V.: DISCS: a distributed coordinate system based on robust nonnegative matrix completion. IEEE/ACM TON **25**(2), 934–947 (2017)
4. Internet connectome project. https://github.com/lisa-lab/pings
5. Ubique nyetworks. http://www.ubiquenet.com/
6. Mohammed, S.A., Shirmohammadi, S., Altamimi, S.: Artificial intelligence-based distributed network latency measurement. In: Proceedings of IEEE I2MTC, no. 6 (2019)
7. Coates, A., Ng, A., Lee, H.: An analysis of single-layer networks in unsupervised feature learning. In: Proceedings of the 14th Conference on Artificial Intelligence and Statistics, pp. 215–223 (2011)

Industrial Track Papers

Exploring Optimal Trading Rules in a High-Frequency Portfolio

Cédric Poutré[1,2]([✉]) and Manuel Morales[1,2]

[1] Department of Mathematics and Statistics,
University of Montreal, Montreal, QC, Canada
{poutre,morales}@dms.umontreal.ca
[2] IVADO, Montreal, QC, Canada

Abstract. Given a set of financial instruments with inherent characteristics at different time intervals, we are interested in finding an optimal trading rule in a high-frequency trading context. The proposed trading rule makes use of a Recurrent Neural Network (RNN) with the profits generated by the strategy as its objective function. This recurrent function incorporates multiple non-linear operations, which make its gradient complex to estimate using a numerical approach. This yields a non-linear optimization problem on a multidimensional parameter space that cannot be resolved by the gradient descent algorithms usually used for training deep learning models. We implement a novel way of training RNNs based on a multivariate continuous-state Markov chain annealing algorithm and show its effectiveness in the context of the trading problem.

Keywords: Recurrent Neural Network · High-frequency trading · Stochastic optimization

1 Data and Systematic Trading Rule on a Portfolio

The data was provided by a hedge fund, therefore the profits generated by the proposed methodology will be censored. It has around 600 trading days for 200 stocks and indexes listed on the Australian Securities Exchange. The data contains: technical indicators, the direction of the trade generated by the hedge fund's proprietary system, the nominals available, the liquidity, the cost of trading, the maximum position allowed on the instruments, and the trade return at each trade time.

The systematic strategy is represented by the graph model in Fig. 1, where $\Theta_{t,d}$ is the concatenation of the aforementioned variables at trade time t and trade day d, $\mathbf{P}_{t,d} \in \mathbb{R}^N$ is the positions vector of each of the N instruments, and $\mathbf{O}_{t,d} \in \mathbb{R}^N$ is the profits vector generated by the strategy for each instrument. The function f, given the network parameters $\mathbf{W}$, returns the new state of the network, i.e. $\mathbf{P}_{t,d} = f(\mathbf{P}_{t-1,d}, \Theta_{t,d}|\mathbf{W})$, and g computes the observation of the network, i.e. $\mathbf{O}_{t,d} = g(\mathbf{P}_{t-1,d}, \Theta_{t,d}|\mathbf{W})$. The positions are reset at each trading day, meaning that $\mathbf{P}_{0,d} = \mathbf{0} \; \forall d$. It is a recurrent neural network with a

© Springer Nature Switzerland AG 2019
M.-J. Meurs and F. Rudzicz (Eds.): Canadian AI 2019, LNAI 11489, pp. 615–616, 2019.
https://doi.org/10.1007/978-3-030-18305-9

lagged output of period one, and the objective function to maximize is given by $J(\mathbf{W}) = \sum_d \sum_t \sum_i O_{i,t,d}$, which is the total profit generated by the network.

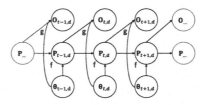

Fig. 1. Graph representation of the proposed systematic trading rule on a portfolio of instruments for one day of trading.

Because of the operations used in f, the gradient of J is difficult to compute numerically, so we implement an optimization algorithm based on simulated annealing algorithms [1], as well as multiple dependant synchronous Markov chains simulated annealing [2]. We compare the performance of our multivariate continuous-state Markov chain annealing algorithm to the random brute-force optimization algorithm, which is the initial benchmark.

Table 1. Relative distribution of the profits as a function of iterations used by our algorithm compared to the optimum value found by the benchmark algorithm, X, based on 100 independent runs with 20 Markov chains

Stats	Iteration of our algo				Iteration of benchmark	
	500	1 000	1 500	2 000	500	5 000
5th Percentile	−12.43%	−7.68%	−5.54%	−4.16%	−14.00%	−6.00%
Median	−7.82%	−3.77%	−2.10%	−1.03%	−8.48%	−4.50%
95th Percentile	−5.07%	−1.11%	+0.97%	+1.60%	−5.00%	−1.00%
Optimum	+1.91%			X (0.00%)		

We can observe that with 1 500 iterations, our algorithm surpasses the distribution generated by 5 000 iterations of the benchmark algorithm. It also finds a better optimum in 2.5 times less iterations.

References

1. Aguiar e Oliveira Jr., H., Ingber, L., Petraglia, A., Rembold Petraglia, M., Augusta Soares Machado, M.: Stochastic Global Optimization and Its Application with Fuzzy Adaptive Simulated Annleaing, 1st edn. Springer, Heidelberg (2012)
2. Ferreiro, A.M., Garcia, J.A., Lopez-Salas, J.G., Vazquez, C.: An efficient implementation of parallel simulated annealing algorithm in GPUs. J. Global Optim. **57**(3), 863–890 (2012)

Auto-Adaptive Learning for Life Science Knowledge

Mickaël Camus[(⊠)]

My Intelligent Machines, Montreal, QC, Canada
mickael@mims.ai

Abstract. Industries and academic research in the field of life science have to deal with massive distributed knowledge from multiple databases with various models. It takes time and lot of collaborations for a life scientist to process, in a secure manner, a project using multi-omic data. We provide an unsupervised auto-adaptive learning system to make available all new knowledge in real-time. We re-engineered the knowledge from these databases to build a distributed symbolic ontology updating in real-time. The system adapts itself to integrate new symbols and consolidate the ontology that may be used directly with natural language.

Keywords: Information extraction · Information retrieval · Knowledge representation

1 Introduction

Aggregating heterogeneous data in a common model is an open problem in industry and research. In the case of the field of life science, several databases, all with different models, are used to save massive data coming from the community [1, 2]. The issue of transforming biological data into knowledge to leverage the work of life scientists is still an important challenge [3]. Moreover, collecting these data from distributed and heterogeneous data warehouse is even harder.

2 Towards a No Query Language Model

As shown in Fig. 1, to aggregate new heterogeneous data in a specific model, we must first understand the model of the source database in order to perform a primary transformation that will be the core of the integration work of new data in a homogeneous ontology. The use of graphs allows working at a high level of abstraction while ensuring that any source model will be transposable. Then, graph aggregation works to integrate and transform the data in a homogeneous ontological graph representing a specific domain of knowledge.

It is important to save each step to avoid restarting the entire operation each time the source database is updated. Given the massive size of the data, the goal is to make the necessary transformations only on the data that has

© Springer Nature Switzerland AG 2019
M.-J. Meurs and F. Rudzicz (Eds.): Canadian AI 2019, LNAI 11489, pp. 617–618, 2019.
https://doi.org/10.1007/978-3-030-18305-9

The Knowledge Engineering Pipeline

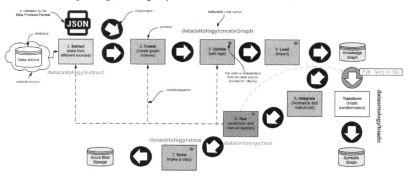

Fig. 1. The knowledge engineering pipeline (schema by Ahmed Halioui)

been modified. This is possible only if the source database offers the appropriate services to check the updates. If this is not the case, then it is necessary to keep a dictionary of updates so as not to repeat the entire import. However, this second method is longer because the entire source database must be scanned.

Once all the data are integrated in a homogeneous graph, it can be transformed in order to obtain a semantic graph. This type of graph makes it possible to represent the data in semantic maps, which makes it possible to put forward the notion of concepts. These concepts are used to answer questions asked in natural language. This amounts to working at the level of knowledge according to the definition of Allen Newell [4]. Thanks to this representation and transposition to a multiagent system, there is no longer any need for a query language to retrieve information; we go from NoSQL to NoQL. Moreover, according to [5], this kind of representation makes it possible to set up actionable knowledge according to an input flow. Which means that it is possible for this kind of system to make decisions at the level of knowledge. This will allow the user to give objectives to the system so that it processes tasks in the background using the regular update of knowledge.

References

1. Thessen, A., Patterson, D.: Data issues in the life sciences. ZooKeys **150**, 15–51 (2011)
2. Karcher, S., et al.: Integration among databases and data sets to support productive nanotechnology: challenges and recommendations. NanoImpact **9** (2017)
3. Livingston, K.M., Bada, M., Baumgartner Jr., W.A., Hunter, L.E.: KaBOB: ontology-based semantic integration of biomedical databases. BMC Bioinf. **16**, 126:1–126:21 (2015)
4. Newell, A.: The knowledge level. Artif. Intell. **18**(1), 87–127 (1982)
5. Camus, M.: Morphology programming with an auto-adaptive system. In: Proceedings of the 2008 International Conference on Artificial Intelligence, ICAI 2008, 14–17 July 2008, Las Vegas, Nevada, USA, vol. 2, pp. 924–930 (2008)

Credit Card Delinquency, Who to Contact? A Machine Learning Based Approach

Nizar Ghoula$^{(\boxtimes)}$, Josée Frappier, and Annie Lemieux

National Bank of Canada, 600 rue de La Gauchetière Ouest Niveau A,
Montreal, QC, Canada
{Nizar.Ghoula,josee.frappier,annie.lemieux}@bnc.ca

Abstract. Credit card delinquency is an issue that most banks and credit card providers face. As an indicator of credit risk, the delinquency rates are monitored quarterly by regulators. Definitions of delinquency might differ but commonly a delinquent account is an account where the minimum payment have not been settled within the given payment period. Statistical models and logistic regression are commonly used approaches for analyzing the delinquent accounts to assess the risk and establish collection strategies. In this talk, we show preliminary results about a modeling experiment that aims to predict if a delinquent account is safe enough to be excluded from the collection process.

Keywords: Credit card delinquency · Machine learning · Collection

1 Motivations

Credit card delinquency is a risk that most banks and credit card providers need to manage. As an indicator of credit risk, the delinquency rates are monitored quarterly by regulators. Definitions of delinquency might differ but commonly a delinquent account is an account where the minimum payment have not been settled within the given payment period. There are different types of delinquency for instance, early stage delinquency = 0–90 days; late stage delinquency = 90–180 days. In Canada, the latest report of the Canadian Bankers Association, shows a delinquency rate of 0.83[1]

Statistical models and logistic regression are commonly used approaches to analyze the delinquent accounts to assess the risk and establish collection strategies. These models are only used at the start of a delinquency process and are used to measure the risk of the signal. Based on the risk score, rules are used for

---

[1] As of April 30, 2018. These rates are average and there might be a variation from a financial institution to another.

© Springer Nature Switzerland AG 2019
M.-J. Meurs and F. Rudzicz (Eds.): Canadian AI 2019, LNAI 11489, pp. 619–621, 2019.
https://doi.org/10.1007/978-3-030-18305-9

segmentation of delinquent accounts to trigger an action. Mainly, to improve the collection process, the models to build are intended to: (1) predict if an account will enter the delinquency cycle accurately, (2) segment the delinquent accounts into multiple segments, (3) and recommend the best action for each segment. Using machine learning can bring new capabilities for advanced analytics by including more variables than the current models. In this talk, we show preliminary results about a modeling experiment that aims to predict if a delinquent account is safe enough to be excluded from the collection process.

2   Experiment and Findings

An account that falls into cycle 1 of delinquency, means that either the client is experiencing difficulties to pay his dept or that this is due to absence of mind or scheduling issues (which is called self-cure). Identifying accurately the accounts that self-cure, saves time and resources since the customer is not going to be contacted by call centers. We propose a model that predicts if a delinquent account at cycle 1 is going to self-cure.

In order to train the model, we managed to import the data sample of 250k clients accounts, the account information are anonymous and do not contain any personal data. We mainly used the variables that are related to the client's behavior score, current balance, utilisation rate, payment habits, transaction volume, and call center data like the number of attempts to contract a customer etc. The corpus was collected over a period of 3 years:

To train the model we first identified the clients that entered a delinquency cycle at most once and did not need to be contacted (by any means of contact) while they were in the first cycle of delinquency. We labeled these accounts as self-cured. The remaining of accounts were labeled as risky. We pre-processed the data to manage NA values, manage rare observations, encode categorical variables and then we split the corpus into train and test subsets. We trained different models and used GridSearch with multiple classifiers to identify the hyper parameters. Figure 1 presents the preliminary results of the model. The model's accuracy = 0.86 but this model needs work to improve the precision and recall.

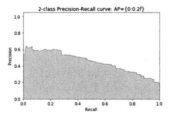

Fig. 1. Results and precision recall curve

Active risk management using machine learning for delinquent accounts helps save costs and also focus on the real bad clients to set the right strategy to recover the dept. Acting fast based on the model's outcome can increase collector capacity by 5–10 percent according to this report[2].

---

[2] Source: https://www.mckinsey.com/business-functions/risk/our-insights/the-analytics-enabled-collections-model.

Extending Recent Advancements in Reinforcement Learning to High Frequency Trading in the Canadian Equities Market

B. Rockwell[1]([✉]), L. Charlin[2], and M. Morales[1,3]

[1] National Bank of Canada, Toronto, ON, Canada
brittany.rockwell@nbc.ca
[2] HEC Montréal, Montreal, QC, Canada
[3] Université de Montréal, Montreal, QC, Canada

Abstract. This talk will discuss the extension of Deep Q-Network (DQN), along with its recent successors, Implicit Quantile Network (IQN) and Rainbow, to a simplified low-latency stock trading environment. Following similar game dynamics from past works, we discretize the action-space by following 'top-of-book' pricing, where the agent manages inventory only by way of issuing market orders. Order fill rates are proportional to the elapsed time before a change in the state occurs.

Keywords: Reinforcement learning · Order execution · Deep learning

1 Description

In 2013, researchers at DeepMind published the first example of a deep neural network learning control policies directly from sensory input data, coined as a Deep Q-Network (DQN) [1]. Impressively, the DQN consistently surpassed human-level performance in a number of experiments all while maintaining a fixed architecture from game-to-game. This meant that while the constructs and constraints could change dramatically from one game to another, the DQN could maintain the same architecture, randomly re-initialize its parameters, and still attain human-level performance in several of the experiments. The paper has since laid the bedrock for numerous advancements in the field of deep RL.

1.1 Dopamine

In 2018, researchers at DeepMind released Dopamine, a reinforcement learning framework intended to encourage consistency and rapid prototyping when using the Atari experiments [2]. With proper trading game design and minor adjustments to the source code, Dopamine could serve as a tool for testing several recent advancements in the context of optimal order execution.

M.-J. Meurs and F. Rudzicz (Eds.): Canadian AI 2019, LNAI 11489, pp. 622–623, 2019.
https://doi.org/10.1007/978-3-030-18305-9

1.2 Data

The study uses real tick-by-tick market data on 5 stocks listed on the Toronto Stock Exchange (TSX) in 2018.

1.3 Contribution

From our knowledge, this paper is the first to extend the IQN and Rainbow to an optimal order execution problem.

2 Method

We follow a similar game design as [3], where a Double Deep Q-Network was used instead of the variants discussed here. This paper discretizes the action-space by only permitting the agent to submit or fill market orders at the leading market price. State transitions occur when the leading bid or ask price of the order book changes. In an attempt to simulate a more realistic low latency environment, the fill rate is proportional to the amount of time until a new state is initiated. To limit large variations in costs, 'shorting' an asset is not permitted. The incremental reward was measured as the sum of any gains or losses realized from engaging in a trade, the market value of any existing stock on-hand and the remaining budget.

Hyper-parameters are mostly left at the values from the original works, although domain specific changes are sometimes made [4]. Consecutive frames are stacked on top of one another to make up one training example. In the trading game, consecutive time steps of stock prices and other relevant features are stacked to create each action-state pair. For further details on each model's implementation within the framework, please refer to [5]. We use two strategies as preliminary benchmarks: randomized action selection and buy/hold. Results of various game designs will be shared in more detail at the talks.

References

1. Mnih, V., et al.: Playing Atari with Deep Reinforcement Learning. arXiv:1312.5602 [cs], December 2013
2. Dabney, W., Ostrovski, G., Silver, D., Munos, R.: Implicit Quantile Networks for Distributional Reinforcement Learning. arXiv:1806.06923 [cs, stat], June 2018
3. Ning, B., Ling, F.H.T., Jaimungal, S.: Double Deep Q-Learning for Optimal Execution. arXiv:1812.06600 [cs, q-fin, stat], December 2018
4. Huang, C.Y.: Financial Trading as a Game: A Deep Reinforcement Learning Approach. arXiv:1807.02787 [cs, q-fin, stat], July 2018
5. Castro, P.S., Moitra, S., Gelada, C., Kumar, S., Bellemare, M.C.: Dopamine: A Research Framework for Deep Reinforcement Learning. arXiv:1812.06110 [cs], December 2018

Author Index